Applied Wave Mathematics

Ewald Quak · Tarmo Soomere
Editors

Applied Wave Mathematics

Selected Topics in Solids, Fluids, and Mathematical Methods

CENS

CMA Centre of Mathematics for Applications

Springer

Ewald Quak
Tarmo Soomere
Centre for Nonlinear Studies
Institute of Cybernetics
Tallinn University of Technology
Akadeemia tee 21
12618 Tallinn
Estonia
ewald.quak@cs.ioc.ee
tarmo.soomere@cs.ioc.ee

ISBN 978-3-642-00584-8 e-ISBN 978-3-642-00585-5
DOI 10.1007/978-3-642-00585-5
Springer Heidelberg Dordrecht London New York

Library of Congress Control Number: 2009935071

Mathematics Subject Classification (2000): 35-XX, 35Q, 35C, 35L, 35Q, 49S, 58A, 65-XX, 65M, 65N, 65T, 74-XX, 74A, 74J, 74L, 74N, 76A, 76B, 76E, 76S, 78M, 80M, 82C, 82D, 86A

Cover design: WMXDesign GmbH, Heidelberg

Printed on acid-free paper

Springer is part of Springer Science+Business Media (www.springer.com)

Dedicated to Jüri Engelbrecht,

the Founder and Leader of CENS,

on the Occasion of his 70th Birthday

Preface

This edited volume consists of twelve contributions related to the EU Marie Curie Transfer of Knowledge Project *Cooperation of Estonian and Norwegian Scientific Centres within Mathematics and its Applications*, CENS-CMA (2005-2009), under contract MTKD-CT-2004-013909, which financed exchange visits to and from CENS, the *Centre for Nonlinear Studies* at the Institute of Cybernetics of Tallinn University of Technology in Estonia.

Seven contributions describe research highlights of CENS members, two the work of members of CMA, the *Centre of Mathematics for Applications*, University of Oslo, Norway, as the partner institution of CENS in the Marie Curie project, and three the field of work of foreign research fellows, who visited CENS as part of the project. The structure of the book reflects the distribution of the topics addressed:

Part I Waves in Solids
Part II Mesoscopic Theory
Part III Exploiting the Dissipation Inequality
Part IV Waves in Fluids
Part V Mathematical Methods

The papers are written in a tutorial style, intended for non-specialist researchers and students, where the authors communicate their own experiences in tackling a problem that is currently of interest in the scientific community. The goal was to produce a book, which highlights the importance of applied mathematics and which can be used for educational purposes, such as material for a course or a seminar.

To ensure the scientific quality of the contributions, each paper was carefully reviewed by two international experts. Special thanks go to all authors and referees, without whom making this book would not have been possible. We also thank Heiko Herrmann for his technical and TeXnical help. The friendly and effective collaboration with Springer Verlag through Martin Peters is kindly appreciated.

Tallinn,
April 2009

Ewald Quak
Tarmo Soomere

Contents

List of Contributors

Jüri Engelbrecht, Project Coordinator
e-mail: je@ioc.ee

Arkadi Berezovski, Project Fellow in Oslo
e-mail: arkadi.berezovski@cs.ioc.ee

Mihhail Berezovski
e-mail: misha@cens.ioc.ee

Irina Didenkulova
e-mail: ira@cs.ioc.ee

Heiko Herrmann, Project Fellow in Tallinn
e-mail: hh@cens.ioc.ee

Christina Papenfuß, Project Fellow in Tallinn
e-mail: c.papenfuss@gmx.de

Ewald Quak, Project Fellow in Tallinn
e-mail: ewald.quak@cs.ioc.ee

Arvi Ravasoo, Project Fellow in Oslo
e-mail: arvi@ioc.ee

Andrus Salupere, Project Fellow in Oslo
e-mail: salupere@ioc.ee

Tarmo Soomere, Project Fellow in Oslo
e-mail: tarmo.soomere@cs.ioc.ee

Centre for Nonlinear Studies
Institute of Cybernetics at Tallinn University of Technology
Akadeemia tee 21
EE–12618 Tallinn
Estonia

Snorre Christiansen
Centre of Mathematics for Applications
University of Oslo
P.O. Box 1053 Blindern
NO–0316 Oslo
Norway
e-mail: snorrec@math.uio.no

Wolfgang Muschik
Institut für Theoretische Physik
Technische Universität Berlin
Hardenbergstr. 36
10623 Berlin
Germany
e-mail: muschik@physik.tu-berlin.de

Nils Henrik Risebro
Centre of Mathematics for Applications
University of Oslo
P.O. Box 1053 Blindern
NO–0316 Oslo
Norway
e-mail: nilshr@math.uio.no

Tomas Torsvik, Project Fellow in Tallinn
Bergen Center for Computational Science
UNIFOB
Thormøhlensgate 55
NO–5008 Bergen
Norway
e-mail: tomas.torsvik@bccs.uib.no

Péter Ván, Project Fellow in Tallinn
Department of Theoretical Physics
KFKI Research Institute of Particle and Nuclear Physics
Konkoly Thege Miklós út 29–33
1525 Budapest
Hungary
e-mail: vpet@rmki.kfki.hu

Ragnar Winther
Centre of Mathematics for Applications
University of Oslo
P.O. Box 1053 Blindern
NO–0316 Oslo
Norway
e-mail: ragnar.winther@cma.uio.no

CENS, CMA and the CENS-CMA Project

Jüri Engelbrecht, Ragnar Winther & Ewald Quak

CENS 1999-2009

The Institute of Cybernetics (IoC) was founded in 1960 within the Estonian Academy of Sciences. Its legendary first director Nikolai Alumäe understood the importance of computer science and control theory but did not forget mechanics, his own field of studies. The needs of mechanics were then one of the driving forces to foster computational methods. So the Department of Mechanics and Applied Mathematics (the present name), shortly DMAM, can look back over almost 50 years in its history. Photoelasticity has always been a topic in the DMAM because of the importance of this powerful method in the analysis of residual stresses. In other fields of studies, however, there has been a shift of focus - from the theory of shells, stability and vibration, the attention was turned step-by-step to the fast-moving front of science in nonlinear dynamics. This opened up new areas of research with surprising ideas and has stimulated many young researchers to join the research teams in DMAM of the Institute of Cybernetics, which since 1997 is part of Tallinn University of Technology (TUT).

About ten years ago it became clear that the research had to be organised in a more flexible way, which could enhance the synergy between strong teams united by basic ideas. The natural concept was to invite other teams sharing similar understandings and working in related fields to join forces and to stress the importance of international co-operation. This was the starting point in 1999 to found the Centre for Nonlinear Studies (CENS) within the IoC. The founders were DMAM with the Laboratory of Photoelasticity of the IoC, the Biomedical Engineering Centre of TUT and the Chair of Geometry of the Institute of Pure Mathematics at the University of Tartu. The idea was to bring under one umbrella the scientific potential in Estonia engaged in interdisciplinary studies of complex nonlinear processes. It was extremely important that CENS invited an International Advisory Board to guide and monitor its activities.

E. Quak, T. Soomere (eds.), *Applied Wave Mathematics*,
DOI 10.1007/978-3-642-00585-5_1, © Springer-Verlag Berlin Heidelberg 2009

The principles of CENS can be formulated briefly: to be at the frontier of science, to participate in international research, to react to national interests, to keep a good research atmosphere and to disseminate knowledge. The spearheads of research were:

- nonlinear waves in solids: complexity of wave motion in solids, solitonics, coherent wave fields, mechanics of microstructured materials, acoustodiagnostics;
- nonlinear integrated photoelasticity: stress field tomography (tensor tomography) and complexity of interference fringes;
- fractality and biophysics: turbulent diffusion, statistical topography and flooding, econophysics, complexity in biophysics, in silico modelling of cardiac mechanics and cell energetics;
- water waves: marine physics, multimodal waves, wave-wave and soliton interactions, anomalies of wave fields, ship wakes, extreme waves, coastal processes;
- nonlinear signal processing: analysis of physiological signals (EKG, EEG), applications in cardiology and brain research;
- geometric methods: Lie-Cartan methods, flows of vector fields on tensor fields.

The synergy between the fields has brought about a new quality of the activities. All the studies are characterized by a strong influence of nonlinearities, a wide range of scales (in space and time), and essential interaction between the constituents of processes. The results are described in the CENS Annual Reports (`http://cens.ioc.ee`) over the period 1999-2008. The lists of refereed papers became steadily longer, several monographs and textbooks have been published together with special issues of international and national journals. It would take too much space here to list all the activities (a summary of best results is given in the Annual Report 2007) but it is worth to mention that 9 fellows of CENS received Estonian Science Awards during the 10 years of CENS, there are 4 fellows of the German Alexander von Humboldt Foundation working in CENS, one Wellcome Trust fellow, etc. In 2002, the Estonian Ministry of Education and Research has included CENS into its list of 10 Centres of Excellence in Research in Estonia (2002-2007).

The decade of research in CENS coincides with a growing interest in the world in complexity and our research is clearly at the frontier of science. Indeed, the study of complex systems investigates collective properties of processes in systems, which are intrinsically nonlinear. While the origins of complexity lie in nonlinear dynamics and physics, today's complexity research is spreading into the medical, economic, and social sciences, as well as into key theories and enabling technologies of the artificial world. The studies listed above formed an excellent basis for complexity studies, and CENS has used this basis to develop its structure and topics in a flexible way. In 2008, CENS was reorganised and includes now three larger units: besides DMAM also the Department of Control Systems (DCS) of the Institute of Cybernetics and the Laboratory of Proactive Technologies (LPT) of TUT. There are now three subunits within DMAM beside its main body: the Laboratory of Photoelasticity, the Laboratory of Systems Biology and the Laboratory of Wave Engineering. The new topics and structures provide better synergy between analysis, synthesis and control. The main spearheads of research from 2008 on are: nonlinear dynamics and complex systems, proactivity and situation awareness, and complex nonlinear control systems. CENS is also the leading centre in Estonia for the EU's Complexity-NET, which unites efforts of 11 European countries.

CENS is an international node of research. For example in 2008, out of around 50 researchers in CENS, 17 were from abroad, representing 12 different countries. Presently CENS has about 30 graduate students, including 3 PhD students and 5 post-docs from abroad. The number of international grants and projects is growing. As one of those projects, the Marie Curie Transfer of Knowledge Development Scheme CENS-CMA (Cooperation of Estonian and Norwegian Scientific Centres within Mathematics and its Applications) linked CENS to the Centre of Mathematics for Applications (CMA) at the University of Oslo, to be described in some detail below. Other projects like the EU Research and Training Network SEAMOCS and a Wellcome Trust grant, etc., have also enhanced the capacity of CENS considerably.

Where do we stand now? Have we followed the main goals formulated at the launching of CENS? Yes, we are well placed according to the international standards in research such as cell energetics, mechanisms of interaction, solitons and turbulent diffusion, analysis of phase-transformation fronts and photoelastic tomography, just to list some results. Several applications are important for Estonia, let it be warnings to diminish the influence of waves from fast ferries in Tallinn Bay, the analysis of risks in financial time series, advice to cardiologists to understand the heart rate variability and the explanation of the influence of microwaves on the nervous system, the optimization of piano scales, etc. We have many graduate students, even though in CENS there is no direct obligation for teaching. The many international and national meetings organised by CENS follow the best practice of research communication. The dissemination of results to the wider public is constantly improving, for example the story of CENS is described in a booklet "The Beauty of a Complex World" (in Estonian), which has been distributed to all Estonian high schools. All that is a result of the good research climate in CENS.

There should always be a look forward. The new ideas for the next decade include linking mesoscopic physics to continuum mechanics, the analysis of emergent behaviour in pervasive computing systems, the unification of discrete- and continuous-time control systems, enlarging the range of applications of the analysis of time-series using physical methods in various fields of sciences, determining optimal ship routes based on the analysis of oil pollution transport, and so on (for more details, see our Annual Report 2008). All these studies are spiced with nonlinearity, emergence, irreversibility and multilevel approaches over space and time scales. These core complexity studies are carried on in CENS despite the small size of the scientific community in Estonia. By placing ourselves into the large European Research Area, we can overcome the problems of critical mass, and make CENS an international node of complexity studies.

Returning to the starting years of the Institute of Cybernetics, the attitude towards research was set up by Nikolai Alumäe: "Take a difficult problem to solve, then you have something to think about. And if you think then you are a happy person". These ideas have been followed and in 2009 CENS is full of happy people including many young researchers.

CMA

The Centre of Mathematics for Applications (CMA) is a Norwegian Centre of Excellence funded by the Research Council of Norway. The centre was established in 2003 as an international research centre in mathematics. The focus of the research is mathematics motivated from applications, with emphasis on problems arising from modern scientific computing. The goal is to create significant development in modern mathematics based on an interplay between theory, computations, and applications.

CMA is located at the University of Oslo as a part of the Faculty of Mathematics and Natural Sciences. Researchers from four departments, Mathematics, Informatics, Physics and Theoretical Astrophysics, are participating in the centre. The link to industrial research is ensured by including researchers from SINTEF as members of the centre. SINTEF is the largest independent research organization in Scandinavia, with a focus on industrial problems. There are also centre members from other Norwegian universities in Bergen and Trondheim. The CMA counts approximately 80 members at any time, 20 senior full-time researchers, 10 in part-time positions, and approximately 40 PhD students or post doc fellows. Furthermore there are a number of international guests, and a significant activity of seminars, workshops, and so on.

The research at CMA is built upon four main disciplines: *geometry*, *stochastic analysis*, *differential equations*, and *applications in the physical sciences*, where the focus of the latter is on astrophysics and computational quantum mechanics. It is our vision that strong interaction between these fields will lead to significant research in mathematics for applications.

The activity in geometry focuses on geometric modeling. Many scientific and industrial problems require a digital description of geometry. The CMA research in this area is based on combining techniques from splines and mesh based modelling, with the more classical tools from algebraic and differential geometry.

Stochastic analysis is fundamental for the development of modern mathematical finance. It also serves as a main mathematical tool in various other fields, like for instance biology, medicine and physics. The stochastic analysis group has focused on developing and applying stochastic calculus to model and manage financial risk. The activity is focused around differentiation and integration with respect to random processes, and various applications like for instance electricity markets and insurance.

Partial differential equations are one of the most fundamental tools in the construction of mathematical models in science and technology. The activity in differential equations at CMA is devoted to theoretical aspects of partial differential equations and to the numerical treatment of such problems.

The physical description of the outer stellar atmospheres results in large sets of coupled partial differential equations. There are major difficulties in constructing numerical methods for these equations, for example related to highly nonlinear reaction terms and in devising proper boundary conditions. Research in these areas is pursued at the CMA. In addition, the activity in cosmology is focused on developing improved algorithms for studying stochastic fields related to data on the Cosmic Microwave Background.

The quantum mechanics group investigates the development of appropriate numerical techniques for studying systems of many interacting particles. The systems of interest span most of the fields in physics covered by non-relativistic quantum mechanics, such as atomic, molecular, nuclear and solid-state physics and the physics of quantum liquids.

The CENS-CMA Project

The Marie Curie actions - intended to enhance the careers of researchers by enabling transnational mobility - have become one of the most popular parts of the Framework Programmes for research and technological development, as successively run by the European Commission.

As part of these actions in the Sixth Framework Programme, the CENS-CMA project (under contract MTKD-CT-2004-013909) lasted for 48 months from May 2005 to April 2009, and its scope is aptly summarized by its full title: Cooperation of Estonian and Norwegian Scientific Centres within Mathematics and its Applications. As a Transfer of Knowledge Development Scheme, the project financed visiting fellows to spend some longer research stay (between two months and two years) at CENS in Tallinn, and also allowed members of CENS to visit our collaboration partner CMA in Oslo for an extended period of time (typically three months). There were two categories of project fellows: experienced researchers (with 4-9 years of research experience, typically Post Docs) and senior researchers (of 10 years or more research experience).

With a budget of 853 333 Euros, a total of 130.5 researcher months were realized. 9 outgoing fellows were sent from Tallinn to Oslo, and 12 incoming fellows recruited to Tallinn: seniors from Norway, Hungary, Australia, Ukraine and Germany; experienced researchers from USA, Germany, Norway, France, UK, and Russia.

Beyond the ongoing scientific work enabled by these research stays, the project concept as a development scheme for the internationalization of CENS has indeed born fruit: two visiting fellows were recruited for longer stays at CENS after their project fellowships were over; helped along by the international collaboration through the project, a new research unit in Wave Engineering was established in early 2009; two international summer schools were held in Tallinn (2006 in shape modelling, 2007 in waves and coastal research, each with ca. 50 participants); two large-scale field experiments in Tallinn Bay were initialized (one held in June/July 2008 and the next one scheduled in June 2009, with participants from 9 countries); application-related international invited speakers from a large variety of countries were invited for the schools or for individual lectures (for example from the SINTEF research institute in Norway, the MATHEON research centre in Germany, and the Boeing Company).

Another major goal of the project was to further develop the skills in obtaining international research funding, and specifically to help define and start a new international research project with a broad partner consortium under the leadership of CENS. The latter goal was achieved through a successful application to the EC-sponsored BONUS-169 initiative. The project BalticWay (The potential of currents for environmental management of the Baltic Sea maritime industry) started in January 2009, as one of only 16 chosen from 149 applications, with 8 partners from 5 Baltic Sea countries, and is being coordinated by the new Wave Engineering group of CENS. Other research funding obtained at CENS recently includes a Wellcome Trust grant, a European Economic Area (EEA) grant involving Estonia, Norway and Russia, a German Feodor Lynen fellowship of the Alexander von Humboldt Foundation, and two FP7 re-integration grants for former project fellows.

A very concrete outcome of the project is of course also this book, meant to document some of the scientific achievements by members of CENS (often prepared during their fellowships in Oslo) and CMA, and by some foreign project fellows visiting Tallinn.

Finally, this is the right place to express our heartfelt thanks to the European Commission for the support provided through this project, which has made a lot of difference for the researchers involved. Specifically we thank the unit T3 of the Marie Curie Actions-Directorate General for Research for all their help in running this project. Among the commission staff, our very special thanks go to Frederic Olsson-Hector for his help at the start in negotiating the contract, and to Marcela Groholova as our long-time project officer, who was always helpful and even found the time to come to Estonia for our midterm meeting, which we think was very important for the project partners and visiting fellows to give and get direct feedback about the project.

Tallinn, April 2009 — *Jüri Engelbrecht*, Leader of CENS
Oslo, April 2009 — *Ragnar Winther*, Leader of CMA
Tallinn, April 2009 — *Ewald Quak*, Project Manager of the CENS-CMA Project

Part I
Waves in Solids

Overview

Arkadi Berezovski

The behaviour of many materials of interest in engineering (e.g., metals, alloys, granular materials, composites, liquid crystals, polycrystals) is often influenced by an existing or emergent microstructure. In general, the components of such a microstructure have different material properties, resulting in a highly complex macroscopic material behaviour.

Among numerous applications, the use of wave energy as a probe is one of the most promising non-destructive means of diagnosing the properties of complex materials. A detailed understanding of how signals evolve in the material of interest is required in this case.

If we know all the details of a given microstructure, namely, size, shape, composition, location, and properties of inclusions as well as properties of the carrier medium, the classical wave theory is sufficient for the description of wave propagation. The less we know concerning the microstructure, the more complicated models are needed for the equivalent description of waves in the carrier medium, which may be considered to be as simple as possible.

It is certainly true that for many types of materials it is sufficient to approximate the real behaviour by some linear model. However, when materials or structures are pushed to extreme conditions, nonlinear effects are always encountered. Such extreme conditions are of major importance in determining the reliability or failure limits of engineering structures.

The standard treatment of wave propagation in solids involves the investigation of the material response to sinusoidal disturbances of infinite extent. That is, we simply obtain the dispersion relation, which gives the relationship between the wave number and the angular frequency. These results apply only to linearly elastic homogeneous materials. Inhomogeneity leads to dispersion. Dispersion of the average composite motion results in a distortion of the stress pulse as it propagates, and the effects of dispersion increase with the duration.

It is important to realize that solutions to partial differential equations, even of linear material models, at infinitesimal strains, describing the response of bodies

E. Quak, T. Soomere (eds.), *Applied Wave Mathematics*,
DOI 10.1007/978-3-642-00585-5_2,

containing a few heterogeneities, are still open problems. This means that complete solutions are in general impossible.

For this reason, the use of homogenized material models is commonplace in the mechanics of heterogeneous materials. The usual approach is to determine the so-called effective properties: averaged or smoothened properties that reflect in some global sense the response of specimens of the material to external loads.

Alternatively, a generalized continuum theory can be applied for the description of complex materials. This approach is illustrated in the paper *Deformation Waves in Solids* by J. Engelbrecht, where the basic theory of waves in homogeneous materials is briefly reviewed, and advanced theories are introduced, focusing on microstructured materials and the concept of internal variables. The governing equations of a microstructure model are presented, with special attention devoted to wave hierarchies and nonlinearities. In addition to standard wave models involving the 2nd order wave operator, the idea of one-wave models is also briefly discussed.

Asymptotic methods for weakly nonlinear wave propagation in solids are presented in the paper *The Perturbation Technique for Wave Interaction in Prestressed Material* by A. Ravasoo. Here the basic notions of continuum mechanics and perturbation methods are introduced and applied to wave propagation and interaction in nonlinear elastic material undergoing inhomogeneous plane prestrain. The five-constant weakly nonlinear theory of elasticity is used in the case of counterpropagating longitudinal waves.

One of the objectives in investigations of nonlinear wave propagation in solids is the development of methods for predicting the effects of dynamic events such as high-velocity impacts and detonation of explosives. Impact loads involve factors, which are not considered in statics. In static problems the deformation energy can be distributed throughout the structure, but in impact loading the volume of energy storage is limited by the speed of the propagation of the waves in the material. For short time impact loads, a small amount of energy in a small volume can result in stresses, which can fracture or otherwise damage the material.

Among many other potential applications, shock wave propagation in multilayer composite materials is relevant to the design and optimization of armor systems for ballistic protection. Layered structures have been a significant contender for developing armor systems and, in many of these, plates of materials with different acoustic impedances are stacked and joined together.

Shock-compression experiments are used to validate the prediction of the material response obtained numerically on the basis of a weakly nonlinear elastic model in the paper *Waves in Inhomogeneous Solids* by A. Berezovski, M. Berezovski and J. Engelbrecht. Hyperbolic conservation laws in the form of a first-order hyperbolic system are applied to the description of wave propagation problems. The emphasis in the paper is on the nonlinearities in these problems, especially those that lead to the development of propagating shock waves in laminates.

The role of the scale effects is clearly demonstrated. When the wavelength of a travelling signal decreases and becomes comparable with the characteristic size of the heterogeneities, successive reflections and refractions of the local waves at the component interfaces lead to dispersion and attenuation of the global wave field.

There are the following issues concerning the existing numerical methods for elastodynamic wave propagation and impact problems:

a) the selection of an effective numerical method among the known ones;
b) due to spurious high-frequency oscillations, the lack of reliable numerical techniques that yield an accurate solution of wave propagation in solids;
c) the treatment of the error accumulation for long-term integration;
d) the increase in accuracy and the reduction of computation time for real-world dynamic problems.

In the paper, it is shown that linear and non-linear wave propagation in media with rapidly-varying properties as well as in functionally graded materials can be successfully simulated by means of the modification of the wave-propagation algorithm based on the non-equilibrium jump relation for true inhomogeneities. The presented algorithm is conservative, stable up to Courant number equal to 1, high-order accurate, and thermodynamically consistent.

Deformation Waves in Solids

Jüri Engelbrecht

Abstract The basic theoretical concepts are analysed on the basis of the continuum theory for modelling deformation waves in solids. First a brief description of modelling for homogeneous solids is presented, which is widely known in practice. Special attention is paid to advanced theories focusing on microstructured materials. Several approaches are described: the separation of macro- and microstructure, the balance of pseudomomentum, and the concept of internal variables. Characteristically, the advanced models describe the hierarchy of waves, which includes the dependence on the internal scale(s). The resulting dispersive effects are often accompanied by nonlinearities and in this case solitary waves may emerge. Finally, some challenges in the theory of waves are briefly listed.

1 Introduction

1.1 General ideas

In order to describe the propagation of mechanical waves in solids, one needs mathematical models to be built based on sound definitions. First, the observable variables, such as displacement and deformation, are sometimes called state variables. A rather general definition, advocated in continuum mechanics by Truesdell and Noll, says [39]:

A wave is a state moving into another state with a finite velocity.

We may also see a wave as *a disturbance which propagates from one point in a medium to other points without giving the medium as a whole any permanent displacement.*

Centre for Nonlinear Studies, Institute of Cybernetics at Tallinn University of Technology, Akadeemia tee 21, 12618 Tallinn, Estonia, e-mail: je@ioc.ee

E. Quak, T. Soomere (eds.), *Applied Wave Mathematics*,
DOI 10.1007/978-3-642-00585-5_3,

Although this definition is widely used, the case of plastic waves (not considered here) needs a more detailed description.

In this paper, we focus on solids. A *solid* is

- *a substance that has a definite volume and shape and resists forces that tend to alter its volume or shape;*
- *a crystalline material in which the constituent atoms are arranged in a 3D lattice with certain symmetries.*

The first definition of a solid is the basis for the theory of continuous media (for example see Eringen [13]) and the second — for the theory of discrete media (for example see Maugin [26]).

From those definitions it is clear that a solid is deformed at a certain point and this disturbance is transmitted from one point to the next, etc.

So, in general, waves in this context correspond to continuous variations of the states of the material points that constitute the solid. The resistance to deformation and the resistance to motion (i.e. inertia) must be overcome during the wave propagation. Consequently, waves can only occur in media in which energy can be stored in both kinetic and potential forms. Again, as mentioned above for waves, some more sophisticated cases like thermoelasticity need a more detailed analysis.

There is a great interest in wave phenomena in solids. First, one has to understand how materials (structures, details, specimens, etc.) resist to dynamical loads. This is not only a problem for technology or engineering, but also in seismology. Second, waves carry information about the source and the material. This property can be used for the nondestructive testing of materials and is closely related to acoustics (ultrasound range). Clearly, given the wide scale of material properties, and the intensity and frequency of excitations, the problems can be very complicated and, therefore, traditional linear theories cannot describe the processes with the needed accuracy. That is why one should pay close attention to the proper modelling of wave motion.

The mathematical models describing waves in continua are based on the conservation laws complemented by suitably chosen constitutive laws. These models should reflect the features given in the definitions above. In pure mathematical terms a finite speed refers to the existence of a real eigenvalue of the corresponding mathematical models. These models are called hyperbolic and are of fundamental importance in wave motion [40]. Apart from strict hyperbolicity, waves may be characterized just by their dispersive relations, i.e., the models should possess certain harmonic solutions with fixed wave numbers and frequencies. These waves are called dispersive [40] and, as easily understood, not described by the definitions above.

However, the physical world with its multiple scales and different processes is rich and the constitutive laws are often based on simplified assumptions, especially when other fields apart from pure mechanical stress are accounted for. Then hyperbolicity in the strict mathematical sense may be lost but still be preserved in some asymptotical sense.

1.2 Notes from history

The theory of wave propagation in solids may be traced back to the 19th century and studies of Cauchy [7], Poisson [33], Lamé [20] a.o. During that time, these studies were simply an extension of the theory of elasticity. Poisson, in fact, was the first to recognize that elastic disturbance was in general composed of two types of fundamental waves, i.e. dilatational and equivolumetrical ones. Many studies followed: for example, those of Rayleigh [34], Love [21], a.o. Numerous objects, such as an elastic 3D medium, a halfspace, two half-spaces in contact, waveguides, etc. were being studied. A wide range of waves were described and special mathematical methods for analysis were derived. The firm framework for classical continuum mechanics has actually been created by Truesdell and Toupin [38] in their monumental treatise "Classical Field Theory". More recently, excellent overviews on wave motion were presented by Kolsky [18], Bland [3], Achenbach [1], and Miklowitz [28]. One has also to mention the studies of waves in fluids, like Lamb [19] or general studies on waves like Brillouin [5] and Whitham [40]. The contemporary understanding of wave theory has pillars in both large areas (i.e. solids and fluids), which is even more important when dealing with nonlinear waves.

1.3 Description of what follows

The brief overview in this paper is based on earlier studies of Engelbrecht ([8], [9], [10]), Jeffrey and Engelbrecht [16], and recent results of CENS.

First, in Section 2 we briefly discuss the concept of the basic theory of waves in homogeneous materials. Section 3 is devoted to advanced theories focusing on microstructured materials and the concept of internal variables. In Section 4, model governing equations are briefly presented. Special attention is devoted to wave hierarchies and nonlinearities. In addition to standard wave models involving the 2nd order wave operator, the idea of the one-wave model is also briefly presented. Section 5 includes final remarks.

2 Basic theory

The conceptual approach in constructing the mathematical models of wave motion is based on the following sequence:

1. basic principles (initial assumptions and conservation laws);
2. constitutive theory (constitutive equations added together with auxiliary postulates in order to formulate closed systems);
3. mathematical models (auxiliary assumptions about the character of field variables and approximations of the constitutive laws).

The details of modelling can be found in monographs by Eringen [13], Engelbrecht [8], [10], Maugin [26], and others. Here we present only a brief description of basic steps.

After fixing the initial assumptions on time, space and medium (Engelbrecht [8]), the conservation laws are formulated. Here we follow, first, Eringen [13] and his notations: T^{KL} – Piola-Kirchhoff stress tensor, E_{KL} – Green deformation tensor, ρ_0 and ρ – initial and current densities, V and v – initial and current volumes, f_k – the components of the body force, A_k – the components of the acceleration, E – internal energy, Q^K – components of the heat, h – the supply of the energy, θ – temperature, S – entropy, $W = E - TS$ – Helmholtz free energy. Space (Euler) coordinates are denoted by x_k, material (Lagrange) coordinates by X_K and indices run over 1, 2, 3. The comma indicates the differentiation with respect to the coordinate and the dot – the differentiation with respect to time. The rule of summation over the diagonally repeated index is used. The conservation laws thus are the following (in a Descartes system):

(i) conservation of mass:

$$\int_V \rho_0 \, dV = \int_v \rho \, dv; \tag{1}$$

(ii) balance of momentum:

$$\left(T^{KL} x_{k,L}\right)_{,K} + \rho_0 \left(f_k - A_k\right) = 0; \tag{2}$$

(iii) balance of moment of momentum (also known as angular momentum) for non-polar materials:

$$T^{KL} = T^{LK}; \tag{3}$$

(iv) conservation of energy:

$$\rho_0 \dot{E} = T^{KL} \dot{E}_{KL} + Q^K_{,K} + \rho_0 h; \tag{4}$$

(v) entropy inequality:

$$T^{KL} \dot{E}_{KL} + \frac{1}{\theta} Q^K \theta_{,K} - \rho_0 \dot{W} - \rho_0 \dot{\theta} S \geq 0. \tag{5}$$

Second, after fixing the auxiliary postulates on the initial state and the character of constitutive equations, a closed system is formulated. The auxiliary postulates on constitutive equations can also be presented verbally as:

- the stress may be determined from the strain alone (perfectly elastic body);
- the stress may be determined from the stretching alone (perfectly plastic body), etc.

According to conventional continuum theory [13], the elastic stress tensor is related to the potential energy and especially to the free energy W, i.e.

$$T^{KL} = \rho_0 \frac{\partial W}{\partial E_{KL}}. \tag{6}$$

In more complicated cases, the stress tensor contains reversible $({}_E T^{KL})$ and irreversible $({}_D T^{KL})$ parts.

Next, auxiliary assumptions involve estimates of possible strains and temperature, such as $E_{KL} \ll 1$, $(\theta - \theta_0)/\theta_0 \ll 1$. These assumptions are rather restrictive and the following models are to be used within these ranges. It is important that they permit the representation of the Helmholtz free energy W in terms of Taylor series.

The final mathematical model on the basis of the balance of momentum (2) is usually written in terms of displacements U_K and temperature θ. Note that the strain tensor is given by $E_{KL} = \frac{1}{2}\left(U_{K,L} + U_{L,K} + U_{I,L} \cdot U_{K,I}\right)$. A compact description in matrix notation is then:

$$\mathbf{I}\frac{\partial \mathbf{U}}{\partial t} + \mathbf{A}^K \frac{\partial \mathbf{U}}{\partial X_K} + \sum_{p=2}^{q} \mathbf{B}_M^{rs} \frac{\partial\, p\mathbf{U}}{\partial \left(X_M\right)^r \partial t^s} + \mathbf{H} = 0, \tag{7}$$

where

$$\mathbf{U} = \begin{vmatrix} U_{N,t} \\ U_{K,L} \\ \theta \\ Q_K \end{vmatrix}, N, K, L = 1, 2, 3, \tag{8}$$

$$\mathbf{A}^K = \mathbf{A}^K(\mathbf{U}),\ \mathbf{B}_M^{rs} = \mathbf{B}_M^{rs}(\mathbf{U}),\ \mathbf{H} = \mathbf{H}(\mathbf{U}), \tag{9}$$

and $\mathbf{I}$ is the unit matrix, $r + s = p$.

Note that in principle, the matrices $\mathbf{A}^K$, $\mathbf{B}_M^{rs}$, and the vector $\mathbf{H}$ may also depend on X_M. The solutions of Eq. (7) are waves $\mathbf{U}(X_M, t)$ and they are sought to satisfy initial and boundary conditions

$$\mathbf{U}(X^K, t)\left.\right|_{t=0} = \psi(X_K), \quad \mathbf{U}(X^K, t)\left.\right|_B = \phi\left(X_K, t\right), \tag{10}$$

where B denotes a certain boundary.

Equation (7) is the governing equation of the wave motion. With $\mathbf{B}_M^{ps} = 0$, it is clearly hyperbolic, but in the general case $\mathbf{B}_M^{rs} = \theta(\varepsilon)$, and hyperbolicity is preserved in the asymptotic sense [8], [10].

Let us note, again, the importance of the existence of kinetic and potential energies for motion. The Lagrangian formalism reflects this property explicitly. Shortly, we define the Lagrangian $\mathcal{L} = \mathrm{K} - \mathrm{W}$, where K is the kinetic energy and W – the potential energy.

Then the Euler-Lagrange equations are

$$\left(\frac{\partial \mathfrak{L}}{\partial \mathbf{r}_t}\right)_t + \left(\frac{\partial \mathfrak{L}}{\partial \mathbf{r}_{,i}}\right)_{,i} - \frac{\partial \mathfrak{L}}{\partial \mathbf{r}} = 0, \qquad (11)$$

where $\mathbf{r}$ is the position vector $\mathbf{r} = \mathbf{r}(X_K, t)$. From Eq. (11), the governing equations of motion can be derived in the full consistence of Eq. (2) (or Eq. (7)).

Now, a detailed description of the types of waves should follow. However, we restrict ourselves only to a description of the two types of waves that can propagate in an unbounded elastic medium. These two types of waves are characterized by comparing the particle motion with the direction of propagation:

- if the particle motion is along the direction of propagation, then the wave is longitudinal;
- if the particle motion is perpendicular to the direction of propagation, then the wave is transverse.

Both of them are frequently referred to as body waves. One should also notice that longitudinal waves are sometimes called dilatational, irrotational, and extension waves, whilst transversal waves are called shear, rotational, distortion, and equivoluminal waves. Within the framework of the linear theory of elasticity, these waves are uncoupled, which is not true according to more advanced theories. If a solid has a free surface, then surface (Rayleigh) waves are possible. The situation is even more complicated in bounded media like rods, plates, shells, etc. For more details, the reader is referred to several monographs in this field (Kolsky [18]; Achenbach [1]; Miklowitz [28]; Maugin [24]; Yerofeyev et al. [41]).

3 Advanced theories

3.1 General ideas

The classical theory of continuous media is developed using the assumption of smoothness of continua. Materials used in contemporary advanced technologies are often characterized by their complex structure satisfying many requirements in practice. This concerns polycrystalline solids, ceramic composites, alloys, functionally graded materials, granular materials, etc. Often the damage effects should also be accounted for, i.e. materials are still usable when then have microcracks. In all these materials there exists an intrinsic space-scale, like the lattice period, the size of a crystallite or a grain, or the distance between the microcracks. Clearly the complex dynamical behaviour of such microstructured materials cannot be explained by the classical theory of continua.

Within the theories of continua the problems of irregularities of media were actually predicted already by the Cosserats and Voigt, and more recently by Mindlin [29], Eringen [14] and others. The elegant mathematical theories of continua with voids or with vector microstructure, of continua with spins, of micromorphic continua, ferroelectric crystals, etc., have been elaborated on since, see the overviews by Capriz [6] and Eringen [15]. Recently, an excellent overview on the complexity of wave motion was presented by Pastrone [32].

The straightforward modelling of microstructured solids leads to the assignment of all physical properties to every volume element dV in a solid, thus introducing the dependence on the material coordinates X_K. Then, the governing equations implicitly include space-dependent parameters, but due to the complexity of the system, can be solved only numerically. Another probably much more effective method is to separate macro- and microstructure in continua. The conservation laws for both structures should then be separately formulated (Mindlin [29]; Eringen [14]; [15]), or the microstructural quantities are separately taken into account in one set of conservation laws (Maugin [24]). In the first case, macrostress and microstress together with the interactive force between macro- and microstructure need to be determined. The last case uses the concept of pseudomomentum and material inhomogeneity force.

In Section 3.2, we shall analyze the first case in more detail. However, both cases, when consistently treated, should give the same result, which will also be demonstrated in Section 3.3.

3.2 Separation of macro- and microstructure

Here we follow Mindlin [29] who has interpreted the microstructure "as a molecule of a polymer, a crystallite of a polycrystal or a grain of a granular material". This microelement is taken as a deformable cell. Note that if this cell is rigid, then the Cosserat model applies. The displacement $\mathbf{U}$ of a material particle in terms of macrostructure is defined by its components $U_I = x_I - X_I$, where $x_I, X_I\,(I = 1,2,3)$ are the components of the spatial and material position vectors, respectively. Within each material volume, there is a microelement and the microdisplacement $\mathbf{U}'$ is defined by its components $U_I' \equiv \widehat{x}_I - \widehat{X}_I$, where the origin of the coordinates $\widehat{X}_I$ moves with the displacement $\mathbf{U}$. The displacement gradient is assumed to be small. This leads to the basic assumption of Mindlin [29] that "the microdisplacement can be expressed as a sum of products of specified functions of $\widehat{X}_I$ and arbitrary functions of x_I and t". The first approximation is then

$$U_J' = \widehat{x}_K\, \varphi_{KJ}\,(x_{I,t})\,. \tag{12}$$

The microdeformation is then

$$\partial\, U_J' / \partial\, \widehat{x}_I = \partial_I'\, U_J' = \varphi_{IJ}\,. \tag{13}$$

Now we consider the simplest 1D case and drop the indices I,J, dealing with U and φ only. The indices X,t used in the sequel denote differentiation.

The fundamental balance laws for microstructured material can be formulated separately for the macroscopic and microscopic scales (see Section 3.1). We show here how the balance laws can be derived from the Lagrangian (Mindlin [29]; Pastrone [31]) $\mathcal{L} = \mathrm{K} - \mathrm{W}$ formed from the kinetic and potential energies

$$\mathrm{K} = \frac{1}{2}\rho_0 U_t^2 + \frac{1}{2} I \varphi_t^2, \quad \mathrm{W} = \mathrm{W}(U_X, \varphi, \varphi_X), \tag{14}$$

where I is the microinertia related to a microelement.

The corresponding Euler–Lagrange equations have the general form (cf. Eq. (11))

$$\left(\frac{\partial \mathcal{L}}{\partial U_t}\right)_t + \left(\frac{\partial \mathcal{L}}{\partial U_X}\right)_X - \frac{\partial \mathcal{L}}{\partial U} = 0, \tag{15}$$

$$\left(\frac{\partial \mathcal{L}}{\partial \varphi_t}\right)_t + \left(\frac{\partial \mathcal{L}}{\partial \varphi_X}\right)_X - \frac{\partial \mathcal{L}}{\partial \varphi} = 0. \tag{16}$$

Inserting the partial derivatives

$$\frac{\partial \mathcal{L}}{\partial U_t} = \rho_0 U_t, \quad \frac{\partial \mathcal{L}}{\partial U_X} = -\frac{\partial \mathrm{W}}{\partial U_X}, \quad \frac{\partial \mathcal{L}}{\partial U} = 0, \tag{17}$$

$$\frac{\partial \mathcal{L}}{\partial \varphi_t} = I \varphi_t, \quad \frac{\partial \mathcal{L}}{\partial \varphi_X} = -\frac{\partial \mathrm{W}}{\partial \varphi_X}, \quad \frac{\partial \mathcal{L}}{\partial \varphi} = -\frac{\partial W}{\partial \varphi}, \tag{18}$$

into Eqs. (15), (16), we obtain the equations of motion

$$\rho_0 U_{tt} - \left(\frac{\partial \mathrm{W}}{\partial U_X}\right)_X = 0, \quad I \varphi_{tt} - \left(\frac{\partial \mathrm{W}}{\partial \varphi_X}\right)_X + \frac{\partial \mathrm{W}}{\partial \varphi} = 0. \tag{19}$$

Denoting

$$T = \frac{\partial \mathrm{W}}{\partial U_X}, \quad P = \frac{\partial \mathrm{W}}{\partial \varphi_X}, \quad R = \frac{\partial \mathrm{W}}{\partial \varphi}, \tag{20}$$

we recognise $T = T^{11}$ as the macrostress (the first Piola-Kirchhoff stress), P as the microstress and R as the interactive force. The equations of motion (19) now take the form

$$\rho_0 U_{tt} = T_X, \quad I \varphi_{tt} = P_X - R. \tag{21}$$

The simplest potential energy function describing the influence of a microstructure is a quadratic function

$$\mathrm{W} = \frac{1}{2}\alpha U_X^2 + A \varphi U_X + \frac{1}{2} B \varphi^2 + \frac{1}{2} C \varphi_X^2, \tag{22}$$

with α, A, B, C denoting material constants. Inserting it into Eqs (20), and the result into Eqs. (21), the governing equations take the form

$$\rho_0 U_{tt} = \alpha U_{XX} + A \varphi_X, \tag{23}$$

$$I\varphi_{tt} = C\varphi_{XX} - AU_X - B\varphi. \tag{24}$$

This is the sought-after mathematical model for 1D longitudinal waves in microstructured materials of the Mindlin type.

3.3 Balance of pseudomomentum

For heterogeneous materials Maugin [24] has introduced the concept of pseudomomentum in the material manifold that leads to a physically transparent presentation of the governing equation.

The core of the governing equation is the balance of momentum. We rewrite Eq. (2) in a more compact form

$$\frac{\partial}{\partial t}\mathbf{p}\,|_{\mathbf{X}} - \nabla_R \mathbf{T} = \mathbf{f}, \tag{25}$$

where $\mathbf{p} = \rho_0 \mathbf{v}$ is the linear momentum at any regular point $\mathbf{X}$, $\mathbf{T}$ is the first Piola-Kirchhoff stress tensor and $\mathbf{f}$ is the body force.

The motion is $\mathbf{x} = \chi(\mathbf{X}, t)$, and the physical velocity $\mathbf{v}$ and the direct–motion deformation gradient $\mathbf{F}$ are defined by

$$\mathbf{v} := \left.\frac{\partial \chi}{\partial t}\right|_{\mathbf{X}}, \quad \mathbf{F} := \left.\frac{\partial \chi}{\partial \mathbf{X}}\right|_{t}. \tag{26}$$

Applying $\mathbf{F}$ from the right to Eq. (25), we obtain

$$\frac{d\mathbf{P}}{dt} - \nabla_R \cdot \mathbf{b} = \mathbf{f}^{\mathrm{int}} + \mathbf{f}^{\mathrm{ext}} + \mathbf{f}^{\mathrm{inh}}, \tag{27}$$

where $\mathbf{P} = -\rho_0 \mathbf{v}\cdot\mathbf{F}$, $\mathbf{b}$ is the material Eshelby stress, $\mathbf{f}^{\mathrm{inh}}$ is the material inhomogeneity force, $\mathbf{f}^{\mathrm{ext}}$ is the material external force, and $\mathbf{f}^{\mathrm{int}}$ is the material internal force. For details, the reader is referred to Maugin [27].

We return now to Eqs. (23) and (24). Let us multiply Eq. (23) by U_X and Eq. (24) by φ_X and add the equations. The identical expressions

$$\rho_0 U_t U_{Xt} + I\varphi_t\varphi_{Xt} = \frac{1}{2}\left(\rho_0 U_t^2 + I\varphi_t^2\right)_X, \tag{28}$$

are added on the other side. Then we recognize the 1D pseudomomentum

$$\mathcal{P} = -\left(\rho_0 U_t U_X + I\varphi_t\varphi_X\right), \tag{29}$$

and the Lagrangian $\mathcal{L}$ with its derivative

$$\mathcal{L}_X = \frac{1}{2}\left(\rho_0 U_t^2 + I\varphi_t^2\right)_X - TU_{XX} - \eta\varphi_{XX} - \tau\varphi_X. \tag{30}$$

Consequently, Eqs. (28)–(30) yield

$$\mathcal{P} - b_X = 0, \tag{31}$$

where the Eshelby stress component b is determined by

$$b = -\frac{1}{2}\left(\rho_0 U_t^2 + I\varphi_t^2 + \alpha U_X^2 - B\varphi^2 + C\varphi_X^2\right). \tag{32}$$

For details of the derivation, see Engelbrecht et al. [12].

3.4 Internal variables

Waves occur in media in which energy can be stored in both kinetic and potential forms (see the Introduction). The usual understanding is that the dependent variables all are inertial, i.e. related to the energy. The first question is that of the temperature, however, the possible modifications of the Fourier law may overcome this certain paradox (Müller [30]). Setting aside this extremely interesting problem, let us pose the question whether, besides stress and strain as the main observable wave variables, there also exist other variables that do not possess inertia? It is clear that while in this case energies exist, the wave motion is possible but may be accompanied also by changes of other, uninertial variables. There are indeed many physical phenomena characterized by uninertial, i.e. internal variables. Maugin [22] and Maugin and Muschik [25] have described the formalism of internal variables and applied this formalism for many cases: nematic liquid crystals, localization of damage coupled with elasticity, microstructure in general, etc. The main idea in this formalism is to introduce, besides the kinetic energy K and potential energy W, also a dissipation potential $\mathcal{D}$, from which the governing equation(s) for internal variable(s) is/are derived.

In general terms, this equation reads

$$\frac{\delta \mathrm{W}}{\delta \alpha} + \frac{\partial \mathcal{D}}{\partial \alpha_t} = 0, \quad \frac{\delta}{\delta \alpha} = \frac{\partial}{\partial \alpha} - \nabla \frac{\partial}{\partial \nabla \alpha}, \tag{33}$$

where α is the internal variable, and $\delta/\delta\alpha$ denotes the Euler–Lagrange derivative. For details, see Maugin [22], Maugin and Muschik [25], and also Engelbrecht [10].

What makes the case of internal variables interesting is the fact that the hyperbolic equations of motion are accompanied by the evolution–diffusion type equations governing the internal variables. This means that wave structures and dissipative structures are combined. Undoubtedly, there are interesting physical phenomena governed by such models.

4 Model governing equations

4.1 Basic linear theory

In terms of the Piola-Kirchhoff stress tensor T^{KL}, which here can be written with lower indices (linear theory!) as only T_{KL}, the governing equation of motion in an isotropic elastic body is (cf. Eq. (2))

$$T_{KL,L} - \rho_0 U_{K,tt} = 0 . \tag{34}$$

The linear constitutive law for the same case is

$$T_{KL} = \lambda E_{NN} \delta_{KL} + 2\mu E_{KL} , \tag{35}$$

where λ and μ are the Lamé constants (the second-order elastic moduli). While

$$E_{KL} = \frac{1}{2} (U_{L,K} + U_{K,L}) , \tag{36}$$

then in terms of displacement U_K, Eqs (34)–(36) yield

$$\rho_0 U_{I,tt} - (\lambda + \mu) U_{K,KI} - \mu U_{I,KK} = 0, \tag{37}$$

which describes both longitudinal and transverse waves in the 3D setting. In components, Eq. (37) reads

$$\rho_0 U_{1,tt} - (\lambda + 2\mu) U_{1,11} - (\lambda + \mu) (U_{2,21} + U_{3,31}) - \mu (U_{1,22} + U_{1,33}) = 0 , \tag{38}$$

$$\rho_0 U_{2,tt} - (\lambda + 2\mu) U_{2,22} - (\lambda + \mu) (U_{1,12} + U_{3,32}) - \mu (U_{2,11} - U_{2,33}) = 0 , \tag{39}$$

$$\rho_0 U_{3,tt} - (\lambda + 2\mu) U_{3,33} - (\lambda + \mu) (U_{1,13} + U_{2,23}) - \mu (U_{3,11} - U_{3,22}) = 0 . \tag{40}$$

If we tackle only $U_I = U_I (X_1, t)$, i.e. a 1D case, then we get the system of equations

$$U_{1,tt} - c_0^2 U_{1,11} = 0, \tag{41}$$

$$U_{2,tt} - c_t^2 U_{2,11} = 0, \tag{42}$$

$$U_{3,tt} - c_t^2 U_{3,11} = 0, \tag{43}$$

where c_0, c_t are the velocities of longitudinal and transverse waves, respectively, and $c_0^2 = (\lambda + 2\mu)/\rho_0$, $c_t^2 = \mu/\rho_0$. In this case, all the waves are uncoupled and nondispersive. The governing equations (41)–(43) are typical hyperbolic wave equations possessing d'Alembert-type solutions (see, for example Achenbach [1]; Bland [4]).

4.2 Wave hierarchy

Waves in microstructured solids (see Section 3) are characterized by dispersion and the governing equations are more complicated than those presented in Section 4.1. In the case of the Mindlin-type model, longitudinal waves are described by Eqs. (23), (24). It is possible to present this system in the form of one equation (Engelbrecht et al. [11]):

$$U_{tt} - \left(c_0^2 - c_A^2\right) U_{XX} + p^2 \left(U_{tt} - c_0^2 U_{XX}\right)_{tt} - \\ - p^2 c_1^2 \left(U_{tt} - c_0^2 U_{XX}\right)_{XX} = 0, \tag{44}$$

where $c_0^2 = \alpha/\rho_0 = (\lambda + 2\mu)/\rho_0$, $c_A^2 = A^2/\rho_0 B$, $c_1^2 = C/I$ are velocities and $p^2 = I/B$ is an inherent time constant. By an asymptotic analysis (Engelbrecht et al. [12]), it is possible to present Eq. (38) in an approximated form

$$U_{tt} - \left(c_0^2 - c_A^2\right) U_{XX} + p^2 c_A^2 \left(U_{tt} - c_1^2 U_{XX}\right)_{XX} = 0. \tag{45}$$

This equation displays explicity the hierarchical character of the waves in the sense of Whitham [40]. In the derivation of Eq. (45), the scale parameter $\delta = l^2/L^2$ plays a crucial role. Here, l is the scale of the microstructure and L – is the wavelength. If δ is small (the wavelength is large), then the waves are governed by the properties of the macrostructure, i.e. the operator

$$L_{\text{macro}} = U_{tt} - \left(c_0^2 - c_A^2\right) U_{XX}, \tag{46}$$

has the leading role. If however δ is large (the wavelength is small) then the waves are governed by the properties of the microstructure, i.e. the operator

$$L_{\text{micro}} = U_{tt} - c_1^2 U_{XX}, \tag{47}$$

has the leading role. When both operators are to be accounted for, then we note the importance of the higher derivatives U_{ttXX} and U_{XXXX} that clearly give rise to dispersive effects. The dispersion analysis over a wide range of parameters is to be found in Engelbrecht et al. [11], [12].

It is possible to develop the modelling of microstructured media from the one-scale case like Eqs. (23), (24) and (44), (45) to the multiple scales (the scale within the scale). In this case, every deformable cell of the microstructure includes new deformable cells at a smaller scale. Then two scale parameters are to be determined: $\delta_1 = l_1^2/L^2$, $\delta_2 = l_2^2/L^2$, where l_1 and l_2 are the scales of both microstructures.

The final governing equation takes on the form (for details see Engelbrecht et al. [12])

$$U_{tt} - \left(c_0^2 - c_{A1}^2\right) U_{XX} + p_1^2 c_{A1}^2 \left[U_{tt} - \left(c_1^2 - c_{A2}^2\right) U_{XX}\right]_{XX} - \\ - p_1^2 c_{A1}^2 \, p_2^2 c_{A2}^2 \left(U_{tt} - c_2^2 U_{XX}\right)_{XXXX} = 0, \tag{48}$$

which should be compared to Eq. (44). Here new velocities c_{A2}, c_2 appear together with the new time constant p_2 while $p_1 = p$ and $c_{A1} = c_A$ in Eq. (44). Characteristically, the sixth order derivatives U_{ttXXXX} and U_{XXXXXX} appear, which become important for small wavelengths. This model is a step closer to crystal structures of materials (cf. Maugin [26]).

4.3 Nonlinearities

The principle of equipresence demands that all the effects of the same order should be taken into account to guarantee the best correspondence between the models and reality. Together with dispersion and dissipation , nonlinear effects are extremely important. It should be stressed that linear models are just first approximations, where the assumption of proportionality rules. Contemporary understanding, however, is different and nonlinearities are taken into account in order to explain many interesting phenomena in the real world. The topic of waves generally refers to the coupling of fields, the appearance of solitons and other solitary waves, the existence of shock waves and dissipative structures, the explanation of fracture mechanisms, etc.

There are many sources of nonlinearities influencing wave motion (see Engelbrecht [10]):

- material (physical) nonlinearities, i.e. the constitutive law(s) is/are nonlinear. In terms of stress-strain relations, this means that the potential energy W has terms higher than quadratic (cf. Eq. (6);
- geometrical nonlinearities, i.e. deformation (cf. strain tensor in its full form);
- kinematical nonlinearities, i.e. convectivity, compound motion, etc.;
- structural nonlinearities, for example, due to constraints limiting the motion of structural elements;
- combined nonlinearities, i.e. coupling of fields.

As an example, in the 1D setting, the equation of motion for longitudinal waves involving physical and geometrical nonlinearities, is

$$U_{1,tt} - c_0^2 \left[1 + 3\left(1 + m_0\right) U_{1,1}\right] U_{1,11} = 0, \tag{49}$$

where the physical nonlinearity is based on the potential energy W:

$$\rho_0 W = \frac{1}{2}\lambda I_1^2 + \mu I_2 + \nu_1 I_1^3 + \nu_2 I_1 I_2 + \nu_3 I_3, \tag{50}$$

from which the needed stress tensor component (cf. Eq. (6)) is

$$T^{11} = (\lambda + 2\mu) U_{1,1} + \left(\frac{1}{2}\lambda + \mu + 3\nu_1 + 3\nu_2 + 3\nu_3\right) U_{1,1}^2. \tag{51}$$

The strain tensor component is simply

$$E_{1,1} = U_{1,1} + \frac{1}{2} U_{1,1}^2. \tag{52}$$

Here λ, μ are Lamé parameters, $v_i, i = 1,2,3$, are the third order elastic constants, and $I_i, i = 1,2,3$, are the algebraic invariants of the strain tensor. In addition,

$$m_0 = 2\,(v_1 + v_2 + v_3)\,(\lambda + 2\mu)^{-1}. \tag{53}$$

Clearly, the nonlinear setting needs more physical parameters and the relations between variables are more cumbersome. But Eq. (49) is able to model the formation of shock waves (discontinuities), which the linear model (41) is not able to grasp. Why are nonlinearities important in modelling wave motion? Nonlinear models are able to describe distortion of wave profiles (spectral changes), amplitude-dependent velocities, interaction of waves, spatio-temporal chaos, and other important physical effects, going also beyond the elastic limit. The effects of plastic deformation are of special importance in structural mechanics, where hysteresis, ratcheting, hardening and other specific phenomena are to be accounted for. Here we refer to treatises of Eringen [13], Bland [3], Whitham [40], Engelbrecht [8], Jeffrey and Engelbrecht [16], Engelbrecht [9], Maugin [23], just to mention some from a long list of studies in nonlinear wave motion.

There is one more point to be stressed. It is not only the nonlinearity itself that influences the outcome, but often the possibility to balance between the nonlinearity and another characteristic property. This "other characteristic property"could be dispersion (then solitons or solitary waves may emerge), dissipation (then shock waves or dissipative structures may emerge), forcing (chaotic regimes may emerge), etc. In this sense, nonlinearity is a cornerstone for new phenomena that is characteristic to complex systems.

4.4 One-wave models

The leading wave operators, such as those in (44), (45), (48) are of the second order with respect to time and describe both left- and right- going waves. In the linear theory, longitudinal and shear waves are separated, but in nonlinear theory the coupling can affect both waves considerably. In addition, dispersive and dissipative effects make the governing systems rather complicated. The main question is then to understand to which wave which physical effects are related, both qualitatively and quantitatively.

One of the possibilities to overcome such difficulties in contemporary wave theory is to introduce the notion of evolution equations governing just one single wave. Physically, this means the separation (if possible) of a multi-wave process into separate waves. The waves are then governed by the so-called evolution equations, everyone of which describes the distortion of a single wave along a properly chosen characteristic (ray). The main idea of constructing such evolution equations is the following: a set of small parameters related either to the initial conditions or to the physical and/or geometrical parameters is introduced and the perturbation method with stretched coordinates is then applied. Taniuti and Nishihara [37], who initiated this approach, called it the "reductive perturbation method". Actually, there are several methods, which are used for this purpose (Jeffrey and Kawahara [17]; Engelbrecht [8], [10]). The stretched coordinates in a 1D case are, for example

$$\xi = \varepsilon^k (c_i t - X_1), \quad \tau = \varepsilon^{k+1} X_1, \tag{54}$$

where ε is a small parameter, c_i is the velocity from the main wave operator and k describes the space scale. Leaving aside the details (see Jeffrey and Kawahara [17]; Engelbrecht, [8], [10]), an evolution equation for 1D longitudinal waves in a material where nonlinearity and dispersion are accounted for, is

$$\frac{\partial u}{\partial t} + mu\frac{\partial u}{\partial \xi} + \Omega^{-2}\frac{\partial^3 u}{\partial \xi^3} = 0, \tag{55}$$

and in a material where nonlinearity and dissipation are accounted for, is

$$\frac{\partial u}{\partial \tau} + mu\frac{\partial u}{\partial \xi} - \Gamma^{-1}\frac{\partial^2 u}{\partial \xi^2} = 0. \tag{56}$$

Here $u \sim \partial U_1/\partial t$, while m, Ω, Γ are constants and $k = 0$. Equation (55) is the celebrated Korteweg-de Vries (KdV) equation and Eq. (56) – the Burgers equation. The first permits the emergence of solitons, the second – shock waves. Although these evolution equations are like fundamental "bricks" in contemporary mathematical physics, governing phenomena in fluids, gases, plasmas, transmission lines, etc., the reality is more complicated and the evolution equations turn out to be more complicated (Engelbrecht [10]; Maugin [26]; etc). For example, for waves in martensitic-austenitic alloys, the evolution equation for 1D longitudinal waves reads (Salupere et al. [35])

$$\frac{\partial u}{\partial \tau} + [P(u)]_\xi + \Omega_1^{-2}\frac{\partial^3 u}{\partial \xi^3} + \Omega_2^{-2}\frac{\partial^5 u}{\partial \xi^5} = 0, \tag{57}$$

$$P(u) = -\frac{1}{2}u^2 + \frac{1}{4}u^4, \tag{58}$$

where two dispersive terms with constants Ω_1 and Ω_2, and quadratic and quartic nonlinearities are taken into account.

5 Final remarks

Propagation of stress or deformation waves in solids is a challenging field of studies. Although the conservation laws as a basis for modelling wave motion have been known for a long time, the wide range of material properties and coupled fields generate more and more new problems. The nonlinear character of nature brings in the complexity of motion.

Here we have tackled but few of the problems: the basic ideas for modelling wave motion in solids and some advanced models, where the assumption of the homogeneity of materials must be replaced by the assumptions on the existence of the microstructure, which has a strong influence on wave characteristics. The models described are based mainly on the theory of elasticity, although the conservation laws in Section 2 include thermal effects. In order to include inelastic effects one has to develop more complicated mathematical models, but the essence of wave motion remains, in a general sense, the same. Indeed, speaking about waves in solids, the present overview is just an introduction. As stated in Section 1, one has to understand the waves in bounded solids (a half space, layered medium), where besides longitudinal and transverse waves, also surface waves exist, one has to model waves in structural elements (rods, plates, shells), one has to analyse the influence of coupled fields (thermoelasticity, electroelasticity, etc), one has to understand waves in the plastic range and dynamic effects of fracture, etc. The advanced theories (Section 3) could also include the gradient theories, nonlocal effects, micromorphic and/or micropolar theories, etc. (see, for example, Eringen [15]; Yerofeyev [41]). The present overview is actually a starting point to other presentations in this volume describing methods for and applications of waves.

Why is all this important? There are many reasons:

- wave characteristics explain dynamical stress and/or strain at dynamical loads, the outcome of which may considerably exceed statical values;
- waves carry information about the material properties and/or stress states in solids and this can be used for nondestructive nesting;
- waves can change the structure of materials (phase-transformation boundaries, for example), etc.

In order to effectively use mathematical models developed for various cases, one should apply effective methods for solving them. There are but few analytical solutions. Of course we know the analytical solution of a classical wave equation (cf. Eq. (35)) or the KdV equation (Eq. 55), but most cases need numerical treatment. Of the many existing algorithms, we describe here the finite volume method (Berezovski et al. [2]) and the pseudospectral method (Salupere [36]).

There is a real challenge in studies of wave motion in solids – to bind the physics of materials and the macrobehaviour. The first step in such an analysis is to use mesoscopic continuum physics together with microcontinuum mechanics that may also involve internal variables. Such a compound theory could be used for a wide range of loadings including high frequency excitations and coupled physical fields (stress-strain, temperature, electromagnetic forces, etc.).

Even now, one has to tackle the two-faced behaviour of materials: numerically discretized continuum models of solids, and discrete crystal lattices, which are viewed as continua in the long wavelength approximation. Based on physical analysis and compound theories, building bridges between these two approaches is an important task. Quite probably, nonlinearity will play a significant role in all new theories, reflecting the complexity of the physical world.

Acknowledgements The author would like to thank the referees for their valuable comments, which improved the presentation.

References

1. Achenbach, J.D.: Wave Propagation in Elastic Solids. North-Holland, Amsterdam (1973)
2. Berezovski, A., Berezovski, M., Engelbrecht, J.: Waves in inhomogeneous solids. In: Quak, E., Soomere, T. (eds.) Applied Wave Mathematics, pp. 55–81. Springer, Heidelberg (2009)
3. Bland, D.R.: Nonlinear Dynamic Elasticity. Blaisdell, Waltham (1969)
4. Bland, D.R.: Wave Theory and Applications. Clarendon Press, Oxford (1988)
5. Brillouin, L.: Wave Propagation in Periodic Structures. Dover, Toronto (1953)
6. Capriz, G.: Continua with Microstructure. Springer, New York (1989)
7. Cauchy, A.L.: Sur les équations qui experiment les conditions d'équilibre ou les lois du mouvement intérieur d'un corps solide, élastique ou non élastique. Ex. de Math. **3**, 160–187 (1822) = Oeuvres (2) **8**, 253–277
8. Engelbrecht, J.: Nonlinear Wave Processes of Deformation in Solids. Pitman, London (1983)
9. Engelbrecht, J.: An Introduction to Asymmetric Solitary Waves. Longman, Harlow (1991)
10. Engelbrecht, J.: Nonlinear Waves Dynamics. Complexity and Simplicity. Kluwer, Dordrecht (1997)
11. Engelbrecht, J., Berezovski, A., Pastrone, P., Braun, M.: Waves in microstructured materials and dispersion. Phil. Mag. **85**, 4127–4141 (2005)
12. Engelbrecht, J., Pastrone, F., Braun, M., Berezovski, A.: Hierarchies of waves in nonclassical materials. In: Delsanto, P.-P. (ed.) Universality of Nonclassical Nonlinearity: Application to Non-Destructive Evaluation and Ultrasonics, pp. 29–47. Springer, New York (2007)
13. Eringen, A.C.: Nonlinear Theory of Continuous Media. McGraw-Hill, New York (1962)
14. Eringen, A.C.: Linear theory of micropolar elasticity. J. Math. Mech. **15**, 909–923 (1966)
15. Eringen, A.C.: Microcontinuum Field Theories. Foundations and Solids. Springer, New York (1999)
16. Jeffrey, A., Engelbrecht, J. (eds.): Nonlinear Waves in Solids. Springer, Wien (1994)
17. Jeffrey, A., Kawahara, T.: Asymptotic Methods in Nonlinear Wave Theory. Pitman, Boston (1982)
18. Kolsky, H.: Stress Waves in Solids, 2nd ed. Dover, New York (1963)
19. Lamb, H.: Hydrodynamics. Cambridge University Press (1879); see also the 1997 edition from CUP.
20. Lamé, G.: Leçons sur la Theorie Mathématique de l'Elasticité des Corps Solides. Bachelier, Paris (1852)
21. Love, A.E.N.: A Treatise on the Mathematical Theory of Elasticity. Cambridge University Press (1906)
22. Maugin, G.A.: Internal variables and dissipative structures. J. Non-Equilib. Thermodyn. **15**, 173–192 (1990)
23. Maugin, G.A.: Thermomechanics of Plasticity and Fracture. Cambridge University Press (1992)
24. Maugin, G.A.: Material Inhomogeneities in Elasticity. Chapman & Hall, London (1993)

25. Maugin, G.A., Muschik, W.: Thermodynamics with internal variables, Part I: general concepts, Part II: applications. J. Non-Equilib. Thermodyn. **19**, 217–249, 250–289 (1994)
26. Maugin, G.A.: Nonlinear Waves in Elastic Crystals. Oxford University Press (1999)
27. Maugin, G.A.: Pseudo-plasticity and pseudo-inhomogeneity effects in materials mechanics. J. Elasticity **71**, 81–103 (2003)
28. Miklowitz, J.: The Theory of Elastic Waves and Waveguides, 2nd ed. North-Holland, Amsterdam (1980)
29. Mindlin, R.D.: Micro-structure in linear elasticity. Arch. Rat. Mech. Anal. **16**, 51–78 (1964)
30. Müller, I.: Thermodynamics. Pitman, London (1985)
31. Pastrone, F.: Waves in solids with vectorial microstructure. Proc. Estonian Acad. Sci. Phys. Math. **52**, 21–29 (2003)
32. Pastrone, F.: Nonlinearity and complexity in elastic wave motion. In: Delsanto, P.-P. (ed.) Universality of Nonclassical Nonlinearity: Application to Non-Destructive Evaluation and Ultrasonics, pp. 15–26. Springer, New York (2007)
33. Poisson, S.D.: Mémoire sur les équations générales de l'equilibre et du mouvement des corps élastiques et des fluides. J. École Poly. **13**(20), 1–174 (1829)
34. Rayleigh, L.: On progressive waves. Proc. London Math. Soc. **17**, 21–26 (1887)
35. Salupere, A., Engelbrecht, J., Maugin, G.A.: Solitonic structures in KdV-based higher-order systems. Wave Motion **34**, 51–61 (2001)
36. Salupere, A.: The pseudospectral method and discrete spectral analysis. In: Quak, E., Soomere, T. (eds.) Applied Wave Mathematics, pp. 301–333. Springer, Heidelberg (2009)
37. Taniuti, T., Nishihara, K.: Nonlinear Waves. World Scientific, Singapore (1983) (in Japanese 1977)
38. Truesdell, C.A., Toupin, R.: The classical field theories. In: Flugge's Handbuch der Physik **III/1**, pp. 226–793. Springer, Berlin (1960)
39. Truesdell, C.A., Noll, W.: The nonlinear field theories. In: Flugge's Handbuch der Physik **III/3**, pp. 1–602. Springer, Berlin (1965)
40. Whitham, G.B.: Linear and Nonlinear Waves. Wiley, New York (1974)
41. Yerofeyev, V.I., Kazhayev, V.V., Semerikova, N.P.: Waves in Rods. Dispersion, Dissipation, Nonlinearity. Physmatlit, Moscow (2002) (in Russian)

The Perturbation Technique for Wave Interaction in Prestressed Material

Arvi Ravasoo

Abstract Perturbation theory is a collection of methods for the systematic analysis of the global behavior of solutions to differential and difference equations. It is most useful when the first step reveals the important features of the solution and the remaining ones give small corrections. This is illustrated by the solution to the problem of interaction of longitudinal waves in an elastic material with inhomogeneous plane prestrain. The perturbative solution to the governing equation is expanded in a series with a small parameter that characterizes the small strain. The obtained solution that is global in space and time and local in the small parameter is analyzed in detail. Utilization of the solution in practical applications of ultrasonic nondestructive material characterization is discussed on the basis of numerical simulation data.

1 Introduction

Perturbation theory [1, 2, 5, 6, 7] is the study of the effects of small disturbances. The basic idea in perturbation theory is to obtain an approximate solution of a mathematical problem by using the presence of a small dimensionless parameter. Perturbation theory is used to find the approximate solution to a problem which cannot be solved exactly and if the problem can be formulated by adding a "small term" to the solution of the exactly solvable problem. This leads to the solution in terms of a power series with small parameter ε

$$U = U^{(0)} + \varepsilon U^{(1)} + \varepsilon^2 U^{(2)} + \varepsilon^3 U^{(3)} + \cdots, \qquad (1)$$

where $U^{(0)}$ is the solution to the exactly solvable problem.

Arvi Ravasoo
Centre for Nonlinear Studies, Institute of Cybernetics at Tallinn University of Technology, Akadeemia tee 21, 12618 Tallinn, Estonia, e-mail: arvi@ioc.ee

E. Quak, T. Soomere (eds.), *Applied Wave Mathematics*,
DOI 10.1007/978-3-642-00585-5_4, © Springer-Verlag Berlin Heidelberg 2009

The second and the subsequent terms describe the deviation of the solution caused by the difference between the considered problem and the exactly solvable problem. It is crucial that for small ε the higher order terms are successively less important. The perturbation theory is usually designated according to the order to which the perturbation is carried out. The first order perturbation theory takes the terms with ε into account, the second order perturbation theory adds to these terms the terms with ε^2, and so on.

As an illustration a perturbative solution to the problem of wave propagation, reflection and interaction in an inhomogeneously prestressed material is derived. The solution which is global in space and time and local in the small parameter enables it to determine the relations between the wave characteristics and the parameters of the physically nonlinear elastic material and its prestressed state, and therefore it may be useful in ultrasonic nondestructive testing (NDT).

2 Prelude

An elementary example to introduce the ideas of perturbation theory is presented. The response of an oscillator to an arbitrary nonlinear external effect is described by the equation

$$U_{,tt} + U = f(U, U_{,t}), \tag{2}$$

where t is the dimensionless time and U notes a dimensionless function of time. An index t after a comma indicates differentiation with respect to time t. The *nonlinear function* $f(U, U_{,t})$ determines the arbitrary nonlinear active force. It is essential that Eq. (2) does not involve a small parameter.

Henceforth, weak nonlinear oscillations are studied. The original problem is converted into a perturbation problem by constructing the general solution to Eq. (2) by means of a power series in a small parameter ε

$$U = \varepsilon U^{(1)} + \varepsilon^2 U^{(2)} + \cdots, \tag{3}$$

where the small parameter $(0 < |\,\varepsilon\,| \ll 1)$ determines the strength of the nonlinearity.

Substitution of the power series (3) into Eq. (2) leads to the equation

$$\varepsilon U^{(1)}_{,tt} + \varepsilon^2 U^{(2)}_{,tt} + \cdots + \varepsilon U^{(1)} + \varepsilon^2 U^{(2)} + \cdots = \varepsilon^2 f(U^{(1)}, U^{(1)}_{,t}) + \cdots \tag{4}$$

with the assumption that the nonlinear right-hand side can be expanded into a power series in the small parameter ε.

Separately equating the coefficients of like powers in ε in Eq. (4) to zero gives a sequence of equations, where the first two are for

$O(\varepsilon)$:

$$U^{(1)}_{,tt} + U^{(1)} = 0, \tag{5}$$

and for
$O(\varepsilon^2)$:

$$U^{(2)}_{,tt} + U^{(2)} = f(U^{(1)}, U^{(1)}_{,t}). \tag{6}$$

Above, the Landau order symbol O is used [5].

Equations (5) and (6) can now be solved successively since the inhomogeneous term on the right-hand side of Eq. (6) involves only the previously determined function.

The general solution to Eq. (5), i.e., the time-dependent part of the first term in the solution (3) is

$$U^{(1)} = \alpha \cos(t + \beta), \tag{7}$$

where α and β are arbitrary constants.

The second term in the series (3) is the solution to Eq. (6) that takes the form

$$U^{(2)}_{,tt} + U^{(2)} = f[\alpha \cos(t + \beta), -\alpha \sin(t + \beta)]. \tag{8}$$

The known right-hand side of Eq. (8) is a periodic function with the period equal to 2π. The particular solution to Eq. (8) is determined from this equation with the right-hand side expanded into the Fourier series [7]

$$\begin{aligned} & f[\alpha \cos(t + \beta), -\alpha \sin(t + \beta)] \\ &= f_0(\alpha) + \sum_{n=1}^{\infty} f_n(\alpha) \cos(nt + n\beta) + \sum_{n=1}^{\infty} g_n(\alpha) \sin(nt + n\beta), \end{aligned} \tag{9}$$

where

$$f_0(\alpha) = \frac{1}{2\pi} \int_0^{2\pi} f(\alpha \cos\varphi, -\alpha \sin\varphi) d\varphi, \tag{10}$$

$$f_n(\alpha) = \frac{1}{\pi} \int_0^{2\pi} f(\alpha \cos\varphi, -\alpha \sin\varphi) \cos n\varphi \, d\varphi, \tag{11}$$

$$g_n(\alpha) = \frac{1}{\pi} \int_0^{2\pi} f(\alpha \cos\varphi, -\alpha \sin\varphi) \sin n\varphi \, d\varphi. \tag{12}$$

Equation (8) can now be rewritten as

$$\begin{aligned} & U^{(2)}_{,tt} + U^{(2)} \\ &= f_0 + \sum_{n=1}^{\infty} f_n \cos(nt + n\beta) + \sum_{n=1}^{\infty} g_n \sin(nt + n\beta). \end{aligned} \tag{13}$$

Equation (13) is linear and the principle of superposition may be used to derive the particular solution to it as a sum of particular solutions that correspond to each term on the right-hand side.

The result is

$$U^{(2)} = f_0 + \frac{1}{2} f_1 t \sin(t+\beta) - \frac{1}{2} g_1 t \cos(t+\beta) - \sum_{n=2}^{\infty} \frac{1}{n^2-1} \left[f_n \cos(nt+n\beta) + g_n \sin(nt+n\beta)\right]. \quad (14)$$

Substitution of expressions (7) and (14) into the series (3) leads to the weakly non-linear perturbative solution to Eq. (2)

$$U = \varepsilon[\alpha \cos(t+\beta)] + \varepsilon^2 \left\{ f_0 + \frac{1}{2} f_1 t \sin(t+\beta) - \frac{1}{2} g_1 t \cos(t+\beta) - \sum_{n=2}^{\infty} \frac{1}{n^2-1} \left[f_n \cos(nt+n\beta) + g_n \sin(nt+n\beta)\right] \right\} + \cdots. \quad (15)$$

The obtained solution (15) contains secular terms and it is not valid if $O(t) \geq O(\varepsilon^{-1})$. Methods that enable to transform the solution (15) to the uniform solution are described in [7], for example.

This simple example illustrates the three basic steps of perturbative analysis:

1. Replace the original problem with a perturbation problem by introducing a small parameter ε.
2. Propose an expression for the solution in the form of a perturbation series and determine the terms of that series.
3. Compose the solution to the original problem by summing the perturbation series for the appropriate value of ε.

These three steps are applied below for a more complicated problem of wave interaction in prestressed material.

3 Basic relations in continuum mechanics

For a better understanding of the physical side of the application of the perturbation technique to the wave interaction problem, a very short introduction to the theory of continuum mechanics is presented.

The theory of continuum mechanics is a scientific discipline concerned with the global behavior of substances under the influence of external disturbances. This theory is described in detail in numerous monographs and textbooks. Below, the terminology and the basic concepts follow the description in the books [4, 10].

3.1 Coordinate systems

Two different coordinate systems are used in continuum mechanics: Eulerian and Lagrangian coordinates. The Eulerian coordinates denoted as x_1, x_2, x_3 are bound to the physical space, where the motion and deformation of the material is observed. The Lagrangian coordinates of the material particle X_1, X_2, X_3 remain the same during the deformation, i. e. they are attached to the deformed material.

Here these coordinates are considered as Cartesian (rectangular) coordinates and they coincide at the instant $t = 0$:

$$x_1 = X_1,\, x_2 = X_2,\, x_3 = X_3. \tag{16}$$

The one-parametric coordinate transformations

$$x_k = x_k(X_1, X_2, X_3, t) \;\; (k = 1,2,3), \tag{17}$$

$$X_K = X_K(x_1, x_2, x_3, t) \;\; (K = 1,2,3), \tag{18}$$

are defined as motion. The Lagrangian description of the law of motion (17) transfers the material point with coordinates $X\,(X_1, X_2, X_3, 0)$ to the point $x\,(x_1, x_2, x_3, t)$ in the physical space. The Eulerian description (18) determines which material point $X\,(X_1, X_2, X_3, 0)$ is at the point $x\,(x_1, x_2, x_3, t)$ in physical space at the instant t. The uniqueness of the laws of motion (17) and (18) is guaranteed provided the Jacobian j of the partial derivatives $\partial x_k / \partial X_K$ $(k, K = 1,2,3)$ is non-zero, i. e.,

$$j = \det\left(\frac{\partial x_k}{\partial X_K}\right) \neq 0 \quad (K, k = 1,2,3) \tag{19}$$

holds at every point $x\,(x_1, x_2, x_3, t)$ and in its neighborhood. The Lagrangian description of the material particle displacement is expressed by the assumption $x_k \equiv X_K$ at $t = 0$ in the form

$$U_K = \delta_{Kk}\, x_k(X_1, X_2, X_3, t) - X_K\,,\; (K, k = 1,2,3)\,. \tag{20}$$

3.2 Conservation laws

The fundamental equations in continuum mechanics are the laws of conservation.

The local conservation of mass:

$$\frac{\rho_o}{\rho} = \det\left(\frac{\partial x_k}{\partial X_K}\right) = j\,, \quad (K, k = 1,2,3)\;, \tag{21}$$

where ρ is the mass density in Eulerian description and ρ_o is the initial density at $t = 0$. The uniqueness of the Lagrangian description requires the Jacobian j to be non-zero. This condition expresses the axiom of continuity so that the deformations are non-destructive: a finite volume deforms to a finite volume and different volumes cannot penetrate into each other.

The conservation of momentum:

$$\frac{\partial T_{Kl}}{\partial X_K} = \rho_o \frac{\partial^2 x_l}{\partial^2 t}, \quad (K,l = 1,2,3) , \tag{22}$$

where T_{Kl} is the first Piola-Kirchhoff stress tensor. The tensor T_{Kl} determines the stress at the point $x(x_1,x_2,x_3,t)$ with respect to the initial unit area at $X(X_1,X_2,X_3,0)$.

The conservation of the moment of momentum:
The third conservation law may be written in the compact form

$$T_{KL} = T_{LK} , \quad (K,L = 1,2,3) , \tag{23}$$

by assuming that the material is free from the surface and volume couples. The second Piola-Kirchhoff stress tensor T_{KL} denotes the stress at the point $X\ (X_1,X_2,X_3,t)$ with respect to the initial unit area at $X\ (X_1,X_2,X_3,0)$.

Both Piola-Kirchhoff stress tensors describe stresses computed for the initial area. Therefore they are often called pseudostress tensors. The relation between these two tensors is

$$T_{KL} = T_{Kl} \frac{\partial X_L}{\partial x_l}, \quad (K,L,l = 1,2,3) . \tag{24}$$

The actual stress at the point $x\,(x_1,x_2,x_3,t)$ of the physical space is determined by the Euler stress tensor t_{kl}. It relates to the first Piola-Kirchhoff stress tensor as

$$t_{kl} = \frac{\rho}{\rho_o} T_{Kl} \frac{\partial x_k}{\partial X_K}, \quad (K,k,l = 1,2,3) . \tag{25}$$

Conservation of energy:
This conservation law states that the time rate of the change of the kinetic and internal energy is equal to the sum of the rate of work done by the external forces plus the sum of the energies that enter or leave the material per unit time. The latter may be the mechanical equivalent of heat, the electric charge, the chemical energy, etc.

3.3 The constitutive equation

Conservation laws are valid for any type of media. These laws are inadequate for a unique determination of the unknowns involved. The system of equations may be closed on the basis of additional information about the material properties of the medium.

Henceforth an ideally elastic material is considered, in which the stress is a function of the state of deformation only. This definition enables it to isolate an initial natural state of the material at $t = 0$, where the stress and strain is absent. This initial state is described by the relations

$$t_{kl} = 0\,,\, C_{KL} = \delta_{KL}\,,\, E_{KL} = 0\,,\, t = 0\,,\ (K,k,L,l = 1,2,3), \tag{26}$$

where C_{KL} is the Green deformation tensor, E_{KL} the Green-Lagrange strain tensor, and δ_{KL} denotes the Kronecker delta. In rectangular coordinates the Green deformation tensor has the form

$$C_{KL} = \delta_{kl}\frac{\partial x_k}{\partial X_K}\frac{\partial x_l}{\partial X_L}\,,\ \ (K,k,L,l = 1,2,3). \tag{27}$$

This enables expressing the Green-Lagrange strain tensor as

$$2E_{KL} = C_{KL} - \delta_{KL} = \frac{\partial U_K}{\partial X_L} + \frac{\partial U_L}{\partial X_K} + \frac{\partial U_M}{\partial X_K}\frac{\partial U_M}{\partial X_L}, \ (K,L = 1,2,3). \tag{28}$$

The constitutive equation for an elastic material is derived by the additional assumption that the material is hyperelastic, i. e., the strain energy is a function of the strain only and there exists a strain energy Σ in the form

$$\Sigma = \Sigma\,(X_K, I_1, I_2, I_3)\,,\ \ (K = 1,2,3). \tag{29}$$

The strain energy depends on the invariants I_1, I_2, I_3 of the Green-Lagrange strain tensor E_{KL}:

$$I_1 = E_{KK},\, I_2 = E_{KL}E_{LK},\, I_3 = E_{KL}E_{ML}E_{KM}, (K,L,M = 1,2,3). \tag{30}$$

The constitutive equation may be presented now in a compact form through the second Piola-Kirchhoff stress tensor and the Green-Lagrange strain tensor

$$T_{KL} = \frac{\partial \Sigma}{\partial E_{KL}}\,,\ \ (K,L, = 1,2,3). \tag{31}$$

Using Eq. (29) the second Piola-Kirchhoff stress tensor (31) may be expressed in the form

$$T_{KL} = \left[\frac{\partial \Sigma}{\partial I_1}\frac{\partial I_1}{\partial E_{KL}} + \frac{\partial \Sigma}{\partial I_2}\frac{\partial I_2}{\partial E_{KL}} + \frac{\partial \Sigma}{\partial I_3}\frac{\partial I_3}{\partial E_{KL}}\right]\,,\ (K,L = 1,2,3). \tag{32}$$

Both Piola-Kirchhoff stress tensors describe stresses computed for the initial undeformed area. Therefore it is convenient to use them if deformations correspond to the small but finite strain

$$\left|\frac{\partial U_K}{\partial X_L}\right| \ll 1\,,\ \ (K,L, = 1,2,3). \tag{33}$$

The nonlinear theory of elasticity is formulated by assumptions that initially the strain energy is equal to zero ($\Sigma(0,0,0) = 0$) and the approximation of the strain energy function (29) with respect to the initial state gives [3]

$$\Sigma = \frac{1}{2}\lambda I_1^2 + \mu I_2 + v_1 I_1^3 + v_2 I_1 I_2 + v_3 I_3 + \cdots . \tag{34}$$

Expression (34) characterizes the five-constant nonlinear theory of elasticity. Herc λ and μ are Lamé constants or the elastic constants of the second order and v_1, v_2 and v_3 are the elastic constants of the third order.

Taking into account Eq. (34), the constitutive equation for the nonlinear elastic material may be expressed by the formula

$$\begin{aligned} T_{KL} = (\lambda I_1 + 3v_1 I_1^2 + v_2 I_2)\frac{\partial I_1}{\partial E_{KL}} + (\mu + v_2 I_1)\frac{\partial I_2}{\partial E_{KL}} \\ + v_3 \frac{\partial I_3}{\partial E_{KL}} + \cdots, \quad (K,L,M,N = 1,2,3) . \end{aligned} \tag{35}$$

3.4 Initial and boundary conditions

The above basic equations allow to predict the unique outcome for the physical phenomena. The governing equations are in the form of partial differential equations that may be solved after the determination of initial and boundary conditions. These conditions are very specific and depend on the physical nature of the problem. Two problems with initial and boundary conditions are described below.

3.5 Compatibility conditions

If the displacement U_K is given then substitution into Eq. (28) determines the Green-Lagrange strain tensor E_{KL}. If instead E_{KL} is given then the determination of the free unknowns U_K requires the solution of the six partial differential equations (28). Such a system is overdetermined and for the existence of a single-valued continuous displacement field, restrictions (compatibility conditions) upon E_{KL} must be imposed.

In the problem below U_K is given and that is why the compatibility conditions are not discussed here in detail. The relevant information is presented in [4, 10], for example.

4 Governing equations

The application of the perturbation technique is demonstrated on the governing equations of the nonlinear theory of elasticity that describe the equilibrium of the elastic material (specimen, structural element) under external forces and the interaction of waves in this prestressed material.

The material is assumed to be isotropic and homogeneous. Deformations of the material that correspond to the small but finite strain (see Eq. 33) are described on the basis of the nonlinear theory of elasticity presented above.

In nonlinear problems it is important to choose the appropriate reference frame. The Lagrangian reference frame is preferable for the considered problem because the aim is to apply external forces to the boundaries of the material (specimen, structural element) and to follow the deformation process on these boundaries. Therefore, the deformations of the material are described in Lagrangian Cartesian coordinates X_K, $K = 1,2,3$.

It is assumed that initially the material is in the natural, stress-free state where, at time instant $t = 0$, it will be subjected to external forces. From now on the material is in a prestressed state, which is considered to be static, i.e., independent of time. The displacement vector and the second Piola-Kirchhoff stress tensor at this state are denoted by $U_K^0(X_J)$ and $T_{KL}^0(X_J)$, respectively. At some instant $t > 0$, the longitudinal wave process represented by $U_K(X_J,t)$ and $T_{KL}(X_J,t)$ is excited in the prestressed material. As a result, the displacement at the present state can be expressed as the sum

$$U_K^*(X_J,t) = U_K^0(X_J) + U_K(X_J,t). \tag{36}$$

The equation of motion of the material at the present state

$$[T_{KL}^*(\delta_{kL} + \delta_{kM}\, U_{M,L}^*)]_{,K} - \rho_0\, \delta_{kM}\, U_{M,tt}^* = 0 \tag{37}$$

is derived on the basis of the laws of conservation (22), (23) and expressions (28) and (35). In Eq. (37) ρ_0 denotes the density of the material in the natural prestress-free state, and δ_{kL} and δ_{kM} are Euclidean shifters, which connect the Lagrangian Cartesian coordinates X_K and the Eulerian Cartesian coordinates x_k. Indices K,L and t after a comma indicate differentiation with respect to X_K, X_L and time t, respectively. The usual summation convention is used and all indices, except time t, run over 1, 2 and 3.

The geometrical and the physical nonlinearities are taken into account during the derivation of the equation of motion (37). The geometrical nonlinearity is included by the consideration of the last term in Eq. (28). The utilization of the geometrical nonlinearity is crucial in the considered problem. This enables to describe the interaction between the prestress field and the stress field evoked by the wave motion.

Physical nonlinearity means the nonlinear relation between the stress and the strain. Here, the nonlinear stress-strain relation is described by Eq. (35). The linear part of it is characterized by the Lamé constants λ and μ and the quadratic nonlinear part by the three elastic constants of the third order ν_1, ν_2 and ν_3.

The case of plane strain is considered, i.e., the component $U_3^*(X_J,t)$ of the displacement vector is assumed to be equal to zero. In this case, the equation of motion (37) takes the form of a system of two equations

$$\begin{aligned}
&[1+k_1\,U^*_{I,I}+k_2\,U^*_{J,J}]\,U^*_{I,II}+[2\,k_3\,U^*_{I,J}+2\,k_4\,U^*_{J,I}]\,U^*_{I,IJ}\\
&+[k_7+k_3\,U^*_{I,I}+k_3\,U^*_{J,J}]\,U^*_{I,JJ}+[k_4\,U^*_{I,J}+k_3\,U^*_{J,I}]\,U^*_{J,II}\\
&+[k_3\,U^*_{I,J}+k_4\,U^*_{J,I}]\,U^*_{J,JJ}+[k_6+k_5\,U^*_{I,I}+k_5\,U^*_{J,J}]\,U^*_{J,JI}\\
&-c^{-2}\,U^*_{I,tt}=0,
\end{aligned}\tag{38}$$

where the indices are $I=1$, $J=2$ for the first equation and $I=2$, $J=1$ for the second. The coefficients

$$\begin{aligned}
&k_1=3+6k\,(\nu_1+\nu_2+\nu_3),\quad k_2=k\,(\lambda+6\,\nu_1+2\,\nu_2),\\
&k_3=1+k\,(\nu_2+\frac{3}{2}\,\nu_3),\qquad k_4=k\,(\mu+\nu_2+\frac{3}{2}\,\nu_3),\\
&k_5=k\,[\lambda+\mu+3\,(2\,\nu_1+\nu_2+\frac{1}{2}\,\nu_3)],\\
&k_6=k\,(\lambda+\mu),\quad k_7=k\,\mu,\quad k=(\lambda+2\,\mu)^{-1},\quad c^{-2}=\rho_0\,k
\end{aligned}\tag{39}$$

characterize the properties of the material.

The governing equations for the equilibrium of the material in a prestressed state and for wave propagation in the prestressed material are obtained after substitution of Eq. (36) into Eq. (38).

The equilibrium of the material in the static prestressed state is described by the set of equations

$$\begin{aligned}
&[1+k_1\,U^0_{I,I}+k_2\,U^0_{J,J}]\,U^0_{I,II}+[2\,k_3\,U^0_{I,J}+2\,k_4\,U^0_{J,I}]\,U^0_{I,IJ}\\
&+[k_7+k_3\,U^0_{I,I}+k_3\,U^0_{J,J}]\,U^0_{I,JJ}+[k_4\,U^0_{I,J}+k_3\,U^0_{J,I}]\,U^0_{J,II}\\
&+[k_3\,U^0_{I,J}+k_4\,U^0_{J,I}]\,U^0_{J,JJ}+[k_6+k_5\,U^0_{I,I}+k_5\,U^0_{J,J}]\,U^0_{J,JI}=0,
\end{aligned}\tag{40}$$

where the indices are $I=1$, $J=2$ for the first equation and $I=2$, $J=1$ for the second.

The counterpropagation of one-dimensional longitudinal waves in the material (structural element) with two parallel boundaries is considered. Two waves are excited simultaneously on the surfaces $X_1=0$ and $X_1=h$ of the material in correspondence with the loading scheme presented in Fig. 1. The prestressed state and the ratio of the width of the excitation zone to the thickness of the material are assumed to assure that the following order relations that characterize the wave propagation along the axis X_1 are satisfied:

$$\begin{aligned}
&|\,U_{2,K}\,|\ll|\,U_{1,K}\,|,\quad |\,U_{2,KL}\,|\ll|\,U_{1,KL}\,|,\quad |\,U_{1,2}\,|\ll|\,U_{1,1}\,|,\\
&|\,U_{1,12}\,|=|\,U_{1,21}\,|\ll|\,U_{1,11}\,|,\quad |\,U_{1,22}\,|\ll|\,U_{1,11}\,|.
\end{aligned}\tag{41}$$

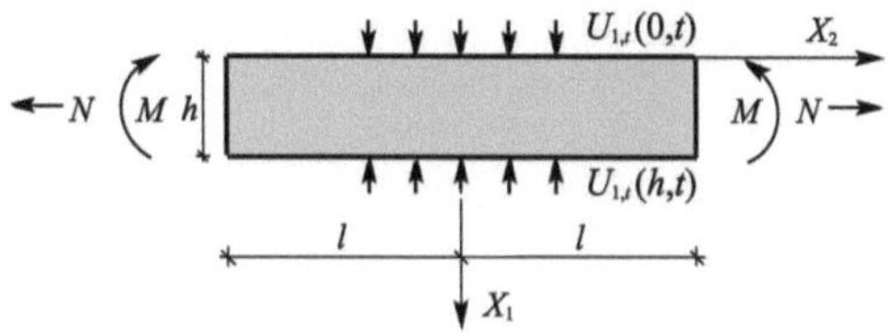

Fig. 1 Loading scheme

The indices K and L in Eq. (41) run over 1 and 2. Physically, this means that the spatial derivatives of the displacements due to the propagating waves are much larger in the direction of propagation X_1 than in the orthogonal direction X_2. Substituting Eq. (36) into Eq. (38) and taking the equations of equilibrium (40) and inequalities (41) into account, the governing equation

$$[1+k_1\,U^0_{1,1}+k_2\,U^0_{2,2}]\,U_{1,11}+[k_1\,U^0_{1,11}+k_3\,U^0_{1,22}+k_5\,U^0_{2,12}]\,U_{1,1} +k_1\,U_{1,11}\,U_{1,1}-c^{-2}\,U_{1,tt}=0 \qquad (42)$$

is obtained, which describes longitudinal wave propagation in a material undergoing inhomogeneous plane strain.

5 The perturbation technique

The intention is to solve the problem of nondestructive evaluation of the state of plain strain in elastic material on the basis of wave propagation data. The wave process in the prestressed material is governed by the equation (42). To solve this equation with respect to wave motion it is necessary to have some preliminary information about the prestressed state, i.e., about the coefficients of the equation. This information may be obtained from the solution to the set of equations (40).

The problem (40), (42) is solved under the assumption that the strain evoked in the material by the prestress and wave motion is small but finite and plastic deformations are not allowed. This leads to the idea to solve Eqs. (40) and (42) making use of the perturbation theory. Thus a small parameter $|\,\varepsilon\,|\ll 1$ that has the physical meaning of a small strain is introduced.

Solutions of equations (40) and (42) are sought assuming that the displacement of the prestressed state can be expressed by the series

$$U^0_K=\sum_{m=1}^{\infty}\varepsilon^m\,U_K^{0\,(m)}, \qquad (43)$$

and the displacement due to the wave motion can be expressed by the series

$$U_1 = \sum_{n=1}^{\infty} \varepsilon^n U_1^{(n)}. \tag{44}$$

In principle, the small parameters in series (43) and (44) that describe displacements caused by the prestress and the wave motion, respectively, may be of different order. Here, the idea is to use nonlinear effects of wave propagation in the nondestructive characterization of the prestressed state of the material. As it is shown in [8], the nonlinear effects of wave motion contain maximum information about the prestressed state provided these displacements are of the same order. This case is considered henceforth.

The result is that the original problem is converted into a perturbation problem by introducing the small parameter ε. The peculiarity of the considered problem is that the small parameter is introduced by boundary conditions.

As it will be seen below, a regular perturbation problem is discussed, where the order of the equations and the number of initial and boundary conditions remain the same for $\varepsilon = 0$. If the order of the equations is reduced when $\varepsilon = 0$, or if one or more initial or boundary conditions have to be discarded the perturbation problem is called singular [1].

5.1 *The prestressed state*

The solution of the equilibrium equations (40) is sought in the form of series (43). Inserting series (43) into Eq. (40) and equating to zero the terms of equal powers of ε, the first two terms of series (43) may be determined from the sets of equations

$O(\varepsilon)$:

$$U_{I,II}^{0(1)} + k_7 U_{I,JJ}^{0(1)} + k_6 U_{J,JI}^{0(1)} = 0, \tag{45}$$

$O(\varepsilon^2)$:

$$\begin{aligned} U_{I,II}^{0(2)} + k_7 U_{I,JJ}^{0(2)} + k_6 U_{J,JI}^{0(2)} + [k_1 U_{I,I}^{0(1)} + k_2 U_{J,J}^{0(1)}]\, U_{I,II}^{0(1)} \\ + [2 k_3 U_{I,J}^{0(1)} + 2 k_4 U_{J,I}^{0(1)}]\, U_{I,IJ}^{0(1)} + [k_3 U_{I,I}^{0(1)} + k_3 U_{J,J}^{0(1)}]\, U_{I,JJ}^{0(1)} \\ + [k_4 U_{I,J}^{0(1)} + k_3 U_{J,I}^{0(1)}]\, U_{J,II}^{0(1)} + [k_3 U_{I,J}^{0(1)} + k_4 U_{J,I}^{0(1)}]\, U_{J,JJ}^{0(1)} \\ + [k_5 U_{I,I}^{0(1)} + k_5 U_{J,J}^{0(1)}]\, U_{J,JI}^{0(1)} = 0. \end{aligned} \tag{46}$$

In each of these sets the indices are $I = 1$, $J = 2$ for the first equation and $I = 2$, $J = 1$ for the second one.

Equations (45) and (46) are solved in the special case illustrated in Fig. 1. A two-dimensional structural element with a thickness h, width $2l$ and unit length is subjected to external symmetrical normal forces N and couples M on the surfaces $X_2 = \pm l$, while the surfaces $X_1 = 0$ and $X_1 = h$ are traction-free in the prestressed state.

The constant normal forces N and couples M on the boundaries can be expressed in terms of stress or displacement on the basis of the theory of elasticity [11].

Using the polynomials

$$P_k^{m,n} = \sum_{p=0}^{m} \sum_{s=0}^{n} r_{p,s} X_1^p X_2^s, \; k = 1, 2, ..., 4$$

with constant coefficients $r_{p,s}$, solution (43) of equation (40) is sought in the form

$$U_1^0 = \varepsilon P_1^{2,2} + \varepsilon^2 P_2^{5,5}, \qquad U_2^0 = \varepsilon P_3^{1,1} + \varepsilon^2 P_4^{5,5}. \tag{47}$$

Substitution of the first terms of equation (47) into the linear equations of equilibrium (45) and the boundary conditions gives the solution

$$P_1^{2,2} = \frac{-\lambda X_1 (hN^{(1)} - 6M^{(1)})}{\mu(\lambda + \mu)h^2} - \frac{-3M^{(1)}[\lambda X_1^2 + (\lambda + 2\mu)X_2^2]}{2\mu(\lambda + \mu)h^3},$$

$$P_3^{1,1} = \frac{\lambda + 2\mu}{\mu(\lambda + \mu)h^2}\left[\frac{(hN^{(1)} - 6M^{(1)})X_2}{4} + \frac{3M^{(1)} X_1 X_2}{h}\right]. \tag{48}$$

Here, constant normal forces N and couples M on the boundaries of the specimen (Fig. 1) are presented in the form

$$N = \varepsilon N^{(1)}, \; M = \varepsilon M^{(1)}.$$

Similarly, the polynomials $P_2^{5,5}$ and $P_4^{5,5}$ in equation (47) are determined from the system of linear differential equations (46) with known right-hand sides under zero boundary conditions. The analytical solution to this system is derived using the symbolic manipulation software Maple. This cumbersome solution is too involved to be presented here.

5.2 Counterpropagating waves

Two longitudinal waves with arbitrary smooth initial profiles that propagate simultaneously in the prestressed material are excited and the initial stage of the wave profile distortion is investigated. It is supposed that in this initial stage the distortion of wave profiles is weak and shock waves are not generated.

Substitution of series (43) and (44), with the small parameter ε, into equation (42) gives an equation which governs wave propagation in the prestressed elastic material.

The first three terms of series (44) are solutions to the following equations
$O(\varepsilon)$:

$$U^{(1)}_{1,11} - c^{-2}\, U^{(1)}_{1,tt} = 0, \tag{49}$$

$O(\varepsilon^2)$:

$$\begin{aligned} U^{(2)}_{1,11} - c^{-2}\, U^{(2)}_{1,tt} = -k_1\, U^{(1)}_{1,11}\, U^{(1)}_{1,1} - \Big[k_1\, U^{0(1)}_{1,11} + k_3\, U^{0(1)}_{1,22} \\ +k_5\, U^{0(1)}_{2,12}\Big]\, U^{(1)}_{1,1} - \Big[k_1\, U^{0(1)}_{1,1} + k_2\, U^{0(1)}_{2,2}\Big]\, U^{(1)}_{1,11}\,, \end{aligned} \tag{50}$$

$O(\varepsilon^3)$:

$$\begin{aligned} U^{(3)}_{1,11} - c^{-2}\, U^{(3)}_{1,tt} = -k_1\, U^{(1)}_{1,11}\, U^{(2)}_{1,1} - k_1\, U^{(2)}_{1,11}\, U^{(1)}_{1,1} - \Big[k_1\, U^{0(1)}_{1,1} \\ +k_2\, U^{0(1)}_{2,2}\Big]\, U^{(2)}_{1,11} - \Big[k_1\, U^{0(2)}_{1,1} + k_2\, U^{0(2)}_{2,2}\Big]\, U^{(1)}_{1,11} - \Big[k_1\, U^{0(1)}_{1,11} + k_3\, U^{0(1)}_{1,22} \\ +k_5\, U^{0(1)}_{2,12}\Big]\, U^{(2)}_{1,1} - \Big[k_1\, U^{0(2)}_{1,11} + k_3\, U^{0(2)}_{1,22} + k_5\, U^{0(2)}_{2,12}\Big]\, U^{(1)}_{1,1}\,. \end{aligned} \tag{51}$$

In order to investigate the interaction of longitudinal waves in an elastic material subjected to inhomogeneous plane prestrain, the structural element in Fig. 1 with two boundaries subjected to a prescribed external affect is considered.

Two longitudinal waves with smooth arbitrary initial profiles are excited on traction free parallel surfaces. The resulting wave propagation process is described by the equations of motion (49) to (51) and the initial and boundary conditions

$$U_1(X_1,X_2,0) = U_{1,t}(X_1,X_2,0) = 0\,, \tag{52}$$

$$U_{1,t}(0,X_2,t) = \varepsilon\, a_0\, \varphi(t)\, H(t)\,, \tag{53}$$

$$U_{1,t}(h,X_2,t) = \varepsilon\, a_h\, \psi(t)\, H(t)\,, \tag{54}$$

where $H(t)$ denotes Heaviside's unit step function, and a_0 and a_h are constants. The smooth arbitrary initial wave profiles are determined by functions $\varphi(t)$ and $\psi(t)$ with $\max \mid \varphi(t) \mid = 1$ and $\max \mid \psi(t) \mid = 1$.

It is assumed that the properties and the prestressed state of the material are known. From the mathematical point of view this means that equations (49) to (51) can be solved as one-dimensional hyperbolic equations with constant coefficients and with known right-hand sides. The coordinate X_2 may be regarded as a parameter.

The first term of series (44) is the solution of the wave equation (49) under the initial and boundary conditions

$$U^{(1)}_1(X_1,X_2,0) = U^{(1)}_{1,t}(X_1,X_2,0) = 0\,, \tag{55}$$

$$U_{1,t}^{(1)}(0,X_2,t) = a_0\, \varphi(t)\, H(t)\,, \tag{56}$$

$$U_{1,t}^{(1)}(h,X_2,t) = a_h\, \psi(t)\, H(t)\,. \tag{57}$$

This solution

$$\begin{aligned} U_1^{(1)}(X_1,X_2,t) &= a_0\, H(\xi) \int_0^{\xi} \varphi(\tau)d\tau + a_h\, H(\eta) \int_0^{\eta} \psi(\tau)d\tau \\ &\quad - a_0\, H(\theta) \int_0^{\theta} \varphi(\tau)d\tau - a_h\, H(\zeta) \int_0^{\zeta} \psi(\tau)d\tau\,, \end{aligned} \tag{58}$$

$$\xi = t - \frac{X_1}{c}\,, \qquad \eta = t - \frac{h - X_1}{c}\,,$$

$$\zeta = t - \frac{h + X_1}{c}\,, \quad \theta = t - \frac{2h - X_1}{c}\,,$$

represents the propagation of two waves in a homogeneous isotropic elastic material in the positive and negative directions of the axis X_1.

The subsequent terms in series (44) can be determined from corresponding equations under the initial and boundary conditions

$$U_1^{(n)}(X_1,X_2,0) = U_{1,t}^{(n)}(X_1,X_2,0) = 0\,, \tag{59}$$

$$U_{1,t}^{(n)}(0,X_2,t) = U_{1,t}^{(n)}(h,X_2,t) = 0, \qquad n = 2,3,\ldots. \tag{60}$$

Equations (50) and (51) may be expressed in the form

$$U_{1,11}^{(n)}(X_1,X_2,t) - c^{-2}U_{1,tt}^{(n)}(X_1,X_2,t) = \sum_{j=1}^{m} G_j^{(n)}(X_1,X_2)F_j^{(n)}(\vartheta_j^{(n)}), \tag{61}$$

$$\vartheta_j^{(n)} = t - g_j^{(n)}(X_1)\,, \quad g_j^{(n)}(X_1) \geq 0\,,$$

with known right-hand sides, where it is possible to separate the independent variables X_1 and $\vartheta_j^{(n)}$.

Laplace transformation with respect to time applied to equation (61) gives

$$\begin{aligned} &U_{1,11}^{(n)L}(X_1,X_2,p) - c^{-2}\, p^2\, U_1^{(n)L}(X_1,X_2,p) \\ &= e^{-p g_j^{(n)}(X_1)} \sum_{j=1}^{m} G_j^{(n)}(X_1,X_2)\; F_j^{(n)L}(p)\,. \end{aligned} \tag{62}$$

Here p is the transform parameter, and the upper index L denotes the Laplace transforms of the corresponding functions.

The inhomogeneous ordinary differential equation (62) with constant coefficients has the solution

$$U_1^{(n)L}(X_1,X_2,p) = \sum_{j=1}^{m} \left\{ \frac{c}{2p} F_j^{(n)L}(p) \left[P_j^{(n)}(X_1,X_2,p) - e^{-ph/c} \left(V_j^{(n)}(X_1,X_2,p) + W_j^{(n)}(X_1,X_2,p) \right) \right] \right\}, \tag{63}$$

where

$$V_j^{(n)}(X_1,X_2,p) = e^{p(X_1-h)/c} \left[-e^{-ph/c} P_j^{(n)}(0,X_2,p) + P_j^{(n)}(h,X_2,p) \right], \tag{64}$$

$$W_j^{(n)}(X_1,X_2,p) = e^{-p(X_1+h)/c} \left[-e^{ph/c} P_j^{(n)}(0,X_2,p) + P_j^{(n)}(h,X_2,p) \right]. \tag{65}$$

The function $F_j^{(n)L}(p)$ depends on the initial profiles of the waves. The function

$$P_j^{(n)}(X_1,X_2,p) = e^{-pX_1/c} \left[e^{2pX_1/c} \int e^{-p(X_1/c+g_j^{(n)}(X_1))} G_j^{(n)}(X_1,X_2)\, dX_1 - \int e^{p(X_1/c-g_j^{(n)}(X_1))} G_j^{(n)}(X_1,X_2)\, dX_1 \right] \tag{66}$$

depends on the inhomogeneous properties of the material through the functions $G_j^{(n)}(X_1,X_2)$ and $g_j^{(n)}(X_1)$.

Now, the solution to equation (61) may be determined through inverse Laplace transformation as

$$U_1^{(n)}(X_1,X_2,t) = \lim_{Y\to\infty} \frac{1}{2\pi i} \int_{\alpha-iY}^{\alpha+iY} e^{\tau p}\, U_1^{(n)L}(X_1,X_2,p)\, dp\,. \tag{67}$$

This solution determines all terms in series (44) except the first and it is valid in the time interval

$$0 \le t < 2h/c. \tag{68}$$

As a result, solution (44) of the one-dimensional problem (42), which describes the propagation and interaction of longitudinal waves with smooth arbitrary initial profiles $\varphi(t)$ and $\psi(t)$ in the prestressed elastic material, has been derived.

6 Harmonic waves

The purpose is (i) to investigate the propagation and interaction of longitudinal waves in the inhomogeneously prestressed elastic material and (ii) to study the possibility of using wave propagation and interaction data in nondestructive evaluation of the prestressed state of the material. It is convenient to investigate the wave profile distortion on the basis of sine-wave propagation data and to determine the initial profiles of waves in boundary conditions (53) and (54) by letting

$$\varphi(t) = \psi(t) = \sin \omega t, \tag{69}$$

where ω denotes the frequency.

The parameters of the sine-wave excitation are determined by boundary conditions (53) and (54). Since the boundary conditions are presented in terms of the derivative of the function U_1 with respect to time, it is convenient to analyze the wave propagation process on the basis of solution (44) in the form

$$U_{1,t} = \sum_{n=1}^{\infty} \varepsilon^n U_{1,t}^{(n)} . \tag{70}$$

The wave propagation process is considered with exactness of the three first terms in solution (70). The first term in series (70)

$$\begin{aligned} U_{1,t}^{(1)} &= a_0 H(\xi) \sin \omega\xi + a_h H(\eta) \sin \omega\eta \\ &\quad - a_0 H(\theta) \sin \omega\theta - a_h H(\zeta) \sin \omega\zeta \end{aligned} \tag{71}$$

is determined from solution (58) of equation (49), and it describes sine-wave propagation in a linear homogeneous elastic material.

The subsequent terms in solution (70) improve the description of the wave process by taking the effects caused by inhomogeneity (inhomogeneous prestress) and nonlinearity into account. The second term, the solution of equation (50) under initial and boundary conditions (59) and (60), has the form

$$\begin{aligned} U_{1,t}^{(2)} &= A_0^{(2)} + \sum_{j=1}^{m_{21}} A_{1j}^{(2)} \sin \omega \vartheta_j + \sum_{j=1}^{m_{22}} A_{2j}^{(2)} \cos \omega \vartheta_j \\ &\quad + \sum_{j=1}^{m_{23}} A_{3j}^{(2)} \sin 2\omega \vartheta_j + \sum_{j=1}^{m_{24}} A_{4j}^{(2)} \cos 2\omega \vartheta_j , \end{aligned} \tag{72}$$

where $\vartheta_j = t + c_{1j} h/c + c_{2j} X_1/c$, and c_{1j}, c_{2j} are constants. This expression consists of a non-periodic term $A_0^{(2)}$, terms with argument $\omega \vartheta_j$, which represent the influence of inhomogeneity (inhomogeneous prestress) on the first harmonic, and terms with argument $2\omega \vartheta_j$, which represent the evolution of the second harmonic in a homogeneous elastic material.

The third term in series (70), the solution to equation (51) under initial and boundary conditions (59) and (60),

$$
\begin{aligned}
U_{1,t}^{(3)} = A_0^{(3)} &+ \sum_{j=1}^{m_{31}} A_{1j}^{(3)} \sin \omega \vartheta_j + \sum_{j=1}^{m_{32}} A_{2j}^{(3)} \cos \omega \vartheta_j \\
&+ \sum_{j=1}^{m_{33}} A_{3j}^{(3)} \sin 2\omega \vartheta_j + \sum_{j=1}^{m_{34}} A_{4j}^{(3)} \cos 2\omega \vartheta_j \\
&+ \sum_{j=1}^{m_{35}} A_{5j}^{(3)} \sin 3\omega \vartheta_j + \sum_{j=1}^{m_{36}} A_{6j}^{(3)} \cos 3\omega \vartheta_j ,
\end{aligned} \tag{73}
$$

further improves the solution. It describes the influence of inhomogeneity on the second harmonic and the propagation of the third harmonic in a homogeneous material.

7 Nondestructive characterization of plane strain

Solution (70) describes the simultaneous propagation of two longitudinal waves in the elastic material (structural element) subjected to plane prestrain (Fig. 1). The sine-wave propagation and interaction process in the material, represented by the first term of solution (70) is illustrated in Fig. 2. There are several possibilities to use this process in nondestructive testing of the prestressed material (structural element). In this paper the idea presented in [9] is used. The wave propagation is excited in the material on the boundaries $X_1 = 0$ and $X_1 = h$ in accordance with the boundary conditions (53) and (54), i.e., in terms of particle velocity. The stress induced by the wave motion is recorded on the same surfaces.

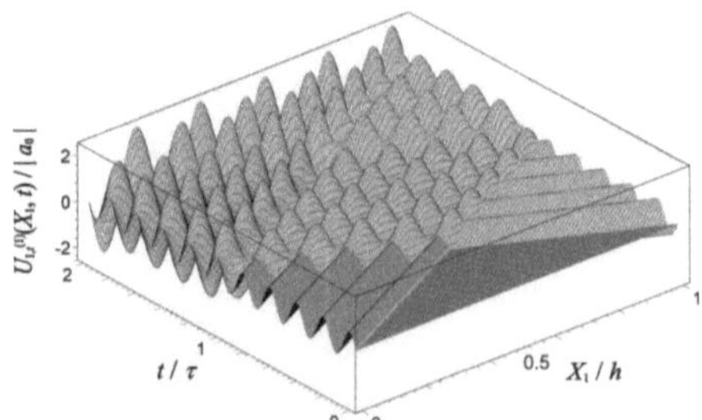

Fig. 2 Counterpropagation of harmonic waves in homogeneous elastic material.

This stress is characterized by the second Piola-Kirchhoff stress tensor component T_{11}. The five-constant nonlinear theory of elasticity describes this stress as a function of the derivatives of the particle displacement with respect to the spatial coordinate:

$$T_{11} = (\lambda + 2\mu)\, U_{1,1} + [\lambda/2 + \mu + 3(\nu_1 + \nu_2 + \nu_3)]\, U_{1,1}^2. \tag{74}$$

In our case this stress is determined from the derivative $U_{1,1}$ which is obtained from solution (44). The expression for the stress assumes the form:

$$\begin{aligned} T_{11} &= \varepsilon\, T_{11}^{(1)} + \varepsilon^2\, T_{11}^{(2)}, \\ T_{11}^{(1)} &= (\lambda + 2\mu)\, U_{1,1}^{(1)}, \\ T_{11}^{(2)} &= (\lambda + 2\mu)\, U_{1,1}^{(2)} + [\lambda/2 + \mu + 3(\nu_1 + \nu_2 + \nu_3)]\, U_{1,1}^{(1)2}. \end{aligned} \tag{75}$$

The stress T_{11} is a function of $U_{1,1}^{(1)}$ and $U_{1,1}^{(2)}$. The function $U_{1,1}^{(1)}$, specified by the first term in solution (44), describes wave motion in the linear homogeneous elastic material. This motion is determined by boundary conditions (56), (57) and by the physical properties of the material. Information about the prestressed state of the material is contained in $U_{1,1}^{(2)}$. Provided the physical properties of the material are known, the function $U_{1,1}^{(1)}$ may be regarded as a known function in the acoustodiagnostics problem of plane strain. The plane strain diagnostics problem turns to the extraction of the stress component $T_{11}^{(2)}$ from the recorded data, i.e., to the analysis of the function $U_{1,1}^{(2)}$.

The following numerical experiment has been implemented. Let the properties of the material (structural element) correspond to duralumin with density $\rho_0 = 2800\,\mathrm{kg/m^3}$, the constants of elasticity $\lambda = 50\,\mathrm{GPa}$, $\mu = 27.6\,\mathrm{GPa}$, $\nu_1 = -136\,\mathrm{GPa}$, $\nu_2 = -197\,\mathrm{GPa}$, $\nu_3 = -38\,\mathrm{GPa}$, and dimensions $h = 0.1\,\mathrm{m}$ and $l = 1\,\mathrm{m}$. The prestressed state of the material corresponds to the plane strain. The strain is characterized by the dimensionless constant ε that is proposed to be equal to $\varepsilon = 10^{-4}$.

The idea is to study the influence of the prestress on the recorded wave induced stress data, i.e., on the value of the function $U_{1,1}^{(2)}$ in equation (75). The recorded data contain maximum information about the prestressed state provided the strain intensities caused by the prestress and the wave motion are of the same order [8]. The experimenter has the possibility to assure this by choosing the amplitudes of the excited waves. In the considered numerical experiment the sine-wave amplitudes are determined by the constants $a_0 = -a_h = c$ in correspondence with boundary conditions (53) and (54). The constant c is defined by equation (39). As the result, the amplitudes of the excited particle velocity at the boundaries $X_1 = 0$ and $X_1 = h$ have opposite signs and equal absolute values $|\,\varepsilon a_0\,| = 0.6130\,\mathrm{m/s}$. The absolute value of the strain amplitude caused by the boundary excitation is equal to $\varepsilon = 10^{-4}$.

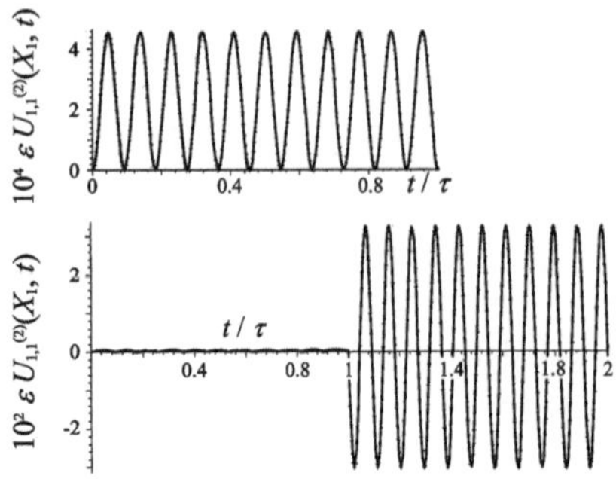

Fig. 3 Second order effects of boundary oscillations in prestress-free material.

The results of the computational simulation are presented in Fig. 2 to Fig. 5, where $\tau = h/c$. In Fig. 3 to Fig. 5 the function $\varepsilon^2\, U_{1,1}^{(2)} / \mid \varepsilon\, a_0/c \mid$ is plotted on the vertical axis and the equality $\varepsilon\, a_0/c = \varepsilon$, valid for the considered special case, is taken into account. This enables to compare the values of the functions $\varepsilon\, U_{1,1}^{(1)}$ and $\varepsilon^2\, U_{1,1}^{(2)}$ with the value of the strain induced on the boundaries by the excitation of the wave process (equations (53) and (54)). Since the amplitudes within the intervals $0 \le t/\tau < 1$ and $1 \le t/\tau < 2$ differ by about 10^2 in magnitude, two plots with different scales are provided in each of the figures.

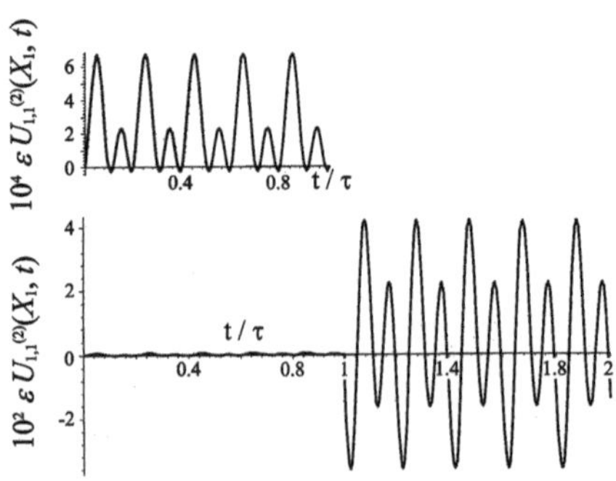

Fig. 4 Influence of uniform homogeneous prestress on the second order effects of boundary oscillations.

It is interesting that the initial value of the wave frequency ω has great influence on the amplitude of the wave-induced stress in the zone of wave interaction. If the number n of wave periods in the interval $0 \le t/\tau < 1$ at $X_1 = 0$ and $X_1 = h$ is equal to an integer ($n = 5$, $\omega = 1.9256 \cdot 10^6$ rad/s) the amplitude of the function $U_{1,1}^{(1)}$ is three times greater in the interval of wave interaction $1 \leq t/\tau < 2$ than in $0 \leq t/\tau < 1$ at $X_1 = 0$ and $X_1 = h$. If the number n is equal to an integer plus a half, there is no amplification of the amplitude in the interval of wave interaction at all. For intermediate values of n, the amplification is less than three times.

From the point of view of nondestructive testing it is important to repeat that the first term in the solution (44) describes the wave motion in a homogeneous elastic material, while the second term is affected by inhomogeneity caused by the external loading. Below, the behavior of the second term is studied for homogeneously and inhomogeneously prestressed material.

The interaction of waves described by the term $U_{1,1}^{(2)}$ is not sensitive to the variation of the sine-wave frequency, but there is a very strong amplification of the wave amplitudes in the interval of wave interaction (Fig. 3).

The profile of the function $U_{1,1}^{(2)}$ recorded at the surfaces $X_1 = 0$ and $X_1 = h$ of a prestress-free nonlinear elastic material is plotted in Fig. 3. This harmonic function is characterized by different constant amplitudes in the intervals $0 \leq t/\tau < 1$ and $1 \leq t/\tau < 2$.

The prestress, induced by the external load and characterized by the stress component T_{22}^0 calculated on the basis of the theory of elasticity on the boundaries $X_2 = \pm l$, modulates the profile of the function $U_{1,1}^{(2)}$ recorded at the surfaces $X_1 = 0$ and $X_1 = h$. The homogeneous prestress caused by the normal force modulates the measured data on both surfaces in the same way (Fig. 4). The depth and profile of modulation are sensitive to the value and sign of the prestress.

The inhomogeneity of the prestress is characterized by different values of the function $U_{1,1}^{(2)}$ at $X_1 = 0$ and $X_1 = h$. The influence of the pure bending caused by simultaneous action of symmetric couples on the surfaces $X_2 = \pm l$ on the value of

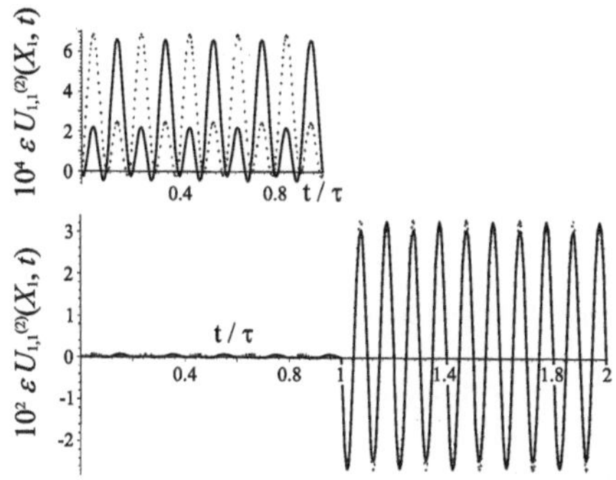

Fig. 5 Second order effects of boundary oscillations in elastic material undergoing pure bending (dotted line: $X_1 = 0$, solid line: $X_1 = h$).

the function $U_{1,1}^{(2)}$ is plotted in Fig. 5. The modulation of the wave profile is intensive in the interval $0 \leq t/\tau < 1$ and less intensive in the interval of wave interaction.

It is essential that the wave profile in the interval of wave interaction is intensively modulated under the influence of simultaneous action of normal forces and couples.

8 Conclusions

The perturbation theory has been successfully applied to solve the problem of nonlinear wave propagation and interaction in nonlinear elastic material undergoing inhomogeneous plane prestrain. The original problem has been converted into a perturbation problem, assuming (i) small strain and (ii) the strains evoked in the material by prestress and by wave motion to be of the same order. The weakly nonlinear perturbative solution about the known exact solution of the linear problem has been derived using the symbolic manipulation software Maple. The perturbation technique has turned out to be most useful for the considered problem since the first few steps reveal the important features of the solution and the remaining ones provide small corrections. The analyses of the solution versus the prestrain verifies the assumption that the extraction of information from the wave interaction data enables to enhance the efficiency of nondestructive testing. Simultaneous recording of the stress on two opposite boundaries makes it possible to distinguish different states of the prestressed material which is important for practical applications. The presented treatment is recommended to be employed for nondestructive prestress evolution in wide-spread materials, structural elements, machinery, etc. The validity is guaranteed if the material is described by the five-constant nonlinear theory of elasticity and if the excited strain is small but finite, i.e., no large strain, no plastic deformations, no shock waves.

Acknowledgements The research was supported by the Estonian Science Foundation through the Grant No. 7728 and by the EU FP6 Transfer of Knowledge Project CENS-CMA.

References

1. Ames, W.F.: Nonlinear Partial Differential Equations in Engineering. Academic Press, New York (1965)
2. Bender, C.M., Orszag, S.A.: Advanced Mathematical Methods for Scientists and Engineers. McGraw-Hill, New York (1978)
3. Bland, D.R.: Nonlinear Dynamic Elasticity. Blaisdell, Waltham (1969)
4. Eringen, A.C.: Nonlinear Theory of Continuous Media. McGraw-Hill, New York (1962)
5. Jeffrey, A., Kawahara, T.: Asymptotic Methods in Nonlinear Wave Theory. Pitman, Boston (1982)
6. Kevorkian, J., Cole, J.D.: Perturbation Methods in Applied Mathematics. Springer, New York (1981)
7. Nayfeh, A.H.: Introduction to Perturbation Techniques. Wiley & Sons, New York (1981)

8. Ravasoo, A.: Nonlinear longitudinal waves in inhomogeneously predeformed elastic media. J. Acoust. Soc. Am. **106**, 3143–3149 (1999)
9. Ravasoo, A.: Non-linear interaction of waves in prestressed material. Int. J. Non-Linear Mechanics **42**, 1162–1169 (2007)
10. Reddy, J.N.: An Introduction to Continuum Mechanics with Applications. Cambridge University Press (2008)
11. Rogers, G.R.: Mechanics of Solids. Wiley & Sons, New York (1982)

Waves in Inhomogeneous Solids

Arkadi Berezovski, Mihhail Berezovski and Jüri Engelbrecht

Abstract The paper aims at presenting a numerical technique used in simulating the propagation of waves in inhomogeneous elastic solids. The basic governing equations are solved by means of a finite-volume scheme that is faithful, accurate, and conservative. Furthermore, this scheme is compatible with thermodynamics through the identification of the notions of numerical fluxes (a notion from numerics) and of excess quantities (a notion from irreversible thermodynamics). A selection of one-dimensional wave propagation problems is presented, the simulation of which exploits the designed numerical scheme. This selection of exemplary problems includes (i) waves in periodic media for weakly nonlinear waves with a typical formation of a wave train, (ii) linear waves in laminates with the competition of different length scales, (iii) nonlinear waves in laminates under an impact loading with a comparison with available experimental data, and (iv) waves in functionally graded materials.

1 Introduction

Waves correspond to continuous variations of the states of material points representing a medium. The characteristic feature of waves is their motion. In mechanics the motion of waves is governed by the conservation laws for mass, linear momentum, and energy. These conservation laws, complemented by constitutive relations, are the basis of the theory of thermoelastic waves in solids [1, 3, 9, 19].

Inhomogeneous solids include layered and randomly reinforced composites, multiphase and polycrystalline alloys, functionally graded materials, ceramics and polymers with certain microstructure, etc. Therefore, it is impossible to present a complete theory of linear and nonlinear wave propagation for the full diversity of

Centre for Nonlinear Studies, Institute of Cybernetics at Tallinn University of Technology, Akadeemia tee 21, 12618 Tallinn, Estonia, e-mail: arkadi.berezovski@cs.ioc.ee

E. Quak, T. Soomere (eds.), *Applied Wave Mathematics*,
DOI 10.1007/978-3-642-00585-5_5,

possible situations, in so far as geometry, contrast of multiphase properties and loading conditions are concerned.

From a practical point of view, we need to perform numerical calculations. Many numerical methods have been proposed to compute wave propagation in heterogeneous solids, among them, the stiffness matrix recursive algorithm [33, 38] and the spectral layer element method [10, 11] should be mentioned, in addition to more common finite-element, finite-difference, and finite-volume methods.

Here the general idea is the following: division of a body into a finite number of computational cells requires the description of all fields inside the cells as well as the interaction between neighboring cells. Approximation of wanted fields inside the cells leads to discontinuities of the fields at the boundaries between cells. This also leads to the appearance of excess quantities, which represent the difference between the exact and approximate values of the fields. Interaction between neighboring cells is described by means of fluxes at the boundaries of the cells. These fluxes correspond to the excess quantities and, therefore, can be calculated by means of jump relations at the boundaries between cells.

In this paper, we demonstrate how the finite-volume wave-propagation algorithm developed in [27] can be reformulated in terms of the excess quantities and then applied to the wave propagation in inhomogeneous solids. Both original and modified algorithms are stable, high-order accurate, thermodynamically consistent, and applicable both to linear and nonlinear waves.

1.1 Governing equations

The simplest example of heterogeneous media is a periodic medium composed by materials with different properties. One-dimensional wave propagation in the framework of linear elasticity is governed by the conservation of linear momentum [1]

$$\rho(x)\frac{\partial v}{\partial t} - \frac{\partial \sigma}{\partial x} = 0, \tag{1}$$

and the kinematic compatibility condition

$$\frac{\partial \varepsilon}{\partial t} = \frac{\partial v}{\partial x}. \tag{2}$$

Here t is time, x is the space variable, the particle velocity $v = u_t$ is the time derivative of the displacement u, the one-dimensional strain $\varepsilon = u_x$ is the space derivative of the displacement, σ is the Cauchy stress, and ρ is the material density. The compatibility condition (2) follows immediately from the definitions of the strain and the particle velocity.

The two equations (1) and (2) contain three unknowns: v, σ and ε.

The closure of the system of equations (1) and (2) is achieved by a constitutive relation, which in the simplest case is Hooke's law

$$\sigma = \rho(x)c^2(x)\,\varepsilon, \tag{3}$$

where $c(x) = \sqrt{(\lambda(x)+2\mu(x))/\rho(x)}$ is the corresponding longitudinal wave velocity, and $\lambda(x)$ and $\mu(x)$ are the so-called Lamé coefficients. The indicated explicit dependence on the point x means that the medium is materially inhomogeneous.

The system of equations (1)–(3) can be expressed in the form of a conservation law

$$\frac{\partial}{\partial t}q(x,t) + \frac{\partial}{\partial x}f(q(x,t)) = 0, \tag{4}$$

with

$$q(x,t) = \begin{pmatrix} \varepsilon \\ \rho v \end{pmatrix} \quad \text{and} \quad f(x,t) = \begin{pmatrix} -v \\ -\rho c^2 \varepsilon \end{pmatrix}. \tag{5}$$

In the linear case, equation (4) can be rewritten in the form

$$\frac{\partial}{\partial t}q(x,t) + A\frac{\partial}{\partial x}q(x,t) = 0, \tag{6}$$

where the matrix A is given by

$$A = \begin{pmatrix} 0 & -1/\rho \\ -\rho c^2 & 0 \end{pmatrix}. \tag{7}$$

We will solve the system of equations (1)–(3) numerically. Although a numerical solution can be difficult with standard methods, high-resolution finite volume methods based on solving Riemann problems have been found to perform very well on linear hyperbolic systems modeling wave propagation in rapidly-varying heterogeneous media [16].

2 The wave-propagation algorithm

Standard methods cannot give high accuracy near discontinuities in the material parameters and will often fail completely in problems where the parameters vary drastically on the grid scale. By contrast, solving the Riemann problem at each cell interface properly resolves the solution into waves, taking into account every discontinuity in the parameters, and automatically handling the reflection and transmission of waves at each interface. This is crucial in developing the correct macroscopic behavior. As a result, Riemann-solver methods are quite natural for this application. Moreover, the methods extend easily from linear to nonlinear problems. Expositions of such methods and pointers to the rich literature base can be found in many sources [17, 20, 27, 36, 37].

2.1 Averaged quantities

Let us introduce a computational grid of cells $C_n = [x_{n-1/2}, x_{n+1/2}]$ with interfaces $x_{n-1/2} = (n-1)/2\Delta x$ and time levels $t_k = k\Delta t$. For simplicity, the grid size Δx and time step Δt are assumed to be constant. Integrating equation (4) over $C_n \times [t_k, t_{k+1}]$ gives

$$\int_{x_{n-1/2}}^{x_{n+1/2}} q(x,t_{k+1})dx = \int_{x_{n-1/2}}^{x_{n+1/2}} q(x,t_k)dx - \tag{8}$$

$$-\left(\int_{t_k}^{t_{k+1}} f(q(x_{n+1/2},t))dt - \int_{t_k}^{t_{k+1}} f(q(x_{n-1/2},t))dt\right).$$

Introducing the average Q_n of the exact solution on C_n at time $t = t_k$ and the numerical flux F_n that approximates the time average of the exact flux taken at the interface between the cells C_{n-1} and C_n, i.e.

$$Q_n \approx \frac{1}{\Delta x}\int_{x_{n-1/2}}^{x_{n+1/2}} q(x,t_k)dx, \quad F_n \approx \frac{1}{\Delta t}\int_{t_k}^{t_{k+1}} f(q(x_{n-1/2},t))dt, \tag{9}$$

we can rewrite equation (8) in the form of a numerical method in the flux-differencing form

$$Q_n^{k+1} = Q_n^k - \frac{\Delta t}{\Delta x}(F_{n+1}^k - F_n^k). \tag{10}$$

In general, however, we cannot evaluate the time integrals on the right-hand side of equation (8) exactly, since $q(x_{n\pm 1/2},t)$ varies with time along each edge of the cell, and we do not have the exact solution to work with. If we can approximate this average flux based on the values Q^k, then we will have a fully-discrete method.

2.2 Numerical fluxes

Numerical fluxes are determined by means of the solution of the Riemann problem at interfaces between cells. The solution of the Riemann problem (at the interface between cells $n-1$ and n) consists of two waves, which we denote, following [27], $\mathcal{W}_n^I$ and $\mathcal{W}_n^{II}$. The left-going wave $\mathcal{W}_n^I$ moves into cell $n-1$, and the right-going wave $\mathcal{W}_n^{II}$ moves into cell n. The state between the two waves must be continuous across the interface (Rankine-Hugoniot condition) [27]:

$$\mathcal{W}_n^I + \mathcal{W}_n^{II} = Q_n - Q_{n-1}. \tag{11}$$

In the linear case, the considered waves are determined by eigenvectors of the matrix A [27]:

$$\mathcal{W}_n^I = \gamma_n^I r_{n-1}^I, \quad \mathcal{W}_n^{II} = \gamma_n^{II} r_n^{II}. \tag{12}$$

This means that equation (11) is represented as

$$\gamma_n^I r_{n-1}^I + \gamma_n^{II} r_n^{II} = Q_n - Q_{n-1}. \tag{13}$$

Considering the definition of eigenvectors $Ar = \lambda r$, we see that the eigenvector

$$r^I = \begin{pmatrix} 1 \\ \rho c \end{pmatrix} \tag{14}$$

corresponds to the eigenvalue $\lambda^I = -c$ (left-going wave). Similarly, the eigenvector

$$r^{II} = \begin{pmatrix} 1 \\ -\rho c \end{pmatrix} \tag{15}$$

corresponds to the eigenvalue $\lambda^{II} = c$ (right-going wave). Substituting the eigenvectors into equation (13), we have

$$\gamma_n^I \begin{pmatrix} 1 \\ \rho_{n-1} c_{n-1} \end{pmatrix} + \gamma_n^{II} \begin{pmatrix} 1 \\ -\rho_n c_n \end{pmatrix} = Q_n - Q_{n-1}, \tag{16}$$

or, more explicitly,

$$\begin{pmatrix} 1 & 1 \\ \rho_{n-1} c_{n-1} & -\rho_n c_n \end{pmatrix} \begin{pmatrix} \gamma_n^I \\ \gamma_n^{II} \end{pmatrix} = \begin{pmatrix} \bar{\varepsilon}_n - \bar{\varepsilon}_{n-1} \\ \rho \bar{v}_n - \rho \bar{v}_{n-1} \end{pmatrix}. \tag{17}$$

Solving the system of linear equations (17), we obtain the amplitudes of the left-going and right-going waves. Then the numerical fluxes in the Godunov-type numerical scheme are determined as follows:

$$F_{n+1}^k = -\lambda_{n+1}^I \mathcal{W}_{n+1}^I = -c_{n+1} \gamma_{n+1}^I r_n^I, \tag{18}$$

$$F_n^k = \lambda_n^{II} \mathcal{W}_n^{II} = -c_n \gamma_n^{II} r_n^{II}. \tag{19}$$

Finally, the Godunov-type scheme is expressed in the form

$$Q_n^{k+1} = Q_n^k + \frac{\Delta t}{\Delta x} \left(c_{n+1} \gamma_{n+1}^I r_n^I - c_n \gamma_n^{II} r_n^{II} \right). \tag{20}$$

This is the standard form for the wave-propagation algorithm [27].

Within the wave-propagation algorithm, every discontinuity in parameters is taken into account by solving the Riemann problem at each interface between discrete elements. The reflection and transmission of waves at each interface are handled automatically for the considered inhomogeneous media.

2.3 Second-order corrections

The scheme considered above is formally first-order accurate only. To increase the order of accuracy, we rewrite the numerical scheme as

$$Q_n^{k+1} = Q_n^k + \Delta_n^{up} - \frac{\Delta t}{\Delta x}(\tilde{F}_{n+1}^k - \tilde{F}_n^k), \tag{21}$$

where Δ_n^{up} equals the upwind flux (or Godunov flux) obtained from equation (20).

The term $\tilde{F}_n$ is used to update the solution so that second order accuracy is achieved. The flux for the second-order Lax-Wendroff scheme may be written as the Godunov flux plus a correction [27],

$$F_n = \frac{1}{2}A(Q_n + Q_{n-1}) - \frac{\Delta t}{2\Delta x}A(Q_n - Q_{n+1}) = F_n^G + \frac{1}{2}|A|\left(1 - \frac{\Delta t}{\Delta x}|A|\right)\Delta Q_n, \tag{22}$$

where $|A| = A^+ - A^-$. Hence, a natural choice for $\tilde{F}$ is

$$\tilde{F}_n = \frac{1}{2}|A|\left(1 - \frac{\Delta t}{\Delta x}|A|\right)\Delta Q_n = \frac{1}{2}\sum_p |\lambda^p|\left(1 - \frac{\Delta t}{\Delta x}|\lambda^p|\right)\mathcal{W}_n^p. \tag{23}$$

The Godunov-type scheme exhibits strong numerical dissipation, and discontinuities in the solution are smeared, causing low accuracy. The Lax-Wendroff scheme, on the other hand, is more accurate in smooth parts of the solution. However, near discontinuities, numerical dispersion generates oscillations, also reducing the accuracy. A successful approach to suppress these oscillations is to apply flux limiters [16, 23, 24, 25].

2.4 The conservative wave propagation algorithm

For the conservative wave-propagation algorithm [2], the solution of the generalized Riemann problem is obtained by using the decomposition of the flux difference $f_n(Q_n) - f_{n-1}(Q_{n-1})$ instead of the decomposition (11):

$$\mathcal{L}_n^I + \mathcal{L}_n^{II} = f_n(Q_n) - f_{n-1}(Q_{n-1}). \tag{24}$$

The waves $\mathcal{L}^I$ and $\mathcal{L}^{II}$ are still proportional to the eigenvectors of the matrix A

$$\mathcal{L}_n^I = \beta_n^I r_{n-1}^I, \quad \mathcal{L}_n^{II} = \beta_n^{II} r_n^{II}, \tag{25}$$

and the corresponding numerical scheme has the form

$$Q_n^{l+1} - Q_n^l = -\frac{\Delta t}{\Delta x}\left(\mathcal{L}_n^{II} + \mathcal{L}_{n+1}^I\right). \tag{26}$$

The coefficients β^I and β^{II} are determined from the solution of the system of linear equations

$$\begin{pmatrix} 1 & 1 \\ \rho_{n-1}c_{n-1} & -\rho_n c_n \end{pmatrix} \begin{pmatrix} \beta_n^I \\ \beta_n^{II} \end{pmatrix} = \begin{pmatrix} -(\bar{v}_n - \bar{v}_{n-1}) \\ -(\rho c^2 \bar{\varepsilon}_n - \rho c^2 \bar{\varepsilon}_{n-1}) \end{pmatrix}. \tag{27}$$

As it is shown in [2], the obtained algorithm is conservative and second-order accurate on smooth solutions.

3 Excess quantities and numerical fluxes

We could simply apply the numerical scheme described in the previous sections to simulate the wave propagation in periodic media. However, the splitting of the body into a finite number of computational cells and averaging all the fields over the cell volumes leads to a situation known in thermodynamics as "endoreversible system" [22]. This means that even if the state of each computational cell can be associated with a corresponding local equilibrium state (and, therefore, temperature and entropy can be defined as usual), the state of the whole body is a non-equilibrium one. The computational cells interact with each other, which leads to the appearance of excess quantities.

In the admitted non-equilibrium description [32], both stress and velocity are represented as the sum of the averaged (local equilibrium) and excess parts:

$$\sigma = \bar{\sigma} + \Sigma, \quad v = \bar{v} + \mathcal{V}. \tag{28}$$

Here $\bar{\sigma}$ and $\bar{v}$ are averaged fields and Σ and $\mathcal{V}$ are the corresponding excess quantities.

Therefore, we rewrite a first-order Godunov-type scheme (10) in terms of the excess quantities

$$(\rho \bar{v})_n^{k+1} - (\rho \bar{v})_n^k = \frac{\Delta t}{\Delta x} \left(\Sigma_n^+ - \Sigma_n^- \right), \tag{29}$$

$$\bar{\varepsilon}_n^{k+1} - \bar{\varepsilon}_n^k = \frac{\Delta t}{\Delta x} \left(\mathcal{V}_n^+ - \mathcal{V}_n^- \right). \tag{30}$$

Here an overbar denotes averaged quantities, a superscript k denotes a time step, a subscript n denotes the number of the computational cell, while Δt and Δx are time step and space step, respectively.

Though excess quantities are determined formally everywhere inside computational cells, we need to know only their values at the boundaries of the cells, where they play the role of numerical fluxes. To determine the values of the excess quantities at the boundaries between computational cells, we apply the jump relation for the linear momentum [6], which is reduced in the isothermal case to

$$[\bar{\sigma} + \Sigma] = 0. \tag{31}$$

Similarly, the jump relation following from the kinematic compatibility (2) reads

$$[\bar{v} + \mathcal{V}] = 0. \tag{32}$$

It should be noted that the two last jump conditions can be considered as the *continuity of genuine unknown fields* at the boundaries between computational cells, which is illustrated in Fig. 1.

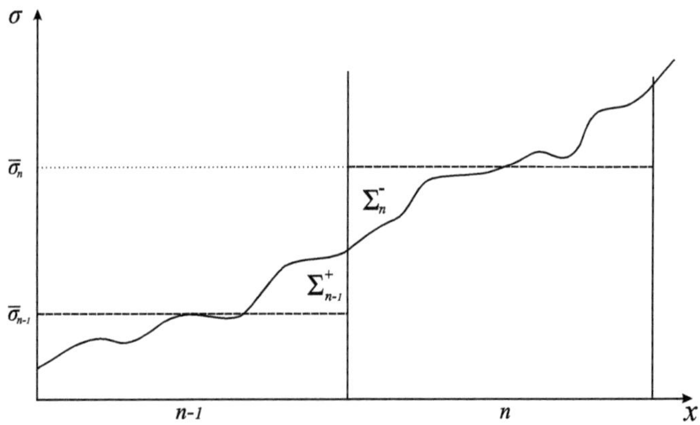

Fig. 1 Stresses in the bulk.

The values of the excess stresses and excess velocities at the boundaries between computational cells are not independent [8]. Considering Riemann invariants at the interface between computational cells, one can see that

$$\rho_n c_n \mathcal{V}_n^- + \Sigma_n^- \equiv 0, \tag{33}$$

$$\rho_{n-1} c_{n-1} \mathcal{V}_{n-1}^+ - \Sigma_{n-1}^+ \equiv 0, \tag{34}$$

i.e., the excess quantities depend on each other at the cell boundary.

3.1 Excess quantities at the boundaries between cells

Rewriting the jump relations (31), (32) in the form

$$(\Sigma^+)_{n-1} - (\Sigma^-)_n = (\bar{\sigma})_n - (\bar{\sigma})_{n-1}, \tag{35}$$

$$(\mathcal{V}^+)_{n-1} - (\mathcal{V}^-)_n = (\bar{v})_n - (\bar{v})_{n-1}, \tag{36}$$

and using the dependence between excess quantities (equations (33) and (34)), we obtain then the system of linear equations for the determination of the excess velocities

$$\mathcal{V}_{n-1}^{+} - \mathcal{V}_{n}^{-} = \bar{v}_n - \bar{v}_{n-1}, \tag{37}$$

$$\mathcal{V}_{n-1}^{+}\rho_{n-1}c_{n-1} + \mathcal{V}_{n}^{-}\rho_n c_n = \rho_n c_n^2 \bar{\varepsilon}_n - \rho_{n-1} c_{n-1}^2 \bar{\varepsilon}_{n-1}. \tag{38}$$

In matrix notation the latter system of equations has the form

$$\begin{pmatrix} 1 & 1 \\ \rho_{n-1}c_{n-1} & -\rho_n c_n \end{pmatrix} \begin{pmatrix} -\mathcal{V}_{n-1}^{+} \\ \mathcal{V}_{n}^{-} \end{pmatrix} = \begin{pmatrix} -(\bar{v}_n - \bar{v}_{n-1}) \\ -(\rho c^2 \bar{\varepsilon}_n - \rho c^2 \bar{\varepsilon}_{n-1}) \end{pmatrix}. \tag{39}$$

Comparing the obtained equation with equation (30), we conclude that

$$\beta_n^{I} = -\mathcal{V}_{n-1}^{+}, \quad \beta_n^{II} = \mathcal{V}_{n}^{-}. \tag{40}$$

This means that the excess quantities following from non-equilibrium jump relations at the boundary between computational cells correspond to the numerical fluxes in the conservative wave-propagation algorithm.

The representation of the wave-propagation algorithm in terms of the excess quantities given here is formally identical to its conservative form [2]. The advantage of the new representation manifests itself at discontinuities, for which jump relations cannot be reduced to the continuity of true values, e.g., at phase-transition fronts or cracks.

4 One-dimensional waves in periodic media

As the first example, we consider the propagation of a pulse in a periodic medium. The initial form of the pulse is given in Fig. 2, where the periodic variation in density is also shown by dashed lines. For the test problem, the materials are chosen as polycarbonate ($\rho = 1190$ kg/m^3, $c = 4000$ m/s) and Al 6061 ($\rho = 2703$ kg/m^3, $c = 6149$ m/s).

We apply the numerical scheme (29) and (30) for the solution of the system of equations (1)–(3). The corresponding excess quantities are calculated by means of equations (35)–(38).

As it was noted, we can exploit all the advantages of the wave-propagation algorithm, including second-order corrections and transversal propagation terms [24]. However, no limiters are used in the calculations. Suppressing spurious oscillations is achieved by means of using a first-order Godunov step after each three second-order Lax-Wendroff steps. This idea of composition was invented in [29].

Calculations are performed with Courant-Friedrichs-Levy number equal to 1. The simulation result for 4000 time steps is shown in Fig. 3.

We observe a distortion of the pulse shape and a decrease in the velocity of the pulse propagation in comparison to the maximal longitudinal wave velocity in the materials. These results correspond to the prediction of the effective media theory [34] both qualitatively and quantitatively [16].

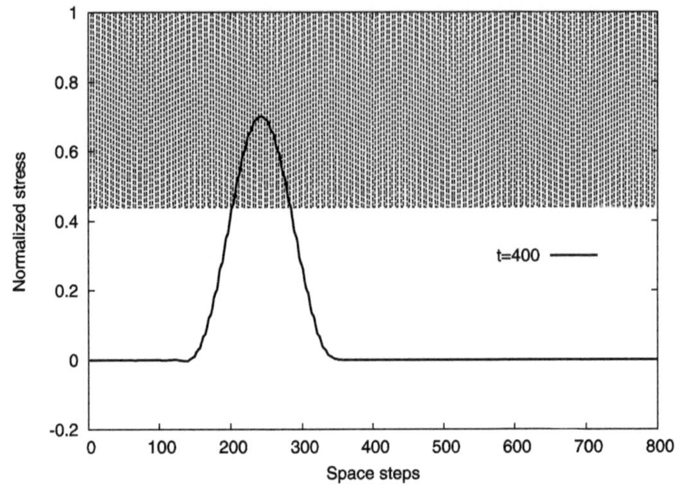

Fig. 2 Initial pulse shape. Reproduced from [5].

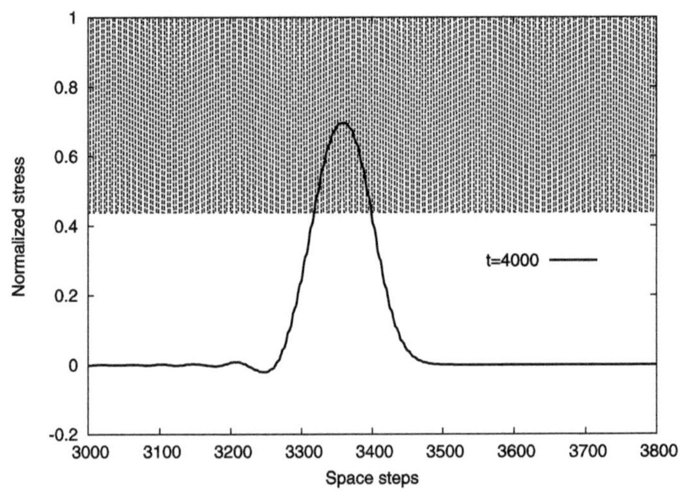

Fig. 3 Pulse shape at time step 4000. Reproduced from [5].

It should be noted that the effective media theory [34] leads to the dispersive wave equation

$$\frac{\partial^2 u}{\partial t^2} = (c^2 - c_a^2)\frac{\partial^2 u}{\partial x^2} + p^2 c_a^2 c_b^2 \frac{\partial^4 u}{\partial x^4}, \tag{41}$$

where u is the displacement, p is the periodicity parameter, and c_a and c_b are parameters of the effective media [15], instead of the wave equation following from equations (1)–(3)

$$\frac{\partial^2 u}{\partial t^2} = c^2 \frac{\partial^2 u}{\partial x^2}. \tag{42}$$

Equation (41) exhibits both *dispersion* (fourth-order space derivative) and the alteration in the longitudinal wave speed.

5 One-dimensional weakly nonlinear waves in periodic media

In the next example, we will see the influence of the materials' nonlinearity on the wave propagation. To close the system of equations (1) and (2) in the case of weakly nonlinear media we apply a simple nonlinear stress-strain relation

$$\sigma = \rho c^2 \varepsilon (1 + B\varepsilon), \tag{43}$$

where B is a parameter of nonlinearity, the values and sign of which are supposed to be different for hard and soft materials.

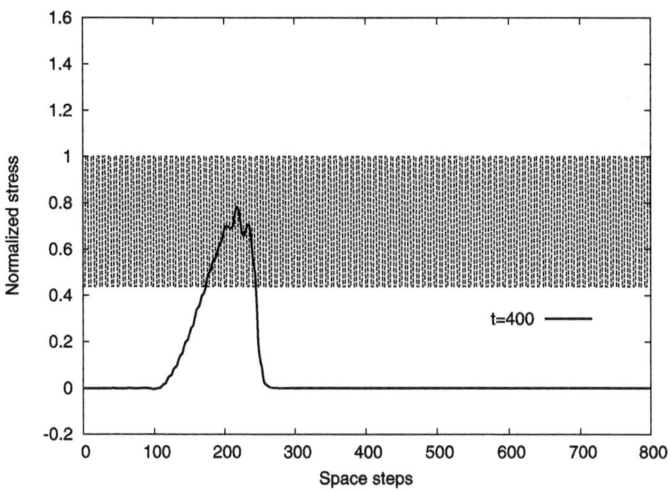

Fig. 4 Pulse shape at time step 400. Nonlinear case.

The solution method is almost the same as before. The approximate Riemann solver for the nonlinear elastic media (equation (43)) is similar to that used in [26, 28]. A modified longitudinal wave velocity $\hat{c}$, following the nonlinear stress-strain relation (43), is applied at each time step in the numerical scheme (29) and (30):

$$\hat{c} = c\sqrt{1 + 2B\varepsilon} \tag{44}$$

instead of the piecewise constant one corresponding to the linear case.

We consider the same pulse shape and the same materials (polycarbonate and Al 6061) as in the case of the linear periodic medium. However, the nonlinear effects appear only for a sufficiently high magnitude of loading. The values of the parameter of nonlinearity B were chosen as 0.24 for Al 6061 and 0.8 for polycarbonate.

The results of the simulations corresponding to 400, 1600, and 5200 time steps are shown in Figs. 4–6.

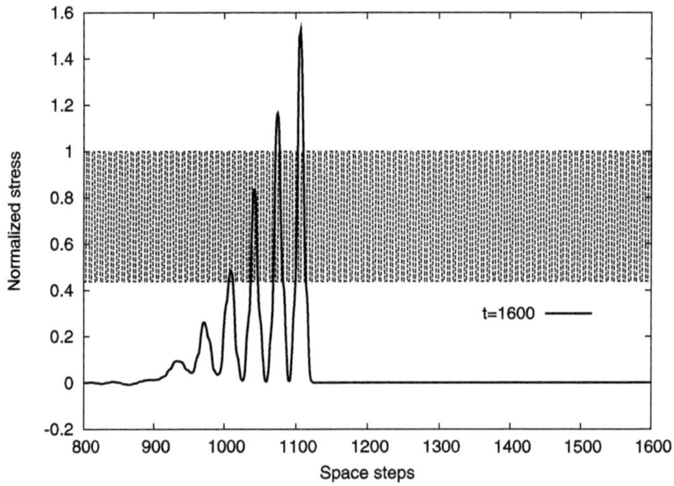

Fig. 5 Pulse shape at time step 1600. Nonlinear case.

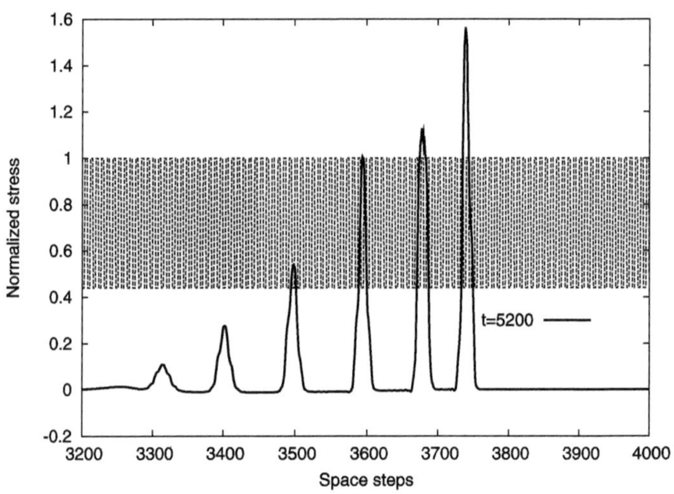

Fig. 6 Pulse shape at time step 5200. Nonlinear case. Reproduced from [5].

We observe that an initial bell-shaped pulse is transformed into a train of soliton-like pulses propagating with amplitude-dependent speeds. Such kind of behavior was first reported in [26], where these pulses were called "stegotons" because their shape is influenced by the periodicity.

In principle, the soliton-like solution could be expected because if we combine the weak nonlinearity (43) with the dispersive wave equation in terms of the effective media theory (41), we arrive at the Boussinesq-type equation

$$\frac{\partial^2 u}{\partial t^2} = (c^2 - c_a^2)\frac{\partial^2 u}{\partial x^2} + \alpha B \frac{\partial u}{\partial x}\frac{\partial^2 u}{\partial x^2} + p^2 c_a^2 c_b^2 \frac{\partial^4 u}{\partial x^4}, \tag{45}$$

which possesses soliton-like solutions.

6 One-dimensional linear waves in laminates

There are three basic length scales in wave propagation phenomena:

- the typical wavelength λ;
- the typical size of the inhomogeneities d;
- the typical size of the whole inhomogeneity domain l.

In the case of infinite periodic media considered above the third length scale was absent. Therefore, it may be instructive to consider wave propagation in a body where the periodic arrangement of layers of different materials is confined within a finite spatial domain.

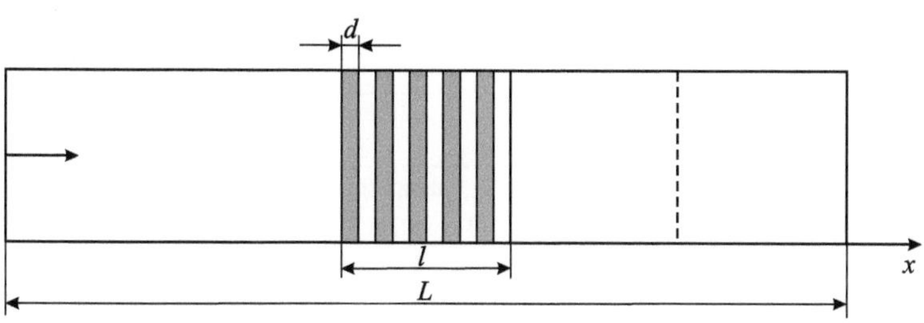

Fig. 7 Length scales in laminate.

To investigate the influence of the size of the inhomogeneity domain, we compare the shape of the pulse in the homogeneous medium with the corresponding pulse transmitted through the periodic array with a different number of distinct layers (Fig. 7).

We use Ti ($\rho = 4510$ kg/m^3, $c = 5020$ m/s) and Al ($\rho = 2703$ kg/m^3, $c = 5240$ m/s) as materials in the distinct layers in the numerical simulations of linear elastic wave propagation.

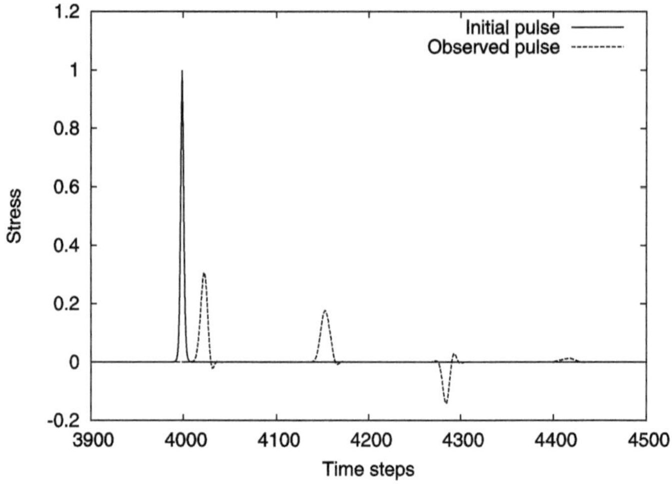

Fig. 8 Pulse shape at 4000 time steps ($d = 64\triangle x, l = 1000\triangle x$).

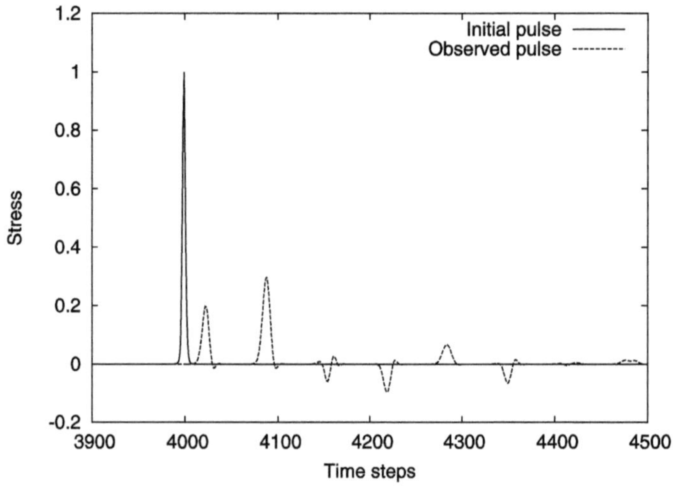

Fig. 9 Pulse shape at 4000 time steps ($d = 32\triangle x, l = 1000\triangle x$).

We apply a stress pulse, the width λ of which corresponds to $30\triangle x$ ($\triangle x$ is the space step)

$$\sigma(t) = \frac{2}{\cosh^2(0.5(t - 15\triangle t))} \tag{46}$$

at the left end of the domain (Fig. 7), and record the resulting pulse at $x = 4000\triangle x$. The location is indicated by the dashed line in Fig. 7.

The results are presented in Figs. 8–10 (dashed lines). The reference pulse calculated for homogeneous media is drawn with a solid line. As can be observed, if

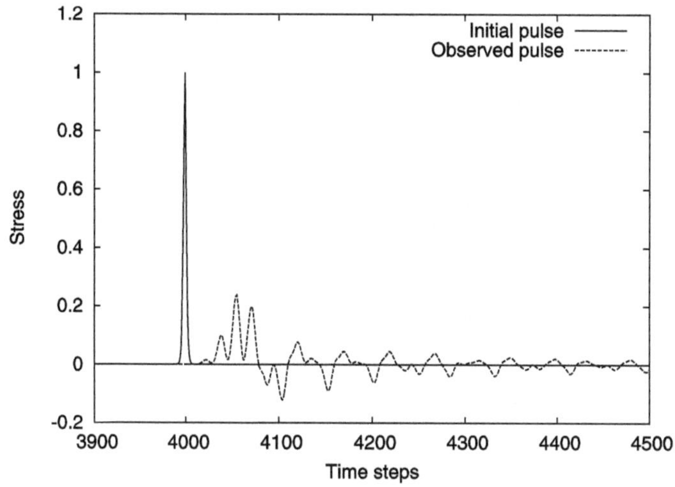

Fig. 10 Pulse shape at 4000 time steps ($d = 8\triangle x, l = 1000\triangle x$).

the wavelength is less then the size of the inhomogeneity ($d \geq \lambda$), we have a strong dispersion of the pulse, i.e., a separation of the wave into components of various frequencies (Figs. 8 and 9). This dispersion is not so strong if, vice versa, the size of the inhomogeneity d is less then the wavelength λ (Fig. 10).

Thus, waves in laminates demonstrate dispersive behavior, which is governed by the relations between the characteristic length scales. Taking into account nonlinear effects, we have seen the soliton-like wave propagation. Both nonlinearity and dispersion effects are observed experimentally in laminates under shock loading.

7 Nonlinear elastic waves in laminates under impact loading

Though the stress response to an impulsive shock loading has been very well understood for homogeneous materials, the same cannot be said for heterogeneous systems. In heterogeneous media, scattering due to interfaces between dissimilar materials plays an important role for shock wave dissipation and dispersion [18].

Diagnostic experiments for the dynamic behavior of heterogeneous materials under impact loading are usually carried out using a plate impact test configuration under a one-dimensional strain state. These experiments were recently reviewed in [12, 13]. For almost all the experiments, the stress response has shown a sloped rising part followed by an oscillatory behavior with respect to a mean value [12, 13]. Such behavior in the periodically layered systems is consistently exhibited in the systematic experimental work [39]. The specimens used in the shock compression experiments [39] were periodically layered two-component composites prepared by repeating a composite unit as many times as necessary to form a specimen with

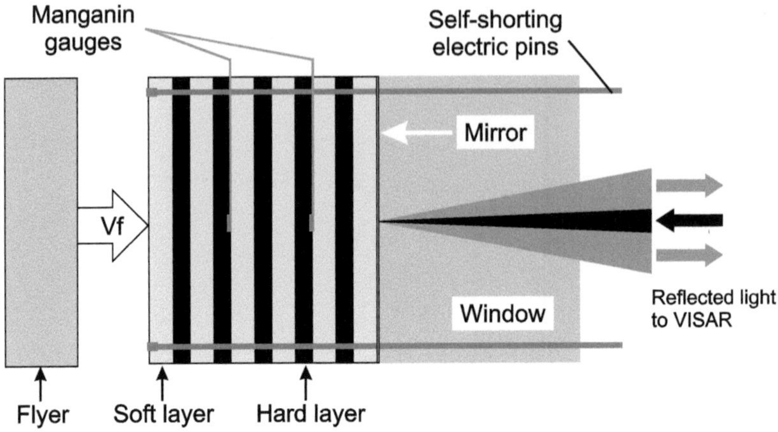

Fig. 11 Experimental setting. Reproduced from [39].

the desired thickness (see Fig. 11). A buffer layer of the same material as the soft component of the specimen was used at the other side of the specimen. A window in contact with the buffer layer was used to prevent the free surface from serious damage due to unloading from shock wave reflection at the free surface. Shock compression experiments were conducted by employing a powder gun loading system, which could accelerate a flat plate flyer to a velocity in the range of 400 m/s to about 2000 m/s. In order to measure the particle velocity history at the specimen window surface, a velocity interferometry system was constructed, and to measure the shock stress history at selected internal interfaces, the manganin stress technique was adopted. Four different materials, polycarbonate, 6061-T6 aluminum alloy, 304 stainless steel, and glass, were chosen as components. The selection of these materials provided a wide range of combinations of shock wave speeds, acoustic impedance and strength levels. The influence of multiple reflections of internal interfaces on shock wave propagation in the layered composites was clearly illustrated by the shock stress profiles measured by manganin gages. The origin of the observed structure of the stress waves was attributed to material heterogeneity at the interfaces. For high velocity impact loading conditions, it was fully realized that material nonlinear effects may play a key role in altering the basic structure of the shock wave.

An approximate solution for layered heterogeneous materials subjected to high velocity plate impact has been developed in [12, 13]. For laminated systems under shock loading, shock velocity, density and volume were related to the particle velocity by means of an equation of state. The elastic analysis was extended to shock response by incorporating the nonlinear effects through computing the shock velocities of the wave trains and superimposing them.

As pointed out in [39], stress wave propagation through layered media made of isotropic materials provides an ideal model to investigate the effect of heterogeneous materials under shock loading, because the length scales, e.g., the thickness of individual layers, and other measures of heterogeneity, e.g., impedance mismatch, are well defined.

Since the impact velocity in shock experiments is sufficiently high, various nonlinear effects may affect the observed behavior. That is why we apply numerical simulations of finite-amplitude nonlinear wave propagation to the study of scattering, dispersion and attenuation of shock waves in layered heterogeneous materials.

The geometry of the problem follows the experimental configuration described in [39] (Fig. 12).

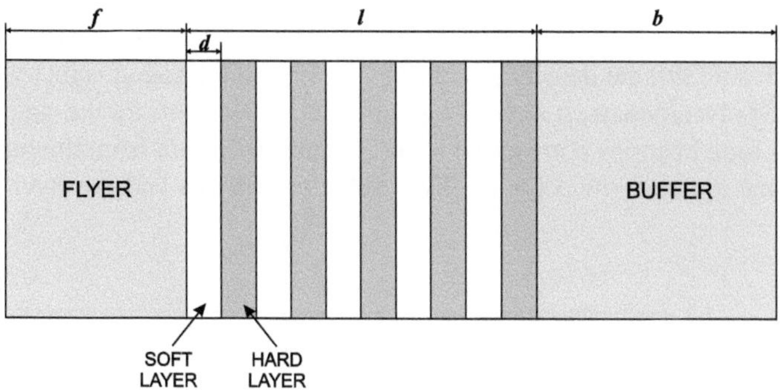

Fig. 12 Geometry of the problem.

We consider the initial-boundary value problem of impact loading of a heterogeneous medium composed of alternating layers of two different materials. The impact is provided by a planar flyer of length L, which has an initial velocity v_0. A buffer of the same material as the soft component of the specimen is used to eliminate the effect of wave reflection at the stress-free surface. The densities of the two materials are different, and the materials' response to compression is characterized by the distinct stress-strain relations $\sigma(\varepsilon)$. Compressional waves propagating in the direction of the layering are modeled by the one-dimensional hyperbolic system of conservation laws (1)–(2).

Initially, stress and strain are zero inside the flyer, the specimen, and the buffer, but the initial velocity of the flyer is nonzero:

$$v(x,0) = v_0, \quad 0 < x < L, \tag{47}$$

where L is the size of the flyer. Both left and right boundaries are stress-free.

Instead of an equation of state like the one used in [12, 13], we apply a simpler nonlinear stress-strain relation $\sigma(\varepsilon,x)$ for each material (43) (cf. [31]):

$$\sigma = \rho c^2 \varepsilon(1 + B\varepsilon), \tag{48}$$

where, as previously, ρ is the density, c is the conventional longitudinal wave speed, and B is a parameter of nonlinearity, the values and signs of which are supposed to be different for hard and soft materials.

We apply the same numerical scheme as in the previous example. The results of the numerical simulations compared with experimental data [39] are presented in the next section.

7.1 Comparison with experimental data

Figure 13 shows the measured and calculated stress time history in the composite, which consists of 8 units of polycarbonate, each 0.74 mm thick, and of 8 units of stainless steel, each 0.37 mm thick. The material properties of the components are extracted from [39]: the density $\rho = 1190$ kg/cm^3 and the sound velocity $c = 1957$ m/s for the polycarbonate; $\rho = 7890$ kg/cm^3 and $c = 5744$ m/s for the stainless steel. The stress time histories correspond to the distance 0.76 mm from the impact face. Calculations are performed for the flyer velocity 561 m/s and the flyer thickness 2.87 mm.

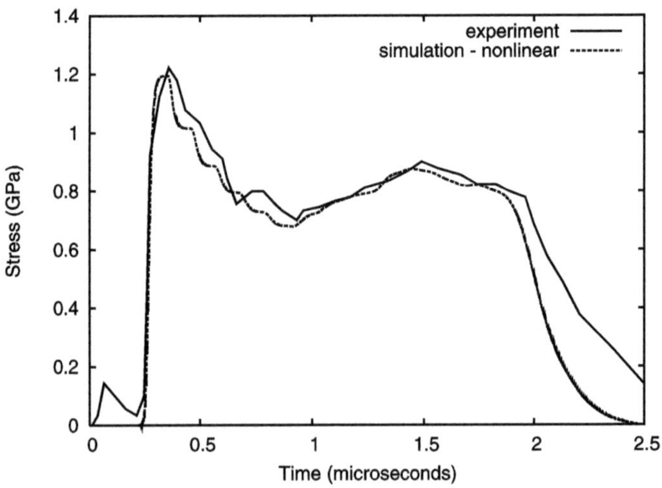

Fig. 13 Comparison of shock stress time histories corresponding to the experiment 112501 [39]. Reproduced from [4].

The results of the numerical calculations depend crucially on the choice of the parameter of nonlinearity B. We choose this parameter from the condition to match the numerical simulations to the experimental results.

Time histories of particle velocity for the same experiment are shown in Fig. 14. It should be noted that the particle velocity time histories correspond to the boundary between the specimen and the buffer. As one can see, both stress and particle velocity time histories are well reproduced by the nonlinear model with the same values of the nonlinearity parameter B.

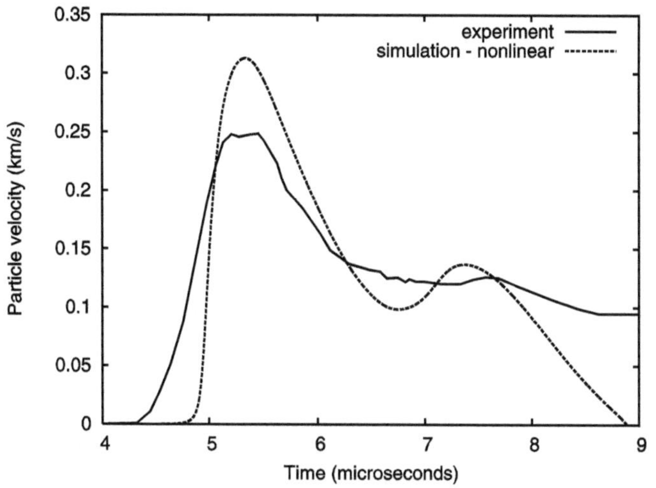

Fig. 14 Comparison of particle velocity time histories corresponding to the experiment 112501 [39]. Reproduced from [4].

As it is pointed out in [39], the influence of multiple reflections of internal interfaces on shock wave propagation in the layered composites is clearly illustrated by the shock stress time histories measured by manganin gages. Therefore, we focus our attention on the comparison of the stress time histories.

Figure 15 shows the stress time histories in the composite, which consists of 16 units of polycarbonate, each 0.37 mm thick, and of 16 units of stainless steel, each 0.19 mm thick. The stress time histories correspond to the distance 3.44 mm from the impact face. Calculations are performed for the flyer velocity 1043 m/s and the flyer thickness 2.87 mm.

The nonlinearity parameter B is chosen here to be 2.80 for polycarbonate and zero for stainless steel. Additionally, the stress time history corresponding to the linear elastic solution (i.e., the nonlinearity parameter is zero for both components) is shown. It can be seen that the stress time history computed by means of the considered nonlinear model is very close to the experimental one. It reproduces three main peaks and decreases with distortion, as it is observed in the experiment [39].

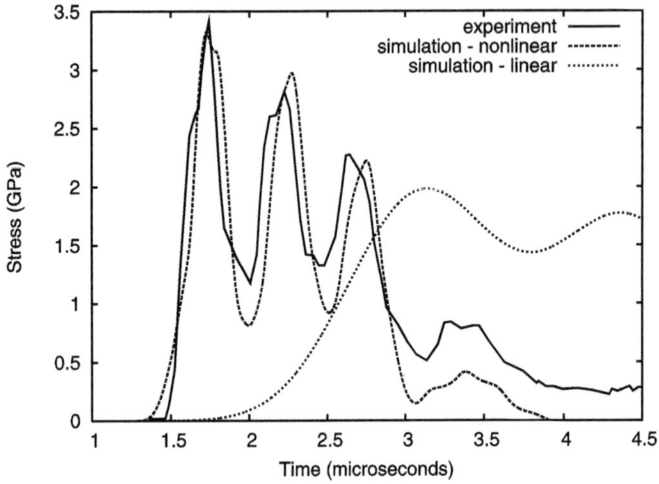

Fig. 15 Comparison of shock stress time histories corresponding to the experiment 110501 [39]. Reproduced from [4].

In Fig. 16 the same comparison is presented for the same composite as in Figure 15, only the flyer thickness is different (5.63 mm). This means that the shock energy is approximately twice as high than that in the previous case. The nonlinearity parameter B is also increased to 4.03 for polycarbonate and remains zero for stainless steel. As a result all 6 experimentally observed peaks are reproduced well.

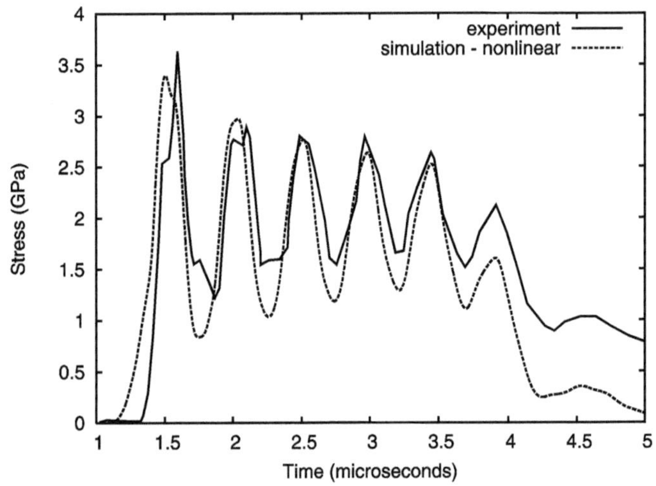

Fig. 16 Comparison of shock stress time histories corresponding to the experiment 110502 [39]. Reproduced from [4].

In Fig. 17 the comparison of stress time histories is presented for the composite consisting of 16 0.37 mm thick units of polycarbonate and 16 0.20 mm thick units of D-263 glass. The material properties of D-263 glass are [39]: the density $\rho = 2510$ kg/cm^3 and the sound velocity $c = 5703$ m/s. The distance between the measurement point and the impact face is 3.41 mm. Corresponding flyer velocity is 1079 m/s and the flyer thickness is 2.87 mm. The nonlinearity parameter B is chosen to be equal 5.025 for polycarbonate and zero for D-263 glass. Again, the stress time history corresponding to the linear elastic solution (i.e., the nonlinearity parameter is zero for both components) is shown. As one can see, the stress time history corresponding to the nonlinear model reproduces all 5 peaks with the same amplitude as observed experimentally.

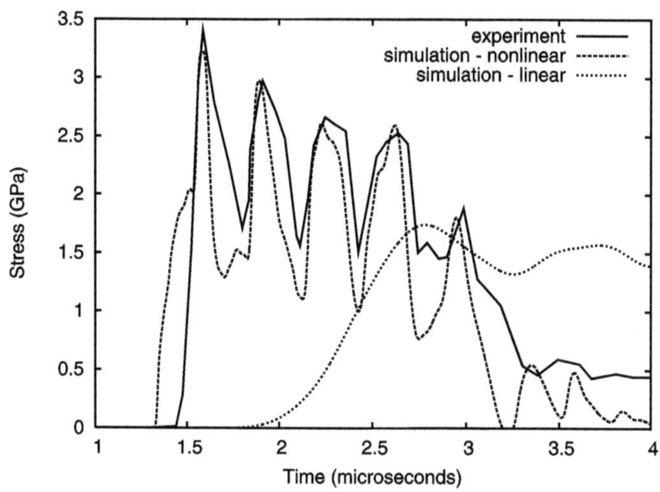

Fig. 17 Comparison of shock stress time histories corresponding to the experiment 112301 [39]. Reproduced from [4].

As it can be seen, the agreement between the results of the calculations and the experiments is achieved by the adjustment of the nonlinearity parameter B.

It follows that the nonlinear behavior of the soft material is affected not only by the energy of the impact, but also by the scattering induced by internal interfaces. It should be noted that the influence of the nonlinearity is not necessarily small. In the numerical simulations, which match with the experiments, the increase of the actual sound velocity of polycarbonate follows. It may be up to two times higher in comparison to the linear case. This conclusion is really surprising, but supported by the stress time histories.

Thus, the application of a nonlinear stress-strain relation for materials in numerical simulations of the plate impact problem of a layered heterogeneous medium shows that a good agreement between computations and experiments can be obtained by adjusting the values of the parameter of nonlinearity [4]. In the numerical simulations of the finite-amplitude shock wave propagation in heterogeneous composites, the flyer size and velocity, the impedance mismatch of hard and soft materials, as well as the number and size of layers in a specimen were the same as in the experiments [39]. Moreover, a nonlinear behavior of materials was also taken into consideration. This means that combining scattering effects induced by internal interfaces and physical nonlinearity in material behavior into one nonlinear parameter, provides the possibility to reproduce the shock response in heterogeneous media observed experimentally. In this context, the parameter B is actually influenced by (i) the physical nonlinearity of the soft material and (ii) the mismatch of the elasticity properties of soft and hard materials. The mismatch effect is similar to the type of nonlinearity characteristic to materials with different moduli of elasticity for tension and compression. The mismatch effect manifests itself due to wave scattering at the internal interfaces, and, therefore, depends on the structure of a specimen. The variation of the parameter of nonlinearity confirms the statement that the nonlinear wave propagation is highly affected by the interaction of the wave with the heterogeneous substructure of a solid [39].

It should be noted that layered media do not exhaust all possible substructures of heterogeneous materials. Another example of a heterogeneous substructure is provided by functionally graded materials.

8 Waves in functionally graded materials

Functionally graded materials (FGMs) are composed of two or more phases that are fabricated so that their compositions vary more or less continuously in some spatial direction and are characterized by nonlinear gradients that result in graded properties. Traditional composites are homogeneous mixtures, and therefore they involve a compromise between the desirable properties of the component materials. Since significant proportions of an FGM contain the pure form of each component, the need for compromise is eliminated. The properties of both components can be fully utilized. For example, the toughness of a metal can be mated with the refractoriness of a ceramic, without any compromise in the toughness of the metal side or the refractoriness of the ceramic side.

Comprehensive reviews of current FGM research may be found in the papers [21] and [30], and in the book [35]. Studies of the evolution of stresses and displacements in FGMs subjected to quasistatic loading [35] show that the utilization of structures and geometry of a graded interface between two dissimilar layers can reduce stresses significantly. Such an effect is also important in the case of dynamical loading, where energy-absorbing applications are of special interest.

We consider the one-dimensional problem in elastodynamics for an FGM slab in which material properties vary only in the thickness direction. It is assumed that the slab is isotropic and inhomogeneous with the following fairly general properties [14]:

$$E'(x) = E'_0 \left(a\frac{x}{l}+1\right)^m, \quad \rho(x) = \rho_0 \left(a\frac{x}{l}+1\right)^n, \tag{49}$$

where ρ is the mass density, l is the thickness, a, m, and n are arbitrary real constants with $a > -1$, while E_0 and ρ_0 are the elastic constant and density at $x = 0$. The elastic constant E_0 is determined under the assumption that $\sigma_{yy} = \sigma_{zz}$ and the slab is fully constrained at infinity. It can thus be shown that

$$E' = \frac{E(1-\nu)}{(1+\nu)(1-2\nu)}, \tag{50}$$

with $E(x)$ and $\nu(x)$ being the Young modulus and the Poisson ratio of the inhomogeneous material.

It is assumed that the slab is at rest for $t \leq 0$, therefore, the following initial conditions are valid:

$$v(x,0) = 0, \quad \sigma(x,0) = 0. \tag{51}$$

The boundary condition at $x = 0$ is

$$v(0,t) = 0, \quad t > 0 \quad (\text{“fixed” boundary}) \tag{52}$$

At $x = l$, the slab is subjected to a stress pulse given by

$$\sigma_{xx}(l,t) = \sigma_0 f(t), \quad t > 0, \tag{53}$$

where the constant σ_0 is the magnitude of the pulse, the function f describes its time profile, and without any loss in generality, it is assumed that $|f| \leq 1$.

Following [14], we consider an FGM slab that consists of nickel and zirconia. The thickness of the slab is $l = 5$ mm. On one surface the medium is pure nickel and on the other surface pure zirconia, while the material properties $E_0(x)$ and $\rho(x)$ vary smoothly in thickness direction. A pressure pulse defined by

$$\sigma_{xx}(l,t) = \sigma_0 f(t) = -\sigma_0 (H(t) - H(t - t_0)) \tag{54}$$

is applied to the surface $x = l$ and the boundary $x = 0$ is “fixed”. Here H is the Heaviside function. The pulse duration is assumed to be $t_0 = 0.2\,\mu s$. The properties of the constituent materials used are given in Table 1 [14].

Material	E (GPa)	ν	ρ (kg/m^3)
ZrO	151	0.33	5331
Ni	207	0.31	8900

Table 1 Properties of materials

The material parameters for the FGMs used are [14]: $a = -0.12354$, $m = -1.8866$, and $n = -3.8866$. The stress is calculated up to $12\,\mu s$ (the propagation time of the plane wave through the thickness $l = 5$ mm is approximately $0.77\,\mu s$ in pure ZrO2 and $0.88\,\mu s$ in Ni).

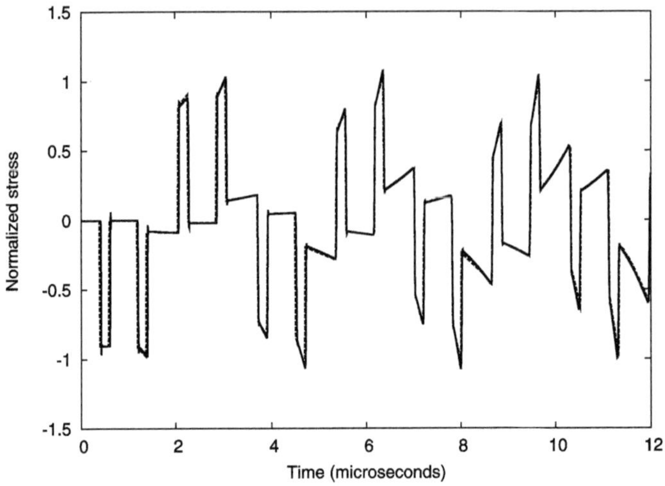

Fig. 18 Variation of stress with time in the middle of the slab. Reproduced from [5].

Numerical simulations were performed by means of the same algorithm as above. The comparison of the results of the numerical simulation and of the analytical solution [14] for the time dependence of the normalized stress σ_{xx}/σ_0 at the location $x/l = 1/2$ is shown in Fig. 18.

As one can see, it is difficult to make a distinction between analytical and numerical results. This means that the applied algorithm is well suited for the simulation of wave propagation in FGM.

A nonlinear behavior for the same materials with the nonlinearity parameter $A = 0.19$ is shown in Figure 19. For the comparison, calculations were performed with the value 0.9 of the Courant number both in the linear and nonlinear case. The amplitude amplification and pulse shape distortion in comparison with the linear case is clearly observed. In addition, the velocity of a pulse in the nonlinear material is increased.

9 Concluding remarks

As we have seen, linear and non-linear wave propagation in media with rapidly-varying properties as well as in functionally graded materials can be successfully simulated by means of the modification of the wave-propagation algorithm based on the non-equilibrium jump relation for true inhomogeneities. It should be emphasized that the used jump relation expresses the continuity of genuine unknown fields at the boundaries between computational cells. The applied algorithm is conservative, stable up to Courant number equal to 1, high-order accurate, and thermodynamically consistent. However, the main advantage of the presented modification of

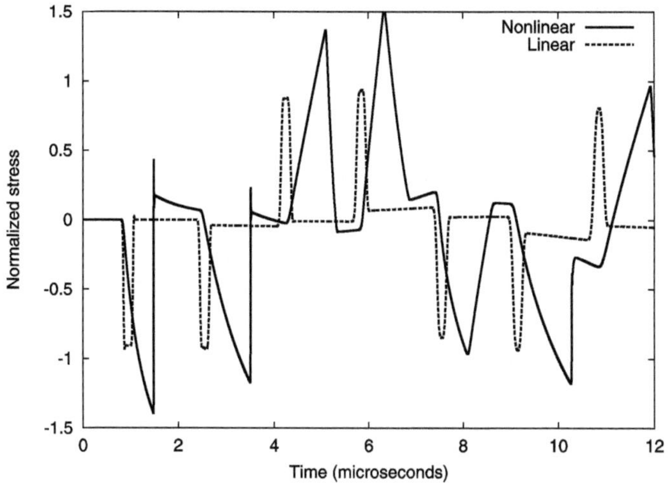

Fig. 19 Variation of stress with time in the middle of the slab. Nonlinear case. Reproduced from [5].

the wave-propagation algorithm is its applicability to the simulation of moving discontinuities. This property is related to the formulation of the algorithm in terms of excess quantities. To apply the algorithm to moving singularities, we simply should change the non-equilibrium jump relation for true inhomogeneities to another non-equilibrium jump relation valid for quasi-inhomogeneities.

Acknowledgements Support of the Estonian Science Foundation (Grant 7037) is gratefully acknowledged.

References

1. Achenbach, J.D.: Wave Propagation in Elastic Solids. North-Holland, Amsterdam (1973)
2. Bale, D.S., LeVeque, R.J., Mitran, S., Rossmanith, J.A.: A wave propagation method for conservation laws and balance laws with spatially varying flux functions. SIAM J. Sci. Comp. **24**, 955–978 (2003)
3. Bedford, A., Drumheller, D.S.: Introduction to Elastic Wave Propagation. Wiley, New York (1994)
4. Berezovski, A., Berezovski, M., Engelbrecht, J.: Numerical simulation of nonlinear elastic wave propagation in piecewise homogeneous media. Mater. Sci. Eng. **A418**, 364–369 (2006)
5. Berezovski A, Berezovski, M., Engelbrecht, J., Maugin, G.A.: Numerical simulation of waves and fronts in inhomogeneous solids. In: Nowacki, W.K., Zhao, H. (eds.) Multi-Phase and Multi-Component Materials under Dynamic Loading, pp. 71-80. Inst. Fundam. Technol. Research, Warsaw (2007)
6. Berezovski, A., Maugin, G.A.: Simulation of thermoelastic wave propagation by means of a composite wave-propagation algorithm. J. Comp. Physics **168**, 249–264 (2001)

7. Berezovski, A., Maugin, G.A.: Thermoelastic wave and front propagation. J. Thermal Stresses **25**, 719–743 (2002)
8. Berezovski, A., Maugin, G.A.: Stress-induced phase-transition front propagation in thermoelastic solids. Eur. J. Mech. A/Solids **24**, 1–21 (2005)
9. Billingham, J., King, A.C.: Wave Motion. Cambridge University Press (2000)
10. Chakraborty, A., Gopalakrishnan, S.: Various numerical techniques for analysis of longitudinal wave propagation in inhomogeneous one-dimensional waveguides. Acta Mech. **162**, 1–27 (2003)
11. Chakraborty, A., Gopalakrishnan, S.: Wave propagation in inhomogeneous layered media: solution of forward and inverse problems. Acta Mech. **169**, 153–185 (2004)
12. Chen, X., Chandra, N.: The effect of heterogeneity on plane wave propagation through layered composites. Comp. Sci. Technol. **64**, 1477–1493 (2004)
13. Chen, X., Chandra, N., Rajendran, A.M.: Analytical solution to the plate impact problem of layered heterogeneous material systems. Int. J. Solids Struct. **41**, 4635–4659 (2004)
14. Chiu, T.-C., Erdogan, F.: One-dimensional wave propagation in a functionally graded elastic medium. J. Sound Vibr. **222**, 453–487 (1999)
15. Engelbrecht, J., Berezovski, A., Pastrone, F., Braun, M.: Waves in microstructured materials and dispersion. Phil. Mag. **85**, 4127–4141 (2005)
16. Fogarthy, T., LeVeque, R.J.: High-resolution finite-volume methods for acoustics in periodic and random media. J. Acoust. Soc. Am. **106**, 261–297 (1999)
17. Godlewski, E., Raviart, P.-A.: Numerical Approximation of Hyperbolic Systems of Conservation Laws. New York, Springer (1996)
18. Grady, D.: Scattering as a mechanism for structured shock waves in metals. J. Mech. Phys. Solids **46**, 2017–2032 (1998)
19. Graff, K.F.: Wave Motion in Elastic Solids. Oxford University Press (1975)
20. Guinot, V.: Godunov-type Schemes: An Introduction for Engineers. Elsevier, Amsterdam (2003)
21. Hirai, T.: Functionally graded materials. In: Processing of Ceramics. Vol. 17B, Part 2, pp. 292-341. VCH Verlagsgesellschaft, Weinheim (1996)
22. Hoffmann, K.H., Burzler, J.M., Schubert, S.: Endoreversible thermodynamics. J. Non-Equil. Thermodyn. **22**, 311–355 (1997)
23. Langseth, J.O., LeVeque, R.J.: A wave propagation method for three-dimensional hyperbolic conservation laws. J. Comp. Physics **165**, 126–166 (2000)
24. LeVeque, R.J.: Wave propagation algorithms for multidimensional hyperbolic systems. J. Comp. Physics **131**, 327–353 (1997)
25. LeVeque, R.J.: Balancing source terms and flux gradients in high-resolution Godunov methods: the quasi-steady wave-propagation algorithm. J. Comp. Physics **148**, 346–365 (1998)
26. LeVeque, R.J.: Finite volume methods for nonlinear elasticity in heterogeneous media. Int. J. Numer. Methods in Fluids **40**, 93–104 (2002)
27. LeVeque, R.J.: Finite Volume Methods for Hyperbolic Problems. Cambridge University Press (2002)
28. LeVeque, R.J., Yong, D.H.: Solitary waves in layered nonlinear media. SIAM J. Appl. Math. **63**, 1539–1560 (2003)
29. Liska, R., Wendroff, B.: Composite schemes for conservation laws. SIAM J. Numer. Anal. **35**, 2250–2271 (1998)
30. Markworth, A.J., Ramesh, K.S., Parks, W.P.: Modelling studies applied to functionally graded materials. J. Mater. Sci. **30**, 2183–2193 (1995)
31. Meurer, T., Qu, J., Jacobs, L.J.: Wave propagation in nonlinear and hysteretic media – a numerical study. Int. J. Solids Struct. **39**, 5585–5614 (2002)
32. Muschik, W., Berezovski, A.: Thermodynamic interaction between two discrete systems in non-equilibrium. J. Non-Equilib. Thermodyn. **29**, 237–255 (2004)
33. Rokhlin, S.I., Wang, L.: Ultrasonic waves in layered anisotropic media: characterization of multidirectional composites. Int. J. Solids Struct. **39**, 5529–5545 (2002)
34. Santosa, F., Symes, W.W.: A dispersive effective medium for wave propagation in periodic composites. SIAM J. Appl. Math. **51**, 984–1005 (1991)

35. Suresh, S., Mortensen, A.: Fundamentals of Functionally Graded Materials. The Institute of Materials, IOM Communications, London (1998)
36. Toro, E.F.: Riemann Solvers and Numerical Methods for Fluid Dynamics. Springer, Berlin (1997)
37. Toro, E.F. (ed.): Godunov Methods: Theory and Applications. Kluwer, New York (2001)
38. Wang, L., Rokhlin, S.I.: Recursive geometric integrators for wave propagation in a functionally graded multilayered elastic medium. J. Mech. Phys. Solids **52**, 2473–2506 (2004)
39. Zhuang, S., Ravichandran, G., Grady, D.: An experimental investigation of shock wave propagation in periodically layered composites. J. Mech. Phys. Solids **51**, 245–265 (2003)

Part II
Mesoscopic Theory

Overview

Wolfgang Muschik

Continuum theory is based on the balance equations of mass, momentum, angular momentum or spin, total or kinetic energy, and internal energy. Additionally one has to consider the balance of entropy for taking into account the Second Law of Thermodynamics. In non-relativistic physics all these balances are defined on time and position $(\mathbf{x},t)$. Beyond the quantities whose balance equations are mentioned above, complex materials need more variables for their unique description. Examples for these additional quantities are internal variables, order and damage parameters, Cosserat triads, and alignment and conformation tensors. In principle there are two possibilities to include these additional quantities into the continuum theoretical description: one can introduce additional fields and their balance equation defined on space-time $(\mathbf{x},t) \in \mathbb{R}^3 \times \mathbb{R}^1$, or the additional quantities are introduced as variables extending space-time to the so-called *mesoscopic space*. The second possibility is called the *mesoscopic concept* [1], which consists of extending the domain of the balance equations by the set of mesoscopic variables

$$(\mathbf{m},\mathbf{x},t) \in \mathcal{M} \times \mathbb{R}^3 \times \mathbb{R}^1. \tag{1}$$

Here $\mathbf{m} \in \mathcal{M}$ is a set of mesoscopic variables in a suitable manifold $\mathcal{M}$, on which an integration can be defined.

An example for $\mathbf{m}$ is the *microscopic director* $\mathbf{n}$ in mesoscopic liquid crystal theory. This microscopic director is defined as a unit vector pointing into the temporary direction of a needle-shaped rigid particle, or, if the particle is of a plane shape, the microscopic director is perpendicular to the particle. Because the microscopic director is defined on a molecular level, it is not a macroscopic field $\mathbf{d}(\mathbf{x},t)$ describing the mean orientation, but a *mesoscopic variable*. Here "mesoscopic" means that the level of description is finer than the macroscopic one, but that no microscopic concepts such as molecular interactions or potentials are used. In this example the manifold $\mathcal{M}$ in (1) is the 2-dimensional unit sphere $\mathbb{S}^2$.

Institut für Theoretische Physik, Technische Universität Berlin, Hardenbergstr. 36, 10623 Berlin, Germany, e-mail: muschik@physik.tu-berlin.de

E. Quak, T. Soomere (eds.), *Applied Wave Mathematics*,
DOI 10.1007/978-3-642-00585-5_6,

Beyond the use of additional variables $\mathbf{m}$, the mesoscopic concept introduces a statistical element, the so-called *mesoscopic distribution function* (MDF) $f(\mathbf{m},\mathbf{x},t)$, generated by the different values of the mesoscopic variables of the molecules in a volume element

$$f(\mathbf{m},\mathbf{x},t) \equiv f(\cdot), \quad (\cdot) \equiv (\mathbf{m},\mathbf{x},t) \in \mathcal{M} \times \mathbb{R}^3 \times \mathbb{R}^1. \tag{2}$$

The MDF is defined on the mesoscopic space $\mathcal{M} \times \mathbb{R}^3 \times \mathbb{R}^1$, describing the distribution of $\mathbf{m}$ in a volume element around $\mathbf{x}$ at time t, and therefore it is always normalized

$$\int f(\mathbf{m},\mathbf{x},t)\, d\mathcal{M} = 1. \tag{3}$$

Now the fields such as mass density, momentum density, specific internal energy, etc. are defined on the mesoscopic space. For distinguishing these fields from the usual, macroscopic ones, the word "mesoscopic" is added. Consequently, the *mesoscopic mass density* is defined by

$$\rho(\cdot) := \rho(\mathbf{x},t) f(\cdot)\,. \tag{4}$$

Here $\rho(\mathbf{x},t)$ is the macroscopic mass density. By use of (3), we obtain

$$\rho(\mathbf{x},t) = \int \rho(\mathbf{m},\mathbf{x},t)\, d\mathcal{M}. \tag{5}$$

This equation shows that the system can be formally treated as a mixture by regarding all particles in a volume element of the same mesoscopic variables as one component of the system having the partial density $\rho(\cdot)$. Thus the MDF results as the fraction of the mass density belonging to one component characterized by the same mesoscopic variables over the total mass density of the mixture. Here the "component index" $\mathbf{m}$ is a continuous one. Because mixture theory is well developed, *mesoscopic balance equations* can be written down very easily.

Starting out with the usual shape of a macroscopic balance in local form

$$\frac{\partial}{\partial t}\mathbf{X}(\mathbf{x},t) + \nabla_x \cdot [\mathbf{v}(\mathbf{x},t)\mathbf{X}(\mathbf{x},t) - \mathbf{S}(\mathbf{x},t)] = \Sigma(\mathbf{x},t), \tag{6}$$

and taking into account that the mesoscopic space (1) includes the mesoscopic variables additional to position and time, we obtain the general shape of *local mesoscopic balances*

$$\frac{\partial}{\partial t}\mathbf{X}(\cdot) + \nabla_x \cdot [\mathbf{v}(\cdot)\mathbf{X}(\cdot) - \mathbf{S}(\cdot)] + \nabla_m \cdot [\mathbf{u}(\cdot)\mathbf{X}(\cdot) - \mathbf{R}(\cdot)] = \Sigma(\cdot). \tag{7}$$

Here the independent field $\mathbf{u}(\cdot)$, called *mesoscopic change velocity*, is the analogue to the mesoscopic material velocity $\mathbf{v}(\cdot)$ and describes the change in time of the set of mesoscopic variables

$$(\mathbf{m},\mathbf{x},t) \longrightarrow (\mathbf{m} + \mathbf{u}(\cdot)\Delta t, \mathbf{x} + \mathbf{v}(\cdot)\Delta t, t + \Delta t). \tag{8}$$

The special balances are obtained by a special identification of $\mathbf{X}$ by which an interpretation of $\mathbf{S}(\cdot)$ and $\mathbf{R}(\cdot)$ follows. We obtain the mesoscopic balances by the subsequent identifications:

Mass

$$\mathbf{X} \equiv \rho, \quad \Sigma(\cdot) \equiv 0, \tag{9}$$

$$\mathbf{S}(\cdot) \equiv \mathbf{0}, \quad \mathbf{R}(\cdot) \equiv \mathbf{0}, \tag{10}$$

$$\frac{\partial}{\partial t}\rho(\cdot) + \nabla_x \cdot \{\rho(\cdot)\mathbf{v}(\cdot)\} + \nabla_m \cdot \{\rho(\cdot)\mathbf{u}(\cdot)\} = 0. \tag{11}$$

Momentum

$$\mathbf{X} \equiv \rho\mathbf{v}, \qquad \Sigma(\cdot) \equiv \rho(\cdot)\mathbf{k}(\cdot), \tag{12}$$

$$\mathbf{S}(\cdot) \equiv \mathbf{T}^\top(\cdot), \qquad \mathbf{R}(\cdot) \equiv \mathcal{T}^\top(\cdot), \tag{13}$$

$$\frac{\partial}{\partial t}[\rho(\cdot)\mathbf{v}(\cdot)] + \nabla_x \cdot \left[\mathbf{v}(\cdot)\rho(\cdot)\mathbf{v}(\cdot) - \mathbf{T}^\top(\cdot)\right] + \\ + \nabla_m \cdot \left[\mathbf{u}(\cdot)\rho(\cdot)\mathbf{v}(\cdot) - \mathcal{T}^\top(\cdot)\right] = \rho(\cdot)\mathbf{k}(\cdot). \tag{14}$$

Here $\mathbf{k}(\cdot)$ is the external acceleration, $\mathbf{T}^\top(\cdot)$ the transposed stress tensor, and $\mathcal{T}^\top(\cdot)$ the transposed stress tensor on $\mathcal{M}$. The other mesoscopic balances of angular momentum, spin, total and internal energy follow by similiar identifications of $\mathbf{X}$.

According to the definition of the mesoscopic mass density (4), we obtain from the mesoscopic mass balance (11) a balance of the MDF $f(\cdot)$ by inserting its definition:

$$\frac{\partial}{\partial t}f(\cdot) + \nabla_x \cdot [\mathbf{v}(\cdot)f(\cdot)] + \nabla_m \cdot [\mathbf{u}(\cdot)f(\cdot)] + \\ + f(\cdot)\left[\frac{\partial}{\partial t} + \mathbf{v}(\cdot) \cdot \nabla_x\right] \ln\rho(\mathbf{x},t) = 0. \tag{15}$$

Because this balance equation of the mesoscopic distribution function includes the macroscopic field of the mass density (5), it is not independent of the macroscopic mass balance defined on $\mathbb{R}^3$. Therefore the mesoscopic rate equation for the MDF contains already macroscopic quantities influencing its time rate. This can be interpreted as an influence of a "mean field" on the mesoscopic motion. As can be seen from (5), macroscopic quantities are defined by integration over the mesoscopic part. Integration of the mesoscopic balances over the mesoscopic variables results in the balances of micropolar media.

One of them is essential: the *entropy balance*. As the Second Law can be formulated only macroscopically, the entropy balance is only interesting in its macroscopic form. It writes

$$\frac{\partial}{\partial t}[\rho(\mathbf{x},t)\eta(\mathbf{x},t)] + \nabla_x \cdot [\rho(\mathbf{x},t)\eta(\mathbf{x},t)\mathbf{v}(\mathbf{x},t) + \phi(\mathbf{x},t)] = \\ = \zeta(\mathbf{x},t) + \sigma(\mathbf{x},t) \tag{16}$$

($\eta(\mathbf{x},t)$ = specific entropy, $\phi(\mathbf{x},t)$ = entropy flux density, $\zeta(\mathbf{x},t)$ = entropy supply, $\sigma(\mathbf{x},t)$ = entropy production density). The Second Law is now expressed by the *dissipation inequality*

$$\sigma(\mathbf{x},t) \geq 0, \tag{17}$$

which has to be taken into account for writing down constitutive equations.

Using the set $\mathbf{m}$ of mesoscopic variables we can introduce the family of the macroscopic *fields of order parameters* which is defined by different moments of the MDF

$$\mathbf{A}(\mathbf{x},t) := \int f(\cdot)\mathbf{m}d\mathcal{M} \tag{18}$$

$$\mathbf{a}(\mathbf{x},t) := \int f(\cdot)\ \overline{\mathbf{mm}}\ d\mathcal{M}, \tag{19}$$

$$\underline{\mathbf{a}}_N(\mathbf{x},t) := \int f(\cdot)\ \overline{\mathbf{m}\text{...N times...}\mathbf{m}}\ d\mathcal{M}, \qquad \text{etc.} \tag{20}$$

Here the symbol $\overline{\quad}$ denotes the traceless symmetric part of the tensor in its argument.

These fields of order parameters describe macroscopically the mesoscopic state of the system introduced by $\mathbf{m}$ and its MDF $f(\cdot)$. Consequently, these fields are the link between the mesoscopic background description of the system and its extended description by additional macroscopic fields.

The contribution of Christina Papenfuss *Dynamics of Internal Variables from the Mesoscopic Background for the Example of Liquid Crystals and Ferrofluids* applies mesoscopic theory to the items mentioned in its title. Here, the mesoscopic variable is the microscopic director with its orientation distribution function and the alignment tensors of different order as order parameters. The method describing liquid crystals is applied to dipolar media and their macroscopic magnetization.

Heiko Herrmann discusses in his contribution *Towards a Description of Twist Waves in Mesoscopic Continuum Physics* how mesoscopic concepts can be applied to wave propagation in liquid crystals and compares mesoscopic methods with the classical ones of internal variables.

References

1. Muschik, W., Ehrentraut, H., Papenfuss, C. Concepts of mesoscopic continuum physics, with application to biaxial liquid crystals. J. Non-Equilib. Thermodyn. **25**, 179–197 (2000)

Dynamics of Internal Variables from the Mesoscopic Background for the Example of Liquid Crystals and Ferrofluids

Christina Papenfuss

Abstract Liquid crystals are an interesting example of complex materials, showing fluid-like flow behavior on one hand and anisotropic solid-like behavior on the other hand. Macroscopically such complex behavior is taken into account by internal variables, and the question of the equations of motion for the internal variables arises. One way to derive such equations of motion is the so-called mesoscopic theory. The general concept of the mesoscopic theory is presented, and it is applied to the examples of liquid crystals and ferrofluids. The internal variables, here the alignment tensor in the case of liquid crystals, and the polarization in case of ferrofluids, are defined from the mesoscopic background. Equations of motion are derived in both cases. The well-known Landau-type equation for the alignment tensor in liquid crystals is recovered.

1 Introduction to liquid crystals

1.1 Some properties of liquid crystals

Liquid crystals were discovered by Reinitzer[1] more than a hundred years ago. Since then the fascinating properties of this "fourth state of matter", which combines the properties of the fluid with those of the solid state, have long been a subject of active research. On one hand, liquid crystalline phases behave like fluids, as they do

Institute of Cybernetics at Tallinn University of Technology, Akadeemia tee 21, 12618 Tallinn, Estonia, e-mail: c.papenfuss@gmx.de
Permanent address: Technische Universität Berlin, Strasse des 17. Juni 135, 10623 Berlin, Germany

[1] These unusual substances were mentioned for the first time in a letter from the botanist F. Reinitzer to the physicist O. Lehmann in 1888. This letter is published in [66]. O. Lehmann was the first one to investigate the physical properties of liquid crystals, especially the morphology of textures under the polarizing microscope.

E. Quak, T. Soomere (eds.), *Applied Wave Mathematics*,
DOI 10.1007/978-3-642-00585-5_7,

not have a well defined shape but flow like highly viscous fluids. On the other hand, they are anisotropic, i.e., material properties depend on the orientation of the sample relative to the measuring device. For example, the electrical conductivity depends on the orientation of the material. The same is true for the dielectric constant. This anisotropy of the dielectric constant leads to optical anisotropy (see Fig. 1). This optical anisotropy led to the discovery of liquid crystals under the polarizing microscope, and it is also the property that is used in the most important application of liquid crystals: optical devices (liquid crystal displays, LCDs).

1.1.1 Liquid crystalline phases

Liquid crystals consist of form-anisotropic molecules. They can be prolate (elongated) or oblate (disk shaped). In the following, it will always be assumed that the effective molecular shape is uniaxial, so that the orientation of the molecule can be described by one direction, the *microscopic director* **n**.

Liquid crystals exhibit a variety of ordered phases between the solid crystalline phase and the isotropic liquid phase. The phase transitions are induced by temperature changes. Heated to above the clearing temperature (the phase transition temperature between the nematic phase and the isotropic phase), the material becomes an isotropic liquid. In most cases the nematic phase occurs upon cooling to below the clearing point; in some cases other liquid crystalline phases occur. In the ne-

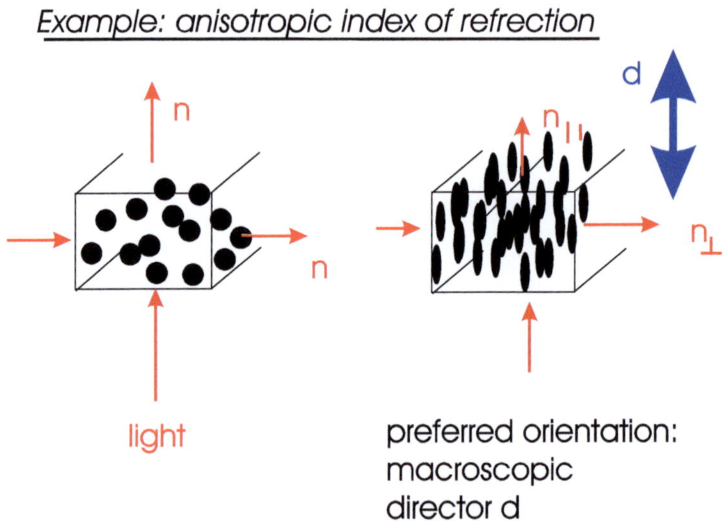

Fig. 1 In the liquid crystalline state, different dielectric constants are measured parallel to the preferred particle orientation and perpendicular to the preferred particle orientation.

matic phase, as well as in the isotropic phase, there is no ordering of the *centers of mass* of the particles. In the nematic phase, in contrast to the isotropic phase, the *molecular orientations* show long range ordering. In addition to this orientational ordering, a one-dimensional positional ordering occurs in the smectic phases. The mass density is modulated periodically, i.e., the centers of mass form layers. The microscopic director can be parallel to or tilted with respect to the normal vector to the smectic layers. In the first case, the liquid crystal is in the smectic A phase, in the second, it is in the smectic C phase. Some liquid crystalline phases are sketched in Fig. 2.

Under special boundary conditions, or the action of electromagnetic fields, it is possible that the liquid crystal is biaxial, i.e., material properties are different in all three perpendicular directions. However, in most cases there is rotation symmetry around one axis, called macroscopic director **d**. Then the phase is called uniaxial. This macroscopic property has to be distinguished from the rotation symmetry of the (microscopic) particles, which is presupposed here in any case.

Macroscopic definition of the alignment tensor

The second order alignment tensor is used as an order parameter in the Landau-theory of phase transitions. Therefore, we define it in such a way that:

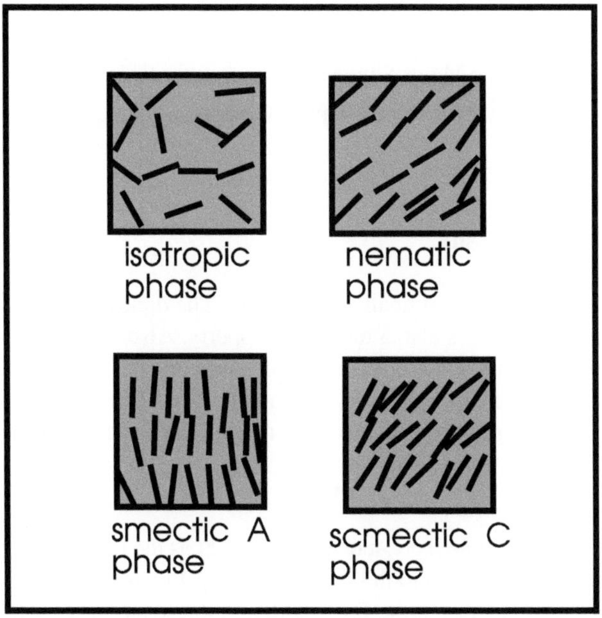

Fig. 2 The most frequent phases in a liquid crystalline material in the order of decreasing temperature: 1. the isotropic (ordinary liquid) phase at high temperature, 2. an anisotropic, but fluid-like nematic phase, 3. the smectic A-phase with a layered structure and particle orientations parallel to the layer normal, 4. the layered smectic C-phase with a tilt angle between the layer normal and the particle orientation.

1. It vanishes in the high temperature phase (the isotropic phase).
2. It is non-zero in the low temperature phase (the nematic phase).
3. It is a dimensionless quantity.

A definition of the alignment tensor with the above properties is:

$$\mathbf{a} = \frac{\varepsilon^e - \frac{1}{3}\,\mathrm{trace}(\varepsilon^e)\delta}{\frac{1}{3}\,\mathrm{trace}(\varepsilon^e)} \tag{1}$$

with the dielectric tensor ε^e ($\mathbf{D} = \varepsilon^e \cdot \mathbf{E}$).

In the isotropic, high-temperature phase the dielectric tensor is proportional to the unit tensor, and the traceless part vanishes, leading to a vanishing alignment tensor, see equation (1). In the liquid crystalline phases, the dielectric tensor is anisotropic, and correspondingly the alignment tensor is non-zero.

In principle, analogous definitions could be given by any other anisotropic property of the liquid crystalline state. However, the dielectric susceptibility is directly related to optical properties and is, therefore, the most important property of liquid crystals.

If the material is (macroscopically) uniaxial (with its symmetry axis denoted by **d**), from symmetry arguments the alignment tensor is of the form:

$$\mathbf{a} = S\,\overline{\mathbf{dd}} \tag{2}$$

with a scalar quantity, the Maier-Saupe order parameter S [54, 55], and unit vector **d**. $\overline{\quad\;}$ denotes the symmetric traceless part of a tensor, in this case: $\overline{\mathbf{dd}} = \mathbf{dd} - \frac{1}{3}\delta$, with the unit tensor δ.

1.1.2 Anisotropic viscosity

Liquid crystals are not only optically anisotropic. The dielectric constant, measured with static electric fields, is also a tensorial property. Also in the measurement of a viscosity coefficient the anisotropic nature of liquid crystals shows up. To illustrate this, consider a simple flow geometry, a Couette flow. The velocity field is in the x-direction, depending linearly on the y-coordinate (a constant gradient):

$$\nabla\mathbf{v} = \gamma\mathbf{e}_y\mathbf{e}_x \tag{3}$$

with shear rate γ. We define the viscosity coefficient as the ratio between the respective stress tensor component and the shear rate t_{yx}/γ. Let us assume that the material has uniaxial symmetry even in the flow field, and the scalar order parameter S does not depend on position. There are different possibilities for the preferred orientation, i.e., the macroscopic director **d**: **d** can be parallel to the flow field, **d** can be parallel to the gradient, **d** can be perpendicular to both the flow direction and the gradient, or **d** can be in the plane of the velocity and the gradient under an angle of 45°

with the velocity. In all these different geometries, different viscosity coefficients are measured. These are the different Miesowicz viscosities [56].

1.1.3 The Franck elastic energy

In several experimental situations, the orientational order is not homogeneous, but there is a gradient of the alignment tensor. This can be enforced by boundary conditions or by external fields. There are even liquid crystalline phases, the so-called blue phases, which are characterized by a periodic lattice of the molecular orientation, i.e. a non-homogeneous, periodic alignment tensor field is inherent to the structure.

The energy density with inhomogeneous alignment is higher than with a homogeneous orientation. The energy difference is called elastic energy, although it is not the elastic energy of a solid but of an orientational order.

A representation theorem for the elastic energy density, depending on the alignment tensor gradient up to second order reads [53]:

$$e_e = \kappa_1 (\nabla \cdot \mathbf{a})^2 + \kappa_2 (\nabla \times \mathbf{a})^2 . \tag{4}$$

In the special case of a uniaxial distribution function, i.e., $\mathbf{a} = S\, \overline{\mathbf{dd}}$, and for a homogenous scalar order parameter $\nabla S = 0$, this assumes the form:

$$e_e = k_1 (\nabla \cdot \mathbf{d})^2 + k_2 (\mathbf{d} \cdot \nabla \times \mathbf{d})^2 + k_3 (\mathbf{d} \times (\nabla \times \mathbf{d}))^2 . \tag{5}$$

All terms in equation (5) satisfy the so called head-tail-symmetry: $\mathbf{d} \leftrightarrow -\mathbf{d}$ is a symmetry operation. Expression (5) is the Franck elastic energy. The first term is the energy of a pure splay deformation, the second one the energy of a pure bend deformation, and the third one the elastic energy of a pure twist deformation.

2 Mesoscopic theory of complex materials

2.1 Complex materials

Let us call substances consisting of spherical particles with interactions depending only on the inter-particle distance *simple materials*. However, most of the interesting and practically important materials have a more complicated internal structure. We call these materials with internal structure *complex materials*. One way to deal with such complex materials is to introduce macroscopic internal variables. Then, equations of motion for these internal variables are needed. We want to sketch now an alternative way to take into account an internal structure of the material within a continuum theory, the so-called mesoscopic theory. First we shall give some further examples of materials with internal structure, in addition to the liquid crystals mentioned in the introduction.

2.2 Examples of internal structure

2.2.1 Steel

Steel consists of micro-crystallites of different orientations of the crystal axes (see Fig. 4).

Under the action of a mechanical load, the micro-crystallites can be reoriented, resulting in history dependent material properties.

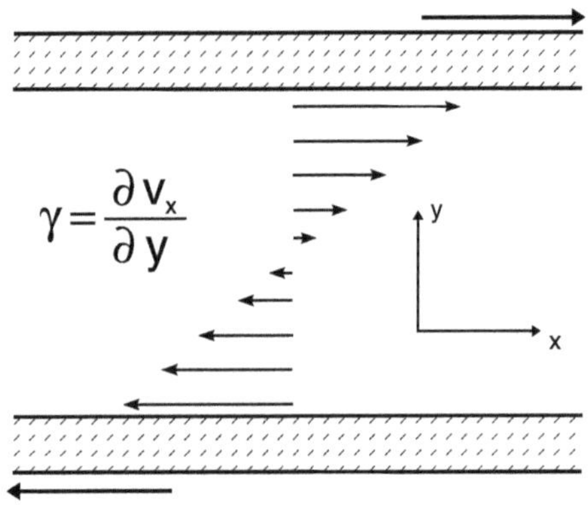

Fig. 3 A simple shear experiment: Couette geometry.

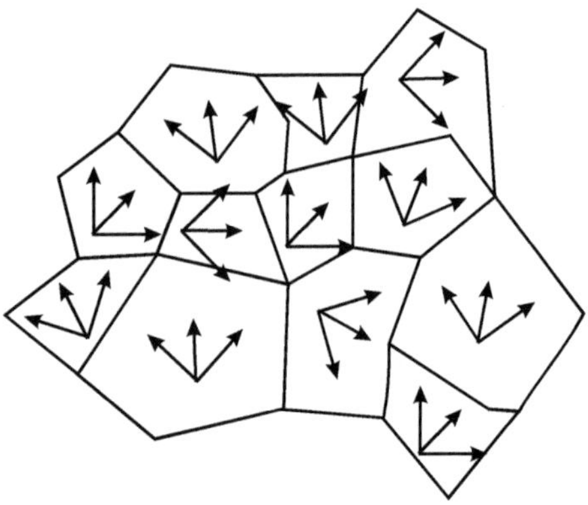

Fig. 4 In a poly-crystalline material the orientation of the crystal axes changes from micro-crystallite to micro-crystallite.

2.2.2 Shape memory alloys

These materials show a very spectacular hysteresis effect: if the original shape is deformed at low temperature, and the material is then heated above a critical temperature, the original un-deformed shape is recovered. There are several applications of these shape memory alloys, such as switches controlled by temperature and even medical applications [61]. The underlying internal structure is a microcrystalline material with different orientations of the micro-crystallites. In addition, the material can undergo a phase transition between martensite and austenite. The phase transition at the critical temperature, and reorientation of crystal axes under mechanical load, can explain the different stress-strain diagrams at different temperatures resulting in the hysteresis effect.

2.2.3 Polymer melts and solutions

Polymer molecules are long chains of organic subunits. Polymer melts or polymer solutions with a low molecular weight solvent show non-Newtonian flow behavior [27, 28, 46]. Let us consider, as an example, a plane Couette flow (see Fig. 3) with the velocity in the x-direction, the gradient in the y-direction, and the shear rate defined by $\gamma = \partial v_x / \partial y$. In a Newtonian fluid, the stress tensor component t_{xy} depends linearly on the shear rate. In polymer melts and solutions, however, we observe a nonlinear dependence, i.e., shear thinning or shear thickening, respectively (see Fig. 5).

This non-Newtonian behavior can be explained assuming stretching and reorientation of the polymer chains in the flow field. The theoretical description of the internal degrees of freedom connected with the conformation of the polymer chain is possible on different levels of accuracy (see Fig. 6). The simplest description is in terms of an end-to-end vector connecting the two ends of the molecule. More information is included in the conformation tensor. The tensor ellipsoid of the con-

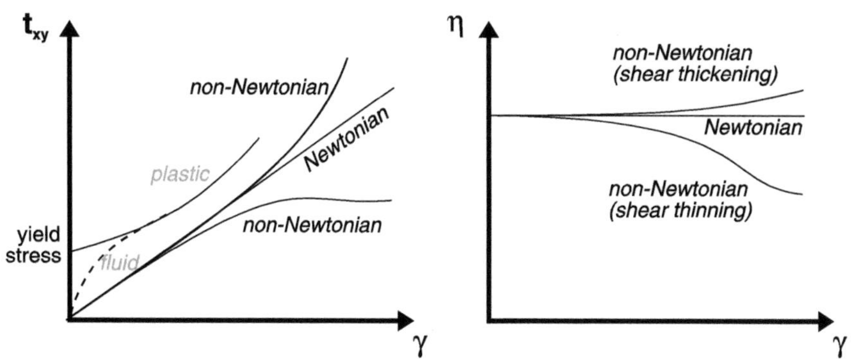

Fig. 5 Non-Newtonian fluid in a shear flow.

formation tensor is the smallest ellipsoid including the polymer chain. The most elevated description is in terms of the positions and orientations of the N subunits of the polymer molecule.

2.2.4 Liquid crystals

Thermotropic liquid crystals consist of rigid non-spherical particles, which are rotationally symmetric. The axis of rotational symmetry is called from now on the microscopic director **n**. The molecules can be rod-like or disc-like. In all liquid crystalline phases, there exists an orientational order of the microscopic directors.

2.2.5 Solids damaged by micro-cracks

An important mechanism of material damage in solids is the growth of micro-cracks under the action of an external load. These microcracks can be modelled as penny-shaped, i.e. flat and rotation symmetric. Then each single crack is characterized by its diameter and orientation of the surface normal [65, 70, 71]. In case of microscopically small cracks there is a large number of cracks in the volume element with a distribution of crack sizes and crack orientations.

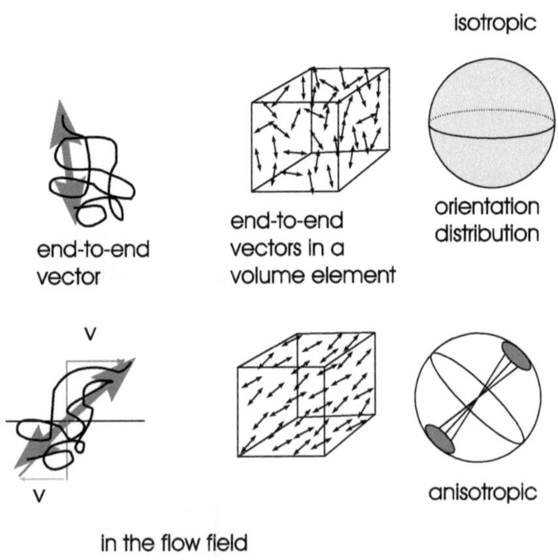

Fig. 6 Polymer chains are stretched and reoriented in the flow field. The orientation distribution of end-to-end vectors becomes anisotropic.

2.3 The mesoscopic concept

There exist many different approaches towards a constitutive theory, and we have to distinguish between macroscopic (phenomenological) theories and theories with a more refined background. The macroscopic quantities of the phenomenological theories are directly measurable, whereas all other theories need some statistical averaging procedure. A wide field of statistical theories is based on the phase space, which is spanned by the particle positions and momenta and, for non-spherical particles, also by orientations and angular momenta. A phase space distribution function is introduced that contains information about the positions and momenta of all, say, 10^{23} particles. This information is not available in experiments. Macroscopic quantities have to be calculated by integrating over particle positions and velocities. The dynamics in this phase space approach is given in terms of an equation of motion for the phase space distribution function. From this dynamics on the phase space the phenomenological equations have to be recovered. For the averaging procedure, a distribution function is needed. In the kinetic theory based on the Boltzmann equation, this is a one-particle distribution function. From this Boltzmann equation, the phenomenological equations of continuum theory [14, 37, 49] can be derived. However, the Boltzmann equation is restricted to dilute gases, where interactions between atoms or molecules are rare and the motions of particles are independent from each other. In dense gases or liquids, the motion of one particle is influenced by the surrounding particles, and a one-particle distribution function is not sufficient to describe the state of the system. A derivation of the phenomenological equations from the level of many-particle distribution functions is much more complicated [29]. The Boltzmann equation is replaced by an infinite coupled set of equations of motion for the one-, two-, three-, ..., N-particle distribution functions, the BBGKY hierarchy [9, 10, 11, 42]. From this theory, as well as from the Boltzmann theory, expressions for constitutive quantities in terms of interaction forces or potentials between molecules are obtained. We will not deal with the microscopic background, but we are concerned with phenomenological theories, the macroscopic theory and the so-called mesoscopic theory. This mesoscopic theory is between microscopic and macroscopic theories in the sense that no microscopic interactions between molecules are introduced; however, mesoscopic fields contain more information than the macroscopic ones. The idea is to enlarge the domain of the field quantities. The new mesoscopic fields are defined on the space $\mathbb{R}_x^3 \times \mathbb{R}_t \times \mathcal{M}$. The manifold $\mathcal{M}$ is given by the set of values the internal degree of freedom can take. Therefore, the choice of $\mathcal{M}$ depends on the complex material under consideration. The domain of the mesoscopic field quantities $\mathbb{R}_x^3 \times \mathbb{R}_t \times \mathcal{M}$ is called mesoscopic space. In Table 1, the physical meaning of the additional mesoscopic variable and of the manifold $\mathcal{M}$ are given for the examples considered in the previous section. The orientation of a triad of crystal axes or molecular axes is determined by a rotation matrix $\in SO(3)$ mapping the actual triad to a reference system.

Beyond the use of additional variables $\mathbf{m}$, the mesoscopic concept introduces a statistical element, the so-called *mesoscopic distribution function* (MDF) $f(\mathbf{m},\mathbf{x},t)$ generated by the different values of the mesoscopic variable of the particles (or subunits) in a volume element:

$$f(\mathbf{m},\mathbf{x},t) \equiv f(\cdot), \ \ (\cdot) \equiv (\mathbf{m},\mathbf{x},t) \in \mathcal{M} \times \mathbb{R}^3 \times \mathbb{R}^1. \tag{6}$$

The MDF describes the distribution of $\mathbf{m}$ in a volume element around $\mathbf{x}$ at time t, and, therefore, it is normalized:

$$\int f(\mathbf{m},\mathbf{x},t)\,d\mathcal{M} \ = \ 1. \tag{7}$$

Now the fields such as mass density, momentum density, etc., are defined on the mesoscopic space. For distinguishing these fields from the macroscopic ones, we add the word "mesoscopic", but we will use the same symbol as for the corresponding macroscopic field. Consequently, the *mesoscopic mass density* is defined by:

$$\rho(\cdot) := \rho(\mathbf{x},t) f(\cdot). \tag{8}$$

Here $\rho(\mathbf{x},t)$ is the macroscopic mass density. By use of (7), we obtain:

$$\rho(\mathbf{x},t) = \int \rho(\mathbf{m},\mathbf{x},t)\,d\mathcal{M}. \tag{9}$$

Other mesoscopic fields defined on the mesoscopic space are the *mesoscopic material velocity* $\mathbf{v}(\cdot)$ of the particles belonging to the mesoscopic variable $\mathbf{m}$ at time t in a volume element around $\mathbf{x}$, the *mesoscopic stress tensor* $\mathbf{t}(\cdot)$, and the *mesoscopic heat flux density* $\mathbf{q}(\cdot)$, etc.

Macroscopic quantities are obtained from mesoscopic ones as averages, with the MDF as probability density:

$$A(\mathbf{x},t) = \int_{\mathcal{M}} A(\cdot) f(\cdot) d\mathbf{m}. \tag{10}$$

material	internal structure	Manifold $\mathcal{M}$
steel	orientation of crystal axes of micro-crystallites	$SO(3)$
shape memory alloys	fraction of different phases and orientation of crystal axes	$[0,1] \times SO(3)$
polymer solutions	orientation and length of end-to-end vector	$S^2 \times [0,r]$
liquid crystals of uniaxial molecules	orientation of particle = microscopic director	S^2
damaged material	size and orientation of micro-cracks	$\mathbb{R}^+ \times S^2$

Table 1 The mesoscopic variable and the manifold $\mathcal{M}$ for different complex materials.

2.4 Mesoscopic balance equations

Let G denote a region in $\mathbb{R}^3 \times \mathcal{M}$ and $\mathbf{X}$ the density of an extensive quantity. Then the global quantity in the region G changes due to a flux over the boundary of G and due to production and supply within G:

$$\frac{d}{dt}\int_G \mathbf{X} d^3x d\mathbf{m} = \int_{\partial G} \phi_{\mathbf{X}}(\cdot) \mathbf{da} + \int_G \Sigma_{\mathbf{X}}(\cdot) d^3x d\mathbf{m}. \tag{11}$$

A generalized Reynolds transport theorem in the mesoscopic space [30] is used to transform the time derivative, and Gauss' theorem is applied to the boundary integral. This is the point where it is necessary to be able to integrate and to differentiate on the manifold $\mathcal{M}$. Then this results in points of the continuum, in which all mesoscopic fields are continuously differentiable, the *local mesoscopic balance* [1]:

$$\frac{\partial}{\partial t}\mathbf{X}(\cdot) + \nabla \cdot [\mathbf{v}(\cdot)\mathbf{X}(\cdot) - \mathbf{S}(\cdot)] + \nabla_m \cdot [\mathbf{u}(\cdot)\mathbf{X}(\cdot) - \mathbf{R}(\cdot)] = \Sigma(\cdot). \tag{12}$$

Here the independent field $\mathbf{u}(\cdot)$, defined on the mesoscopic space, describes the change in time of the set of mesoscopic variables. With respect to $\mathbf{m}$, the *mesoscopic change velocity* $\mathbf{u}(\cdot)$ is the analogue to the mesoscopic material velocity $\mathbf{v}(\cdot)$ referring to $\mathbf{x}$. If a molecule is characterized by $(\mathbf{m},\mathbf{x},t)$, then for $\Delta t \to +0$ it is characterized by $(\mathbf{m}+\mathbf{u}(\cdot)\Delta t, \mathbf{x}+\mathbf{v}(\cdot)\Delta t, t+\Delta t)$. Besides the usual gradient, also the gradient with respect to the set of mesoscopic variables appears as an additional flux term on $\mathcal{M}$ in all balance equations. In general, there is a convective and a non-convective flux on $\mathcal{M}$.

The balance equations to be considered in any case are the balances of mass, momentum, and energy. Depending on the internal degrees of freedom, additional balance equations have to be taken into account. In the example of liquid crystals, this will be the balance of internal angular momentum connected to rotations of the molecules.

The set of balance equations is not a closed system of equations, but requires constitutive equations for mesoscopic quantities. The domain of the constitutive mappings is the state space: here, a mesoscopic one. It is possible that the mesoscopic state space includes *only* mesoscopic quantities, or that it includes mesoscopic *and* macroscopic quantities. An example of the second kind is discussed for liquid crystals in [31].

3 Mesoscopic theory of uniaxial liquid crystals

An example for $\mathbf{m}$ is the *microscopic director* $\mathbf{n}$ in mesoscopic liquid crystal theory [1, 6]. This microscopic director is defined as a unit vector pointing in the temporary direction of a needle-shaped rigid particle. As the microscopic director is defined on a molecular level, it is not a macroscopic field describing the "mean orientation" but rather represents a *mesoscopic variable*, which spans the 2-dimensional unit sphere S^2. The unit sphere is the manifold $\mathcal{M}$ in this example. The MDF in (6) describes the orientation of the molecules in a volume element. In the case of liquid crystals, the MDF is called orientation distribution function (ODF). It is a distribution function on the unit sphere. A so-called head-tail-symmetry is always observed, i.e., one orientation and the reversed one are equally probable:

$$f(\mathbf{x},t,\mathbf{n}) = f(\mathbf{x},t,-\mathbf{n}). \tag{13}$$

Consequently, an inversion symmetry of the graph of the ODF on the unit sphere is observed (see Fig. 7).

The ODF allows the identification of the different phases. In the isotropic phase, all particle orientations are equally probable, and the orientation distribution function is isotropic, i.e., a homogeneous function on the unit sphere S^2. The other extreme is the totally ordered phase, where all particle orientations are the same. The corresponding distribution function has a non-zero value only for this single orientation, i.e., it is δ-shaped. Due to thermal motion, this totally ordered phase does not occur at non-zero temperature. There is partial ordering of orientations, and the corresponding distribution functions show some concentration around a preferred orientation. There are two possibilities: that the ODF is rotationally symmetric around an axis $\mathbf{d}$, or that there is no such rotational symmetry. In the first case, the phase is called uniaxial; in the second case, it is called biaxial. In most cases, nematic liquid crystalline phases are observed to be uniaxial. Biaxiality can be induced by the action of crossed electric and magnetic fields [13], by the action of a flow field [12], or by boundary conditions. This symmetry of the phase has to be distinguished from the symmetry of the particles, which are assumed here to be rotation symmetric in all cases. The different phases are sketched in Fig. 7.

For applications of the mesoscopic concept to liquid crystals, see [3, 4, 5, 6, 7, 8, 31, 57, 58, 59, 62].

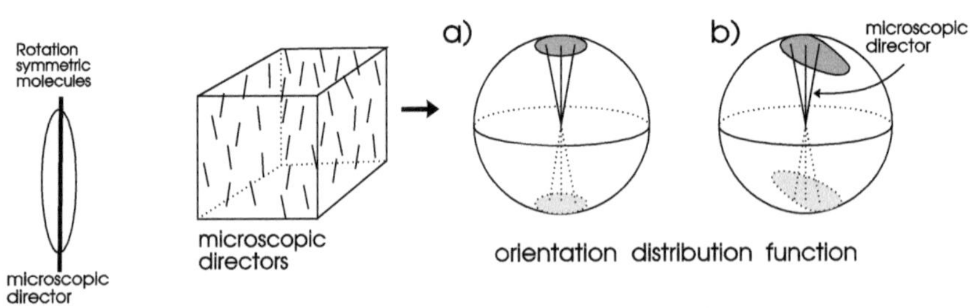

Fig. 7 The orientation distribution function (ODF) in the uniaxial and biaxial liquid crystalline phases. In the isotropic phase, all orientations are equally probable, whereas in the liquid crystalline phases, the ODF is anisotropic.

3.1 Mesoscopic balance equations

The derivative with respect to the mesoscopic variable, the microscopic director $\mathbf{n}$, is denoted by ∇_n. It is the covariant derivative on the unit sphere, and the orientation change velocity $\mathbf{u} = \dot{\mathbf{n}}$ is tangential to the unit sphere.

We obtain the special balances by special identifications of the abstract quantities $\mathbf{X}(\cdot)$, $\Sigma(\cdot)$, $\mathbf{S}(\cdot)$ and $\mathbf{R}(\cdot)$ in (12). These balances are expressed [6]:

Mass:

$$\frac{\partial}{\partial t}\rho(\cdot) + \nabla_x \cdot \{\rho(\cdot)\mathbf{v}(\cdot)\} + \nabla_n \cdot \{\rho(\cdot)\mathbf{u}(\cdot)\} = 0. \tag{14}$$

The last term shows that the density of particles of the particular orientation $\mathbf{n}$ may change due to rotation of particles with velocity $\mathbf{u}$.

Momentum:

$$\frac{\partial}{\partial t}[\rho(\cdot)\mathbf{v}(\cdot)] + \nabla_x \cdot \left[\mathbf{v}(\cdot)\rho(\cdot)\mathbf{v}(\cdot) - \mathbf{t}^\top(\cdot)\right] + \\ + \nabla_n \cdot \left[\mathbf{u}(\cdot)\rho(\cdot)\mathbf{v}(\cdot) - \mathcal{T}^\top(\cdot)\right] = \rho(\cdot)\mathbf{a}(\cdot). \tag{15}$$

Here $\rho\mathbf{a}(\cdot)$ is the external force density, $\mathbf{t}^\top(\cdot)$ the transposed stress tensor, and $\mathcal{T}^\top(\cdot)$ the transposed stress tensor on the unit sphere S^2, i.e., the non-convective momentum flux on the unit sphere.

Angular Momentum:
In addition, we have a balance of angular momentum, which is independent of the balance of momentum, due to rotations of the particles, which are not point like:

$$\mathbf{S}(\cdot) := \mathbf{x} \times \mathbf{v}(\cdot) + \mathbf{s}(\cdot), \tag{16}$$

$$\frac{\partial}{\partial t}[\rho(\cdot)\mathbf{S}(\cdot)] + \nabla_x \cdot \left[\mathbf{v}(\cdot)\rho(\cdot)\mathbf{S}(\cdot) - (\mathbf{x} \times \mathbf{t}(\cdot))^\top - \Pi^\top(\cdot)\right] + \\ + \nabla_n \cdot \left[\mathbf{u}(\cdot)\rho(\cdot)\mathbf{S}(\cdot) - (\mathbf{x} \times \mathcal{T}(\cdot))^\top - \mathcal{W}^\top(\cdot)\right] = \\ = \rho(\cdot)\mathbf{x} \times \mathbf{a}(\cdot) + \rho(\cdot)\mathbf{g}. \tag{17}$$

Here $\mathbf{s}$ is the vector of the specific spin (internal angular momentum due to particle rotations), $\mathbf{g}$ the vector of volume torque density, the second order tensor Π is the surface torque (the non-convective flux of angular momentum in position space), and $\mathcal{W}$ is the analogue to Π with respect to the orientation variable $\mathbf{n}$. All these quantities are mesoscopic ones, depending on position, time, and orientation.

From the balance of total angular momentum, we can derive the balance of internal angular momentum, denoted as spin:

$$\begin{aligned}&\frac{\partial \rho(\cdot)\mathbf{s}(\cdot)}{\partial t}+\nabla\cdot\left(\rho(\cdot)\mathbf{v}(\cdot)\mathbf{s}(\cdot)-\Pi(\cdot)^{T}\right)\\ +\nabla_n\cdot&\left(\rho(\cdot)\mathbf{u}(\cdot)\mathbf{s}(\cdot)-\mathcal{W}^{\top}(\cdot)\right)=\varepsilon:\mathbf{t}(\cdot)+\rho(\cdot)\mathbf{g}(\cdot)\end{aligned} \tag{18}$$

with the totally antisymmetric, third order tensor ε.

Total Energy:

$$e^{\text{tot}}(\cdot) := \frac{1}{2}\mathbf{v}^2(\cdot) + \frac{1}{2}\mathbf{s}(\cdot)\cdot\Theta^{-1}\cdot\mathbf{s}(\cdot) + e(\cdot), \tag{19}$$

$$\begin{aligned}&\frac{\partial}{\partial t}[\rho(\cdot)e^{\text{tot}}(\cdot)] +\\ &+ \nabla_x\cdot\left[\mathbf{v}(\cdot)\rho(\cdot)e^{\text{tot}}(\cdot)-\mathbf{t}(\cdot)^{\top}\cdot\mathbf{v}(\cdot)-\Pi(\cdot)^{\top}\cdot\Theta^{-1}\cdot\mathbf{s}(\cdot)+\mathbf{q}(\cdot)\right]+\\ &+ \nabla_n\cdot\left[\mathbf{u}(\cdot)\rho(\cdot)e^{\text{tot}}(\cdot)-\mathcal{T}(\cdot)^{\top}\cdot\mathbf{v}(\cdot)-\mathcal{W}(\cdot)^{\top}\cdot\Theta^{-1}\cdot\mathbf{s}(\cdot)+\mathbf{Q}(\cdot)\right] =\\ &= \rho(\cdot)\mathbf{a}\cdot\mathbf{v}(\cdot) + \rho(\cdot)\mathbf{g}(\cdot)\cdot\Theta^{-1}\cdot\mathbf{s}(\cdot)+\rho(\cdot)r(\cdot).\end{aligned} \tag{20}$$

Here r is the absorption supply, Θ the moment of inertia tensor of the particles, $\mathbf{q}$ the heat flux density, and $\mathbf{Q}$ the heat flux density on $\mathcal{M}$. Here again, all quantities are mesoscopic ones.

Summary

- Compared to macroscopic balance equations, the mesoscopic equations include additional flux terms in orientation space, the terms with the derivative with respect to the orientational variable $\mathbf{n}$.
- Compared to simple liquids, liquid crystals require consideration of the balance of angular momentum. This is clear from the internal degree of freedom, the rotations of particles.

3.2 Macroscopic balance equations

For the extensive quantities of mass density, momentum density, and spin density, the macroscopic quantities are integrals of the mesoscopic ones over the unit sphere:

$$\rho(\mathbf{x},t) = \oint_{S^2}\rho(\cdot)d^2n, \tag{21}$$

$$\rho(\mathbf{x},t)\mathbf{v}(\mathbf{x},t) = \oint_{S^2}\rho(\cdot)\mathbf{v}(\cdot)d^2n, \tag{22}$$

$$\rho(\mathbf{x},t)\mathbf{s}(\mathbf{x},t) = \oint_{S^2}\rho(\cdot)\mathbf{s}(\cdot)d^2n. \tag{23}$$

Therefore, we obtain the macroscopic balance equations by integrating the mesoscopic ones over all orientations. For the balance of mass, we have:

$$\int_{S^2} \frac{\partial}{\partial t}\rho(\cdot)d^2n + \int_{S^2} \nabla\cdot(\mathbf{v}(\cdot)\rho(\cdot))\,d^2n + \int_{S^2} \nabla_n\cdot(\mathbf{u}(\cdot)\rho(\cdot))\,d^2n \tag{24}$$

or

$$\frac{\partial}{\partial t}\int_{S^2} \rho(\cdot)d^2n + \nabla\int_{S^2} \cdot(\mathbf{v}(\cdot)\rho(\cdot))\,d^2n + \int_{S^2} \nabla_n\cdot(\mathbf{u}(\cdot)\rho(\cdot))\,d^2n \tag{25}$$

or

$$\frac{\partial}{\partial t}\rho(\mathbf{x},t) + \nabla\cdot(\mathbf{v}(\mathbf{x},t)\rho(\mathbf{x},t)) = 0. \tag{26}$$

The last term in equation (25) is zero due to Gauss' theorem on the unit sphere and the fact that S^2 is a closed surface. Accordingly, there is no boundary term on S^2:

$$\int_{S^2} \nabla_n\cdot(\mathbf{u}\rho)d^2n = \int_{\partial S^2} \rho\mathbf{u}\cdot\mathbf{n}dl = 0. \tag{27}$$

Analogously to the macroscopic balance of mass, we obtain the other macroscopic balance equations of momentum and angular momentum.

Balance of momentum:

$$\frac{\partial}{\partial t}(\rho(\mathbf{x},t)\mathbf{v}(\mathbf{x},t)) + \nabla\cdot\left(\mathbf{v}(\mathbf{x},t)\rho(\mathbf{x},t)\mathbf{v}(\mathbf{x},t) - \mathbf{t}^T(\mathbf{x},t)\right) = \rho(\mathbf{x},t)\mathbf{a}(\mathbf{x},t). \tag{28}$$

Balance of angular momentum:

$$\begin{aligned}\frac{\partial}{\partial t}(\rho(\mathbf{x},t)(\mathbf{x}\times\mathbf{v}(\mathbf{x},t)+\mathbf{s}(\mathbf{x},t))) &+ \\ \nabla\cdot\big(\mathbf{v}(\mathbf{x},t)\rho(\mathbf{x},t)(\mathbf{x}\times\mathbf{v}(\mathbf{x},t)+\mathbf{s}(\mathbf{x},t)) &- \mathbf{x}\times\mathbf{t}^T(\mathbf{x},t) \\ -\Pi^T(\mathbf{x},t)\big) &= \rho(\mathbf{x},t)(\mathbf{x}\times\mathbf{a}(\mathbf{x},t)+\mathbf{g}(\mathbf{x},t)).\end{aligned} \tag{29}$$

Along with the balance of energy, these are the balance equations of a micropolar continuum [20, 32, 33, 34, 35]. They have been derived here from the mesoscopic balance equations. The advantage of this derivation is that, on the mesoscopic level, it is very clear from the internal structure which balance equations have to be taken into account: in this instance, the balance of internal angular momentum, in addition to the balances of mass, momentum, and energy.

3.3 Macroscopic constitutive quantities

From the integrated mesoscopic equations, we can identify the macroscopic constitutive quantities in terms of mesoscopic ones. For example, for the stress tensor and the internal energy we find:

$$\mathbf{t}(\mathbf{x},t) = \int_{S^2} (\mathbf{t}(\cdot) + \rho(\mathbf{x},t)\,(\mathbf{v}(\mathbf{x},t)\mathbf{v}(\mathbf{x},t) - \mathbf{v}(\cdot)\mathbf{v}(\cdot)))\,d^2n \tag{30}$$

and

$$e(\mathbf{x},t) = \langle e(\cdot)\rangle + \frac{1}{2}\left(\langle \mathbf{v}(\cdot)^2\rangle - \mathbf{v}(\mathbf{x},t)^2 + \langle \theta \mathbf{u}(\cdot)^2\rangle - \frac{1}{\theta}\mathbf{s}(\mathbf{x},t)^2\right), \tag{31}$$

where $\langle \quad \rangle$ denotes the average of a mesoscopic quantity:

$$\langle \mathbf{v}(\cdot)^2\rangle = \int_{S^2} \mathbf{v}(\cdot)^2 f(\cdot) d^2n. \tag{32}$$

Similar relations can be derived for the heat flux and for the energy supply.

Summary

For constitutive quantities, the macroscopic quantities are in general not simply averages of the corresponding mesoscopic ones. Instead, deviations of the mesoscopic velocity from the macroscopic one, and of the mesoscopic angular velocity from the macroscopic one, make additional contributions.

3.4 Order parameters

We will now give a definition of the alignment tensor based on the orientation distribution function. This alignment tensor is an internal variable.

Using the microscopic director **n** as a mesoscopic variable, we can introduce the family of the macroscopic *fields of order parameters*, which is defined by different moments of the ODF:

$$\mathbf{a}(\mathbf{x},t) := \int_{S^2} f(\cdot)\ \overline{\mathbf{nn}}\ d^2n, \tag{33}$$

$$\mathbf{a}^{(4)}(\mathbf{x},t) := \int_{S^2} f(\cdot)\ \overline{\mathbf{nnnn}}\ d^2n, \tag{34}$$

$$\mathbf{a}^{(6)}(\mathbf{x},t) := \int_{S^2} f(\cdot)\ \underbrace{\overline{\mathbf{n}\ldots\mathbf{n}}}_{6 \text{ times}} d^2n, \qquad \text{etc.} \tag{35}$$

These are tensors of successive order. Only the even order tensors are non-zero, due to the inversion symmetry of the orientation distribution function ($f(\mathbf{x},-\mathbf{n},t) = f(\mathbf{x},\mathbf{n},t)$). The symbol $\overline{\quad}$ denotes the traceless symmetric part of the tensor in its argument [2]. If the ODF is rotationally symmetric with axis of rotational symmetry **d**, the alignment tensors can be written as:

$$\mathbf{a}^{(k)} = S^{(k)} \underbrace{\overline{\mathbf{d}......\mathbf{d}}}_{\text{k times}}, \tag{36}$$

with scalar order parameters $S^{(k)}$, and a unit vector $\mathbf{d}$. The well-known Ericksen-Leslie-theory is the special case, where all particles have exactly the same orientation. In this case all scalar order parameters $S^{(k)}$ are equal to one.

These fields of order parameters describe macroscopically the mesoscopic state of the system introduced by $\mathbf{n}$. Consequently, these fields are the link between the mesoscopic background description of the liquid crystal and its description by additional macroscopic fields (internal variables). In the isotropic phase, all alignment tensors are zero, whereas in the liquid crystalline phases, at least some alignment tensors are non-zero. The alignment tensors are internal variables. In equilibrium, they are determined by the equilibrium variables mass density and temperature. The most important one is the alignment tensor of 2nd order $\mathbf{a}^{(2)} := \mathbf{a}$. We shall see later that it can be interpreted as the order parameter in Landau-theory.

3.5 Differential equation for the distribution function and for the alignment tensors

The orientation distribution function has been defined as the mass fraction:

$$f(\mathbf{x},\mathbf{n},t) = \frac{\rho(\mathbf{x},\mathbf{n},t)}{\rho(\mathbf{x},t)}. \tag{37}$$

For the macroscopic mass density, the balance of mass is applied, and for the mesoscopic mass density, we have the mesoscopic balance of mass. Here, we assume in addition that the flow velocity is independent of the orientation: $\mathbf{v}(\cdot) = \mathbf{v}(\mathbf{x},t)$. We obtain for the distribution function:

$$\frac{\partial f(\mathbf{x},\mathbf{n},t)}{\partial t} + \mathbf{v}(\mathbf{x},\mathbf{n},t)\cdot\nabla f(\mathbf{x},\mathbf{n},t) + \nabla_n\cdot(\mathbf{u}(\mathbf{x},\mathbf{n},t) f(\mathbf{x},\mathbf{n},t)) = 0. \tag{38}$$

The differential equation (38) for the ODF allows the derivation of a system of differential equations for the alignment tensors of successive order, after inserting an expression for the orientation change velocity $\mathbf{u}$. In these equations, the alignment tensors of all orders may be coupled, depending on the expression for $\mathbf{u}$. In general, a closure relation is needed in order to deal with only a limited number of moments (see for instance [64]). A closure relation expresses the higher order alignment tensors $\mathbf{a}^{(k)}$ $(k = 4, 6, \ldots)$ in terms of the second order one. Together with such a closure relation, these equations are the differential equations for the internal variable $\mathbf{a}$. In the next section we will give an example of a derivation of a differential equation for the second order alignment tensor.

3.6 Example of a closed differential equation for the second order alignment tensor

The differential equation for the ODF can be exploited further only if an expression for the orientation change velocity $\mathbf{u}$ is inserted. The orientation change velocity is related to the angular velocity ω and to the spin density $\mathbf{s}$:

$$\omega = \mathbf{n} \times \mathbf{u} , \tag{39}$$

$$\mathbf{s} = \Theta \omega. \tag{40}$$

The aim here is not to solve the differential equation for the spin density on the mesoscopic level, but to make appropriate simplifications, reducing it to an algebraic equation.

3.6.1 Approximations

We now introduce some approximations. The mesoscopic velocity does not depend on the orientation, that means it is equal to the barycentric velocity. The liquid crystal is assumed to be incompressible.

$$\mathbf{v}(\cdot) \equiv \mathbf{v}(\mathbf{x},t), \tag{41}$$

$$\nabla_x \cdot \mathbf{v}(\mathbf{x},t) = 0. \tag{42}$$

We are interested in stationary and uniform states, i.e., there are no spatial gradients, except for possibly a constant velocity gradient, for instance, in a Couette-flow experiment:

$$\frac{\partial}{\partial t}\mathbf{s}(\cdot) = \mathbf{0}, \tag{43}$$

$$\nabla_x \equiv \mathbf{0}, \quad \text{except } \nabla_x \mathbf{v}(\cdot) = \text{const.} \tag{44}$$

There is no non-convective flux of spin on the mesoscopic level:

$$\mathcal{W}(\cdot) = \mathbf{0}. \tag{45}$$

Taking these approximations and the balance of mass (14) into account, the orientational spin balance (18) results in

$$\rho(\cdot)\mathbf{u}(\cdot) \cdot \nabla_n \mathbf{s}(\cdot) = \varepsilon : \mathbf{t}(\cdot). \tag{46}$$

With our assumptions we have basically reduced the balance of spin to an algebraic equation for the orientation change velocity $\mathbf{u}$.

3.6.2 Constitutive Equations

With the above simplifications we obtain from (46)

$$\nabla_n \mathbf{s} = \mathbf{0} \longrightarrow \varepsilon : \mathbf{t} = \mathbf{0} \tag{47}$$

From (47) we see that $\varepsilon : \mathbf{t}$ is homogeneous in $\nabla_n \mathbf{s}$. Thus we have a constitutive equation of the form:

$$\varepsilon : \mathbf{t} = \rho \mathbf{G} \cdot \nabla_n \mathbf{s} \tag{48}$$

with a not yet specified constitutive function $\mathbf{G}$. Taking into account that ∇_n is the covariant derivative on the unit sphere

$$\nabla_n := \partial_n - \mathbf{n}(\mathbf{n} \cdot \partial_n), \tag{49}$$

$\mathbf{u}$ is a solution of the stationary balance of spin, if

$$\mathbf{u} = \mathbf{P} \cdot \mathbf{G}(\cdot), \quad \mathbf{P} = \delta - \mathbf{nn}$$

with the projector $\mathbf{P}$ orthogonal to $\mathbf{n}$.

In fact the simplifying assumptions reduced the differential equation (the balance of spin) for the orientation change velocity $\mathbf{u}$ to an algebraic relation. The orientation change velocity became a constitutive function on the state space, because it is given in terms of the stress tensor (see equation (46)), which itself is a constitutive function.

We now have to define what the domain of the constitutive equations is. This domain is called the *state space*. Here, the state space is chosen in such a way that it consists of a mesoscopic and of a macroscopic part

$$\mathcal{Z} = (\mathcal{Z}^{macro}, \mathcal{Z}^{meso}), \tag{50}$$

$$\mathcal{Z}^{macro} = (\rho(\mathbf{x},t), T(\mathbf{x},t), \overline{\nabla_x \mathbf{v}}\ (\mathbf{x},t), \mathbf{a}(\mathbf{x},t)), \tag{51}$$

$$\mathcal{Z}^{meso} = (\mathbf{n}, \nabla_n \rho(\cdot)), \tag{52}$$

$$\mathcal{Z} = (\rho(\mathbf{x},t), T(\mathbf{x},t), \overline{\nabla_x \mathbf{v}}\ (\mathbf{x},t), \mathbf{a}(\mathbf{x},t), \mathbf{n}, \nabla_n \rho(\cdot)) \,. \tag{53}$$

The state space variables are the macroscopic fields of temperature T, mass density, velocity gradient and alignment tensor, as well as the mesoscopic variables of orientation $\mathbf{n}$ and orientation gradient of the mesoscopic density. The mesoscopic variable $\mathbf{n}$ has been included explicitly. The orientation gradient of mass $\nabla_n \rho(\cdot)$ accounts for the effect of orientation diffusion, because thermic motion alone tends to make the distribution of orientations homogeneous. In the macroscopic part of the state space it is necessary to include the alignment tensor $\mathbf{a}$ to account for the orienting effect of the 'mean field' of the surrounding oriented particles. Spatial gradients are omitted completely, except for the velocity gradient. Spatial gradients can be included also, as well as external fields, like an electric or magnetic field, see for instance, [62].

The dependency of the constitutive function on the state space variables can be arbitrarily non-linear. Here, we make a constitutive assumption for $\mathbf{G}$ on the state space (53), as linear in $\mathbf{a}$ and $\overline{\nabla\mathbf{v}}$, but non-linear in the microscopic director $\mathbf{n}$:

$$\mathbf{u} = \mathbf{P}\cdot\mathbf{G}(\cdot) = \beta_1 \frac{\nabla_n \rho(\cdot)}{\rho(\cdot)} + (\delta - \mathbf{nn})\cdot\left(\beta_2 \overline{\nabla\mathbf{v}} + \beta_3 \mathbf{a}\right)\cdot\mathbf{n}. \tag{54}$$

Inserting this expression into the differential equation for the ODF, we end up with a differential equation of Fokker-Planck type for the orientation distribution function:

$$\begin{aligned}\frac{d}{dt}f + \beta_1 \nabla_n\cdot\nabla_n f - 3f\mathbf{n}\cdot\{\beta_2 \overline{\nabla_x \mathbf{v}} + \beta_3 \mathbf{a}\}\cdot\mathbf{n}&\\ +\mathbf{n}\cdot\{\beta_2 \overline{\nabla_x \mathbf{v}} + \beta_3 \mathbf{a}\}]\cdot\nabla_n f = 0&\end{aligned} \tag{55}$$

with the material time derivative

$$\frac{df}{dt} := \frac{\partial f}{\partial t} + \mathbf{v}\cdot\nabla_x f. \tag{56}$$

As we do not want to deal with this equation on the mesoscopic level, we will now derive from it an equation of motion for the second order alignment tensor, a macroscopic quantity.

3.6.3 Dynamics of the second order alignment tensor

Multiplication of (55) by $\overline{\mathbf{nn}}$, integration over the microscopic directors, and introducing the abbreviation for the non-traceless fourth moment of the ODF:

$$\mathbf{A}^{(4)} := \int_{S^2} f\mathbf{nnnn}d^2n \tag{57}$$

yields the following differential equation for the second order alignment tensor:

$$\begin{aligned}\frac{d}{dt}\mathbf{a} = 6\beta_1\mathbf{a} + \frac{2}{5}\left(\beta_2 \overline{\nabla\mathbf{v}} + \beta_3\mathbf{a}\right)&\\ + \frac{6}{7}\overline{\left(\beta_2 \overline{\nabla\mathbf{v}} + \beta_3\mathbf{a}\right)\cdot\mathbf{a}}&\\ -2\ \overline{\mathbf{a}:\mathbf{A}^{(4)}}.&\end{aligned} \tag{58}$$

This is not yet a closed differential equation for the second order alignment tensor, but a closure relation for the fourth moment $\mathbf{A}^{(4)}$ is needed. One possibility is an algebraic relation between the fourth moment and the second order alignment tensor. A simple relation of this form will be introduced in the next section. Any such choice of algebraic relation leads to a closed equation for the second order tensor, which is of first order in time. Another possibility, leading to a differential equation of second order in time will be discussed in the last subsection.

A simple algebraic closure relation for the fourth moment reads

$$\mathbf{A}^{(4)} = \int_{S^2} f(\cdot)\mathbf{nnnn}d^2n = \int_{S^2} f(\cdot)\mathbf{nn}d^2n \int_{S^2} f(\cdot)\mathbf{nn}d^2n$$
$$= \left(\mathbf{a} + \frac{1}{3}\delta\right)\left(\mathbf{a} + \frac{1}{3}\delta\right). \tag{59}$$

A more sophisticated possibility is to derive a closure relation from a maximization of entropy [64], but this will not be discussed here.

With the closure relation (59), equation (58) leads to

$$\frac{d}{dt}\mathbf{a} = 6\beta_1\mathbf{a} + \frac{2}{5}\left(\beta_2\ \overline{\nabla\mathbf{v}}\ + \beta_3\mathbf{a}\right)$$
$$+\frac{6}{7}\ \overline{\left(\beta_2\ \overline{\nabla\mathbf{v}}\ + \beta_3\mathbf{a}\right)\cdot\mathbf{a}}\ - 2\mathbf{a}:\mathbf{aa}. \tag{60}$$

3.7 *Landau theory of phase transitions as a special case*

The equation of motion for the alignment tensor without a flow field (60) can be interpreted as the equation of motion for the order parameter in the dynamical Landau theory. Landau theory was developed to deal with second order phase transitions, originally with phase transitions in ferromagnetic materials. It has been applied to various kinds of phase transitions, for instance: the transition nematic/ isotropic phase in liquid crystals [21, 22, 23, 53, 73, 74], other transitions between liquid crystalline phases [21, 38, 39], the transition to the superfluid phase of liquid helium, and the transition to the super-conducting phase [47]. In general, it is assumed that there exists an order parameter that is nonzero in one phase (usually the lower temperature phase) and zero in the other phase. The alignment tensor **a** is a candidate of an order parameter here.

Now we will consider a liquid crystal at rest, i.e. $\overline{\nabla\mathbf{v}} = 0$. Then equation (60) simplifies to

$$\frac{d}{dt}\mathbf{a} = 6\beta_1\mathbf{a} + \frac{2}{5}\beta_3\mathbf{a} + \frac{6}{7}\ \overline{\beta_3\mathbf{a}\cdot\mathbf{a}}\ - 2\mathbf{a}:\mathbf{aa} = \overline{\frac{d\Sigma}{d\mathbf{a}}}\ , \tag{61}$$

which shows that the dynamics of the alignment tensor is derived from a potential of the form

$$\Sigma = \frac{1}{2}A(T)\mathbf{a}:\mathbf{a} - \frac{1}{3}B\,\mathrm{trace}\,(\mathbf{a}\cdot\mathbf{a}\cdot\mathbf{a})$$

$$+\frac{1}{4}C_1\,(\mathbf{a}:\mathbf{a})^2 + C_2\,\mathrm{trace}\,(\mathbf{a}\cdot\mathbf{a}\cdot\mathbf{a}\cdot\mathbf{a})\,. \tag{62}$$

This is exactly what is postulated in the dynamical Landau theory with order parameter **a**. The temperature dependence of the parameter $A(T)$ cannot be derived from the mesoscopic theory. For the phase transition to occur at the temperature T_c, the coefficient A must vanish at T_c. From a thermodynamic background it can be argued that

$$A(T) = A_0\,(T - T_c)\,. \tag{63}$$

Especially, in the stationary case we have:

$$\frac{d}{dt}\mathbf{a} = 0 \Rightarrow \overline{\frac{d\Sigma}{d\mathbf{a}}} = 0\,, \tag{64}$$

and the alignment tensor in the stationary state is found from the extremum of the potential. The resulting equation is an ordinary differential equation, because spatial gradients have been excluded from the beginning. On the mesoscopic level, as well as on the macroscopic level, gradients of the alignment tensor can be included in the set of relevant variables. On the macroscopic level, one can derive a Ginzburg-Landau-type partial differential equation [47] for the alignment tensor from thermodynamic arguments. However, on the macroscopic level it is not sufficient to enlarge the state space, but in the exploitation of the dissipation inequality one has to take into account differentiated balance equations as well [67, 68, 69]. A comparison between the results of the mesoscopic and the macroscopic theory in the case that alignment tensor gradients are relevant, is left for a future work.

As the alignment tensor has five independent components, the conditions for a stationary state (64) are still five coupled, nonlinear equations. The situation is considerably simplified, if the orientation distribution is rotation symmetric, i.e., the phase is uniaxial. This symmetry is found in most experimental situations. The axis of rotation symmetry is called macroscopic director and denoted as **d**. Then the alignment tensor has the form $\mathbf{a} = S\,\overline{\mathbf{dd}}$ with the scalar order parameter S and all products of alignment tensors in the potential can be calculated. The Landau potential reduces to a function of the scalar order parameter S alone: $\Sigma(\mathbf{a}) = \Sigma(S,\mathbf{d}) = \Sigma(S)$. The resulting condition for a stationary state reads:

$$A(T)S - \frac{1}{3}BS^2 + \frac{2}{3}C_1S^3 + \frac{1}{3}C_2S^3 = 0. \tag{65}$$

At high temperatures T, there exists only one minimum of the free energy density at $S = 0$, the isotropic phase (see Fig. 8). On lowering the temperature, there occurs a second minimum at temperature $T = T_c^*$, at which the value of the free energy is higher then at the isotropic minimum. This second minimum is meta-stable, and the corresponding ordered phase can be obtained as a meta-stable phase by overheating. At the temperature T_c, both minima have the same value of the free energy. At this temperature, the clearing temperature, there occurs the phase transition. The order parameter jumps between zero (isotropic phase) and a finite value (ordered phase). As the variable S is discontinuous at the phase transition, it is a first order transition. On lowering the temperature further, the second minimum becomes the absolute minimum and the liquid crystalline phase is the stable one. The isotropic phase (the minimum at order parameter zero) becomes unstable at temperature $T = T^*$. Up to this temperature, the isotropic phase can be obtained by supercooling as a meta-stable phase. The temperatures can be expressed in terms of the coefficients. These relations can be used to identify the coefficients from experimental data.

3.8 A remark on constitutive theory and the Second Law of Thermodynamics

Among the principles of material frame indifference and material symmetry, the Second Law of Thermodynamics imposes additional restrictions on constitutive

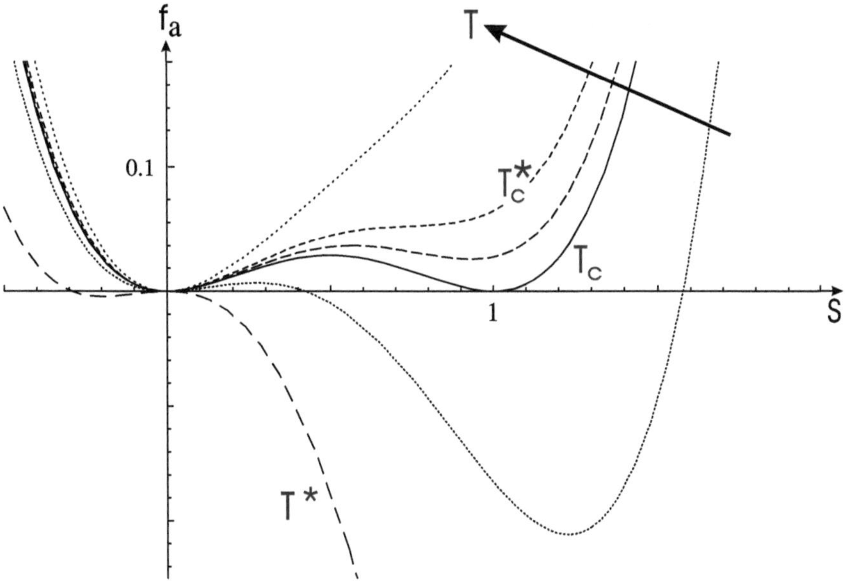

Fig. 8 The free energy density as a function of the scalar order parameter for different temperatures. The number of extremum values depends on temperature, and the absolute minimum corresponds to stable equilibrium phase. At temperature T_c, the phase transition occurs.

functions. There exist different methods of exploiting these restrictions from the Second Law, for instance, the method of Irreversible Thermodynamics [24, 25], the method of Coleman and Noll [17, 18, 19], and the method of Liu [51, 52]. These methods have been applied widely on the macroscopic level to many different kinds of materials. For applications in the field of liquid crystals see, for instance, [3, 40, 41, 50, 60, 63, 72]. In contrast to that, the role of the Second Law on the mesoscopic level is a point to be investigated in the future.

In the last two sections, we derived macroscopic equations for the second order alignment tensor, or a set of alignment tensors, respectively. The constitutive coefficients in these equations will be restricted by the Second Law. In order to derive these restrictions, the entropy inequality should be exploited, taking into account all balance equations, and, in addition, the equation(s) of motion for the alignment tensor(s). This point is left for a future investigation. It should be mentioned that an equation of motion of the form of equation (60) has been derived by Hess in [40, 41] from the exploitation of the Second Law with the method of Irreversible Thermodynamics. From this point of view, the Landau-type equation (60) is simply the linear relation between the thermodynamic force (the derivative of the potential) and the thermodynamic flux (the time derivative of the alignment tensor).

3.9 A set of differential equations for the moments and a second order differential equation for the alignment tensor

The alignment tensors have been defined as traceless tensors of successive order. For the sake of simplicity, we introduce in addition the non-traceless moments

$$\mathbf{A}^{(k)} = \int_{S^2} f \underbrace{\mathbf{n} \ldots \mathbf{n}}_{k} d^2 n. \tag{66}$$

In the following, we use the abbreviation

$$\mathbf{B} = \beta_2 \overline{\nabla \mathbf{v}} + \beta_3 \mathbf{a}. \tag{67}$$

The differential equation for the orientation distribution function allows us to derive a differential equation for any of the moments (66). The resulting equation for the kth order moment reads:

$$\frac{d}{dt}\mathbf{A}^{(k)} = \frac{d}{dt}\int_{S^2} f\underbrace{\mathbf{n}\ldots\mathbf{n}}_{k} d^2n = -\int_{S^2}\nabla_n\cdot(\mathbf{u}f)\underbrace{\mathbf{n}\ldots\mathbf{n}}_{k} d^2n$$

$$= \int_{S^2} f\mathbf{u}\cdot\nabla_n\left(\underbrace{\mathbf{n}\ldots\mathbf{n}}_{k}\right) d^2n = k\int_{S^2} f\left(\mathbf{u}\underbrace{\mathbf{n}\ldots\mathbf{n}}_{k-1}\right)^{sym} d^2n$$

$$= k\int_{S^2} f\beta_1\frac{\nabla_n\rho(\cdot)}{\rho(\cdot)}\underbrace{\mathbf{n}\ldots\mathbf{n}}_{k-1} d^2n + k\int_{S^2} f\mathbf{P}\cdot\mathbf{B}\cdot\mathbf{n}\underbrace{\mathbf{n}\ldots\mathbf{n}}_{k-1} d^2n$$

$$= k\int_{S^2}\frac{\rho(\cdot)}{\rho(\mathbf{x},t)}\beta_1\mathbf{P}\cdot\frac{\nabla_n\rho(\cdot)}{\rho(\cdot)}\underbrace{\mathbf{n}\ldots\mathbf{n}}_{k-1} d^2n + k\int_{S^2} f\mathbf{P}\cdot\mathbf{B}\cdot\mathbf{n}\underbrace{\mathbf{n}\ldots\mathbf{n}}_{k-1} d^2n$$

$$= k\beta_1\frac{1}{\rho(\mathbf{x},t)}\left(\int_{S^2}\nabla_n\cdot\left(\mathbf{P}\rho(\cdot)\underbrace{\mathbf{n}\ldots\mathbf{n}}_{k-1}\right) d^2n - \int_{S^2}(\nabla_n\cdot\mathbf{P})\rho(\cdot)\underbrace{\mathbf{n}\ldots\mathbf{n}}_{k-1} d^2n\right.$$

$$\left. - \int_{S^2}\nabla_n\left(\underbrace{\mathbf{n}\ldots\mathbf{n}}_{k-1}\right)\rho(\cdot)d^2n\right) + k\int_{S^2} f\mathbf{P}\cdot\mathbf{B}\cdot\mathbf{n}\underbrace{\mathbf{n}\ldots\mathbf{n}}_{k-1} d^2n$$

$$= k\beta_1\frac{1}{\rho(\mathbf{x},t)}\left(\int_{S^2} 2\rho(\cdot)\underbrace{\mathbf{n}\ldots\mathbf{n}}_{k} d^2n - \int_{S^2}(k-1)\rho(\cdot)\left(\underbrace{\mathbf{n}\ldots\mathbf{n}}_{k-2}\delta\right)^{sym} d^2n\right.$$

$$\left. + \int_{S^2}\rho(\cdot)(k-1)\underbrace{\mathbf{n}\ldots\mathbf{n}}_{k} d^2n\right) + k\int_{S^2} f(\delta - \mathbf{nn})\cdot\mathbf{B}\cdot\mathbf{n}\underbrace{\mathbf{n}\ldots\mathbf{n}}_{k-1} d^2n$$

$$= k\beta_1\left((k+1)\mathbf{A}^{(k)} - (k-1)\left(\mathbf{A}^{(k-2)}\delta\right)^{sym}\right) + k\left(\mathbf{B}\cdot\mathbf{A}^{(k)} - \mathbf{B}:\mathbf{A}^{(k+2)}\right). \quad (68)$$

Analogously we have the differential equation for the next higher order moment:

$$\frac{d}{dt}\mathbf{A}^{(k+2)} = (k+2)\beta_1\left((k+3)\mathbf{A}^{(k+2)} - (k+1)\left(\mathbf{A}^{(k)}\delta\right)^{sym}\right)$$
$$+(k+2)\left(\mathbf{B}\cdot\mathbf{A}^{(k+2)} - \mathbf{B}:\mathbf{A}^{(k+4)}\right). \quad (69)$$

Our aim is now to eliminate the tensor of order $k+2$ from the coupled set of equations (68) and (69). We will demonstrate this in the case of a uniaxial distribution function and especially for the lowest order tensors, i.e. the tensors of order two and four.

3.9.1 The uniaxial case

In the practically most important case of a rotationally symmetric distribution function, the uniaxial case, all moments of different order are of the form:

$$\mathbf{A}^{(k)} = S^{(k)} \underbrace{\mathbf{d} \ldots \mathbf{d}}_{k}, \tag{70}$$

i.e., we have k scalar parameters $S^{(k)}$ and one unit vector $\mathbf{d}$. The k-fold scalar product of equation (68) with the tensor $\mathbf{A}^{(k)}$ leads to a differential equation for the scalar parameter $S^{(k)}$. Here, thc fact is used that $\mathbf{d} \cdot \dot{\mathbf{d}} = 0$, because $\mathbf{d}$ is a unit vector, so $\mathbf{d} \cdot \mathbf{d} = 1$. For the scalar parameter we find:

$$\dot{S}^{(k)} = k\beta_1 \left((k+1)S^{(k)} - (k-1)S^{(k-2)} \right) + k \left(\mathbf{B} : \mathbf{dd} S^{(k)} - \mathbf{B} : \mathbf{dd} S^{(k+2)} \right). \tag{71}$$

This can be solved for $S^{(k+2)}$:

$$S^{(k+2)} = \left(-\dot{S}^{(k)} + k\beta_1 \left((k+1)S^{(k)} - (k-1)S^{(k-2)} \right) + k\mathbf{B} : \mathbf{dd} S^{(k)} \right) \frac{1}{k\mathbf{B} : \mathbf{dd}}. \tag{72}$$

Analogously to equation (71) we obtain for the order parameter of order $k+2$:

$$\begin{aligned} \dot{S}^{(k+2)} = (k+2)\beta_1 \left((k+3)S^{(k+2)} - (k+1)S^{(k)} \right) + \\ (k+2) \left(\mathbf{B} : \mathbf{dd} S^{(k+2)} - \mathbf{B} : \mathbf{dd} S^{(k+4)} \right). \end{aligned} \tag{73}$$

$S^{(k+2)}$ can be eliminated by differentiating equation (71):

$$\begin{aligned} \ddot{S}^{(k)} = k\beta_1 \left((k+1)\dot{S}^{(k)} - (k-1)\dot{S}^{(k-2)} \right) + k \Big(\dot{\mathbf{B}} : \mathbf{dd} S^{(k)} + 2\mathbf{B} : \dot{\mathbf{d}}\mathbf{d} S^{(k)} \\ + \mathbf{B} : \mathbf{dd} \dot{S}^{(k)} - \dot{\mathbf{B}} : \mathbf{dd} S^{(k+2)} - 2\mathbf{B} : \dot{\mathbf{d}}\mathbf{d} S^{(k+2)} - \mathbf{B} : \mathbf{dd} \dot{S}^{(k+2)} \Big). \end{aligned} \tag{74}$$

These are differential equations for the scalar order parameters of different order. Together with the macroscopic director $\mathbf{d}$ they determine the alignment tensors. The differential equation for the macroscopic director can also be derived from equation (68), by taking the $(k-1)$-fold scalar product with the tensor $\underbrace{\mathbf{d} \ldots \mathbf{d}}_{k-1}$:

$$\dot{\mathbf{d}} = -4\beta_1 \mathbf{d} + (\delta - \mathbf{dd}) \cdot \mathbf{B} \cdot \mathbf{d}. \tag{75}$$

The macroscopic director is reoriented by the flow field $\overline{\nabla \mathbf{v}}$ (the same by an electric field $\mathbf{EE}$) and by the mean field of the surrounding oriented particles $\mathbf{a}$.

Finally, inserting the equations (72), (73) and (75) into equation (74), together with a closure relation for $S^{(k+4)}$, leads to a closed differential equation for $S^{(k)}$, which is of second order in time, and it contains also the order parameters of lower order. Therefore it is coupled to the equations for S^{k-2}, S^{k-4}, and so on. This example shows, that the final set of equations for the relevant order parameters is not limited to first order equations in time, although the differential equation for the distribution function is of first order in time. An interesting point for future research is the inclusion of spatial gradients of the alignment tensor and the derivation of partial differential equations for the order parameters. Then there arises the question of hyperbolicity of the set of equations of motion. For the properties of solutions of hyperbolic equations, see for instance, [36, 48]. In this context it is interesting that the mesoscopically derived equations of the order parameter can be higher order in time, depending on the level at which the set of equations is closed.

Let us consider explicitly the case, where the second and the fourth order alignment tensor are considered as independent variable, i.e. the case $k = 2$. In the uniaxial case, the macroscopic director **d**, the second order scalar parameter $S^{(2)}$ and the fourth order parameter $S^{(4)}$ are the independent variables. After eliminating the fourth order parameter, we end up with a second order differential equation for the second order parameter:

$$\ddot{S}^{(2)} = \dot{S}^{(2)}\left(26\beta_1 + 4\mathbf{B}:\mathbf{dd}\right) - \frac{\dot{\mathbf{B}}:\mathbf{dd} + 2\mathbf{B}:\dot{\mathbf{d}}\mathbf{d}}{\mathbf{B}:\mathbf{dd}}\left(6\beta_1 S^{(2)} - \dot{S}^{(2)}\right)$$
$$-S^{(2)}\left(120\beta_1^2 + 40\beta_1\mathbf{B}:\mathbf{dd}\right) - 8\left(\mathbf{B}:\mathbf{dd}\right)^2\left(S^{(2)} - S^{(6)}\right). \qquad (76)$$

It is coupled to the equation of motion for the macroscopic director **d** given by equation (75). A closure relation for the sixth moment $S^{(6)}$, which is not an independent variable, is needed in addition.

3.9.2 Conclusion

Although the differential equation for the distribution function is of first order in time, as well as the infinite hierarchy of differential equations for the moments, closing this set of equations and eliminating the fourth order moment from the equations, leads to a differential equation of second order in time for the second order moment. This second order time derivative can be understood as a correction term, appearing in the more refined description, taking also the fourth moment as an independent variable. It does not occur at the lowest level of accuracy, namely, taking only the differential equation for the second moment and applying a closure relation for the fourth order alignment tensor, see the previous subsection and [31, 58, 62].

Space derivatives do not appear here in the equations, because we have restricted ourselves to uniform systems. However, gradients can be taken into account in the state space. With gradient dependent constitutive functions, such spatial inhomogeneities are introduced into the differential equation for the distribution function and into the equations for the alignment tensors, via the orientation change velocity **u**. For the simple case of considering the second order alignment tensor as the only variable, such a gradient dependence has been considered in [62].

4 Application of the mesoscopic theory to dipolar media

Electric and magnetic dipolar media have been studied for instance with the methods of irreversible non-equilibrium thermodynamics [43, 44, 45]. Decomposing the vector of electric polarization into a reversible part and an irreversible part (an internal variable), a relaxation equation for the electric polarization **P**

$$\xi^{(0)}_{(EP)}\mathbf{E}+\frac{d\mathbf{E}}{dt}=\xi^{(0)}_{(PE)}\mathbf{P}+\xi^{(1)}_{(PE)}\frac{d\mathbf{P}}{dt} \tag{77}$$

has been obtained. This equation is also known as the Debye equation [26]. The same form of equation can be obtained by replacing the polarization by the magnetization **M**, and the electric field **E** by the magnetic induction **B**. This Debye equation has been generalized to a second order differential equation for the polarization, or magnetization [16, 45], and the cross coupling effects between relaxation of the polarization, heat conduction, and viscous flow have also been taken into account [15].

The aim of the present section is to apply the mesoscopic theory to paramagnetic materials and derive a differential equation for the magnetization in dipolar media from the mesoscopic theory.

Let us denote the orientation of a single dipole by a unit vector **n** with $\mathbf{n}\cdot\mathbf{n}=1$. The orientation of the dipole can take any value on the unit sphere S^2. According to the concept of the mesoscopic theory, we introduce mesoscopic fields, defined on the enlarged (mesoscopic) space $\mathbb{R}^3_x\times\mathbb{R}_t\times S^2$, where the last argument in the domain of the fields is the orientation of the dipole **n**. This mesoscopic space is the same as for liquid crystals, and therefore the mesoscopic balance equations look the same for a dipolar medium as for liquid crystals. The difference between these two materials shows up in the constitutive theory. An important difference is that for liquid crystals, a "head-tail-symmetry" is observed, meaning that there **n** and $-$**n** have to be identified. A similar symmetry does *not* exist for dipoles because the dipole and its reverse are distinguished.

4.1 Orientation distribution function and alignment tensors

Macroscopically, the dipole moments show up as a magnetization only if their orientations are not distributed isotropically, but they are oriented more or less parallel. This orientational order can be described by introducing an orientation distribution function (ODF) (see Fig. 7). In contrast to liquid crystals, there is no inversion symmetry of the ODF because **n** and $-$**n** are distinguished.

The orientation distribution function gives the probability density of finding a dipole of orientation **n** in the continuum element at position **x** and time t. As in the case of liquid crystals it is defined as the mass fraction:

$$f(\mathbf{x},\mathbf{n},t) = \frac{\rho(\mathbf{x},\mathbf{n},t)}{\rho(\mathbf{x},t)}. \tag{78}$$

Under the assumptions of an incompressible material where, in addition, dipoles of different orientations have the same translational velocity ($\mathbf{v}(\mathbf{x},\mathbf{n},t) = \mathbf{v}(\mathbf{x},t)$), we have the differential equation for the ODF (see the section on liquid crystals):

$$\frac{\partial f(\cdot)}{\partial t} + \mathbf{v}(\mathbf{x},t)\cdot\nabla f(\cdot) + \nabla_n\cdot(\mathbf{u}(\cdot)f(\cdot)) = 0. \tag{79}$$

The alignment tensors are again defined as the following traceless tensors:

$$a^{(k)}(\mathbf{x},t) := \int_{S^2} f(\mathbf{x},\mathbf{n},t)\,\underbrace{\overline{\mathbf{n}\ldots\mathbf{n}}}_{k}\,d^2n. \tag{80}$$

They are introduced in such a way that they all vanish if the distribution of dipole orientations is isotropic. Here the odd and even order tensors can be nonzero because the ODF has no inversion symmetry. The first order alignment tensor is a measure of the average orientation of the dipoles. It is proportional to the macroscopic magnetization. Here it is convenient to introduce also alignment tensors $A^{(k)}$, which are not traceless:

$$A^{(k)}(\mathbf{x},t) := \int_{S^2} f(\mathbf{x},\mathbf{n},t)\underbrace{\mathbf{n}\ldots\mathbf{n}}_{k}d^2n. \tag{81}$$

4.2 Exploitation of the balance of spin

The balance of internal angular momentum is exploited exactly the same way as it was done in the case of liquid crystals. The only difference is in the constitu-

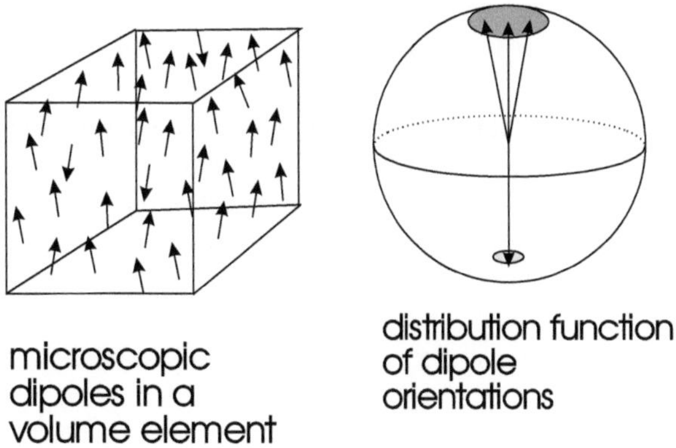

Fig. 9 The dipole moments give rise to an orientation distribution function on the unit sphere.

tive function **G**. The domain of the constitutive mappings, i.e., the state space Z, is chosen to be:

$$Z = \{\rho, T, \mathbf{B}, \dot{\mathbf{B}}, \mathbf{a}^{(1)}, \mathbf{a}^{(2)}, \mathbf{n}\}, \tag{82}$$

where T is temperature and **B** is magnetic induction. It includes macroscopic and mesoscopic variables. The macroscopic variables are temperature, mass density, magnetic induction, its time derivative, and the first and second order alignment tensors. These alignment tensors in the state space account for the fact that the dipoles tend to align parallel, i.e., the surrounding dipoles exert an aligning "mean field". In a simpler model, it would be enough to include only the first order alignment tensor. The first order tensor expresses the tendency of the dipoles to align parallel. The second order alignment tensor accounts for the influence of a quadrupolar ordering. We will discuss the case without the second order alignment tensor as a special case later. The mass density ρ in the state space is the macroscopic one because the dependence on the orientation **n** is written out explicitly.

Then a representation theorem for **G**, linear in all quantities except for the mesoscopic variable **n**, gives:

$$\mathbf{u} = \mathbf{P} \cdot \mathbf{G}(\cdot) = (\delta - \mathbf{nn}) \cdot \left(\beta_1 \mathbf{B} + \beta_2 \dot{\mathbf{B}} + \beta_4 \mathbf{a}^{(1)} + \beta_3 \mathbf{a}^{(2)} \cdot \mathbf{n} \right). \tag{83}$$

The coefficients β_j are functions of the macroscopic mass density $\rho(\mathbf{x},t)$ and the temperature $T(\mathbf{x},t)$. They do not depend on the mesoscopic mass density because the dependence on the mesoscopic variable **n** is written explicitly in the constitutive equation.

4.3 Equation of motion for the magnetization

The first moment of equation (79) reads:

$$\frac{\partial}{\partial t} \int_{S^2} f \mathbf{n} d^2 n + \mathbf{v} \cdot \nabla \int_{S^2} f \mathbf{n} d^2 n + \int_{S^2} \mathbf{n} \nabla_n \cdot (f \mathbf{u}) d^2 n = 0. \tag{84}$$

On the other hand, the variable **n** is proportional to the microscopic magnetization (magnetization per unit mass), i.e., it is the orientation of the microscopic dipole moment: $\mathbf{m} = \alpha \mathbf{n}$ with $\alpha = const$. The first moment of the orientation distribution function is proportional to the average of the microscopic magnetization, i.e., the macroscopic magnetization:

$$\mathbf{M}(\mathbf{x},t) = \alpha \rho(\mathbf{x},t) \int_{S^2} f \mathbf{n} d^2 n = \alpha \rho(\mathbf{x},t) \mathbf{a}^{(1)}. \tag{85}$$

The first two terms in equation (84) are derivatives of the first order alignment tensor. The third term is integrated by parts using Gauss' theorem on the unit sphere. The resulting equation reads:

$$\frac{\partial \mathbf{a}^{(1)}}{\partial t} + \mathbf{v} \cdot \nabla \mathbf{a}^{(1)} = \frac{d\mathbf{a}^{(1)}}{dt} = \int_{S^2} f \mathbf{u} \cdot \nabla_n(\mathbf{n}) d^2 n. \tag{86}$$

Then, inserting the equation for the orientation change velocity equation (83), and taking into account $\nabla_n(\mathbf{n}) = \mathbf{P} = \delta - \mathbf{nn}$ and $\mathbf{n} \cdot \nabla_n(\dots) = 0$ (because ∇_n is the covariant derivative on the unit sphere), we obtain:

$$\frac{d\mathbf{a}^{(1)}}{dt} = \int_{S^2} \Big(\beta_1 \mathbf{B} + \beta_2 \dot{\mathbf{B}} + \beta_4 \mathbf{a}^{(1)} + \beta_5 \mathbf{a}^{(2)} \cdot \mathbf{n}$$

$$-\beta_1 \mathbf{nn} \cdot \mathbf{B} - \beta_2 \mathbf{nn} \cdot \dot{\mathbf{B}} - \beta_4 \mathbf{nn} \cdot \mathbf{a}^{(1)} - \beta_5 \mathbf{a}^{(2)} : \mathbf{nnn}\Big) f d^2 n, \tag{87}$$

using the fact that $\mathbf{P}$ is a projector ($\mathbf{P} \cdot \mathbf{P} = \mathbf{P}$).

The first moment of the dipole distribution function is proportional to the magnetization (see equation (85)). In the resulting equation there enter also the second and the third orientational moments $\mathbf{A}^{(2)}$ and $\mathbf{A}^{(3)}$ of the dipole distribution function:

$$\mathbf{A}^{(2)} = \int_{S^2} f(\cdot) \mathbf{nn} d^2 n, \tag{88}$$

$$\mathbf{A}^{(3)} = \int_{S^2} f(\cdot) \mathbf{nnn} d^2 n, \tag{89}$$

which are denoted by capital $\mathbf{A}$, because they are not traceless. From equation (87), it follows in terms of the magnetization:

$$\alpha^{-1} \frac{d \frac{\mathbf{M}(\mathbf{x},t)}{\rho(\mathbf{x},t)}}{dt} = \beta_1 \mathbf{B}(\mathbf{x},t) + \beta_2 \dot{\mathbf{B}}(\mathbf{x},t)$$

$$+\beta_4 \alpha^{-1} \frac{\mathbf{M}(\mathbf{x},t)}{\rho(\mathbf{x},t)} + \beta_5 \alpha^{-1} \mathbf{a}^{(2)}(\mathbf{x},t) \cdot \frac{\mathbf{M}(\mathbf{x},t)}{\rho(\mathbf{x},t)}$$

$$-\beta_1 \mathbf{A}^{(2)}(\mathbf{x},t) \cdot \mathbf{B}(\mathbf{x},t) - \beta_2 \mathbf{A}^{(2)}(\mathbf{x},t) \cdot \dot{\mathbf{B}}(\mathbf{x},t)$$

$$-\beta_4 \alpha^{-1} \mathbf{A}^{(2)}(\mathbf{x},t) \cdot \frac{\mathbf{M}(\mathbf{x},t)}{\rho(\mathbf{x},t)} - \beta_5 \mathbf{a}^{(2)}(\mathbf{x},t) : \mathbf{A}^{(3)}(\mathbf{x},t). \tag{90}$$

This is a macroscopic equation for the field quantities that depend on position and time. If the material is incompressible, the left hand side simplifies to:

$$\frac{d \frac{\mathbf{M}(\mathbf{x},t)}{\rho(\mathbf{x},t)}}{dt} = \frac{d\mathbf{M}(\mathbf{x},t)}{dt} \frac{1}{\rho(\mathbf{x},t)}. \tag{91}$$

The resulting differential equation for the magnetization is of the type of a relaxation equation. Interestingly, orientation diffusion does not show up in the macroscopic equation for the magnetization.

Let us consider the special case that the second order alignment tensor is not included in the state space, i.e., $\beta_5 = 0$ and the material is incompressible:

$$\frac{1}{\alpha\rho}\frac{d\mathbf{M}}{dt} = \beta_1\mathbf{B} + \beta_2\dot{\mathbf{B}} + \beta_4\frac{1}{\alpha\rho}\mathbf{M}$$
$$-\beta_1\mathbf{A}^{(2)}\cdot\mathbf{B} - \beta_2\mathbf{A}^{(2)}\cdot\dot{\mathbf{B}} - \beta_4\frac{1}{\alpha\rho}\mathbf{A}^{(2)}\cdot\mathbf{M}. \tag{92}$$

In both cases there is a coupling to higher order moments, at least to the second order one, which cannot be avoided, even if the second order alignment tensor is not included in the state space ($\beta_5 = 0$). Therefore, a closure relation is needed, expressing the higher order moments in terms of the second order one. Such a closure relation can be derived from the principle of maximum entropy [64], or it has to be postulated as a constitutive equation. The simplest assumption is that the orientations of the dipoles are statistically independent (which is an approximation only). Then the closure relations are very simple:

$$\mathbf{A}^{(2)} = \int_{S^2} f(\mathbf{x},\mathbf{n},t)\mathbf{n}\mathbf{n}d^2n = \int_{S^2} f(\mathbf{x},\mathbf{n},t)\mathbf{n}d^2n \int_{S^2} f(\mathbf{x},\mathbf{n},t)\mathbf{n}d^2n$$
$$= \mathbf{a}^{(1)}\mathbf{a}^{(1)}, \tag{93}$$

$$\mathbf{A}^{(3)} = \int_{S^2} f(\mathbf{x},\mathbf{n},t)\mathbf{n}\mathbf{n}\mathbf{n}d^2n$$
$$= \int_{S^2} f(\mathbf{x},\mathbf{n},t)\mathbf{n}d^2n \int_{S^2} f(\mathbf{x},\mathbf{n},t)\mathbf{n}d^2n \int_{S^2} f(\mathbf{x},\mathbf{n},t)\mathbf{n}d^2n = \mathbf{a}^{(1)}\mathbf{a}^{(1)}\mathbf{a}^{(1)}. \tag{94}$$

With these assumptions on the higher order alignment tensors, there results a closed equation for the magnetization. For the irreducible alignment tensor, we have to insert $\mathbf{a}^{(2)} = \mathbf{A}^{(2)} - \frac{1}{3}\delta$ with the unit tensor δ. With the second order alignment tensor in the state space, we have:

$$\frac{d\mathbf{M}}{dt} = \beta_1\alpha\rho\mathbf{B} + \beta_2\alpha\rho\dot{\mathbf{B}} + \beta_4\mathbf{M} + \beta_5\left(\frac{1}{\alpha^2\rho^2}\mathbf{M}\mathbf{M} - \frac{1}{3}\delta\right)\cdot\mathbf{M}$$
$$-\beta_1\frac{1}{\alpha\rho}\mathbf{M}\mathbf{M}\cdot\mathbf{B} - \beta_2\frac{1}{\alpha\rho}\mathbf{M}\mathbf{M}\cdot\dot{\mathbf{B}} - \beta_4\frac{1}{\alpha^2\rho^2}\mathbf{M}\mathbf{M}\cdot\mathbf{M} -$$
$$\beta_5\frac{1}{\alpha^2\rho^2}\left(\frac{1}{\alpha^2\rho^2}\mathbf{M}\mathbf{M} - \frac{1}{3}\delta\right) : \mathbf{M}\mathbf{M}\mathbf{M}, \tag{95}$$

and without the second order alignment tensor in the state space:

$$\frac{d\mathbf{M}}{dt} = \beta_1 \alpha\rho \mathbf{B} + \beta_2 \alpha\rho \dot{\mathbf{B}} + \beta_4 \mathbf{M}$$
$$-\beta_1 \frac{1}{\alpha\rho} \mathbf{M}\mathbf{M} \cdot \mathbf{B} - \beta_2 \frac{1}{\alpha\rho} \mathbf{M}\mathbf{M} \cdot \dot{\mathbf{B}} - \beta_4 \frac{1}{\alpha^2\rho^2} \mathbf{M}\mathbf{M} \cdot \mathbf{M}. \tag{96}$$

If the value of the magnetization is small enough, we can neglect quadratic and higher order terms of the magnetization. In this linear limit, the equations (95) and (96) simplify to:

$$\frac{d\mathbf{M}}{dt} = \beta_1 \alpha\rho \mathbf{B} + \beta_2 \alpha\rho \dot{\mathbf{B}} + \beta_4 \mathbf{M} - \frac{1}{3}\beta_5 \mathbf{M},$$
$$\frac{d\mathbf{M}}{dt} = \beta_1 \alpha\rho \mathbf{B} + \beta_2 \alpha\rho \dot{\mathbf{B}} + \beta_4 \mathbf{M}, \tag{97}$$

respectively.

Both are of the form of the well-known Debye equation for dielectric relaxation phenomena, here in the analogous form for magnetic relaxation, equation (77). This fact can be used to identify the coefficients α_1, α_2, β_4, and β_5.

4.4 Summary

Starting with the mesoscopic balance equations, especially the balance of mass, and the balance of internal angular momentum in the stationary case, we derived a differential equation for the orientation distribution function of the dipoles. The first moment of this distribution function is proportional to the magnetization. The resulting equation for the magnetization is of relaxation type, and it is nonlinear, even in the simplest case of a state space. Its linear limit is of the form of the well-known Debye-equation.

5 Summary of the mesoscopic theory

The mesoscopic concept introduces field quantities defined on an enlarged domain, the mesoscopic space, which takes into account the internal structure of the complex material. In addition, a mesoscopic distribution function is introduced as a statistical element. The advantages of such a mesoscopic background theory over a purely macroscopic description are:

- From the internal structure, it is clear which balance equations have to be taken into account. For liquid crystals, for example, the balance of angular momentum is relevant.
- Based on the mesoscopic distribution function order parameters are defined that are the internal variables in a purely macroscopic description. From the meso-

scopic balance equations, there result equations of motion for the internal variables. This is a set of coupled equations for the order parameters of different order, and a closure relation for the highest order parameters is needed. Closing this set of equations is possible on different levels of accuracy. Depending on the number of order parameters considered as independent variables, the final equation for the lowest order parameter is of first or of second, or even of higher order in time.
- The mesoscopic constitutive theory gives the order parameter dependence of macroscopic constitutive quantities.
- Different physical systems can be considered within the same or analogous mesoscopic theories. So, for instance, the orientation of a biaxial molecule and the orientation of the crystal axes of a micro-crystallite are described by the same mesoscopic variable. In addition, this mesoscopic variable can be an element of S^3 [30], which makes the theory for biaxial liquid crystals analogous to the theory of uniaxial liquid crystals.
- On the other hand, there are more constitutive equations needed on the mesoscopic level than on the macroscopic level, and the choice of the relevant variables is less obvious. For instance, the state space can be a purely mesoscopic one, or it can contain mesoscopic and macroscopic quantities, as in our examples. In addition, it is not clear whether restrictions concerning constitutive functions can be obtained by exploitation of a dissipation inequality on the mesoscopic level. The role of the Second Law of Thermodynamics on the mesoscopic level is a question to be investigated in the future.

Acknowledgements Financial support by the European Union through the FP6 Marie Curie Transfer of Knowledge Project CENS-CMA (MC-TK-013909) is gratefully acknowledged.

References

1. Blenk, S., Ehrentraut, H., Muschik, W.: Orientation balances for liquid crystals and their representation by alignment tensors. Mol. Cryst. Liqu. Cryst. **204**, 133–141 (1991)
2. Blenk, S., Ehrentraut, H., Muschik, W.: Statistical foundation of macroscopic balances for liquid crystals in alignment tensor formulation. Physica A **174**, 119–138 (1991)
3. Blenk, S., Ehrentraut, H., Muschik, W.: Macroscopic constitutive equations for liquid crystals induced by their mesoscopic orientation distribution. Int. J. Eng. Sci. **30**(9), 1127–1143 (1992)
4. Blenk, S., Ehrentraut, H., Muschik, W.: A continuum theory for liquid crystals describing different degrees of orientational order. Liquid Crystals **14**(4), 1221–1226 (1993)
5. Blenk, S., Ehrentraut, H., Muschik, W., Papenfuss, C.: Mesoscopic orientation balances and macroscopic constitutive equations of liquid crystals. In: Proc. 7th Intl.Symp. on Continuum Models of Discrete Systems, Materials Science Forum, vol. 123–125, pp. 59–68. Trans Tech, Paderborn (1992)
6. Blenk, S., Muschik, W.: Orientational balances for nematic liquid crystals. J. Non-Equilib. Thermodyn. **16**, 67–87 (1991)
7. Blenk, S., Muschik, W.: Orientational balances for nematic liquid crystals describing different degrees of orientational order. ZAMM **72**(5), T400–T403 (1992)

8. Blenk, S., Muschik, W.: Mesoscopic concepts for constitutive equations of nematic liquid crystals in alignment tensor formulation. ZAMM **73**(4-5), T331–T333 (1993)
9. Bogolyubov, N.N.: Kinetic equations. Acad. Sci. USSR J. Phys. **10**, 265–274 (1946)
10. Born, M., Green, H.S.: A general kinetic theory of liquids I: the molecular distribution functions. Proc. Roy. Soc. Lond. A **188**, 10–18 (1946)
11. Born, M., Green, H.S.: A general kinetic theory of liquids III: dynamical properties. Proc. Roy. Soc. Lond. A **190**, 455–474 (1947)
12. Brand, H., Pleiner, H.: Hydrodynamics of biaxial discotics. Phys. Rev. A **24**(5), 2777–2779 (1981)
13. Carlsson, T., Leslie, F.M.: Behaviour of biaxial nematics in the presence of electric and magnetic fields. Liq. Cryst. **10**(3), 325–340 (1991)
14. Chapman, S., Cowling, T.G.: The mathematical theory of nonuniform gases. Cambridge University Press (1970)
15. Ciancio, V.: On the generalized Debye equation of media with dielectric relaxation phenomena described by vectorial internal thermodynamic variables. J. Non-Equilib. Thermodyn. **14**, 239–250 (1989)
16. Ciancio, V., Dolfin, M., Ván, P.: Thermodynamic theory of dia- and paramagnetic materials. Int. J. of Applied Electromagnetics and Mechanics **7**, 237–247 (1996)
17. Coleman, B.D., Gurtin, M.E.: Thermodynamics with internal state variables. J. Chem. Phys. **47**(2), 597–613 (1967)
18. Coleman, B.D., Mizel, V.J.: Thermodynamics and departures from Fouriers law of heat conduction. Arch. Rational Mech. Anal. **13**, 245–261 (1963)
19. Coleman, B.D., Noll, W.: The thermodynamics of elastic materials with heat conduction and viscosity. Arch. Rat. Mech. Anal. **13**, 167–168 (1963)
20. Cosserat, E., Cosserat, F.: Sur la méchanique générale. Acad. Sci. Paris **145**, 1139–1142 (1907)
21. De Gennes, P.G.: The Physics of Liquid Crystals. Clarendon Press, Oxford (1974)
22. De Gennes, P.G.: Simple Views on Condensed Matter, Series in Modern Condensed Matter Physics, vol. 4. World Scientific, Singapore (1992)
23. De Gennes, P.G., Prost, J.: The Physics of Liquid Crystals, 2nd ed. Monographs on Physics. Clarendon Press, Oxford (1995)
24. De Groot, S.R.: Thermodynamics of Irreversible Processes. North-Holland, Amsterdam (1951)
25. De Groot, S.R., Mazur, P.: Non-equilibrium Thermodynamics. North-Holland, Amsterdam (1962)
26. Debye, P.: Polar Molecules. Dover, New York (1945)
27. Doi, M.: Molecular dynamics and rheological properties of concentrated solutions of rodlike polymers in isotropic and liquid crystalline phases. J. Polym. Sci. Polym. Phys. **19**(2), 229–243 (1981)
28. Doi, M., Edwards, S.F.: The Theory of Polymer Dynamics. Clarendon, Oxford (1986)
29. Dreyer, W.: Molekulare Erweiterte Thermodynamik Realer Gase. Habilitationsschrift, Technische Universität Berlin (1990)
30. Ehrentraut, H.: A Unified Mesoscopic Continuum Theory of Uniaxial and Biaxial Liquid Crystals. Wissenschaft und Technik, Berlin (1996)
31. Ehrentraut, H., Muschik, W., Papenfuss, C.: Mesoscopically derived orientation dynamics of liquid crystals. J. Non-Equilib. Thermodyn. **22**, 285–298 (1997)
32. Eringen, A.C.: Simple microfluids. Int. J. Eng. Sci. **2**, 205–217 (1964)
33. Eringen, A.C.: Theory of micropolar fluids. J. Math. Mech. **16**, 1–18 (1966)
34. Eringen, A.C.: Micropolar theory of liquid crystals. In: J.F. Johnson, R.S. Porter (eds.) Liquid Crystals and Ordered Fluids, Vol. 3, pp. 443–474. Plenum Press, New York (1978)
35. Eringen, A.C., Lee, J.D.: Relations of two continuum theories of liquid crystals. In: Johnson, J.F., Porter, R.S. (eds.) Liquid Crystals and Ordered Fluids, Vol. 2, pp. 315–330. Plenum Press, New York (1974)
36. Friedrichs, K.O., Lax, P.D.: Systems of conservation equations with a convex extension. Proc. Nat. Acad. Sci USA **68**, 1686–1688 (1971)

37. Grad, H.: Principles of the kinetic theory of gases. In: Flügge, S. (ed.) Handbuch der Physik XII. Springer, Berlin (1958)
38. Grebel, H., Hornreich, R.M., Shtrikman, S.: Landau theory of cholesteric blue phases. Phys. Rev. A **28**, 1114–1138 (1983)
39. Grebel, H., Hornreich, R.M., Shtrikman, S.: Landau theory of cholesteric blue phases: The role of higher harmonics. Phys. Rev. A **30**, 3264–3278 (1984)
40. Hess, S.: Irreversible thermodynamics of nonequilibrium alignment phenomena in molecular liquids and in liquid crystals. Z. Naturforsch. **30a**, 728–733 (1975)
41. Hess, S.: Irreversible thermodynamics of nonequilibrium alignment phenomena in molecular liquids and in liquid crystals ii. Z. Naturforsch. **30a**, 1224–1232 (1975)
42. Kirkwood, J.G.: The statistical mechanical theory of transport processes I: general theory. J. Chem. Phys. **14**, 180–201 (1946)
43. Kluitenberg, G.A.: On dielectric and magnetic relaxation phenomena and non-equilibrium thermodynamics. Physica **68**, 75–92 (1973)
44. Kluitenberg, G.A.: On dielectric and magnetic relaxation phenomena and vectorial internal degrees of freedom in thermodynamics. Physica **87A**, 302–330 (1977)
45. Kluitenberg, G.A.: On vectorial internal variables and dielectric and magnetic relaxation phenomena. Physica **109A**, 91–122 (1981)
46. Kröger, M.: Rheologie und Struktur von Polymerschmelzen. W&T Verlag, Berlin (1995). PhD Thesis, Technische Universität Berlin
47. Landau, L.D., Ginzburg, V.L.: K theorii sverkhrovodimosti. Zh. Eksp. Teor. Fiz. **20**, 1064–1082 (1950). English translation: On the theory of superconductivity. In: ter Haar, D. (ed.) Collected Papers of L.D. Landau, pp. 626–633. Pergamon, Oxford (1965)
48. Lax, P.D.: Hyperbolic Systems of Conservation Laws and the Mathematical Theory of Shock Waves. Society for Industrial and Applied Mathematics, Philadelphia (1973)
49. Reichl, L.E.: A Modern Course in Statistical Physics. Edward Arnold, London (1980)
50. Leslie, F.M.: Some constitutive equations for liquid crystals. Arch. Rat. Mech. Anal. **28**, 265–283 (1968)
51. Liu, I.S.: Method of Lagrange multipliers for exploitation of the entropy principle. Arch. Rat. Mech. Anal. **46**, 131–148 (1972)
52. Liu, I.S.: Continuum Mechanics. Springer, New York (2002)
53. Longa, L., Monselesan, D., Trebin, H.R.: An extension of the Landau-Ginzburg-de Gennes theory for liquid crystals. Liquid Crystals **2**(6), 769–796 (1987)
54. Maier, W., Saupe, A.: Eine einfache molekular-statistische Theorie der nematischen kristallinflüssigen Phase. Teil 1. Z. Naturforsch. **14a**, 882–889 (1959)
55. Maier, W., Saupe, A.: Eine einfache molekular-statistische Theorie der nematischen kristallinflüssigen Phase. Teil 2. Z. Naturforsch. **15a**, 287–292 (1960)
56. Miesowicz, M.: The three coefficients of viscosity of anisotropic liquids. Nature **158**(4001), 27–27 (1946)
57. Muschik, W., Ehrentraut, H., Papenfuss, C., Blenk, S.: Mesoscopic theory of liquid crystals. In: Brey, J.J., Marro, J., Rubi, J.M., San Miguel, M. (eds.) 25 Years of Non-Equilibrium Statistical Mechanics, Proceedings of the XIII Sitges Conference 1994, Lecture Notes in Physics, vol. 445, pp. 303–311. Springer, New York (1995)
58. Muschik, W., Papenfuss, C., Ehrentraut, H.: Alignment tensor dynamics induced by the mesoscopic balance of the orientation distribution function. Proc. Estonian Acad. Sci. Phys. Math. **46**, 94–101 (1997)
59. Muschik, W., Papenfuss, C., Ehrentraut, H.: Sketch of the mesoscopic description of nematic liquid crystals. J. Non-Newton. Fluid Mech. **119**(1-3), 91–104 (2004)
60. Müller, I.: Thermodynamics. Pitman, Boston (1985)
61. Müller, I.: Grundzüge der Thermodynamik. Springer, Berlin (1994)
62. Papenfuss, C.: Nonlinear dynamics of the alignment tensor in the presence of electric fields. Arch. Mech. **50**(3), 529–536 (1998)
63. Papenfuss, C., Muschik, W.: Constitutive theory for two dimensional liquid crystals. Mol. Cryst. Liq. Cryst. **262**, 473–484 (1995)

64. Papenfuss, C., Muschik, W.: Orientational order in free standing liquid crystalline films and derivation of a closure relation for higher order alignment tensors. Mol. Cryst. Liq. Cryst. **330**, 541–548 (1999)
65. Papenfuss, C., Ván, P., Muschik, W.: Mesoscopic theory of microcracks. Archive of Mechanics **55**, 481–499 (2003)
66. Stegemeyer, H., Kelker, H.: 100-jähriges Jubiläum: Flüssigkristalle-Blaue Phasen. Nachr. Chem. Tech. Lab. **36**(4), 360–364 (1988)
67. Ván, P.: Weakly nonlocal irreversible thermodynamics. Annalen der Physik (Leipzig) **12**, 142–169 (2003)
68. Ván, P.: Exploiting the second law in weakly non-local continuum physics. Periodica Polytechnica Ser. Mech. Eng. **49**(1), 79–94 (2005)
69. Ván, P.: The Ginzburg-Landau equation as a consequence of the second law. Continuum Mech. Thermodyn. **17**, 165–169 (2005)
70. Ván, P., Papenfuss, C., Muschik, W.: Mesoscopic dynamics of microcracks. Phys. Rev. E **62**(5), 6206–6215 (2000)
71. Ván, P., Papenfuss, C., Muschik, W.: Griffith cracks in the mesoscopic microcrack theory. J. Phys. A **37**(20), 5315–5328 (2004). Published online: Condensed Matter, abstract, cond-mat/0211207; 2002.
72. Verhás, J.: Irreversible thermodynamics of nematic liquid crystals. Acta Phys. Hung. **55**, 275–291 (1984)
73. Vertogen, G., Jeu, de W.: Thermotropic Liquid Crystals: Fundamentals. Series in Chemical Physics Vol. 45. Springer, Berlin (1988)
74. Virga, E.G.: Variational Theories for Liquid Crystals. Chapman and Hall, London (1994)

Towards a Description of Twist Waves in Mesoscopic Continuum Physics

Heiko Herrmann

Abstract An introduction to Mesoscopic Continuum Physics is given with a focus on liquid crystals. Mesoscopic Continuum Physics introduces variables describing the microstructure – like orientation of crystals – into the domain of the fields, thus treating them equivalently to space. The theory of Mesoscopic Continuum Physics has been reformulated, resulting in more compact equations. In this formulation the balance of spin shows up naturally as component equations of the balance of momentum. This is an advantage over the standard formulation, in which it seems to be postulated separately. Starting from this, a wave equation for twist waves has been derived on the mesoscopic space. Twist waves are one of the fundamental modes of orientation waves in liquid crystals. A short repetition of twist waves of liquid crystals in the Ericksen-Leslie theory is given and compared to a description using the mesoscopic theory.

1 Introduction

The requirements to contemporary materials are very complicated due to the wide range of loadings including high frequency inputs and coupled physical fields (deformation, temperature, electromagnetic forces, etc.). The concept of homogeneity of materials used in classical theories and algorithms cannot be used any more, and the internal structure of materials must be described by far more exact theories based on physics of materials. This is a hot topic in contemporary theoretical and applied mechanics. The main theme of modern materials science and engineering is to optimize micro-structure for desired properties through advanced processing. However, our ability to characterize and quantitatively predict micro-structural evolution, and hence to yield an unambiguous processing-property relationship is rather limited

Feodor Lynen Fellow of the Alexander von Humboldt Foundation, Institute of Cybernetics at Tallinn University of Technology, Akadeemia tee 21, 12618 Tallinn, Estonia, e-mail: hh@cens.ioc.ee

E. Quak, T. Soomere (eds.), *Applied Wave Mathematics*,
DOI 10.1007/978-3-642-00585-5_8, © Springer-Verlag Berlin Heidelberg 2009

because of the extreme complexity of micro-structure and the nonlinear interaction of its elements. Both understanding and taking advantage of the complexity and the underlying nonlinear dynamic processes are two of the biggest challenges for scientists and engineers today.

The nonlinearities present in micro-structured materials, though they might be small, are often the key to understanding the system's properties. Specifically phenomena like self-organization, pattern formation and the existence of localized structures can only exist in the presence of nonlinearities.

Liquid crystals consist of rigid rod- or lens-shaped "particles" (form-anisotropic molecules), see Fig. 1. In special ranges of temperature these systems show a spontaneous alignment in connection with a long-range order of orientation but disordered centers of mass. This represents a phase of matter which is between liquid and crystal. Liquid crystals combine typical properties of liquids (flow properties) with those of solids (anisotropy).

The order of the phase is dependant on the temperature. The simplest liquid crystalline phase is the nematic phase, which is characterized by an orientational order – which means the molecules are more or less aligned in the same direction. This orientation – and therefore the optical properties of the phase – can be influenced by electrical fields.

At the moment two competitive theories for complex media exist, these are *continuum mechanics with internal variables* and *mesoscopic continuum physics*. Both theories have advantages – and disadvantages – over the other.

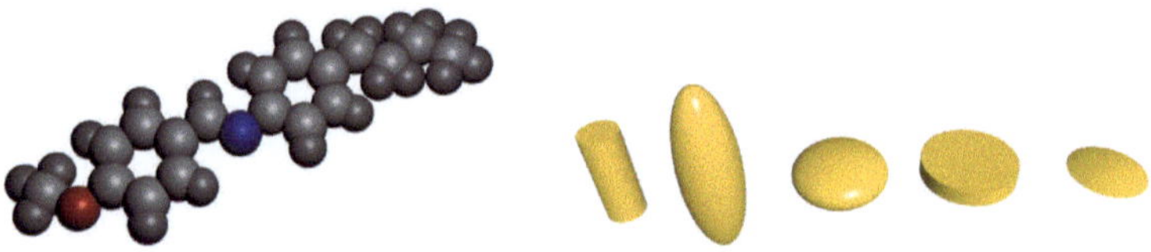

(a) (b)

Fig. 1 (left) A well-known liquid crystal: MBBA (n-4'-MethoxyBenzylidene-n-ButylAnilin). **(right)** Several models for uniaxial liquid crystals.

Continuum mechanics with internal variables introduces new fields on the space-time to describe the internal structure of the considered material. This method is a very natural and intuitive procedure. A crucial point are the evolution equations for these new fields, which cannot be derived but have to be postulated. Quite often the Second Law of Thermodynamics is not exploited in the sense of constitutive theory, but constitutive functions and a Gibbs' relation for the entropy are postulated.

For the description of the liquid crystalline order and the resulting properties several categories of precision can be found in the literature: macroscopic director, scalar order parameter and alignment tensors:

Macroscopic director: The macroscopic director has been introduced as a unity vector describing the main direction in the liquid crystals [9, 10, 11, 18, 19, 20]. According to the normalization the macroscopic director cannot vanish. Therefore the phase-transition nematic–isotropic cannot be described, because in the isotropic case there is no distinguished direction.

Scalar order parameter: The scalar order parameter was originally introduced in the framework of a statistical mean-field theory [23, 24]. It assumes values between 0 for the isotropic case and 1 for total alignment. By use of the macroscopic director together with the order parameter, one can describe multiple degrees of nematic order as well as the phase-transition nematic–isotropic.

Alignment tensors: Alignment tensors are the moments of the orientation distribution function, the alignment tensor of second rank for example is the second moment of the distribution function. It has been introduced into irreversible thermodynamics as an internal variable [15, 16, 17, 29] for describing phenomena of flow alignment, of streaming birefringence and of non-linear viscosity. The alignment tensor is the order parameter of the Landau-Theory of phase transitions [22], or the internal variable in the derivation of equations of motion by a variational principle [31].

2 Mesoscopic Continuum Physics

2.1 *Generalization of vector fields to the mesoscopic space*

Mesoscopic continuum physics does not introduce new fields on the space-time, but expands the space-time by introducing new variables. This results in a higher-dimensional mesoscopic space on which the "usual" fields are defined. This method has similarities to mixture theory in chemical physics. This procedure is not very intuitive, but has the advantage that no new evolution equations are necessary, but the old ones are given on the mesoscopic space in a natural and automatic way.

The orientation of a liquid crystal is given by a unit-vector in $\mathbb{R}^3$, this vector has 2 independent components. If one uses spherical polar coordinates the length of the vector is constant and only the angles remain as variables. Thus it is suitable to describe the orientation in generalized coordinates on S^2.

Mesoscopic continuum physics has been introduced for liquid crystals and has been applied to different problems [5, 25]. In each volume element particles of different orientation are assumed. An orientation distribution function is introduced by which all macroscopic quantities can be defined by averaging. The macroscopic quantities are relevant for comparison with experiments and therefore the interesting quantities. The introduction of a fine description enables us to put some knowledge of the internal structure into the model.

phys. object	generalization	math. object
Space	$\mathbf{x} \to (\mathbf{x},\mathbf{n}) = \tilde{\mathbf{x}}$	$\mathbf{x} \in \mathbb{R}^3$, $\mathbf{n} \in S^2$, $\tilde{\mathbf{x}} \in \mathbb{R}^3 \times S^2$
velocity	$\mathbf{v} \to (\mathbf{v},\mathbf{u}) = \tilde{\mathbf{v}}$	$\tilde{\mathbf{v}} : \mathbb{R}^3 \times S^2 \to T_{\tilde{\mathbf{x}}}\mathbb{R}^3 \times T_{\tilde{\mathbf{x}}}S^2$
mass density	$\rho(\mathbf{x}) \to \tilde{\rho}(\tilde{\mathbf{x}})$	$\tilde{\rho} : \mathbb{R}^3 \times S^2 \to \mathbb{R}$
internal energy density	$\varepsilon(\mathbf{x}) \to \tilde{\varepsilon}(\tilde{\mathbf{x}})$	$\tilde{\varepsilon} : \mathbb{R}^3 \times S^2 \to \mathbb{R}$
gradient	$\nabla \to (\nabla_{\mathbf{x}}, \nabla_{\mathbf{n}}) = \nabla_{\tilde{\mathbf{x}}}$	
stress tensor	$\mathbf{t}(\mathbf{x}) \to \tilde{\mathbf{t}}(\tilde{\mathbf{x}})$	$\tilde{\mathbf{t}} : (\mathbb{R}^3 \times S^2) \to (\mathbb{R}^3 \times T_{\tilde{\mathbf{x}}}S^2) \times (\mathbb{R}^3 \times T_{\tilde{\mathbf{x}}}S^2)$
force density	$\mathbf{f}(\mathbf{x}) \to (f,k)(\tilde{\mathbf{x}}) = \tilde{\mathbf{f}}(\tilde{\mathbf{x}})$	$\tilde{\mathbf{f}} : \mathbb{R}^3 \times S^2 \to \mathbb{R}^3 \times T_{\tilde{\mathbf{x}}}S^2$
heat flux density	$\mathbf{q}(\mathbf{x}) \to (q,o)(\tilde{\mathbf{x}}) = \tilde{\mathbf{q}}(\tilde{\mathbf{x}})$	$\tilde{\mathbf{q}} : \mathbb{R}^3 \times S^2 \to \mathbb{R}^3 \times T_{\tilde{\mathbf{x}}}S^2$
heat supply	$r(\mathbf{x}) \to \tilde{r}(\tilde{\mathbf{x}})$	$\tilde{r} : \mathbb{R}^3 \times S^2 \to \mathbb{R}$

Table 1 Generalization of vector fields for uniaxial liquid crystals.

The macroscopic balances are derived from the mesoscopic balances by averaging with respect to the mesoscopic variables. Further macroscopic variables – as a measure for the orientational order in the liquid crystalline phase – are the alignment tensors, which can be constructed from the moments of the distribution function [2, 3, 7]. The alignment tensor of second rank is also used in the macroscopic theory as an internal variable.

In addition to the mesoscopic balances, constitutive functions must be used, like in the macroscopic theory. On the macroscopic level the constitutive functions are restricted by the Second Law, this is also true for the mesoscopic constitutive functions in an indirect way, because the macroscopic quantities are constructed from the mesoscopic ones by averaging [4, 6]:

$$\rho(\mathbf{x},t) = \oint_{S^2} \tilde{\rho}(\tilde{\mathbf{x}},t), d^2n, \tag{1}$$

$$F(\tilde{\mathbf{x}},t) := \frac{\tilde{\rho}(\tilde{\mathbf{x}},t)}{\rho(\mathbf{x},t)}, \tag{2}$$

$$\mathbf{v}(\mathbf{x},t) = \oint_{S^2} F(\tilde{\mathbf{x}},t)\tilde{\mathbf{v}}_{\mathbf{x}}(\tilde{\mathbf{x}},t)\, d^2n, \tag{3}$$

$$\mathbf{u}(\mathbf{x},t) = \oint_{S^2} F(\tilde{\mathbf{x}},t)\tilde{\mathbf{v}}_{\mathbf{n}}(\tilde{\mathbf{x}},t)\, d^2n, \tag{4}$$

where F is the *orientation distribution function*. The rotation velocity $\mathbf{u}(\mathbf{x},t)$ is still given on S^2, in macroscopic theory it is nevertheless standard to define the rotation velocity ω as an axial vector on $\mathbb{R}^3$ and the spin $\mathbf{s}(\mathbf{x},t) = \Theta\omega(\mathbf{x},t)$. This can be done by embedding S^2 in R^3 and using spherical polar coordinates with constant radius $r \stackrel{!}{=} 1$, then defining $\omega := \mathbf{e}_r \times (u_1\mathbf{e}_\phi + u_2\mathbf{e}_\theta)$.

2.2 Mesoscopic balances and a transport theorem

Starting out from motivating global balance equations on the mesoscopic space completely analogous to the macroscopic balances, it is possible to derive local mesoscopic balance equations by the use of an extended Reynolds transport theorem.

A balance equation for an arbitrary extensive quantity $\tilde{\phi}(\tilde{\mathbf{x}},t)$ is given by

$$\frac{d}{dt}\int_{\mathcal{B}_t} \tilde{\rho}(\tilde{\mathbf{x}},t)\tilde{\phi}(\tilde{\mathbf{x}},t)d\tilde{\mathbf{x}} = -\oint_{\partial\mathcal{B}_t} \mathbf{J}^{\tilde{\phi}}(\tilde{\mathbf{x}},t)\cdot d\tilde{\mathbf{a}} + \int_{\mathcal{B}_t} \sigma^{\tilde{\phi}}(\tilde{\mathbf{x}},t) + \pi^{\tilde{\phi}}(\tilde{\mathbf{x}},t)d\tilde{\mathbf{x}}. \quad (5)$$

As in general the control volume is time-dependent, the following theorem is needed to exchange time derivative and integration.

Theorem 2.1 (Extended Reynolds transport theorem). *If $\Phi_t : \mathbb{R}^3 \times S^2 \to \mathbb{R}$ is a C^1-function and $\mathcal{B}_t \subset \mathbb{R}^3 \times S^2$ is a sufficiently smooth region and differentiable with respect to t in the sense that the evolution of $\mathcal{B}_t$*

$$(\xi,\eta) : \mathcal{B}_0 \times \mathbb{R} \to \mathcal{B}_t, \qquad (\mathbf{x}_0,\mathbf{n}_0,t) \mapsto (\mathbf{x},\mathbf{n})$$

provides "velocity fields"

$$(\mathbf{w},\mathbf{u}) := \frac{d}{dt}(\xi,\eta),$$

then the proposition

$$\frac{d}{dt}\int_{\mathcal{B}_t} \Phi_t(\mathbf{x},\mathbf{n})\,d(\mathbf{x},\mathbf{n}) = \int_{\mathcal{B}_t} \left\{\frac{\partial}{\partial t}\Phi_t + \nabla_x\cdot(\mathbf{w}\Phi_t) + \nabla_n\cdot(\mathbf{u}\Phi_t)\right\} d(\mathbf{x},\mathbf{n}) \quad (6)$$

holds true. Here ∇_x and ∇_n denote the covariant derivatives on $\mathbb{R}^3$ and S^2, respectively.

Consider a body[1] G_t in nematic space $\mathbb{R}^3 \times S^2$ and formulate the global balance of mass for particles in a given spatial region having orientations within a given solid angle. Since the mass in such a region of the nematic space is conserved, we obtain the equation

$$\frac{d}{dt}\int_{G_t} \tilde{\rho}(\tilde{\mathbf{x}},t)\,d(\tilde{\mathbf{x}}) = 0. \quad (7)$$

Using the extended Reynolds theorem we conclude that

$$\int_{G_t} \left\{\frac{\partial}{\partial t}\tilde{\rho}(\tilde{\mathbf{x}},t) + \nabla_{\tilde{\mathbf{x}}}\cdot(\tilde{\mathbf{v}}(\tilde{\mathbf{x}},t)\tilde{\rho}(\tilde{\mathbf{x}},t))\right\} d(\mathbf{x},\mathbf{n}) = 0, \quad (8)$$

[1] One has to follow the particles in the orientation space (when they change orientation) the same way one follows them in position space (when they change position).

where $\tilde{\mathbf{v}}(\tilde{\mathbf{x}},t)$ is the material velocity in the nematic space, holds true for all regions G_t and so the term within the $\{\dots\}$ has to vanish. Following the same procedure for the mesoscopic momentum $\tilde{\rho}\tilde{\mathbf{v}}$ and the mesoscopic energy $\tilde{\rho}\tilde{\varepsilon}+\tilde{\rho}\tilde{\mathbf{v}}^2$ one gets the set of local mesoscopic balance equations:

Mesoscopic Mass

$$\frac{\partial}{\partial t}\tilde{\rho}(\tilde{\mathbf{x}},t)+\nabla_{\tilde{\mathbf{x}}}\cdot(\tilde{\mathbf{v}}(\tilde{\mathbf{x}},t)\tilde{\rho}(\tilde{\mathbf{x}},t))=0. \tag{9}$$

Mesoscopic Momentum

$$\begin{aligned}\frac{\partial}{\partial t}(\tilde{\rho}(\tilde{\mathbf{x}},t)\tilde{\mathbf{v}}(\tilde{\mathbf{x}},t))+\\ +\nabla_{\tilde{\mathbf{x}}}\cdot\Big(\tilde{\mathbf{v}}(\tilde{\mathbf{x}},t)\tilde{\rho}(\tilde{\mathbf{x}},t)\tilde{\mathbf{v}}(\tilde{\mathbf{x}},t))-\tilde{\mathbf{t}}^{\top}_{\tilde{\mathbf{x}},\tilde{\mathbf{x}}}(\tilde{\mathbf{x}},t)\Big)=\tilde{\rho}(\tilde{\mathbf{x}},t)\tilde{\mathbf{f}}(\tilde{\mathbf{x}},t).\end{aligned} \tag{10}$$

Mesoscopic Energy

$$\begin{aligned}\frac{\partial}{\partial t}(\tilde{\rho}(\tilde{\mathbf{x}},t)\tilde{\varepsilon}(\tilde{\mathbf{x}},t)+\frac{1}{2}\tilde{\rho}(\tilde{\mathbf{x}},t)\tilde{\mathbf{v}}^2(\tilde{\mathbf{x}},t))+\\ +\nabla_{\tilde{\mathbf{x}}}\cdot\Big(\tilde{\mathbf{v}}(\tilde{\mathbf{x}},t)\tilde{\rho}(\tilde{\mathbf{x}},t)\tilde{\varepsilon}(\tilde{\mathbf{x}},t)+\\ +\tilde{\mathbf{v}}(\tilde{\mathbf{x}},t)\frac{1}{2}\tilde{\rho}(\tilde{\mathbf{x}},t)\tilde{\mathbf{v}}^2(\tilde{\mathbf{x}},t)+\\ +\tilde{\mathbf{q}}(\tilde{\mathbf{x}},t)-\tilde{\mathbf{v}}(\tilde{\mathbf{x}},t)\cdot\tilde{\mathbf{t}}(\tilde{\mathbf{x}},t)\Big)=\tilde{\rho}(\tilde{\mathbf{x}},t)\tilde{r}(\tilde{\mathbf{x}},t).\end{aligned} \tag{11}$$

It is common in macroscopic theory to simplify the balance of energy by subtracting the with $\tilde{\mathbf{v}}$ multiplied balance of momentum and adding the with $\frac{1}{2}\tilde{\mathbf{v}}\cdot\tilde{\mathbf{v}}$ multiplied balance of mass, this also works in the mecoscopic theory. The result of this is the balance of internal energy.

Mesoscopic Internal Energy

$$\begin{aligned}\frac{\partial}{\partial t}(\tilde{\rho}(\tilde{\mathbf{x}},t)\tilde{\varepsilon}(\tilde{\mathbf{x}},t))+\\ +\nabla_{\tilde{\mathbf{x}}}\cdot(\tilde{\mathbf{v}}(\tilde{\mathbf{x}},t)\tilde{\rho}(\tilde{\mathbf{x}},t)\tilde{\varepsilon}(\tilde{\mathbf{x}},t)+\tilde{\mathbf{q}}(\tilde{\mathbf{x}},t))-\\ -\tilde{\mathbf{t}}(\tilde{\mathbf{x}},t):\nabla_{\tilde{\mathbf{x}}}\tilde{\mathbf{v}}(\tilde{\mathbf{x}},t)=\tilde{\rho}(\tilde{\mathbf{x}},t)\tilde{r}(\tilde{\mathbf{x}},t)-\\ -\tilde{\rho}(\tilde{\mathbf{x}},t)\tilde{\mathbf{v}}(\tilde{\mathbf{x}},t)\cdot\tilde{\mathbf{f}}(\tilde{\mathbf{x}},t)\end{aligned} \tag{12}$$

with $\tilde{\mathbf{t}}:\nabla_{\tilde{\mathbf{x}}}\tilde{\mathbf{v}}$ defined as $t_{\mu\nu}\nabla^{\nu}v^{\mu}$, using Einstein's sum convention.

An example for the spatial part of a mesoscopic stress tensor can be found in [28]:

$$\tilde{\mathbf{t}}(\tilde{\mathbf{x}},t) = \frac{\tilde{\rho}(\tilde{\mathbf{x}},t)}{\rho(\mathbf{x},t)}\Big(\hat{t}(\rho,T) + \alpha_1 \mathbf{nnnn} : (\nabla\mathbf{v})^{sym} + \alpha_2 \mathbf{nN} + \alpha_3 \mathbf{Nn} + \\ + \alpha_4 (\nabla\mathbf{v})^{sym} + \alpha_5 \mathbf{nn} : (\nabla\mathbf{v})^{sym} + \alpha_6 \mathbf{n}\cdot(\nabla\mathbf{v})^{sym}\cdot\mathbf{n} + \\ + \xi_1 \mathbf{nn} : (\nabla\mathbf{v})^{sym}\delta + \xi_2 \mathbf{nn}\nabla\cdot\mathbf{v} + \xi_3 \nabla\cdot\mathbf{v}\delta\Big), \tag{13}$$

$$\mathbf{N} := \dot{\mathbf{n}} - (\nabla \times \mathbf{v}(\mathbf{x},t)) \times \mathbf{n}. \tag{14}$$

Examples for the stress tensor in classical director theory are given in [30] and [21], which also provide examples for the couple stress.

3 Orientation waves

In addition to surface elevation and density/pressure shock waves liquid crystals can show another type of wave – orientation waves. These waves are propagating changes of the orientation of the liquid crystals. One can distinguish two types of orientation waves, twist and tilt waves, which differ in the change of orientation being in perpendicular to the direction of propagation or aligned with it, see Fig. 2.

(a) Tilt wave (b) Twist wave (c) Tilt wave

Fig. 2 Schematic representation of orientation waves in liquid crystals. (propagation from left to right)

One of the first descriptions of twist waves was given by Ericksen in 1968 [12] using the theory of Leslie [19]. Since then orientation waves have been of continuous interest, e.g. experiments are described in [14] and recent analysis of director waves using variational theory can be found in [1].

3.1 Twist waves in classical macroscopic theory

In this section the main ideas presented in the aforementioned articles will be repeated to give an overview of twist waves in the Ericksen-Leslie-Theory.

The first step is to introduce a (macroscopic) director d_i and the total time derivative

$$\frac{\mathrm{d}}{\mathrm{d}t}\Box := \frac{\partial}{\partial t}\Box + v_i \nabla_i \Box, \tag{15}$$

$$w_i = \frac{\partial}{\partial t} d_i. \tag{16}$$

in this section the index notation will be used as it is done by Ericksen and Leslie. Although in general one should clearly distinguish between co- and contra-variant tensors, this will be omitted and lower indices will be used for both. Furthermore Einstein's sum convention will be used and a short form for derivatives is introduced, where "$X_{j,i}$" denotes $\partial X_j / \partial x_i$:

$$\frac{\mathrm{d}}{\mathrm{d}t}\rho + \rho v_{i,i} = 0, \tag{17}$$

$$\rho \frac{\mathrm{d}}{\mathrm{d}t} v_i = \rho F_i + \sigma_{ji,j}, \tag{18}$$

$$\rho \frac{\mathrm{d}}{\mathrm{d}t} w_i = \rho G_i + g_i + \pi_{ji,j}. \tag{19}$$

Next the symmetric and anti-symmetric part of the velocity gradient are introduced

$$A_{ij} := \frac{1}{2}(v_{i,j} + v_{j,i}), \tag{20}$$

$$\omega_{ij} := \frac{1}{2}(v_{i,j} - v_{j,i}), \tag{21}$$

$$N_i := w_i + \omega_{ki} d_k, \tag{22}$$

$$N_{ij} := w_{i,j} + \omega_{ki} d_{k,j}. \tag{23}$$

Without thermal effects the free energy is only dependent on the director and its gradient $F \stackrel{!}{=} F(d_i, d_{i,j})$ (Franck Energy [13]):

$$F \stackrel{!}{=} \frac{1}{2\rho}\left(k_{22} d_{i,j} d_{i,j} + (k_{11} - k_{22} - k_{24}) d_{i,j}^2 + (k_{33} - k_{22}) d_k d_j d_{k,j} + k_{24} d_{i,j} d_{j,i}\right). \tag{24}$$

Following the constitutive considerations given in [19], one finds

$$\sigma_{ji} = -p\delta_{ij} - \rho \frac{\partial F}{\partial d_{k,j}} d d_{k,i} + \hat{\sigma}'_{ji}, \tag{25}$$

$$\hat{\sigma}'_{ji} = \mu_1 d_k d_p A_{kp} d_i d_j + \mu_2 d_j N_i + \mu_3 d_i N_j + \mu_4 A_{ij} + \mu_5 d_j d_k A_{ki} + \mu_6 d_i d_k A_{kj}, \tag{26}$$

$$\tau_{ij} = \beta_j d_i + \rho \frac{\partial F}{\partial d_{k,j}}, \tag{27}$$

$$g_i = \gamma d_i - \beta_{i,j} - \rho \frac{\partial F}{\partial d_i} + \hat{g}'_i, \tag{28}$$

$$\hat{g}'_i = \lambda_1 N_i + \lambda_2 d_j A_{ji}, \tag{29}$$

where β_i is an arbitrary vector and γ an arbitrary constant. A necessary condition is

$$\lambda_1 = \mu_2 - \mu_3, \tag{30}$$

$$\lambda_2 = \mu_5 - \mu_6. \tag{31}$$

These considerations start out from the dissipation inequality and use representation theorems. It is necessary to consider first a static deformation and then a dynamic case (shear flow).

When assuming incompressibility, absence of extrinsic body forces or couples, neglecting thermal effects, and assuming a fluid at rest, which means putting the velocity equal to zero, one gets the following equations [12]:

$$\sigma_{ji,i} = 0, \tag{32}$$

$$\rho_1 \frac{\partial^2 d_i}{\partial t^2} = g_i + \tau_{ji,j}, \tag{33}$$

$$d_i d_i = 1, \tag{34}$$

$$\sigma_{ji} = -p\delta_{ij} - \tau_{jk} d_{k,i} + \mu_2 d_j \frac{\partial d_i}{\partial t} + \mu_3 d_i \frac{\partial d_j}{\partial t}, \tag{35}$$

$$\tau_{ij} = \rho \frac{\partial F}{\partial d_{j,i}}, \tag{36}$$

$$g_i = \gamma d_i - \rho \frac{\partial F}{\partial d_i} + \lambda_1 \frac{\partial d_i}{\partial t}, \tag{37}$$

where p is an arbitrary pressure, γ is a constant corresponding to the constraint (34), $\rho_1 > 0$, $\rho > 0$, $\lambda = \mu_2 - \mu_3 \leq 0$, μ_2 and μ_3 are constants. F, the free energy, is an objective function of d_i and $d_{i,j}$:

$$F = F(d_i, d_{j,k}), \tag{38}$$

furthermore F is a symmetric function of its arguments, i.e., if one (or both) of d_i and $d_{j,k}$ change its sign, F does not change:

$$F(d_i, d_{j,k}) = F(-d_i, d_{j,k}) = F(-d_i, -d_{j,k}) = F(d_i, -d_{j,k}). \tag{39}$$

By combination of these equations it follows that

$$(p + \rho F)_{,i} = \left(\lambda_1 \frac{\partial d_k}{\partial t} - \rho_1 \frac{\partial^2 d_k}{\partial t^2}\right) d_{k,i} + \left(\mu_2 d_j \frac{\partial d_i}{\partial t} + \mu_3 d_i \frac{\partial d_j}{\partial t}\right)_{,j}. \tag{40}$$

Assuming special cases in which $\mathbf{d}$ has the following form

$$\mathbf{d} = (\cos\varphi, \sin\varphi, 0), \tag{41}$$

$$\varphi = \varphi(x_3, t), \tag{42}$$

one finds that

$$\rho F = W(\varphi_{,3}), \tag{43}$$

$$W = \frac{1}{2} k_{22} \varphi_{,3}^2, \tag{44}$$

$$\frac{\partial F}{\partial d_3} = \frac{\partial F}{\partial d_{3,3}} = 0, \tag{45}$$

$$\rho \left(d_1 \frac{\partial F}{\partial d_{2,3}} - d_2 \frac{\partial F}{\partial d_{1,3}} \right) = \frac{\mathrm{d}W}{\mathrm{d}\varphi_{,3}}, \tag{46}$$

$$d_1 \frac{\partial F}{\partial d_2} - d_2 \frac{\partial F}{\partial d_1} = d_{2,3} \frac{\partial F}{\partial d_{1,3}} - d_{1,3} \frac{\partial F}{\partial d_{2,3}}. \tag{47}$$

By combining these equations it is possible to derive a wave equation

$$\rho_1 \frac{\partial^2}{\partial t^2} \varphi = \lambda_1 \frac{\partial}{\partial t} \varphi + k_2 2 \varphi_{,33}. \tag{48}$$

3.2 Twist waves in mesoscopic theory

In the case of twist waves, the mesoscopic space can be reduced to $\mathbb{R}^1 \times S^1$, if one assumes translation invariance in the space direction perpendicular to the direction of propagation and fixes the movement of the microscopic director also in this plane. W.l.o.g. the propagarion direction shall be x_3. This choice is compatible with the one made in the previous section.

The effect of the necessary assumptions can best be seen when splitting the tensors and divergence into two parts, one for the "real" space and one for the mesoscopic part. It is also common to separate the balance of momentum into a balance of "linear" momentum and angular momentum. In the literature (e.g. [8, 27]) the mesoscopic balances are presented using this split.

Starting out from Eqs. (9), (10) and (12), and assumimg vanishing heat flux density (i.e. $\tilde{\mathbf{q}} = 0$), no external forces and no external torque (i.e. $\tilde{\mathbf{f}} = 0$), no absorption of radiation (i.e. $r = 0$), and in addition constant velocity in the x_3 direction and co-moving observer (i.e. $v_{x_3} \stackrel{!}{=} 0$), one arrives at the following equation:

Mesoscopic Mass

$$\frac{\partial}{\partial t} \tilde{\rho}(\tilde{\mathbf{x}}, t) + \nabla_x \cdot (\tilde{\mathbf{v}}_{\mathbf{x}}(\tilde{\mathbf{x}}, t) \tilde{\rho}(\tilde{\mathbf{x}}, t)) + \nabla_n \cdot (\tilde{\mathbf{v}}_{\mathbf{n}}(\tilde{\mathbf{x}}, t) \tilde{\rho}(\tilde{\mathbf{x}}, t)) = 0 \tag{49}$$

$$\rightarrow \qquad \frac{\partial}{\partial t} \tilde{\rho}(x_3, \varphi, t) + \frac{\partial}{\partial \varphi} \left(\tilde{v}_\varphi(x_3, \varphi, t) \tilde{\rho}(x_3, \varphi, t) \right) = 0. \tag{50}$$

Mesoscopic Momentum

$$\begin{aligned}&\frac{\partial}{\partial t}(\tilde{\rho}(\tilde{\mathbf{x}},t)\tilde{\mathbf{v}}(\tilde{\mathbf{x}},t))+\\+\nabla_x\cdot&\Big(\tilde{\mathbf{v}}_{\mathbf{x}}(\tilde{\mathbf{x}},t)\tilde{\rho}(\tilde{\mathbf{x}},t)\tilde{\mathbf{v}}(\tilde{\mathbf{x}},t)-\tilde{\mathbf{t}}^{\top}_{\tilde{\mathbf{x}},\mathbf{x}}(\tilde{\mathbf{x}},t)\Big)+\\+\nabla_n\cdot&\Big(\tilde{\mathbf{v}}_{\mathbf{n}}(\tilde{\mathbf{x}},t)\tilde{\rho}(\tilde{\mathbf{x}},t)\tilde{\mathbf{v}}(\tilde{\mathbf{x}},t)-\tilde{\mathbf{t}}^{\top}_{\tilde{\mathbf{x}},\mathbf{n}}(\tilde{\mathbf{x}},t)\Big)=\tilde{\rho}(\tilde{\mathbf{x}},t)\tilde{\mathbf{f}}(\tilde{\mathbf{x}},t).\end{aligned} \tag{51}$$

This balance can be split into two balances:

$$\begin{aligned}&\frac{\partial}{\partial t}(\tilde{\rho}(\tilde{\mathbf{x}},t)\mathbf{v}(\tilde{\mathbf{x}},t))+\\+\nabla_x\cdot&\Big(\tilde{\mathbf{v}}_{\mathbf{x}}(\tilde{\mathbf{x}},t)\tilde{\rho}(\tilde{\mathbf{x}},t)\tilde{\mathbf{v}}_{\mathbf{x}}(\tilde{\mathbf{x}},t)-\tilde{\mathbf{t}}^{\top}_{\mathbf{x},\mathbf{x}}(\tilde{\mathbf{x}},t)\Big)+\\+\nabla_n\cdot&\Big(\tilde{\mathbf{v}}_{\mathbf{n}}(\tilde{\mathbf{x}},t)\tilde{\rho}(\tilde{\mathbf{x}},t)\tilde{\mathbf{v}}_{\mathbf{x}}(\tilde{\mathbf{x}},t)-\tilde{\mathbf{t}}^{\top}_{\mathbf{x},\mathbf{n}}(\tilde{\mathbf{x}},t)\Big)=\tilde{\rho}(\tilde{\mathbf{x}},t)\tilde{\mathbf{f}}_{\mathbf{x}}(\tilde{\mathbf{x}},t)\end{aligned} \tag{52}$$

$$\rightarrow \qquad -\frac{\partial}{\partial x_3}\tilde{\mathbf{t}}^{\top}_{x_3x_3}(x_3,\varphi,t)-\frac{\partial}{\partial\varphi}\tilde{\mathbf{t}}^{\top}_{x_3\varphi}(x_3,\varphi,t)=0. \tag{53}$$

$$\begin{aligned}&\frac{\partial}{\partial t}(\tilde{\rho}(\tilde{\mathbf{x}},t)\tilde{\mathbf{v}}_{\mathbf{n}}(\tilde{\mathbf{x}},t))+\\+\nabla_x\cdot&\Big(\tilde{\mathbf{v}}_{\mathbf{x}}(\tilde{\mathbf{x}},t)\tilde{\rho}(\tilde{\mathbf{x}},t)\tilde{\mathbf{v}}_{\mathbf{n}}(\tilde{\mathbf{x}},t)-\tilde{\mathbf{t}}^{\top}_{\mathbf{n},\mathbf{x}}(\tilde{\mathbf{x}},t)\Big)+\\+\nabla_n\cdot&\Big(\tilde{\mathbf{v}}_{\mathbf{n}}(\tilde{\mathbf{x}},t)\tilde{\rho}(\tilde{\mathbf{x}},t)\tilde{\mathbf{v}}_{\mathbf{n}}(\tilde{\mathbf{x}},t)-\tilde{\mathbf{t}}^{\top}_{\mathbf{n},\mathbf{n}}(\tilde{\mathbf{x}},t)\Big)=\tilde{\rho}(\tilde{\mathbf{x}},t)\tilde{\mathbf{f}}_{\mathbf{n}}(\tilde{\mathbf{x}},t)\end{aligned} \tag{54}$$

$$\begin{aligned}\rightarrow\qquad &\frac{\partial}{\partial t}(\tilde{\rho}(x_3,\varphi,t)\tilde{v}_{\varphi}(x_3,\varphi,t))-\frac{\partial}{\partial x_3}\tilde{\mathbf{t}}^{\top}_{\varphi x_3}(x_3,\varphi,t)+\\+\frac{\partial}{\partial\varphi}&\Big(\tilde{v}_{\varphi}(x_3,\varphi,t)\tilde{\rho}(x_3,\varphi,t)\tilde{v}_{\varphi}(x_3,\varphi,t)-\tilde{\mathbf{t}}^{\top}_{\varphi\varphi}(x_3,\varphi,t)\Big)=0.\end{aligned} \tag{55}$$

The first one corresponds to the macroscopic balance of momentum, the second one corresponds to the macroscopic spin balance.

Mesoscopic Internal Energy

$$\begin{aligned}&\frac{\partial}{\partial t}(\tilde{\rho}(\tilde{\mathbf{x}},t)\tilde{\varepsilon}(\tilde{\mathbf{x}},t))+\\+\nabla_x\cdot&\Big(\tilde{\mathbf{v}}_{\mathbf{x}}(\tilde{\mathbf{x}},t)\tilde{\rho}(\tilde{\mathbf{x}},t)\tilde{\varepsilon}(\tilde{\mathbf{x}},t)+\tilde{\mathbf{q}}_{\mathbf{x}}(\tilde{\mathbf{x}},t)\Big)+\\+\nabla_n\cdot&\Big(\tilde{\mathbf{v}}_{\mathbf{n}}(\tilde{\mathbf{x}},t)\tilde{\rho}(\tilde{\mathbf{x}},t)\tilde{\varepsilon}(\tilde{\mathbf{x}},t)+\tilde{\mathbf{q}}_{\mathbf{n}}(\tilde{\mathbf{x}},t)\Big)-\\&-\tilde{\mathbf{t}}_{\mathbf{x},\mathbf{x}}(\tilde{\mathbf{x}},t):\nabla_{\mathbf{x}}\tilde{\mathbf{v}}_{\mathbf{x}}(\tilde{\mathbf{x}},t)-\\&-\tilde{\mathbf{t}}_{\mathbf{x},\mathbf{n}}(\tilde{\mathbf{x}},t):\nabla_{\mathbf{n}}\tilde{\mathbf{v}}_{\mathbf{x}}(\tilde{\mathbf{x}},t)-\\&-\tilde{\mathbf{t}}_{\mathbf{n},\mathbf{x}}(\tilde{\mathbf{x}},t):\nabla_{\mathbf{n}}\tilde{\mathbf{v}}_{\mathbf{n}}(\tilde{\mathbf{x}},t)-\\&-\tilde{\mathbf{t}}_{\mathbf{n},\mathbf{n}}(\tilde{\mathbf{x}},t):\nabla_{\mathbf{n}}\tilde{\mathbf{v}}_{\mathbf{n}}(\tilde{\mathbf{x}},t)=\tilde{\rho}(\tilde{\mathbf{x}},t)\tilde{r}(\tilde{\mathbf{x}},t)-\\&\qquad-\tilde{\rho}(\tilde{\mathbf{x}},t)\tilde{\mathbf{v}}(\tilde{\mathbf{x}},t)\cdot\tilde{\mathbf{f}}(\tilde{\mathbf{x}},t)\end{aligned} \tag{56}$$

$$\rightarrow \quad \frac{\partial}{\partial t}(\tilde{\rho}(x_3,\varphi,t)\tilde{\varepsilon}(x_3,\varphi,t))+$$
$$+\frac{\partial}{\partial \varphi}\left(\tilde{v}_\varphi(x_3,\varphi,t)\tilde{\rho}(x_3,\varphi,t)\tilde{\varepsilon}(x_3,\varphi,t)\right)-$$
$$-\tilde{\mathbf{t}}_{\varphi x_3}(x_3,\varphi,t)\frac{\partial}{\partial \varphi}\tilde{v}_\varphi(x_3,\varphi,t)-$$
$$-\tilde{\mathbf{t}}_{\varphi\varphi}(x_3,\varphi,t)\frac{\partial}{\partial \varphi}\tilde{v}_\varphi(x_3,\varphi,t)=0. \tag{57}$$

The following assumptions have been made: Vanishing heat flux density (i.e. $\tilde{\mathbf{q}}=0$), no external forces and no external torque (i.e. $\tilde{\mathbf{f}}=0$), no absorption of radiation (i.e. $r=0$), and in addition constant velocity in the x_3-direction and a co-moving observer are assumed (i.e. $v_{x_3}\stackrel{!}{=}0$).

The resulting set of equations is:

$$\frac{\partial}{\partial t}\tilde{\rho}(x_3,\varphi,t)+\frac{\partial}{\partial \varphi}\left(\tilde{v}_\varphi(x_3,\varphi,t)\tilde{\rho}(x_3,\varphi,t)\right)=0, \tag{58}$$

$$-\frac{\partial}{\partial x_3}\tilde{\mathbf{t}}^{\top}_{x_3x_3}(x_3,\varphi,t)-\frac{\partial}{\partial \varphi}\tilde{\mathbf{t}}^{\top}_{x_3\varphi}(x_3,\varphi,t)=0, \tag{59}$$

$$\frac{\partial}{\partial t}(\tilde{\rho}(x_3,\varphi,t)\tilde{v}_\varphi(x_3,\varphi,t))-\frac{\partial}{\partial x_3}\tilde{\mathbf{t}}^{\top}_{\varphi x_3}(x_3,\varphi,t)+$$
$$+\frac{\partial}{\partial \varphi}\left(\tilde{v}_\varphi(x_3,\varphi,t)\tilde{\rho}(x_3,\varphi,t)\tilde{v}_\varphi(x_3,\varphi,t)-\tilde{\mathbf{t}}^{\top}_{\varphi\varphi}(x_3,\varphi,t)\right)=0, \tag{60}$$

$$\frac{\partial}{\partial t}(\tilde{\rho}(x_3,\varphi,t)\tilde{\varepsilon}(x_3,\varphi,t))+$$
$$+\frac{\partial}{\partial \varphi}\left(\tilde{v}_\varphi(x_3,\varphi,t)\tilde{\rho}(x_3,\varphi,t)\tilde{\varepsilon}(x_3,\varphi,t)\right)-$$
$$-\tilde{\mathbf{t}}_{\varphi x_3}(x_3,\varphi,t)\frac{\partial}{\partial \varphi}\tilde{v}_\varphi(x_3,\varphi,t)-\tilde{\mathbf{t}}_{\varphi\varphi}(x_3,\varphi,t)\frac{\partial}{\partial \varphi}\tilde{v}_\varphi(x_3,\varphi,t)=0. \tag{61}$$

These equations can either be solved directly (usually by numerical approximation) or can be transformed into a wave-like equation by standard methods. The necessary steps are:

1. differentiation of the continuity equation with respect to time;
2. taking the divergence of the balances of linear and angular momentum;
3. subtraction of these equations.

Taking the time derivative of Eq. (58) yields

$$\frac{\partial^2}{\partial t^2}\tilde{\rho}(x_3,\varphi,t)+\frac{\partial}{\partial t}\frac{\partial}{\partial \varphi}\left(\tilde{v}_\varphi(x_3,\varphi,t)\tilde{\rho}(x_3,\varphi,t)\right)=0, \tag{62}$$

the divergence of Eq. (59)

$$-\frac{\partial^2}{\partial x_3^2}\tilde{\mathbf{t}}^\top_{x_3x_3}(x_3,\varphi,t)-\frac{\partial}{\partial x_3}\frac{\partial}{\partial\varphi}\tilde{\mathbf{t}}^\top_{x_3\varphi}(x_3,\varphi,t)=0 \tag{63}$$

and the divergence of Eq. (60)

$$\frac{\partial}{\partial\varphi}\frac{\partial}{\partial t}(\tilde{\rho}(x_3,\varphi,t)\tilde{v}_\varphi(x_3,\varphi,t))-\frac{\partial}{\partial\varphi}\frac{\partial}{\partial x_3}\tilde{\mathbf{t}}^\top_{\varphi x_3}(x_3,\varphi,t)+$$
$$+\frac{\partial^2}{\partial\varphi^2}\left(\tilde{v}_\varphi(x_3,\varphi,t)\tilde{\rho}(x_3,\varphi,t)\tilde{v}_\varphi(x_3,\varphi,t)-\tilde{\mathbf{t}}^\top_{\varphi\varphi}(x_3,\varphi,t)\right)=0. \tag{64}$$

Equation (63) is only of interest if

$$\frac{\partial}{\partial x_3}\frac{\partial}{\partial\varphi}\tilde{\mathbf{t}}^\top_{x_3\varphi}(x_3,\varphi,t)+\frac{\partial}{\partial\varphi}\frac{\partial}{\partial x_3}\tilde{\mathbf{t}}^\top_{\varphi x_3}(x_3,\varphi,t)\;=\;0,$$

i.e., the "mixed" part of the mesoscopic stress tensor is skew-symmetric. In this case Eqs. (63) and (64) are subtracted from (62), and the resulting equation is:

$$\frac{\partial^2}{\partial t^2}\tilde{\rho}(x_3,\varphi,t)+\frac{\partial^2}{\partial x_3^2}\tilde{\mathbf{t}}^\top_{x_3x_3}(x_3,\varphi,t)-$$
$$-\frac{\partial^2}{\partial\varphi^2}\left(\tilde{v}_\varphi(x_3,\varphi,t)\tilde{\rho}(x_3,\varphi,t)\tilde{v}_\varphi(x_3,\varphi,t)-\tilde{\mathbf{t}}^\top_{\varphi\varphi}(x_3,\varphi,t)\right)=0. \tag{65}$$

Otherwise only Eq. (64) is subtracted from (62), and the resulting equation is:

$$\frac{\partial^2}{\partial t^2}\tilde{\rho}(x_3,\varphi,t)+\frac{\partial}{\partial\varphi}\frac{\partial}{\partial x_3}\tilde{\mathbf{t}}^\top_{\varphi x_3}(x_3,\varphi,t)-$$
$$-\frac{\partial^2}{\partial\varphi^2}\left(\tilde{v}_\varphi(x_3,\varphi,t)\tilde{\rho}(x_3,\varphi,t)\tilde{v}_\varphi(x_3,\varphi,t)-\tilde{\mathbf{t}}^\top_{\varphi\varphi}(x_3,\varphi,t)\right)=0, \tag{66}$$

where $\tilde{t}_{\varphi\varphi}=\tilde{t}_{\varphi\varphi}(\tilde{\rho},v_\varphi,\dots)$, $\tilde{\mathbf{t}}_{x_3\varphi}$, $\tilde{\mathbf{t}}_{\varphi x_3}$ and $\tilde{\mathbf{t}}_{\varphi\varphi}$ are usually at least functions of the density and velocity. Therefore these terms will produce the expected $\partial^2/\partial\varphi^2\tilde{\rho}$ and $\partial^2/\partial x_3^2\tilde{\rho}$. Note that orientation waves transform into waves of the mesoscopic density.

4 Comparison

For a comparison of both ways to describe twist waves in liquid crystals it is useful to know how mesoscopic balances are related to the macroscopic ones. Furthermore, one needs relations between the mesoscopic density and macroscopic order parameters, like the macroscopic director or alignment tensors.

4.1 Mesoscopic mass density, orientation distribution function and macroscopic director

To introduce macroscopic fields that represent the orientation distribution function $F(\tilde{\mathbf{x}},t)$ on $\mathbb{R}^3$, one can use a multi-pole expansion in [8] a suitable tensor basis is constructed and "Fourier coefficients" with respect to that basis are used to represent $F(\tilde{\mathbf{x}},t)$ on $\mathbb{R}^3$.

This multi-pole expansion produces, however, an infinite hierarchy of alignment tensors. From experiments usually only one or two are available for comparison. If the orientation distribution function $F(\tilde{\mathbf{x}},t)$ possesses a symmetry axis such that

$$F(\mathbf{x},\mathbf{n},t) = g(\mathbf{x},\mathbf{n}\cdot\mathbf{d},t) \quad \text{with an even function } g \text{ and a suitable } \mathbf{d}\in S^2 \tag{67}$$

is valid, one calls the vector **d** the macroscopic director.

It is worth to mention that many different mesoscopic distributions will result in the same macroscopic director – this is illustrated in Figs. 3 & 4 and also that not always a macroscopic director will exist. Therefore it is obvious that the mesoscopic description is more accurate and more general applicable.

4.2 Macroscopic balance equations

The balances in section 2.2 have been formulated on the mesoscopic space. For applications it is often necessary to formulate the balances on $\mathbb{R}^3$. In order to get these, one integrates over the mesoscopic variables.

The balance of mesoscopic momentum splits into two balances, the balance of momentum and the balance of spin (internal angular momentum). Here the problem arises that the spin is still given as velocity in the tangential space on S^2, but traditionally the spin is defined as an axial vector in $\mathbb{R}^3$.

A proposition, which is very useful, is the following:

Proposition 4.1. *Let* **U** *be a tangential C^1-vector field on S^2, i.e.* $\mathbf{U}\cdot\mathbf{n}=0$ *for all* $\mathbf{n}\in S^2$. *Then the identity*

$$\oint_{S^2} \nabla_n \cdot \mathbf{U}\, d^2n = 0 \tag{68}$$

is valid.

Proposition (4.1) implies that the terms with $\nabla_\mathbf{n}$ do not contribute to the macroscopic balances. The resulting equations are those of a micro-polar medium with the wanted fields ρ, **v** and **s** obtained as averages of their mesoscopic counterparts [8, 26, 27]:

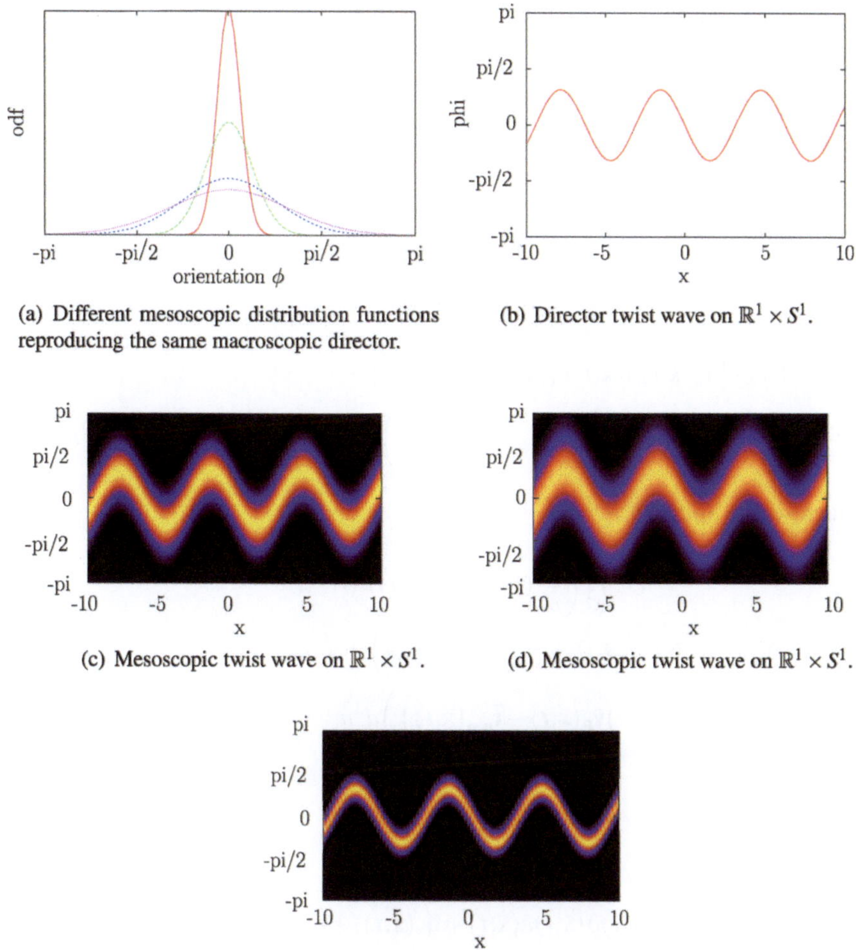

(a) Different mesoscopic distribution functions reproducing the same macroscopic director.

(b) Director twist wave on $\mathbb{R}^1 \times S^1$.

(c) Mesoscopic twist wave on $\mathbb{R}^1 \times S^1$.

(d) Mesoscopic twist wave on $\mathbb{R}^1 \times S^1$.

(e) Mesoscopic twist wave on $\mathbb{R}^1 \times S^1$.

Fig. 3 Comparison of different mesoscopic distribution functions reproducing the same macroscopic director.

Macroscopic Mass

$$\int_{S^2} \left(\frac{\partial}{\partial t} \tilde{\rho}(\tilde{\mathbf{x}},t) + \nabla_x \cdot (\tilde{\mathbf{v}}_\mathbf{x}(\tilde{\mathbf{x}},t)\tilde{\rho}(\tilde{\mathbf{x}},t)) + \nabla_n \cdot (\tilde{\mathbf{v}}_\mathbf{n}(\tilde{\mathbf{x}},t)\tilde{\rho}(\tilde{\mathbf{x}},t)) \right) d^2n = 0, \quad (69)$$

$$\rightarrow \qquad \frac{\partial}{\partial t}\rho(\mathbf{x},t) + \nabla_x \cdot (\mathbf{v}(\mathbf{x},t)\rho(\mathbf{x},t)) = 0. \quad (70)$$

Macroscopic Momentum

$$\begin{aligned} \int_{S^2} \Bigg(&\frac{\partial}{\partial t}(\tilde{\rho}(\tilde{\mathbf{x}},t)\mathbf{v}(\tilde{\mathbf{x}},t)) + \\ &+\nabla_x \cdot \left(\tilde{\mathbf{v}}_\mathbf{x}(\tilde{\mathbf{x}},t)\tilde{\rho}(\tilde{\mathbf{x}},t)\tilde{\mathbf{v}}_\mathbf{x}(\tilde{\mathbf{x}},t) - \tilde{\mathbf{t}}^\top_{\mathbf{x},\mathbf{x}}(\tilde{\mathbf{x}},t) \right) + \\ &+\nabla_n \cdot \left(\tilde{\mathbf{v}}_\mathbf{n}(\tilde{\mathbf{x}},t)\tilde{\rho}(\tilde{\mathbf{x}},t)\tilde{\mathbf{v}}_\mathbf{x}(\tilde{\mathbf{x}},t) - \tilde{\mathbf{t}}^\top_{\mathbf{x},\mathbf{n}}(\tilde{\mathbf{x}},t) \right) \Bigg) d^2n = \int_{S^2} \left(\tilde{\rho}(\tilde{\mathbf{x}},t)\tilde{\mathbf{f}}_\mathbf{x}(\tilde{\mathbf{x}},t) \right) d^2n. \end{aligned} \quad (71)$$

Macroscopic Spin

$$\begin{aligned} \int_{S^2} \Bigg(&\frac{\partial}{\partial t}(\tilde{\rho}(\tilde{\mathbf{x}},t)\tilde{\mathbf{v}}_\mathbf{n}(\tilde{\mathbf{x}},t)) + \\ &+\nabla_x \cdot \left(\tilde{\mathbf{v}}_\mathbf{x}(\tilde{\mathbf{x}},t)\tilde{\rho}(\tilde{\mathbf{x}},t)\tilde{\mathbf{v}}_\mathbf{n}(\tilde{\mathbf{x}},t) - \tilde{\mathbf{t}}^\top_{\mathbf{n},\mathbf{x}}(\tilde{\mathbf{x}},t) \right) + \\ &+\nabla_n \cdot \left(\tilde{\mathbf{v}}_\mathbf{n}(\tilde{\mathbf{x}},t)\tilde{\rho}(\tilde{\mathbf{x}},t)\tilde{\mathbf{v}}_\mathbf{n}(\tilde{\mathbf{x}},t) - \tilde{\mathbf{t}}^\top_{\mathbf{n},\mathbf{n}}(\tilde{\mathbf{x}},t) \right) \Bigg) d^2n = \int_{S^2} \left(\tilde{\rho}(\tilde{\mathbf{x}},t)\tilde{\mathbf{f}}_\mathbf{n}(\tilde{\mathbf{x}},t) \right) d^2n. \end{aligned} \quad (72)$$

Macroscopic Energy

$$\begin{aligned} \int_{S^2} \Bigg(&\frac{\partial}{\partial t}(\tilde{\rho}(\tilde{\mathbf{x}},t)\tilde{\varepsilon}(\tilde{\mathbf{x}},t)) + \\ &+\nabla_x \cdot (\tilde{\mathbf{v}}_\mathbf{x}(\tilde{\mathbf{x}},t)\tilde{\rho}(\tilde{\mathbf{x}},t)\tilde{\varepsilon}(\tilde{\mathbf{x}},t) + \tilde{\mathbf{q}}_\mathbf{x}(\tilde{\mathbf{x}},t)) + \\ &+\nabla_n \cdot (\tilde{\mathbf{v}}_\mathbf{n}(\tilde{\mathbf{x}},t)\tilde{\rho}(\tilde{\mathbf{x}},t)\tilde{\varepsilon}(\tilde{\mathbf{x}},t) + \tilde{\mathbf{q}}_\mathbf{n}(\tilde{\mathbf{x}},t)) - \\ &-\tilde{\mathbf{t}}_{\mathbf{x},\mathbf{x}}(\tilde{\mathbf{x}},t) : \nabla_\mathbf{x}\tilde{\mathbf{v}}_\mathbf{x}(\tilde{\mathbf{x}},t) - \tilde{\mathbf{t}}_{\mathbf{x},\mathbf{n}}(\tilde{\mathbf{x}},t) : \nabla_\mathbf{n}\tilde{\mathbf{v}}_\mathbf{x}(\tilde{\mathbf{x}},t) - \\ &-\tilde{\mathbf{t}}_{\mathbf{n},\mathbf{x}}(\tilde{\mathbf{x}},t) : \nabla_\mathbf{n}\tilde{\mathbf{v}}_\mathbf{n}(\tilde{\mathbf{x}},t) - \tilde{\mathbf{t}}_{\mathbf{n},\mathbf{n}}(\tilde{\mathbf{x}},t) : \nabla_\mathbf{n}\tilde{\mathbf{v}}_\mathbf{n}(\tilde{\mathbf{x}},t) \Bigg) d^2n = \int_{S^2} (\tilde{\rho}(\tilde{\mathbf{x}},t)\tilde{r}(\tilde{\mathbf{x}},t) - \\ &\qquad - \tilde{\rho}(\tilde{\mathbf{x}},t)\tilde{\mathbf{v}}(\tilde{\mathbf{x}},t) \cdot \tilde{\mathbf{f}}(\tilde{\mathbf{x}},t)) d^2n. \end{aligned} \quad (73)$$

A local macroscopic balance equation has the form

$$\frac{\partial}{\partial t}(\rho(\mathbf{x},t)\phi(\mathbf{x},t)) + \nabla_x \cdot \left(\mathbf{v}(\mathbf{x},t)\phi(\mathbf{x},t) + \mathbf{J}^\phi_{\mathrm{cond}}(\mathbf{x},t) \right) = \pi^\phi(\mathbf{x},t) + \sigma^\phi(\mathbf{x},t). \quad (74)$$

These balances contain *macroscopic* constitutive functions, which are in general *not* the averages of the mecoscopic constitutive functions, but contain additional terms.

The constituents of the macroscopic constitutive functions are identified by comparison of the general form with the averaged balance equations. This is left as an exercise to the reader.

In addition to the balance equations one has to take into account the Second Law of Thermodynamics, which is represented by the dissipation inequality. Using the procedures of Liu or Coleman and Noll, one can derive restrictions for the macroscopic constitutive functions. A typical state space for liquid crystals contains some quantities describing the orientational order in the fluid. The problem is that the dissipation inequality is formulated in the macroscopic space, while the constitutive functions are defined on the mesoscopic space and are transformed into macroscopic ones by averaging.

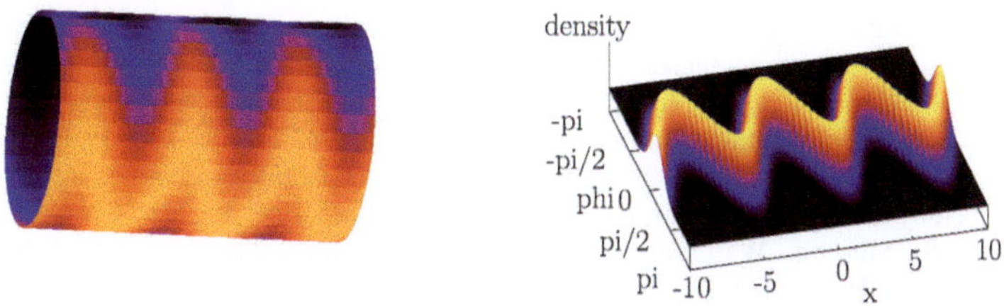

(a) Twist wave on $\mathbb{R}^1 \times S^1$ embedded in $\mathbb{R}^3$. (b) Mesoscopic twist wave on $\mathbb{R}^1 \times S^1$.

Fig. 4 Representation of twist waves in liquid crystals. The symmetric wave, caused by a possible "head-tail" symmetry of the liquid crystals has not been plotted.

5 Conclusions and outlook

Mesoscopic Continuum Physics has been reviewed using generalized coordinates. The use of generalized coordinates has the advantage that one only has to deal with 5 component equations for the velocity and 5 coordinates instead of 6 coordinates, 6 equations and 1 constraint. This makes quite a difference for numerics.

It has been shown that the balance of spin (internal angular momentum) follows naturally from the balance of momentum in the higher dimensional space. Here the macroscopic balance of spin is derived from a mesoscopic description (background theory).

A short repetition of twist-waves of liquid crystals in the Ericksen-Leslie theory has been presented and it has been shown that in the mesoscopic description orientation waves transform into density waves. It is worth mentioning that many different mesoscopic distributions will result in the same macroscopic director, and also that a macroscopic director will not always exist. Therefore it is obvious that the mesoscopic description is more accurate and more generally applicable.

Future work will include proposing an mesoscopic stress tensor also for the angular components and actual numerical calculations.

Acknowledgements Support by the European Union through the FP6 Marie Curie Transfer of Knowledge Project CENS-CMA (MC-TK-013909) is gratefully acknowledged. Furthermore support by the Alexander von Humboldt Foundation in form of a Feodor-Lynen-Fellowship is gratefully acknowledged.The author thanks Jüri Engelbrecht, host for the Feodor-Lynen-Fellowship, for valuable discussions and support.

Data for MBBA in Fig. 1 from http://pubchem.ncbi.nlm.nih.gov.

Almost countless open source tools, including Debian Linux (etch and lenny), gnuplot and LaTeX 2_ε, have been used to prepare this document.

References

1. Ali, G., Hunter, J.K.: Orientation Waves in a Director Field With Rotational Inertia. arXiv math/0609189 (2007)
2. Blenk, S., Ehrentraut, H., Muschik, W.: Orientation balances for liquid crystals and their representation by alignment tensors. Mol. Cryst. Liqu. Cryst. **204**, 133–141 (1991)
3. Blenk, S., Ehrentraut, H., Muschik, W.: Statistical foundation of macroscopic balances for liquid crystals in alignment tensor formulation. Physica A **174**, 119–138 (1991)
4. Blenk, S., Ehrentraut, H., Muschik, W.: Macroscopic constitutive equations for liquid crystals induced by their mesoscopic orientation distribution. Int. J. Eng. Sci. **30**(9), 1127–1143 (1992)
5. Blenk, S., Muschik, W.: Orientational balances for nematic liquid crystals. J. Non-Equilib. Thermodyn. **16**, 67–87 (1991)
6. Blenk, S., Muschik, W.: Mesoscopic concepts for constitutive equations of nematic liquid crystals in alignment tensor formulation. ZAMM **73**(4–5), T331–T333 (1993)
7. Ehrentraut, H., Muschik, W.: On symmetric irreducible tensors in d-dimensions. ARI **51**, 149–159 (1998)
8. Ehrentraut, H., Muschik, W., Papenfuss, C.: Concepts of Continuum Thermodynamics - 5 Lectures on Fundamentals, Methods and Examples. Sekcja Poligrafi Politechniki Swietokrzyskiej, Kielce (1997)
9. Ericksen, J.L.: Conservation laws for liquid crystals. Trans. Soc. Rheol. **5**, 23–34 (1961)
10. Ericksen, J.L.: Twisting of liquid crystals. J. Fluid Mech. **27**, 59–64 (1967)
11. Ericksen, J.L.: Equilibrium theory of liquid crystals. Adv. Liqu. Cryst. **2**, 233–298 (1976)
12. Ericksen, S.L.: Twist waves in liquid crystals. Quart. J. Mech. Appl. Math. **XXI** (1968)
13. Frank, F.C.: On the theory of liquid crystals. Discuss. Faraday Soc. **25**, 19–28 (1958). DOI 10.1039/DF9582500019
14. Guozhen, Z.: Experiments on director waves in nematic liquid crystals. Phys. Rev. Lett. **49**(18), 1332–1335 (1982). DOI 10.1103/PhysRevLett.49.1332
15. Hess, S.: Irreversible thermodynamics of nonequilibrium alignment phenomena in molecular liquids and in liquid crystals. Z. Naturforsch. **30a**, 728–733 (1975)
16. Hess, S.: Irreversible thermodynamics of nonequilibrium alignment phenomena in molecular liquids and in liquid crystals II. Z. Naturforsch. **30a**, 1224–1232 (1975)
17. Hess, S.: Pre- and post-transitional behaviour of the flow alignment and flow-induced phase transition in liquid crystals. Z. Naturforsch. **31a**, 1507–1513 (1976)
18. Leslie, F.M.: Some constitutive equations for anisotropic fluids. Quart. J. Mech. Appl. Math. **19**, 357–370 (1966). DOI 10.1093/qjmam/19.3.357
19. Leslie, F.M.: Some constitutive equations for liquid crystals. Arch. Rat. Mech. Anal. **28**, 265–283 (1968)
20. Leslie, F.M.: Theory of flow phenomena in liquid crystals. Adv. Liqu. Cryst. **4**, 1–81 (1979)
21. Lhuillier, D., Rey, A.D.: Liquid-crystalline nematic polymers revisited. J. Non-Newtonian Fluid Mech. **120**(1–3), 85–92 (2004). DOI 10.1016/j.jnnfm.2004.01.016.
22. Longa, L., Monselesan, D., Trebin, H.R.: An extension of the Landau-Ginzburg-de Gennes theory for liquid crystals. Liq. Crys. **2**(6), 769–796 (1987)

23. Maier, W., Saupe, A.: Eine einfache molekular-statistische Theorie der nematischen kristallinflüssigen Phase. Teil 1. Z. Naturforsch. **14a**, 882–889 (1959)
24. Maier, W., Saupe, A.: Eine einfache molekular-statistische Theorie der nematischen kristallinflüssigen Phase. Teil 2. Z. Naturforsch. **15a**, 287–292 (1960)
25. Muschik, W., Ehrentraut, H., Blenk, S.: Ericksen-Leslie liquid crystal theory revisited from a mesoscopic point of view. J. Non-Equilib. Thermodyn. **20**, 92–101 (1995)
26. Muschik, W., Papenfuss, C., Ehrentraut, H.: Mesoscopic theory of liquid crystals. J. Non-Equilib. Thermodyn. pp. 75–106 (2004)
27. Muschik, W., Papenfuss, C., Ehrentraut, H.: Sketch of the mesoscopic description of nematic liquid crystals. J. Non-Newtonian Fluid Mech. **119**(1–3), 91–104 (2004). DOI 10.1016/j.jnnfm.2004.01.011.
28. Papenfuß, C.: Thermodynamic constitutive theory. Habilitation, TU Berlin (2005)
29. Pardowitz, I., Hess, S.: On the theory of irreversible processes in molecular liquids and liquid crystals, nonequilibrium phenomena assosiated with the second and fourth rank alignment-tensors. Physica **100A**, 540–562 (1980)
30. Shahinpoor, M.: On the stress tensor in nematic liquid crystals. Rheol. Acta **15**, 99–103 (1976). DOI 10.1007/BF01517500
31. Sonnet, A.M., Maffetone, P.L., Virga, E.G.: Continuum theory for nematic liquid crystals with tensorial order. J. Non-Newtonian Fluid Mech. **119**(1–3), 51–59 (2004)

Part III
Exploiting the Dissipation Inequality

Overview

Wolfgang Muschik

In field formulation, a thermodynamical system is described in general by n time-dependent fields $\mathbf{a}(\mathbf{x},t)$, which satisfy the following balances

$$\partial_t(\rho\mathbf{a}) + \nabla\cdot(\mathbf{v}\rho\mathbf{a}+\boldsymbol{\Phi}) = \Sigma. \tag{1}$$

Here $\rho(\mathbf{x},t)$ is the field of the mass density, $\mathbf{v}(\mathbf{x},t)$ the field of the material velocity, $\boldsymbol{\Phi}(\mathbf{x},t)$ the fields of the conductive fluxes belonging to $\mathbf{a}$ and $\Sigma(\mathbf{x},t)$ the sum of the production and of the supply terms of $\mathbf{a}$. In general, not all of the $\mathbf{a}$ are *wanted* (or *basic) fields*, but some of them may be constitutive equations or other variables such as internal ones. In hydrodynamics, these basic fields are the following five ones (5-field theory)

$$(\rho,\varepsilon,\mathbf{v})(\mathbf{x},t). \tag{2}$$

Here $\varepsilon(\mathbf{x},t)$ is the field of the internal specific energy.

The balance equations (1) are underdetermined: they represent n equations for more than n quantities. Consequently, we need additional equations for obtaining a determined system of differential equations. These additional equations are the *constitutive equations*, which describe the considered material of which the system consists. For performing the derivatives in (1), we need a *(constitutive) state space*, which is defined as the domain of the constitutive equations. In general the state space is different from the wanted fields (2). In non-extended thermodynamics, it includes also derivatives of the wanted fields

$$\mathbf{z} = (\rho,\varepsilon,\mathbf{v},\partial_k\rho,\partial_k\varepsilon,\partial_k\mathbf{v},\partial_t\varepsilon,....). \tag{3}$$

In *extended thermodynamics*, the state space does not contain derivatives of the basic fields, but instead the traceless stress tensor $\mathbf{P}^0$ and dissipative fluxes, such as the heat flux density $\mathbf{q}$

$$\mathbf{z} = (\rho,\varepsilon,\mathbf{v},\mathbf{P}^0,\mathbf{q},...). \tag{4}$$

Institut für Theoretische Physik, Technische Universität Berlin, Hardenbergstr. 36, 10623 Berlin, Germany, e-mail: muschik@physik.tu-berlin.de

E. Quak, T. Soomere (eds.), *Applied Wave Mathematics*,
DOI 10.1007/978-3-642-00585-5_9, © Springer-Verlag Berlin Heidelberg 2009

Besides the balance equations (1), we have to take into consideration the *dissipation inequality* by which the *Second Law of Thermodynamics* is introduced

$$\partial_t(\rho s) + \nabla\cdot(\mathbf{v}\rho s+\phi) - \varphi = \sigma \geq 0. \tag{5}$$

Here s is the specific entropy, ϕ the entropy flux, φ the entropy supply and σ the entropy production density, which is non-negative according to the Second Law. By introducing an amendment to the Second Law, one can prove its *Coleman-Mizel formulation*

> *All solutions of the balance equations (1) have to satisfy the dissipation inequality (5),*

which describes the task to perform.

This can be done in two different ways: The balances (1) are solved by assuming constitutive ansatzes, and then the resulting entropy is checked, if it satisfies the dissipation inequality (5). If not, the procedure begins again until solving the problem. The second approach does not start out with a solution of the balances, but is asking for all possible constitutive equations – belonging to a chosen constitutive state space – which satisfy both the balances and the dissipation inequality. Here, the second way is shortly discussed.

The constitutive equations of the balance equations (1) and of the dissipation inequality (5) are

$$\mathbf{M} \equiv \{\mathbf{a}(\mathbf{z}), \boldsymbol{\Phi}(\mathbf{z}), \Sigma(\mathbf{z}), s(\mathbf{z}), \phi(\mathbf{z}), \varphi(\mathbf{z})\}, \tag{6}$$

and the derivatives occuring in (1) become by use of the chain rule

$$\partial_t\,\mathbf{a} = \frac{\partial\mathbf{a}}{\partial\mathbf{z}}\cdot\partial_t\mathbf{z}, \qquad \nabla\mathbf{a} = \frac{\partial\mathbf{a}}{\partial\mathbf{z}}\cdot\nabla\mathbf{z}. \tag{7}$$

If we now introduce (7) into the balances (1), we obtain by using the balance of mass

$$\frac{\partial}{\partial t}\rho(\mathbf{x},t) + \nabla\cdot\{\rho(\mathbf{x},t)\mathbf{v}(\mathbf{x},t)\} = 0, \tag{8}$$

$$\rho\frac{\partial\mathbf{a}}{\partial\mathbf{z}}\cdot\partial_t\mathbf{z} + \rho\mathbf{v}\frac{\partial\mathbf{a}}{\partial\mathbf{z}}:\nabla\mathbf{z} + \frac{\partial\boldsymbol{\Phi}}{\partial\mathbf{z}}:\nabla\mathbf{z} = \Sigma(\mathbf{z}). \tag{9}$$

These balance equations are called *balances on the state space*. Dependent on the special constitutive equations, this system of differential equations is often highly non-linear.

The entropy balance (5) on the state space becomes

$$\rho\frac{\partial s}{\partial\mathbf{z}}\cdot\partial_t\mathbf{z} + \rho\mathbf{v}\frac{\partial s}{\partial\mathbf{z}}:\nabla\mathbf{z} + \frac{\partial\phi}{\partial\mathbf{z}}:\nabla\mathbf{z} \geq \varphi(\mathbf{z}). \tag{10}$$

Introducing the so-called *higher derivatives* of the state space variables (3)

$$\mathbf{y} := (\partial_t \mathbf{z}, \partial_k \mathbf{z}), \tag{11}$$

Eqs. (9) and (10) are linear in the higher derivatives, and they can be written in the following form

$$\mathbf{A}(\mathbf{z}) \cdot \mathbf{y} = \mathbf{C}(\mathbf{z}), \qquad \mathbf{B}(\mathbf{z}) \cdot \mathbf{y} \geq D. \tag{12}$$

Using the *Liu procedure*, one can get rid of the higher derivatives, and one obtains the *Liu equations*

$$\Lambda(\mathbf{z}) \cdot \mathbf{A} = \mathbf{B} \tag{13}$$

and the *residual dissipation inequality*

$$\Lambda(\mathbf{z}) \cdot \mathbf{C} \geq D. \tag{14}$$

Because there are more state space variables than balance equations, $\mathbf{A}$ has a right-hand reciprocal

$$\mathbf{A} \cdot \bar{\mathbf{A}} = \mathbf{1}, \tag{15}$$

and we obtain from (13) and (14):

$$\Lambda = \mathbf{B} \cdot \bar{\mathbf{A}}, \qquad \mathbf{B} \cdot \bar{\mathbf{A}} \cdot \mathbf{C} \geq D. \tag{16}$$

The inequality $(16)_2$ represents the constraints to the constitutive equations $\mathbf{A}, \mathbf{C}, \mathbf{B}$ and D induced by the Second Law.

An other contraint is induced by the *principle of material frame indifference*, which states that the constitutive equations (6) do not depend on the material velocity

$$\frac{\partial \mathbf{M}}{\partial \mathbf{v}} \equiv \mathbf{0}. \tag{17}$$

In his contribution *Weakly Nonlocal Non-equilibrium Thermodynamics – Variational Principles and Second Law*, Péter Ván investigates the influence of the dissipation inequality on the evolution equations of the internal variables spanning constitutive state spaces of first and second order of non-locality in resting media. Subsequently, the question is discussed whether these evolution equations are of Hamiltonian structure allowing variational principles. Using the balance of total energy (and the velocity as a state space variable), Ván applies the Liu procedure to obtain the constitutive constraints of a viscous liquid.

References

1. Muschik, W., Ehrentraut, H. An amendment to the second law. J. Non-Equilib. Thermodyn. **21**, 175–192 (1996)

Weakly Nonlocal Non-equilibrium Thermodynamics – Variational Principles and Second Law

Péter Ván

Abstract A general, uniform, rigorous and constructive thermodynamic approach to weakly nonlocal non-equilibrium thermodynamics is reviewed. A method is given to construct and restrict the evolution equations of physical theories according to the Second Law of Thermodynamics and considering weakly nonlocal constitutive state spaces. The evolution equations of internal variables, the classical irreversible thermodynamics and Korteweg fluids are treated.

1 Introduction

Weakly nonlocal, coarse grained, phase field and gradient are attributes of theories from different fields of physics indicating that in contradistinction to the traditional treatments, the governing equations of the theory depend on higher order space derivatives of the state variables. The origin of the idea goes back to the square gradient model of van der Waals for phase interfaces [100], where it is extensively applied [3, 38, 39, 42]. Later applications go far beyond phase boundaries or thermodynamics. Nowadays weakly nonlocal is a nomination in continuum physics dealing with internal structures [10, 27, 46, 50, 61, 66], coarse grained or phase field appears in statistically motivated thermodynamics [1, 3, 5, 34, 80], and gradient is frequently used in mechanics in different context [2, 11, 37, 43, 44, 57, 58, 59, 79, 83, 84, 101]. The simplest way to demonstrate the meaning of weakly nonlocal extensions can be exemplified by the Ginzburg-Landau equation, which is not only a specific equation in superconductivity, as it was introduced originally by Landau and Khalatnikov [45], but a first weakly nonlocal extension of a homogeneous relaxation equation of an internal variable. The traditional derivation of the Ginzburg-Landau equation is

Department of Theoretical Physics, KFKI, Research Institute of Particle and Nuclear Physics, Konkoly Thege Miklós út 29-33., 1525 Budapest, Hungary, and
Department of Energy Engineering, Budapest University of Technology and Economics
Bertalan Lajos u. 4-6. 1111 Budapest, Hungary, e-mail: vpet@rmki.kfki.hu

E. Quak, T. Soomere (eds.), *Applied Wave Mathematics*,
DOI 10.1007/978-3-642-00585-5_10,

based on a characteristic mixing of variational and thermodynamic considerations. One applies a variational principle for the static part, and the functional derivatives are introduced as thermodynamic forces into a relaxation type equation. A clear variational derivation to obtain a first order differential equation is impossible without any further ado (e.g., without introducing new variables to avoid the first order time derivative, which is not a symmetric operator) [98]. One can apply these kinds of arguments in continuum theories in general, preserving the doubled theoretical framework separating reversible and irreversible parts of the equations [28, 77, 78]. However, there are also other attempts to unify the two parts with different additional hypotheses and to eliminate this inconsistency of the traditional approach [6, 30, 50].

The ultimate aim is to find a unified, general, rigorous and predictive theoretical framework that makes it possible to extend the governing equations of physics with higher order gradients of the continuum fields, beyond the traditional terms. The method should be uniformly applicable from classical systems in local equilibrium up to relativistic systems beyond local equilibrium; should be general to incorporate most of the mentioned classical examples of weakly nonlocal theories without specific assumptions; should reduce the independent additional assumption to a minimum and should be constructive to give calculational methods for systematic higher order extensions of the constitutive space.

This paper is a general tutorial to the mathematical framework of such a theory. In the first section a general methodology of exploiting the Second Law for weakly nonlocal systems is given. The Second Law is considered as a constrained inequality, where the constraints are the evolution equations of the system and their derivatives, depending on the order of the nonlocality. In the third section, evolution equations of internal variables and their different weakly nonlocal extensions are treated. A first order weakly nonlocal theory leads to relaxation type ordinary differential equations, a second order nonlocality leads to the Ginzburg-Landau equation, and a second order nonlocal theory to dual internal variables, unifying the evolution equations of internal variables derived by mechanical methods (by variational principles and dissipation potentials) and by thermodynamics (by heuristic application of the Second Law). In the fourth section we show that classical irreversible thermodynamics can be incorporated naturally in our treatment, a first order weakly nonlocal theory of balance type evolution equations leads to the thermodynamic flux-force relations of classical irreversible thermodynamics with gradients of the intensives as thermodynamic forces. Finally, we demonstrate the applicability of the method to one component heat conducting Korteweg fluids that are first order weakly nonlocal in the energy and in the velocity, and second order weakly nonlocal in the density. In that case nontrivial forms of the pressure tensor ensure the compatibility to the Second Law . As a particular example we derive the constitutive functions of the Schrödinger-Madelung fluids. Finally a summary and discussions follow.

2 Second law and weakly nonlocal constitutive spaces

In this section we shortly summarize some methodological specialities of exploiting the Second Law in weakly nonlocal systems. One can find some further details in [86, 87], where the role of the entropy flux is treated, and in [90], where the key idea is introduced, the derivative of the constraints as additional constraint.

There are several interpretations of the Second Law in non-equilibrium thermodynamics (see e.g. [36, 68]). If we wanted to ensure that the Second Law is a consequence of material properties, then we should postulate that the entropy was an increasing function along the solutions of the evolution equationss. With this assumption we are compatible with the classical heuristic method that puts the evolution equation in the balance of the entropy, and exploits the consequences of the inequality. In the classical approach a quadratic expression is recognized and then a relation is sought between the introduced thermodynamic fluxes and forces [18, 29]. Here the evolution equations are considered as constraints for the inequality of the entropy production [15]. We arrive at the Coleman-Noll procedure recognizing the algebraic part of the problem where the different derivatives are independent. When instead of putting constraints into the evolution equations, the constrained algebraic inequality is solved by multipliers then the calculation is called the Liu procedure [47]. The Coleman-Noll and Liu procedures are equivalent in simple systems [82], but the later one preserves the symmetries of the evolution equations and the constraints. The Liu procedure is based on a linear algebraic theorem, called Liu's theorem in the thermodynamic literature [69, 71], and an interpretation of the role of entropy inequality. Hauser and Kirchner recognized that Liu's theorem is a consequence of a famous statement of optimization theory and linear programming, the so called Farkas's lemma [33]. That theorem was proved first by Farkas in 1894 and independently by Minkowski in 1896 [65]. In the Appendix we have formulated and proved the classical Farkas lemma, the affine Farkas lemma and Liu's theorem.

Originally Farkas developed his lemma to formulate correctly Fourier's principle of mechanics of mass-points, which is the generalization of d'Alembert's principle in case of inequality constraints [21]. The role of Liu's theorem in continuum physics is similar in some sense: we want to give the correct form of the evolution equations taking into account the requirement of the Second Law, that is the entropy inequality. In fact only the material part of the evolution equations – the constitutive functions – is/are restricted. The mathematical formulation introduces a kind of dual point of view, because in the Liu procedure the evolution equations – the partial or ordinary differential equations determining the evolution of the system – form a condition, a constraint for the entropy inequality. The variables (e.g., fields) in the evolution equation form the *basic state space*. The constitutive quantities depend on these functions, on the basic state and some of its derivatives. To make the problem algebraically manageable the basic state variables and their derivatives are considered as independent quantities. Some of them can be incorporated into the *constitutive state space* (or large state space [69]), into the domain of the constitutive functions. The entropy inequality, our objective function, has a special balance form, and determines the *process directions,* the independent variables of the algebraic problem: those are the derivatives of the constitutive state that are not included in the constitutive space. The choice of the constitutive state space is crucial and can result in different kinds of restricted constitutive functions with the Liu procedure.

As we have already mentioned, weakly nonlocal constitutive spaces contain space derivatives. In these cases some derivatives of the constraints – balances and others – can appear as additional constraints that further restrict the entropy inequality. Whether these additional constraints should be considered or not depends on the order of weak nonlocality, the structure of the constraints and physical considerations. This is the peculiarity of the exploitation of the Second Law for weakly nonlocal systems.

In the following sections we will give several examples to demonstrate the application of the formalism.

3 Thermodynamic evolution of internal variables

In this section we investigate the thermodynamic restrictions on the evolution equation of internal variables in continua at rest. First in a first order weakly nonlocal constitutive state space, then in a second order weakly nonlocal one, and finally considering the peculiarities of dual internal variables. Further details of the related calculations and the physical interpretation can be found in [90, 91, 95].

3.1 *First order nonlocality – relaxation*

Let us investigate the thermodynamic restrictions for the evolution of a classical internal variable field $a(t, \mathbf{r})$, in a continuum at rest related to an inertial observer. There are no constraints or knowledge regarding the form of its evolution equation. Therefore the evolution equation can be given in a general form as

$$\underline{\partial_t a} + \hat{f} = 0, \tag{1}$$

with an arbitrary constitutive function $\hat{f}$. Here and in the following the partial time derivatives are denoted by ∂_t and the constitutive quantities are denoted by a hat ($\hat{}$). First we introduce a *first order weakly nonlocal constitutive state space* spanned by the basic fields a and their gradients $\partial_i a$, where $i = 1,2,3$. The notation with indices with Einstein summation convention is applied by distinguishing covariant and contravariant components (vectors and covectors) by upper and lower indices, e.g. $\partial_i J^i (= Div\mathbf{J} = \nabla \cdot \mathbf{J})$ denotes the divergence of the vector field J^i.

The above evolution equation is not completely arbitrary, it is restricted by the Second Law of thermodynamics. Therefore we assume that there is an entropy balance with a nonnegative production term

$$\partial_t \hat{s} + \partial_i \hat{J}^i \geq 0. \tag{2}$$

Here the entropy density $\hat{s}$ and the entropy flux $\hat{J}^i$ are constitutive quantities, too. Therefore in this case

- the basic state space is spanned by a,
- the constitutive state space is spanned by $(a, \partial_i a)$,
- the constitutive functions are $\hat{s}, \hat{J}^i$ and $\hat{f}$.

We may develop the derivatives according to the constitutive assumptions as

$$\partial_a \hat{s} \underline{\partial_t a} + \partial_{\partial_i a} \hat{s}\, \underline{\partial_{it} a} + \partial_a \hat{J}^i\, \partial_i a + \partial_{\partial_j a} \hat{J}^i\, \underline{\partial_{ij} a} \geq 0. \tag{3}$$

Here the partial derivatives of the constitutive functions are denoted in an abbreviated manner, e.g. $\partial \hat{s}/\partial a = \partial_a \hat{s}$. The underlined partial derivatives of the basic field are not in the constitutive space, they are independent algebraic quantities and span the process direction space $(\partial_t a, \partial_{ti} a, \partial_{ij} a)$. Then we can apply the theorem of Liu identifying the underlined partial derivatives in (1) and (3) by **p** and their coefficients by **a** and **b** respectively. Now we introduce a Lagrange-Farkas multiplier $\hat{\lambda}$, which is a constitutive quantity, for the evolution equation (1) and apply the Liu procedure to determine the form of the constitutive functions, required by the entropy inequality:

$$0 \leq \left(\partial_a \hat{s} - \hat{\lambda}\right) \underline{\partial_t a} + \partial_{\partial_i a} \hat{s}\, \underline{\partial_{it} a} + \partial_{\partial_j a} \hat{J}^i\, \underline{\partial_{ij} a} + \partial_a \hat{J}^i\, \partial_i a - \hat{\lambda} \hat{f}. \tag{4}$$

Here the existence of the multiplier $\hat{\lambda}$ follows from Liu's theorem, and we will exploit that the process directions are independent of the constitutive space in the sense that the values of these derivatives can be different for the same values of its coefficients depending, e.g., on the initial conditions of (1). Therefore the multipliers of the underlined terms in (4) are zero and give the Liu equations:

$$\partial_t a : \partial_a \hat{s} = \hat{\lambda}, \tag{5}$$

$$\partial_{it} a : \partial_{\partial_i a} \hat{s} = 0^i, \tag{6}$$

$$\partial_{ij} a : \partial_{\partial_j a} \hat{J}^i = 0^{ij}. \tag{7}$$

The first equation determines the Lagrange-Farkas multiplier and the last two ones show that both the entropy and the entropy flux are local, independent of the gradient of a. This is a complete solution of the system. The residual inequality is

$$0 \leq \partial_i \hat{J}^i(a) - \partial_a \hat{s}(a)\, \hat{f}(a, \partial_i a). \tag{8}$$

Now we assume that the residual entropy flux is zero: $\hat{J}^i \equiv 0^i$. This is the situation in isotropic materials according to the representation theorem of isotropic vector functions. Therefore the entropy inequality reduces to a flux-force system. If we want to determine the form of the evolution equation then the entropy should be considered as a given function and $\hat{f}$ as an undetermined constitutive quantity. The classical solution of the above inequality is that the constitutive function $\hat{f}$ is proportional to the entropy derivative by a nonnegative constitutive multiplier:

$$\hat{f} = -\hat{l}\,\partial_a \hat{s}, \qquad \hat{l} > 0. \tag{9}$$

This is a general solution of the inequality if we assume two times differentiability of $\hat{f}$ (due to the mean value theorem of Lagrange, see e.g. [30]). In practice we rarely exploit this generality and restrict ourselves to the case of constant $\hat{l}$. This remark applies also in (18), (36), (37), and (53). Therefore the evolution equation of an internal variable restricted by the Second Law in isotropic materials is a relaxation type differential equation:

$$\partial_t a = \hat{l}\partial_a \hat{s}. \tag{10}$$

Now let us summarize the most important steps of the procedure:

- Identification of the basic inequality and the basic constraints and the domain of the corresponding functions (constitutive state space).
- Performing the partial derivations and the identification of the process direction space by the partial derivatives of the basic state space and introducing the additional (derivative) constraints, if necessary.
- Application of Liu's theorem.
- Solution of the Liu equations and the dissipation inequality.

In this example there was no need to apply additional derivative constraints, but it can be necessary in case of more extended constitutive state spaces as one can see in the next problem.

3.2 Second order nonlocality – the Ginzburg-Landau equation

In this case we face a similar problem as we are to determine the thermodynamic restrictions on the general evolution equation (1), but now we assume that there is a second order weakly nonlocal state space spanned by the basic field a and its first and second space derivatives $\partial_i a$ and $\partial_{ij} a$.

Therefore in our second example

- the basic state space is spanned by a,
- the constitutive state space is spanned by $(a, \partial_i a, \partial_{ij} a)$,
- the constitutive functions are $\hat{s}, \hat{J}^i$ and $\hat{f}$.

The corresponding process direction space is spanned by the next derivatives $(\partial_t a, \partial_{ti} a, \partial_{tij} a, \partial_{ijk} a)$. Let us observe that these are not independent any more, the gradient of (1) is a linear relation on the process direction space. Therefore we should consider

$$\partial_{ti} a + \partial_i \hat{f} = 0_i \tag{11}$$

as a further constraint to the entropy inequality (2). We introduce the Lagrange-Farkas multipliers $\hat{\lambda}$ for (1) and $\hat{\Lambda}^i$ for (11). Let us apply the Liu procedure again, but in this case not separating the different steps:

$$
\begin{aligned}
0 &\leq \partial_t \hat{s} + \partial_i \hat{J}^i - \hat{\lambda}\left(\partial_t a + \hat{f}\right) - \hat{\Lambda}^i\left(\partial_{ti} a + \partial_i \hat{f}\right) \\
&= \partial_a \hat{s}\, \partial_t a + \partial_{\partial_i a} \hat{s}\, \partial_{it} a + \partial_{\partial_{ij} a} \hat{s}\, \partial_{ijt} a + \partial_a \hat{J}^i\, \partial_i a + \partial_{\partial_j a} \hat{J}^i\, \partial_{ij} a + \partial_{\partial_{jk} a} \hat{J}^i\, \partial_{ijk} a \\
&\quad - \hat{\lambda}\left(\partial_t a + \hat{f}\right) - \hat{\Lambda}^i\left(\partial_{ti} a + \partial_a \hat{f}\, \partial_i a + \partial_{\partial_j a} \hat{f}\, \partial_{ij} a + \partial_{\partial_{jk} a} \hat{f}\, \partial_{ijk} a\right) \\
&= \left(\partial_a \hat{s} - \hat{\lambda}\right) \underline{\partial_t a} + \left(\partial_{\partial_i a} \hat{s} - \hat{\Lambda}^i\right)\, \underline{\partial_{it} a} + \partial_{\partial_{ij} a} \hat{s}\, \underline{\partial_{ijt} a} + \\
&\quad \left(\partial_{\partial_{jk} a} \hat{J}^i - \hat{\Lambda}^i \partial_{\partial_{jk} a} \hat{f}\right) \underline{\partial_{ijk} a} + \partial_a \hat{J}^i\, \partial_i a + \partial_{\partial_j a} \hat{J}^i\, \partial_{ij} a - \\
&\quad \hat{\Lambda}^i \left(\partial_a \hat{f}\, \partial_i a + \partial_{\partial_j a} \hat{f}\, \partial_{ij} a\right) - \hat{\lambda} \hat{f}.
\end{aligned} \tag{12}
$$

We can see that the degenerations are different than in the previous case. In the following we do not give the detailed expositions of the theorem, one can reconstruct that from the detailed presentation of the calculations.

The multipliers of the underlined partial derivatives in (12), the process direction space, give the Liu equations:

$$\partial_t a : \partial_a \hat{s} = \hat{\lambda}, \tag{13}$$

$$\partial_{it} a : \partial_{\partial_i a} \hat{s} = \hat{\Lambda}^i, \tag{14}$$

$$\partial_{ijt} a : \partial_{\partial_{ij} a} \hat{s} = 0^{ij}, \tag{15}$$

$$\partial_{ijk} a : \partial_{\partial_{jk} a} \hat{J}^i = \hat{\Lambda}^i \partial_{\partial_{jk} a} \hat{f}. \tag{16}$$

The first two equations determine the Lagrange-Farkas multipliers by the entropy derivatives, and the third equation shows that the entropy is independent of the second gradient of a. As a consequence, the Lagrange-Farkas multiplier $\hat{\Lambda}^i$ is independent of that derivative, therefore the last equation can be integrated and gives

$$\hat{J}^i(a, \partial_i a, \partial_{ij} a) = \partial_{\partial_i a} \hat{s}(a, \partial_i a)\ \hat{f}(a, \partial_i a, \partial_{ij} a) + \hat{\mathfrak{J}}^i(a, \partial_i a), \tag{17}$$

where the residual entropy flux $\hat{\mathfrak{J}}^i$ is an arbitrary constitutive function, and the variables of the constitutive functions are explicitly written. This is a complete solution of the system of the Liu equations (13)–(16). Therefore the dissipation inequality simplifies to the following form

$$0 \leq \partial_i \hat{\mathfrak{J}}^i + \left(\partial_i(\partial_{\partial_i a} \hat{s}) - \partial_a \hat{s}\right) \hat{f}. \tag{18}$$

Assuming that the residual entropy flux is zero, $\hat{\mathfrak{J}}^i \equiv 0^i$, the entropy inequality reduces to a flux-force system. In this case the classical solution of the inequality is

$$\hat{f} = \hat{l}\ \left(\partial_i(\partial_{\partial_i a} \hat{s}) - \partial_a \hat{s}\right), \qquad \hat{l} > 0. \tag{19}$$

Therefore the form of the evolution equation of an internal variable in a second order weakly nonlocal constitutive state space is the Ginzburg-Landau equation:

$$\partial_t a = \hat{l}\ \left(\partial_a \hat{s} - \partial_i(\partial_{\partial_i a} \hat{s})\right). \tag{20}$$

We can get the classical form of the equation with a particular form of the entropy density $\hat{s}(a, \partial_i a) = s_o(a) - \gamma(\partial_i a)^2$, with quadratic dependence on the gradient and with a constant coefficient γ. We have $\gamma > 0$ because of the concavity of the entropy. Other nonlocal thermodynamic potentials also can be used in the above derivation, e.g., the free energy, but one should be careful to use a correct thermodynamic structure when coupling to other thermodynamic interactions beyond the one related to the internal variable [80].

The Ginzburg-Landau equation and its variants appear in different fields of physics and are applied to several phenomena. Beyond their original appearance in superconductivity, they play an important role in pattern formation and they are the prototypical phase field models [1, 5].

As we have mentioned in the introduction, the traditional derivation of the Ginzburg-Landau equation has two main ingredients:

- The static, equilibrium part is derived from a variational principle.
- The dynamic part is added by stability arguments (relaxational form).

The physical content of the two ingredients of the classical derivation is sound and transparent [34]. On the other hand, the origin of the variational principle, the coupling of the two parts and the role of the Second Law of thermodynamics is ad-hoc and is not compatible with the general balance and constitutive structure of continuum physics. Here we unified the ingredients in a thermodynamic derivation, and we derived the form of the entropy flux, too. We did not refer to any kind of variational principle, however, the derived static part has a complete Euler-Lagrange form. The dynamic part contains a first order time derivative, therefore one cannot hope to derive it from a variational principle of Hamiltonian type [98]. The static part belongs to the zero entropy production, in this sense it is reversible. In our approach we get the "reversible" part as a specific case of the thermodynamic, irreversible thinking.

There are several alternate derivations based on different concepts [30, 50]. Here we have demonstrated that the physical assumptions to get the Ginzburg-Landau equation are very moderate. The typical Ginzburg-Landau structure is a straightforward consequence of the entropy inequality without any further additional set of concepts like the microforce balance or a variational principle.

3.3 Dual internal variables – Hamiltonian structure

There are two classical approaches to determine the evolution of internal variables.

When the evolution equations of internal variables are constructed exploiting the entropy inequality, using exclusively thermodynamic principles, then the corresponding variables are called *internal variables of state* [61]. This frame has the advantage of operating with familiar thermodynamic concepts (thermodynamic force, entropy), however, no inertial effects are considered. The thermodynamic theory of internal variables has a rich history (see the historical notes in [63]). A first more or less complete thermodynamic theory was suggested by Coleman and Gurtin [14], and a clear presentation of the general ideas of the theory was given by Muschik [67]. Internal variables of state were applied for several phenomena in different areas of physics, biology, and material sciences. A complete description of the thermodynamic theory with plenty of applications based on this concept of internal variables of state can be found in [104].

There is a second method that generates the kinetic relations through the Hamiltonian variational principle and suggests that inertial effects are unavoidable. This approach has a mechanical flavor, and the corresponding variables are called *internal degrees of freedom.* Dissipation is added by dissipation potentials. This theoretical frame has the advantage of operating with familiar mechanical concepts (force, energy). The method was suggested by Maugin [59], and it has also a large number of applications [19, 60]. A clear distinction between these two methods with a number of application areas is given by Maugin and Muschik [63, 64] and Maugin [61].

We call internal variables of state those physical field quantities – beyond the classical ones – whose evolution is determined by thermodynamical principles. We call internal degrees of freedom those physical quantities – beyond the classical ones – whose dynamics is determined by mechanicalprinciples.

One of the questions concerning this doubled theoretical frame is related to the joint application of variational principles and thermodynamics. Basic physical equations of thermodynamical origin do not have variational formulations, at least not without any further ado [98]. That is well reflected by the appearance of dissipation potentials as separate theoretical entities in variational models dealing with dissipation.

On the other hand, with pure thermodynamical methods – in the internal variables approach – inertial effects are not considered. Therefore, the coupling to simplest mechanical processes seemingly requires some additional assumptions; those are usually new principles of mechanical origin.

In the following we show that the mechanical structure arises from thermodynamic principles. Our suggestion requires dual internal variables and a particular generalization of the usual postulates of non-equilibrium thermodynamics: we do not require reciprocity relations. With dual internal variables we are able to get inertial effects and to reproduce the evolution of internal degrees of freedom. This would be impossible with a single internal variable. This is the price we pay for the generalization. In other words, instead of the doubling of the theoretical structure we suggest the doubling of the number of internal variables.

Let us consider a thermodynamic system where the state space is spanned by two scalar internal variables a and b. Then the evolution of these variables is determined by the following differential equations

$$\partial_t a + \hat{f} = 0, \tag{21}$$
$$\partial_t b + \hat{g} = 0. \tag{22}$$

The functions $\hat{f}$ and $\hat{g}$ are constitutive functions, restricted by the Second Law of Thermodynamics The entropy inequality, the main ingredient of the Second Law, can be written in the same form as previously in (2). The domain of the constitutive functions (our constitutive space) is spanned by the state space variables and by their first and second gradients. Therefore in this case

- the basic state space is spanned by (a, b),
- the constitutive state space is spanned by $(a, \partial_i a, \partial_{ij} a, b, \partial_i b, \partial_{ij} b)$,
- the constitutive functions are $\hat{s}, \hat{J}^i, \hat{f}$ and $\hat{g}$.

This is a weakly nonlocal constitutive space with second order weak nonlocality in both variables. The corresponding process direction space is spanned by the next derivatives $(\partial_t a, \partial_{ti} a, \partial_{tij} a, \partial_{ijk} a, \partial_t b, \partial_{ti} b, \partial_{tij} b, \partial_{ijk} b)$. The gradients of the evolution equations (21)-(22) are constraints for the entropy inequality in the framework of a second order constitutive state space for both of our variables

$$\partial_{ti} a + \partial_i \hat{f} = 0_i, \tag{23}$$
$$\partial_{ti} b + \partial_i \hat{g} = 0_i. \tag{24}$$

We introduce the Lagrange-Farkas multipliers $\hat{\lambda}_a, \hat{\lambda}_b$ for (21)–(22) and $\hat{\Lambda}^i_a, \hat{\Lambda}^i_b$ for (23)–(24), respectively. The Liu procedure results in the Ginzburg-Landau structure of the previous subsection in a doubled form

$$\begin{aligned}
0 \leq\ & \partial_t \hat{s} + \partial_i \hat{J}^i - \hat{\lambda}_a (\partial_t a + \hat{f}) - \hat{\Lambda}^i_a (\partial_{ti} a + \partial_i \hat{f}) - \\
& \qquad \hat{\lambda}_b (\partial_t b + \hat{g}) - \hat{\Lambda}^i_b (\partial_{ti} b + \partial_i \hat{g}) \\
=\ & \partial_a \hat{s}\, \partial_t a + \partial_{\partial_i a} \hat{s}\, \partial_{it} a + \partial_{\partial_{ij} a} \hat{s}\, \partial_{ijt} a + \partial_a \hat{J}^i\, \partial_i a + \partial_{\partial_j a} \hat{J}^i\, \partial_{ij} a + \partial_{\partial_{jk} a} \hat{J}^i\, \partial_{ijk} a - \\
& \hat{\lambda}_a (\partial_t a + \hat{f}) - \hat{\Lambda}^i_a \Big(\partial_{ti} a + \partial_a \hat{f}\, \partial_i a + \partial_{\partial_j a} \hat{f}\, \partial_{ij} a + \partial_{\partial_{jk} a} \hat{f}\, \partial_{ijk} a + \\
& \partial_b \hat{f}\, \partial_i b + \partial_{\partial_j b} \hat{f}\, \partial_{ij} b + \partial_{\partial_{jk} b} \hat{f}\, \partial_{ijk} b \Big) \\
& + \partial_b \hat{s}\, \partial_t b + \partial_{\partial_i b} \hat{s}\, \partial_{it} b + \partial_{\partial_{ij} b} \hat{s}\, \partial_{ijt} b + \partial_b \hat{J}^i\, \partial_i b + \partial_{\partial_j b} \hat{J}^i\, \partial_{ij} b + \partial_{\partial_{jk} b} \hat{J}^i\, \partial_{ijk} b - \\
& \hat{\lambda}_b (\partial_t b + \hat{g}) - \hat{\Lambda}^i_b \Big(\partial_{ti} b + \partial_b \hat{g}\, \partial_i b + \partial_{\partial_j b} \hat{g}\, \partial_{ij} b + \partial_{\partial_{jk} b} \hat{g}\, \partial_{ijk} b + \\
& \partial_a \hat{g}\, \partial_i a + \partial_{\partial_j a} \hat{g}\, \partial_{ij} a + \partial_{\partial_{jk} a} \hat{g}\, \partial_{ijk} a \Big).
\end{aligned} \tag{25}$$

After some rearrangements one can get

$$\begin{aligned}
0 \leq\ & (\partial_a \hat{s} - \hat{\lambda}_a)\partial_t a + (\partial_{\partial_i a}\hat{s} - \hat{\Lambda}_a^i)\partial_{it} a + \partial_{\partial_{ij} a}\hat{s}\, \partial_{ijt} a + \\
& (\partial_{\partial_{jk} a}\hat{J}^i - \hat{\Lambda}_a^i \partial_{\partial_{jk} a}\hat{f} - \hat{\Lambda}_b^i \partial_{\partial_{jk} a}\hat{g})\partial_{ijk} a - \\
& \hat{\lambda}_a \hat{f} - \hat{\Lambda}_a^i \left(\partial_a \hat{f}\, \partial_i a + \partial_{\partial_j a}\hat{f}\, \partial_{ij} a + \partial_b \hat{f}\, \partial_i b + \partial_{\partial_j b}\hat{f}\, \partial_{ij} b\right) + \\
& \partial_a \hat{J}^i\, \partial_i a + \partial_{\partial_j a}\hat{J}^i\, \partial_{ij} a \\
+\ & (\partial_b \hat{s} - \hat{\lambda}_b)\partial_t b + (\partial_{\partial_i b}\hat{s} - \hat{\Lambda}_b^i)\partial_{it} b + \partial_{\partial_{ij} b}\hat{s}\, \partial_{ijt} b + \\
& (\partial_{\partial_{jk} b}\hat{J}^i - \hat{\Lambda}_a^i \partial_{\partial_{jk} b}\hat{f} - \hat{\Lambda}_b^i \partial_{\partial_{jk} b}\hat{g}\,)\partial_{ijk} b - \\
& \hat{\lambda}_b \hat{g} - \hat{\Lambda}_b^i \left(\partial_b \hat{g}\, \partial_i b + \partial_{\partial_j b}\hat{g}\, \partial_{ij} b + \partial_a \hat{g}\, \partial_i a + \partial_{\partial_j a}\hat{g}\, \partial_{ij} a\right) + \\
& \partial_b \hat{J}^i\, \partial_i b + \partial_{\partial_j b}\hat{J}^i\, \partial_{ij} b.
\end{aligned}$$

Here the multipliers of the process direction space give the Liu equations:

$$\partial_t a : \partial_a \hat{s} = \hat{\lambda}_a, \tag{26}$$

$$\partial_{it} a : \partial_{\partial_i a}\hat{s} = \hat{\Lambda}_a^i, \tag{27}$$

$$\partial_t b : \partial_b \hat{s} = \hat{\lambda}_b, \tag{28}$$

$$\partial_{it} b : \partial_{\partial_i b}\hat{s} = \hat{\Lambda}_b^i, \tag{29}$$

$$\partial_{ijt} a : \partial_{\partial_{ij} a}\hat{s} = 0^{ij}, \tag{30}$$

$$\partial_{ijt} b : \partial_{\partial_{ij} b}\hat{s} = 0^{ij}, \tag{31}$$

$$\partial_{ijk} a : \partial_{\partial_{jk} a}\hat{J}^i = \hat{\Lambda}_a^i \partial_{\partial_{jk} a}\hat{f} + \hat{\Lambda}_b^i \partial_{\partial_{jk} a}\hat{g}, \tag{32}$$

$$\partial_{ijk} b : \partial_{\partial_{jk} b}\hat{J}^i = \hat{\Lambda}_b^i \partial_{\partial_{jk} b}\hat{g} + \hat{\Lambda}_a^i \partial_{\partial_{jk} b}\hat{f}. \tag{33}$$

The first four equations determine the Lagrange-Farkas multipliers by the entropy derivatives. The fifth and the sixth one show that the entropy is independent of the second gradient of a and b. Consequently, the Lagrange-Farkas multipliers $\hat{\Lambda}_a^i$ and $\hat{\Lambda}_b^i$ are independent of those derivatives, therefore the last two equations can be integrated and give

$$\hat{J}^i = \partial_{\partial_i a}\hat{s}\, \hat{f} + \partial_{\partial_i b}\hat{s}\, \hat{g} + \hat{\mathfrak{J}}^i(a, \partial_i a, b, \partial_i b). \tag{34}$$

Here the variables of the the residual entropy flux $\hat{\mathfrak{J}}^i$, an arbitrary constitutive function, are explicitly written. This is a complete solution of the system of Liu equations (26)–(33). Therefore the dissipation inequality simplifies to the following form

$$0 \leq \partial_i \hat{\mathfrak{J}}^i + \left(\partial_i(\partial_{\partial_i a}\hat{s}) - \partial_a \hat{s}\right) \hat{f} + \left(\partial_i(\partial_{\partial_i b}\hat{s}) - \partial_b \hat{s}\right) \hat{g}. \tag{35}$$

Now assuming that the residual entropy flux is zero, $\hat{\mathfrak{J}}^i \equiv 0^i$, the entropy inequality reduces to a flux-force system of the following form:

$$\text{a-force: } \hat{A} = \partial_i(\partial_{\partial_i a}\hat{s}) - \partial_a \hat{s}, \qquad \text{a-flux: } \hat{f},$$

$$\text{b-force: } \hat{B} = \partial_i(\partial_{\partial_i b}\hat{s}) - \partial_b \hat{s}, \qquad \text{b-flux: } \hat{g}.$$

The classical solution of the entropy inequality gives a coupling of fluxes and forces as a system of coupled Ginzburg-Landau equations:

$$\partial_t a = \hat{f} = \hat{l}_1 \left(\partial_i(\partial_{\partial_i a}\hat{s}) - \partial_a \hat{s}\right) + \hat{l}_{12} \left(\partial_i(\partial_{\partial_i b}\hat{s}) - \partial_b \hat{s}\right), \tag{36}$$

$$\partial_t a = \hat{g} = \hat{l}_{21} \left(\partial_i(\partial_{\partial_i a}\hat{s}) - \partial_a \hat{s}\right) + \hat{l}_2 \left(\partial_i(\partial_{\partial_i b}\hat{s}) - \partial_b \hat{s}\right). \tag{37}$$

The constitutive Onsagerian coefficients $\hat{l}_1, \hat{l}_2, \hat{l}_{12}, \hat{l}_{21}$ are restricted by the Second Law. For the sake of generality we do not assume any kind of reciprocity here. We decouple the symmetric and antisymmetric parts introducing $\hat{l} = (\hat{l}_{12} + \hat{l}_{21})/2$ and $\hat{k} = (\hat{l}_{12} - \hat{l}_{21})/2$. The entropy production is nonnegative if

$$\hat{l}_1 > 0, \quad \hat{l}_2 > 0 \quad \text{and} \quad \hat{l}_1\hat{l}_2 - \hat{l}^2 \geq 0. \tag{38}$$

Now the evolution equations (36)–(37) can be written equivalently as

$$\partial_t a = \hat{k}\hat{B} + \hat{l}_1\hat{A} + \hat{l}\hat{B}, \tag{39}$$

$$\partial_t b = -\hat{k}\hat{A} + \hat{l}\hat{A} + \hat{l}_2\hat{B}. \tag{40}$$

3.3.1 Remark on dissipation potentials

We may introduce dissipation potentials for the dissipative part of the equations if the condition of their existence is satisfied. Dissipation potentials are the children of variational principles, they are artificially added to a set of reversible equations of variational origin to generate some kind of dissipative effects. In our case there is no need for this assumption, our construction gives the most general dissipative system without any further ado. Moreover, here it is clear what belongs to the dissipative part and what belongs to the nondissipative part of the evolution equations. The terms with the symmetric conductivity contribute to the entropy production, and the terms from the skew symmetric part do not. On the other hand, there is no need of potential construction, as we are not looking for a variational formulation. Moreover, the symmetry relations are not sufficient for the existence of dissipation potentials in general. In case of constant coefficients (strict linearity), the dissipation potentials always exist for the dissipative (symmetric) part.

3.3.2 Remark on the reciprocity relations

The reciprocity relations are the main results of the great idea of Lars Onsager connecting fluctuation theory to macroscopic thermodynamics [7, 48, 74, 75, 76]. As it was written by Onsager himself on the validity of his result: "The restriction was stated: on a kinetic model, the thermodynamic variables must be algebraic sums of (a large number of) molecular variables, and must be *even* functions of those molecular variables which are odd functions of time (like molecular velocities)" [48]. The Casimir reciprocity relations are based on microscopic fluctuations, too [7]. We do not have such a microscopic background for most of internal variables. E. g., in the case of damage, the internal variables are reflecting a structural disorder on a mesoscopic scale. The relation between thermodynamic variables and the microscopic structure is hopelessly complicated. On the other hand, the Onsagerian reciprocity is based on time reversal properties of corresponding physical quantities either at the macro or at the micro level. Looking for the form of evolution equations without a microscopic model, we do not have any information on the time reversal properties of our physical quantities neither at the micro- nor at the macroscopic level. Therefore, we can conclude that lacking the conditions of the Onsagerian or Casimirian reciprocity gives no reasons to assume their validity in the internal variable theory.

Let us observe the correspondence of evolution equations for internal variables with the reciprocity relations by means of a few simple examples.

3.3.3 Example 1: internal variables

Let us consider materials with diagonal conductivity matrix **L** ($\hat{l} = 0$, $\hat{k} = 0$). It is clear that the Onsagerian reciprocity relations are satisfied, and we return to the classical situation with fully uncoupled internal variables:

$$\partial_t a = \hat{l}_1 \hat{A},$$
$$\partial_t b = \hat{l}_2 \hat{B}.$$

In this case the evolution equations for dual internal variables a and b are the same as in the case of single internal variable.

3.3.4 Example 2: internal degrees of freedom

We now assume that all conductivity coefficients are constant, and their values are $l_1 = l = 0, k = 1$. This means that $l_{12} = -l_{21}$, i.e., the Casimirian reciprocity relations are satisfied.

For simplicity, we consider a specific decomposition of the entropy density into two parts, which depend on different internal variables

$$\hat{s}(a, \partial_i a, b, \partial_i b) = -K(b) - W(a, \partial_i a).$$

The negative signs are introduced taking into account the concavity of the entropy. Then the thermodynamic forces are represented as

$$\hat{A} = \partial_a W - \partial_i(\partial_{\partial_i a} W), \qquad \hat{B} = d_b K = K'(b),$$

and Eqs. (39)–(40) are simplified to

$$\partial_t a = \hat{B} = K'(b), \tag{41}$$

$$\partial_t b = -\hat{A} + l_2\hat{B} = -\partial_a W + \partial_i(\partial_{\partial_i a} W) + l_2 K'(b). \tag{42}$$

One may recognize that the obtained system of equations corresponds exactly to a Hamiltonian form with the last term of (42) as added dissipation. Here the entropy density $\hat{s}$ plays the role of a Hamiltonian density, a is a Lagrangian variable and b corresponds to the (slightly generalized) conjugated moment. The transformation into a Lagrangian form is trivial if K is quadratic $K(b) = b^2/2m$, where m is a constant. Then the whole system corresponds to the general structure of internal degrees of freedom, as one can compare to Eq. (5.14) in [63] with the Lagrangian

$$\mathfrak{L}(\partial_t a, a, \partial_i a) = m\frac{(\partial_t a)^2}{2} - W(a, \partial_i a), \quad \text{and} \quad D(a, \partial_i a) = \frac{ml_2}{2}(\partial_t a)^2$$

as dissipation potential. Moreover, the entropy flux density (34) in case of our special conditions can be written as

$$J^i = -\partial_{\partial_i a} s K'(b) + \mathfrak{J}_0^i,$$

and one can infer that natural boundary conditions of the variational principle related to the above Lagrangian correspond to the condition of vanishing entropy flow at the boundary, with $\mathfrak{J}_0^i \equiv 0^i$.

Therefore, the variational structure of internal degrees of freedom is recovered in the pure thermodynamic framework. The thermodynamic structure resulted in several sign restrictions of the coefficients, and the form of the entropy flux is also recovered. The natural boundary conditions correspond to a vanishing extra entropy flux.

3.3.5 Example 3: diffusive internal variables [16, 20, 62]

Now we give an additional example to see clearly the reduction of evolution equations of internal degrees of freedom to evolution equations for internal variables and the extension of the latter to the previous one.

We keep the values of conductivity coefficients (i.e., $l_1 = l = 0, k = 1$), but assume that both K and W are quadratic functions

$$K(b) = \frac{\beta}{2}b^2, \quad W(a, \partial_i a) = \frac{\alpha}{2}(\partial_i a)^2,$$

where α and β are positive constants according to the concavity requirement.

In this case, the evolution equations (41)–(42) reduce to

$$\partial_t a = \hat{B} = K'(b) = \beta b, \tag{43}$$

$$\partial_t b = -\partial_a W + \partial_i(\partial_{\partial_i a} W) + l_2 K'(b) = \alpha \partial_{ii} a + l_2 \beta b. \tag{44}$$

Putting b from Eq. (43) into Eq. (44), we have

$$\frac{1}{\alpha\beta}\partial_{tt}a - \frac{l_2}{\alpha}\partial_t a = \partial_{ii}a, \tag{45}$$

which is a Cattaneo-Vernotte type hyperbolic equation (telegraph equation) for the internal variable a. This can be considered as an extension of a diffusion equation by an inertial term or as an extension of a damped Newtonian equation (without forces) by a diffusion term.

3.3.6 Discussion

In the framework of the thermodynamic theory with dual weakly nonlocal internal variables we are able to recover the evolution equations for internal degrees of freedom.

We have seen that the form of the evolution equations depends on the mutual interrelations between the two internal variables. In the special case of internal degrees of freedom, the evolution of one variable is driven by the second one, and vice versa. This can be viewed as a duality between the two internal variables. In the case of pure internal variables of state, this duality is replaced by self-driven evolution for each internal variable. The general case includes all intermediate situations.

It is generally accepted that internal variables are "measurable but not controllable" (see e.g. [40]). Controllability can be achieved by boundary conditions or fields directly acting on the physical quantities. We have seen how natural boundary conditions arise considering nonlocality of the interactions through weakly nonlocal constitutive state spaces.

As we wanted to focus on generic inertial effects, our treatment is simplified from several points of view. Vectorial and tensorial internal variables were not considered and the couplings to traditional continuum fields result in degeneracies and more complicated situations than in our simple examples.

It is important to remark that skew symmetric couplings are not always related directly to inertial effects and indicate two directions, one into mechanics and one into thermodynamics, where our method can be generalized. Let us mention here the related pioneering works of Verhás, where skew symmetric conductivity equations appear in different inspiring contexts [103, 105].

Finally, let us mention that the idea of constructing a unified theoretical frame for reversible and irreversible dynamics has a long tradition. The corresponding research was not restricted to the case of internal variables and was looking for a classical Hamiltonian or a generalized variational principle that would be valid for both dissipative and nondissipative evolution equations (see, e.g., [25, 31, 51, 99] and the references therein).

4 Classical Irreversible Thermodynamics

In this section we investigate the evolution of extensive physical field quantities whose equation of motion is a balance, and we will see that for first order weakly nonlocal state spaces the thermodynamic force-flux system is a consequence of the nonnegative entropy production. A treatment of Extended Irreversible Thermodynamics and further details can be found in [87].

A balance type evolution equation for a conserved quantity $\mathbf{a}$ can be given in a general form as

$$\partial_t \mathbf{a} + \partial_i \hat{\mathbf{j}}^i = 0. \tag{46}$$

Here $i \in \{1,2,3\}$ denotes the spatial coordinates. The conserved quantity can be a Descartes product of densities of extensives of any tensorial order, e.g., $\mathbf{a} = (\rho, e, p^j, ...)$, where ρ is the mass density, e is the internal energy density and p^j is the momentum density. Our continuum is at rest, $\hat{\mathbf{j}}^i$ are the corresponding fluxes, e.g., $\mathbf{j}^i = (\rho v^i, \rho e v^i + j_e^i, \rho v^j v^i + P^{ji}, ...)$, where the first term is the current of the mass, the second is the convective and conductive current of internal energy and the third is the current of momentum. Therefore with this convenient notation we introduce only those free indices that are important from the point of view of the balance structure of the evolution equation. For the balances of particular real physical quantities there are peculiarities that we do not introduce here. For example for a continuum at rest traditionally there is no mass flux as a consequence of barycentric velocities, the source terms can play an important role (for internal energy, chemical production, etc.). Those peculiarities introduce further constraints and additional terms in the final equations that we do not consider in this general calculation to show the core structure of classical irreversible thermodynamics. Moreover, in this case we restrict ourselves to a continuum at rest. Some consequences of the motion of the continua is considered and investigated in the next section. We assume here a first order weakly nonlocal state space.

Therefore in this case

- the basic state space is spanned by $\mathbf{a}$,
- the constitutive state space is spanned by $(\mathbf{a}, \partial_i \mathbf{a})$,
- the constitutive functions are $\hat{s}, \hat{J}^i$ and $\hat{\mathbf{j}}^i$.

The above evolution equation is not completely arbitrary, but restricted by the Second Law of Thermodynamics (2). Let us introduce a Lagrange-Farkas multiplier λ for the evolution equation (46) and apply the Liu procedure to determine the form of the constitutive functions, required by the entropy inequality:

$$
\begin{aligned}
0 &\le \partial_t \hat{s}(\mathbf{a}, \partial_i \mathbf{a}) + \partial_i \hat{J}^i(\mathbf{a}, \partial_i \mathbf{a}) - \hat{\lambda}\left(\partial_t \mathbf{a} + \partial_i \hat{\mathbf{j}}^i(\mathbf{a}, \partial_i \mathbf{a})\right) \\
&= \partial_{\mathbf{a}} \hat{s}\, \partial_t \mathbf{a} + \partial_{\partial_i \mathbf{a}} \hat{s}\, \partial_{it} \mathbf{a} + \partial_{\mathbf{a}} \hat{J}^i\, \partial_i \mathbf{a} + \partial_{\partial_j \mathbf{a}} \hat{J}^i\, \partial_{ij} \mathbf{a} \\
&\quad - \hat{\lambda}\left(\partial_t \mathbf{a} + \partial_{\mathbf{a}} \hat{\mathbf{j}}^i \partial_i \mathbf{a} + \partial_{\partial_j \mathbf{a}} \hat{\mathbf{j}}^i \partial_{ij} \mathbf{a}\right) \\
&= \left(\partial_{\mathbf{a}} \hat{s} - \hat{\lambda}\right) \underline{\partial_t \mathbf{a}} + \partial_{\partial_i \mathbf{a}} \hat{s}\, \underline{\partial_{it} \mathbf{a}} + \left(\partial_{\mathbf{a}} \hat{J}^i - \hat{\lambda} \partial_{\mathbf{a}} \hat{\mathbf{j}}\right) \partial_i \mathbf{a} \\
&\quad + \left(\partial_{\partial_j \mathbf{a}} \hat{J}^i - \hat{\lambda} \partial_{\partial_j \mathbf{a}} \hat{\mathbf{j}}^i\right) \underline{\partial_{ij} \mathbf{a}}.
\end{aligned} \tag{47}
$$

The underlined partial derivatives of the basic field are not in the constitutive space, they are independent algebraic quantities and span the process direction space. The multipliers of those terms are zero according to Liu's theorem and give the Liu equations:

$$\partial_t \mathbf{a} : \partial_{\mathbf{a}} \hat{s} = \hat{\lambda}, \tag{48}$$

$$\partial_{it} \mathbf{a} : \partial_{\partial_i \mathbf{a}} \hat{s} = 0^i, \tag{49}$$

$$\partial_{ij} \mathbf{a} : \partial_{\partial_j \mathbf{a}} \hat{J}^i = \hat{\lambda} \partial_{\partial_j \mathbf{a}} \hat{\mathbf{j}}^i. \tag{50}$$

The first equation determines the Lagrange-Farkas multiplier and the second one shows that the entropy should be local, independent of the gradient of **a**. Therefore one can solve the third equation as

$$\hat{J}^i(\mathbf{a}, \partial_i \mathbf{a}) = \partial_{\mathbf{a}} s(\mathbf{a}) \cdot \hat{\mathbf{j}}^i(\mathbf{a}, \partial_i \mathbf{a}) + \hat{\mathfrak{J}}^i(\mathbf{a}), \tag{51}$$

where we have denoted the variables of the constitutive functions and we have introduced the local residual entropy flux $\hat{\mathfrak{J}}^i$. Several authors suggest this kind of additive supplement to the classical entropy flux (e.g. the **K** vector of Müller [70]). This is a complete solution of the system (48)–(50). The dissipation inequality is

$$0 \le \partial_i \hat{\mathfrak{J}}^i + \hat{\mathbf{j}}^i \partial_i (\partial_{\mathbf{a}} \hat{s}). \tag{52}$$

Assuming that the residual entropy flux is zero, $\hat{\mathfrak{J}}^i \equiv 0^i$, the entropy inequality reduces to the usual flux-force system of Classical Irreversible Thermodynamics, where the thermodynamic forces are the gradients of the intensives and the thermodynamic fluxes are identical to the fluxes of the extensives from the balances.

$$\text{Flux: } \hat{\mathbf{j}}^i, \qquad \text{Force: } \partial_i(\partial_{\mathbf{a}} \hat{s}).$$

The classical solution of the above inequality is that the fluxes are proportional to the forces:

$$\hat{\mathbf{j}}^i = \hat{\mathbf{L}}^{ik} \partial_k (\partial_{\mathbf{a}} \hat{s}), \tag{53}$$

where $\hat{\mathbf{L}}^{ik}$ is symmetric and positive definite. Therefore balances of extensives are reduced to transport equations of the form:

$$\partial_t \mathbf{a} + \partial_i \left(\mathbf{L}^{ij} \partial_j (\partial_{\mathbf{a}} \hat{s})\right) = 0. \tag{54}$$

We have seen that in our calculations the classical form of the entropy production (the first term in the inequality above) *and* the classical form of the entropy flux were consequences of the Second Law, with the assumption of first order nonlocality. A rigorous treatment showed that the seemingly intuitive steps of irreversible thermodynamic modelling are well supported and explained by the Liu procedure. We can see that the second part of the usual phenomenological formulation of the local equilibrium hypothesis [25] is a consequence of the assumption of first order nonlocality: the state functions are those in equilibrium due to the local entropy.

5 One component fluids – second order nonlocal in the density

Up to know we have investigated evolution equations of internal variables and the general structure of Classical Irreversible Thermodynamics in a somewhat abstract manner. In this section we analyze a more particular and less abstract example where the physical meaning of the variables in the basic state space is well known. Our example is a one component heat conducting fluid where the constitutive state space is first order for the energy and the velocity fields, but second order in the density. We have seen that in case of first order weakly nonlocal state spaces our method gives the well known classical structure in a somewhat simplified manner, where the number of independent assumptions are reduced. For example, the form of the entropy flux is calculated and not postulated, as we have demonstrated in the previous section. However, the second order nonlocality in the density variables leads to a surprising result and gives the viable family of Korteweg fluids, those that are compatible with the Second Law. One can find more details in [96, 97] and with alternate methods in [17, 66].

5.1 Fluid mechanics in general

The *basic state space* of one-component fluid mechanics is spanned by the mass density ρ, the velocity v^i and the energy density e of the fluid. Hydrodynamics is based on the balance of mass, energy and momentum [31]. In classical fluid mechanics the constitutive space, the domain of the constitutive functions, is spanned by the basic state space (ρ, v^i, e) and the gradient of the velocity $\partial_i v^j$ and the temperature $\partial_i T$. The pressure/stress tensor and the flux of the internal energy are the constitutive quantities in the theory.

The balance of mass can be written as

$$\partial_t \rho + \partial_i(\rho v^i) = 0. \tag{55}$$

There is only a convective flux for mass. The balance of momentum, i.e., the Cauchy equation, is

$$\partial_t(\rho v^i) + \partial_j\left(\hat{P}^{ij} + \rho v^i v^j\right) = 0^i, \tag{56}$$

where ρv^i is the momentum density and $\hat{P}^{ij}$ is the pressure tensor, the conductive flux of the momentum.

The balance of energy is given as

$$\partial_t e + \partial_i(\hat{q}^i + ev^i) = 0. \tag{57}$$

Here e is the total energy density and $\hat{q}^i$ is the conductive flux of the energy. The Second Law requires that the production of the entropy is nonnegative in insulated and source-free systems, therefore source terms are not considered in the balances. Hence in our case there is no mass production, the external forces are absent and the energy is conserved. We do not need the concept of the internal energy yet, first we analyse the consequences of the Second Law with the balance of the total energy. As we are dealing with a fluid, it is convenient to separate the conductive and convective entropy fluxes. Hence (2) will be written as

$$\partial_t\hat{s} + \partial_i(\hat{J}^i + \hat{s}v^i) \geq 0. \tag{58}$$

Therefore in this case

- the basic state space is spanned by (ρ, v^i, e),
- the constitutive state space is spanned by $(\rho, \partial_i\rho, \partial_{ij}\rho, v^i, \partial_i v^j, e, \partial_i e)$,
- the constitutive functions are $\hat{s}, \hat{J}^i, \hat{q}^i$ and $\hat{P}^{ij}$.

It is somewhat convenient to introduce the relative velocity v^i as basic state variable (instead of the momentum) and the gradient of the energy as constitutive state variable (instead of the temperature gradient). As we have a second order weakly nonlocal extension in the mass density, we need the gradient of (55) as a further constraint in the entropy inequality

$$\partial_{it}\rho + \partial_{ij}(\rho v^j) = 0_i. \tag{59}$$

We introduce the Lagrange-Farkas multipliers $\hat{\lambda}, \hat{\Lambda}^i, \hat{\Gamma}_i, \hat{\gamma}$ for the balances (55), (59), (56), and (57), respectively.

Now we apply the Liu procedure, with the method of Lagrange-Farkas multipliers as in the previous sections

$$\begin{aligned}0 \leq\ & \partial_t\hat{s} + \partial_i\hat{J}^i + \partial_i(\hat{s}v^i) - \hat{\lambda}\left(\partial_t\rho + \partial_i(\rho v^i)\right) - \hat{\Lambda}^i\left(\partial_{it}\rho + \partial_{ij}(\rho v^j)\right) - \\ & \hat{\Gamma}_i\left(\partial_t(\rho v^i) + \partial_j\left(\hat{P}^{ij} + \rho v^i v^j\right)\right) - \hat{\gamma}\left(\partial_t e + \partial_i(\hat{q}^i + ev^i)\right).\end{aligned} \tag{60}$$

Developing the partial derivatives of the constitutive functions gives:

$$
\begin{aligned}
&\partial_\rho \hat{s}\partial_t\rho + \partial_{\partial_i\rho}\hat{s}\partial_{ti}\rho + \partial_{\partial_{ij}\rho}\hat{s}\partial_{tij}\rho + \partial_{v^i}\hat{s}\partial_t v^i + \partial_{\partial_i v^j}\hat{s}\partial_{tj}v^i + \partial_e\hat{s}\partial_t e + \\
&\partial_{\partial_i e}\hat{s}\partial_{ti}e + \partial_\rho\hat{J}^i\partial_i\rho + \partial_{\partial_j\rho}\hat{J}^i\partial_{ij}\rho + \partial_{\partial_{jk}\rho}\hat{J}^i\partial_{ijk}\rho + \partial_{v^j}\hat{J}^i\partial_i v^j + \partial_{\partial_k v^j}\hat{J}^i\partial_{jk}v^i + \\
&\partial_e\hat{J}^i\partial_i e + \partial_{\partial_j e}\hat{J}^i\partial_{ij}e + \\
&\partial_i(\hat{s}v^i) - \hat{\lambda}\left(\partial_t\rho + \rho\partial_i v^i + v^i\partial_i\rho\right) - \\
&\hat{\Lambda}^i\left(\partial_{it}\rho + \rho\partial_{ij}v^j + v^j\partial_{ij}\rho + \partial_i\rho\partial_j v^j + \partial_j\rho\partial_i v^j\right) - \\
&\hat{\Gamma}_i\left(\rho\partial_t v^i + v^i\partial_t\rho + \rho v^i\partial_j v^j + \rho v^j\partial_j v^i + v^j v^i\partial_j\rho + \partial_\rho\hat{P}^{ij}\partial_j\rho + \partial_{\partial_k\rho}\hat{P}^{ij}\partial_{jk}\rho + \right. \\
&\left.\partial_{\partial_{kl}\rho}\hat{P}^{ij}\partial_{jkl}\rho + \partial_{v^k}\hat{P}^{ij}\partial_j v^k + \partial_{\partial_l v^k}\hat{P}^{ij}\partial_{jl}v^k + \partial_e\hat{P}^{ij}\partial_j e + \partial_{\partial_k e}\hat{P}^{ij}\partial_{jk}e\right) - \\
&\hat{\gamma}\left(\partial_t e + e\partial_i v^i + v^i\partial_i e + \partial_\rho\hat{q}^i\partial_i\rho + \partial_{\partial_j\rho}\hat{q}^i\partial_{ji}\rho + \right. \\
&\left.\partial_{\partial_{jk}\rho}\hat{q}^i\partial_{ijk}\rho + \partial_{v^j}\hat{q}^i\partial_i v^j + \partial_{\partial_k v^j}\hat{q}^i\partial_{ik}v^j + \partial_e\hat{q}^i\partial_i e + \partial_{\partial_j e}\hat{q}^i\partial_{ji}e\right) \geq 0.
\end{aligned}
$$

A little rearrangement of the terms results in

$$
\begin{aligned}
&(\partial_\rho\hat{s} - \hat{\lambda} - \hat{\Gamma}_i v^i)\partial_t\rho + (\partial_{\partial_i\rho}\hat{s} - \hat{\Lambda}^i)\partial_{ti}\rho + \partial_{\partial_{ij}\rho}\hat{s}\partial_{tij}\rho + (\partial_{v^i}\hat{s} - \rho\hat{\Gamma}_i)\partial_t v^i + \\
&\partial_{\partial_i v^j}\hat{s}\partial_{tj}v^i + (\partial_e\hat{s} - \hat{\gamma})\partial_t e + \partial_{\partial_i e}\hat{s}\partial_{ti}e + \\
&(\partial_{\partial_{jk}\rho}\hat{J}^i - \hat{\Gamma}_l\partial_{\partial_{jk}\rho}\hat{P}^{li} - \partial_{\partial_{jk}\rho}\hat{q}^i)\partial_{ijk}\rho + \\
&(\partial_{\partial_k v^j}\hat{J}^i - \hat{\Gamma}_l\partial_{\partial_k v^i}\hat{P}^{lj} - \hat{\gamma}\partial_{\partial_k v^i}\hat{q}^j\partial_{ik}v^j)\partial_{jk}v^i - \hat{\Lambda}^i\rho\partial_{ij}v^j + \\
&(\partial_{\partial_j e}\hat{J}^i - \hat{\Gamma}_l\partial_{\partial_i e}\hat{P}^{lj} + \hat{\gamma}\partial_{\partial_j e}\hat{q}^i)\partial_{ij}e + \\
&\partial_\rho\hat{J}^i\partial_i\rho + \partial_{\partial_j\rho}\hat{J}^i\partial_{ij}\rho + \partial_{v^j}\hat{J}^i\partial_i v^j + \partial_e\hat{J}^i\partial_i e + \\
&\partial_i(\hat{s}v^i) - \hat{\lambda}\left(\rho\partial_i v^i + v^i\partial_i\rho\right) - \\
&\hat{\Lambda}^i\left(\rho\partial_{ij}v^j + v^j\partial_{ij}\rho + \partial_i\rho\partial_j v^j + \partial_j\rho\partial_i v^j\right) - \\
&\hat{\Gamma}_i\left(\rho v^i\partial_j v^j + \rho v^j\partial_j v^i + v^j v^i\partial_j\rho + \partial_\rho\hat{P}^{ij}\partial_j\rho + \partial_{\partial_k\rho}\hat{P}^{ij}\partial_{jk}\rho + \right. \\
&\left.\partial_{v^k}\hat{P}^{ij}\partial_j v^k + \partial_e\hat{P}^{ij}\partial_j e\right) - \\
&\hat{\gamma}\left(e\partial_i v^i + v^i\partial_i e + \partial_\rho\hat{q}^i\partial_i\rho + \partial_{\partial_j\rho}\hat{q}^i\partial_{ji}\rho + \partial_{v^j}\hat{q}^i\partial_i v^j + \partial_e\hat{q}^i\partial_i e\right) \geq 0.
\end{aligned}
$$

Then the Liu-equations follow as the multipliers of the members of the process direction space, the derivatives that are out of the constitutive space:

$$\partial_t\rho \; : \partial_\rho \hat{s} - \hat{\lambda} - \hat{\Gamma}_i v^i = 0, \tag{61}$$

$$\partial_{ti}\rho \; : \partial_{\partial_i\rho} \hat{s} - \hat{\Lambda}^i = 0^i, \tag{62}$$

$$\partial_{tij}\rho \; : \partial_{\partial_{ij}\rho} \hat{s} = 0^{ij}, \tag{63}$$

$$\partial_t v^i \; : \partial_{v^i} \hat{s} - \rho\hat{\Gamma}_i = 0_i, \tag{64}$$

$$\partial_{tj} v^i \; : \partial_{\partial_j v^i} \hat{s} = 0_i^j, \tag{65}$$

$$\partial_t e \; : \partial_e \hat{s} - \hat{\gamma} = 0, \tag{66}$$

$$\partial_{ti} e \; : \partial_{\partial_i e} \hat{s} = 0^i, \tag{67}$$

$$\partial_{ijk}\rho \; : \partial_{\partial_{kj}\rho} \hat{J}^i = \hat{\Gamma}_l \partial_{\partial_{kj}\rho} \hat{P}^{li} + \hat{\gamma}\partial_{\partial_{kj}\rho} \hat{q}^i, \tag{68}$$

$$\partial_{jk} v^i \; : \partial_{\partial_k v^i} \hat{J}^j = \hat{\Gamma}_l \partial_{\partial_k v^i} \hat{P}^{lj} + \hat{\gamma}\partial_{\partial_k v^i} \hat{q}_j + \frac{\rho}{2}\hat{\Lambda}^l(\delta_l^j\delta_i^k - \delta_l^k\delta_i^j), \tag{69}$$

$$\partial_{ij} e \; : \partial_{\partial_j e} \hat{J}^i = \hat{\Gamma}_l \partial_{\partial_i e} \hat{P}^{lj} + \hat{\gamma}\partial_{\partial_j e} \hat{q}^i. \tag{70}$$

As a consequence of (63), (65), and (67), the entropy density does not depend on $\partial_{ij}\rho$, $\partial_j v^i$ and $\partial_i e$. (61), (62), (64), and (66) give the Lagrange-Farkas multipliers in terms of the entropy derivatives. Therefore, from a thermodynamic point of view, the Lagrange-Farkas multipliers are the normal and generalized intensive variables [41]. Now, one can give a solution of (68)–(70) as

$$\hat{J}^i = \partial_e \hat{s}\hat{q}^i + \frac{\rho}{2}\left(\partial_{\partial_i\rho}\hat{s}\partial_j v^j + \partial_{\partial_j\rho}\hat{s}\partial_j v^i\right) + \frac{1}{\rho}\partial_{v^j}\hat{s}\hat{P}^{ji} + \hat{\mathfrak{J}}^i(\rho, \partial_i\rho, v^i, e), \tag{71}$$

where the residual entropy flux $\hat{\mathfrak{J}}^i$ is an arbitrary function. Thus Liu's equations can be solved and yield the Lagrange-Farkas multipliers as well as restrictions for the entropy and the entropy flux. Applying these solutions of the Liu equations, the dissipation inequality can be simplified to the following form

$$0 \le \sigma_s = \partial_i \hat{\mathfrak{J}}^i + \hat{q}^i \partial_i(\partial_e(\rho\mathfrak{s})) + \hat{P}^{ij}\partial_i(\partial_{v^j}\mathfrak{s}) +$$
$$\partial_j v^j\left(\hat{s} + e\partial_e\hat{s} - \rho\partial_\rho\hat{s} + \frac{\rho^2}{2}\partial_i\left(\partial_{\partial_i\rho}\mathfrak{s}\right)\right) + \partial_j v^i\left(\frac{\rho^2}{2}\partial_i\left(\partial_{\partial_j\rho}\mathfrak{s}\right)\right).$$

Here we have introduced the specific entropy as $\mathfrak{s} := \hat{s}/\rho$. It is interesting to give this inequality, the entropy production in a more traditional form, without indices, too:

$$\nabla\cdot\mathbf{J}_0 + \mathbf{q}\cdot\nabla\partial_e(\rho\mathfrak{s}) +$$
$$\nabla\partial_{\mathbf{v}}\mathfrak{s} : \mathbf{P} + \left[\frac{\rho^2}{2}\left(\nabla\cdot\partial_{\nabla\rho}\mathfrak{s}\mathbf{I} + \nabla\partial_{\nabla\rho}\mathfrak{s}\right) + (\hat{s} + e\partial_e\hat{s} - \rho\partial_\rho\hat{s})\mathbf{I}\right] : \nabla\mathbf{v} \ge 0.$$

Here $\mathbf{I}$ denotes the second order unit tensor δ_{ij}, ∇ is the gradient and $\nabla\cdot$ is the divergence of the corresponding field quantity. To get the traditional form of the equation we should introduce the internal energy of the fluid as the difference of the total and kinetic energies, and assume that the entropy function does not depend on the total and kinetic energies independently, but only on the internal energy $\hat{s} = \hat{s}(\rho, \partial_i\rho, e - \rho v^2/2)$. Moreover, we want to define the entropy as an extensive quantity, therefore we require that the specific entropy depends on the specific internal energy ε:

$$\hat{s}(\rho, \partial_i\rho, v^i, e) = \rho\mathfrak{s}\left(\rho, \frac{e}{\rho} - \frac{v^2}{2}, \partial_i\rho\right). \tag{72}$$

From this form of the entropy function we get the Gibbs relation

$$d\varepsilon = Td\mathfrak{s} + \frac{p}{\rho^2}d\rho - A^i d\partial_i\rho,$$

where $\varepsilon = e/\rho - v^2/2$ is the specific internal energy. The temperature and pressure are defined by the customary partial derivatives of the entropy. The temperature can be connected both to the derivative by the total energy and the internal energy, because the corresponding derivatives are equal. The quantity A^i is defined as the partial derivative of the entropy by the density gradient, like the traditional intensives:

$$\partial_e(\rho\mathfrak{s}) = \frac{1}{T}, \quad \partial_{v^j}\mathfrak{s} = -\frac{v^j}{T}, \quad \partial_\rho\mathfrak{s} = -\frac{p}{T\rho^2}, \quad \partial_{\partial_i\rho}\mathfrak{s} = \frac{A^i}{T}.$$

With these quantities we can write the dissipation inequality and the entropy flux as

$$0 \le \sigma_s = \partial_i\hat{\mathfrak{J}}^i + (\hat{q}^i - v_j\hat{P}^{ij})\partial_i\frac{1}{T} -$$
$$\frac{1}{T}\left(\hat{P}^{ij} - \left(p + \frac{T\rho^2}{2}\partial_k\left(\partial_{\partial_k\rho}\mathfrak{s}\right)\right)\delta^i_j - \frac{T\rho^2}{2}\partial_j\left(\partial_{\partial_i\rho}\mathfrak{s}\right)\right)\partial_i v^j,$$

and

$$\hat{J}^i = (\hat{q}^i - v_j\hat{P}^{ji})\frac{1}{T} + \frac{\rho}{2}\left(\partial_{\partial_i\rho}\hat{s}\partial_j v^j + \partial_{\partial_k\rho}\hat{s}\partial_k v^i\right) + \hat{\mathfrak{J}}^i. \tag{73}$$

We can see that a flux of the internal energy is introduced as in the case of the traditional, first order weakly nonlocal theories. However, the appearance of temperature inside the expression of the viscous pressure indicates the possibility of alternate, better definitions.

Now we change the notation to a coordinate invariant one, introducing the nabla operator for the space derivatives, as it is customary in fluid mechanics. In this case, e.g., $\partial_i v^i = \nabla \cdot \mathbf{v}$ and $\partial_i v^j = \nabla\mathbf{v}$. In the pure mechanical, reversible case our thermodynamic force for mechanical interactions is zero. Therefore we introduce the *nonlocal reversible pressure* as

$$\hat{\mathbf{P}}^r = \frac{T\rho^2}{2}\left[\left(\nabla\cdot\partial_{\nabla\rho}\mathfrak{s} - 2\partial_\rho\mathfrak{s}\right)\mathbf{I} + \nabla\partial_{\nabla\rho}\mathfrak{s}\right]. \tag{74}$$

If the pressure is equal to the reversible pressure, there is no dissipation, the theory is reversible (conservative). In the case of a local entropy (independent of the gradient of the density) we obtain

$$\hat{\mathbf{P}}^{\mathrm{r}}_{Euler}(\rho) := -T\rho^2\partial_\rho \mathfrak{s}(\rho)\mathbf{I}, \tag{75}$$

therefore, the corresponding equations are of the ideal Euler fluid, where $p(\rho) = -T\rho^2\partial_\rho \mathfrak{s}(\rho)$ is the scalar pressure function. Introducing the viscous pressure $\mathbf{P}^{\nu}$ as usual, we can solve the dissipation inequality and give the corresponding Onsagerian conductivity equation as

$$\hat{\mathbf{P}}^{\nu} := \hat{\mathbf{P}} - \hat{\mathbf{P}}^{\mathrm{r}} = \mathbf{L}_{ONS} \cdot \nabla \mathbf{v}.$$

Here $\mathbf{L}_{ONS}$ is a nonnegative constitutive function. Note that if $\mathfrak{s}$ is independent of the gradient of the density, $\mathbf{L}_{ONS}$ is constant, and $\hat{\mathbf{P}}^{\nu}$ is an isotropic function of only $\nabla \mathbf{v}$, then we obtain the traditional Navier-Stokes fluid (see e.g. [31]).

One can prove easily that the reversible part of the pressure is *potentializable*, i.e., there is a scalar valued function U such that

$$\nabla \cdot \hat{\mathbf{P}}^{\mathrm{r}} = \rho \nabla \hat{U}. \tag{76}$$

$\hat{U}$ can be calculated from the entropy function as

$$\hat{U} = \nabla \cdot (\rho \partial_{\nabla\rho} \mathfrak{s}) - \partial_\rho(\rho \mathfrak{s}). \tag{77}$$

Therefore in case of reversible fluids the momentum balance can be written alternatively as

$$\rho \dot{\mathbf{v}} + \nabla \cdot \hat{\mathbf{P}}^{r} = \mathbf{0} \quad \Longleftrightarrow \quad \dot{\mathbf{v}} + \nabla \hat{U} = \mathbf{0}. \tag{78}$$

Let us give an interesting particular example of a weakly nonlocal fluid.

5.2 Schrödinger-Madelung fluid

Here the entropy is defined as

$$s_{SchM}(\rho, \nabla\rho, \mathbf{v}) = -\frac{\nu}{2}\left(\frac{\nabla\rho}{2\rho}\right)^2 - \frac{\mathbf{v}^2}{2} = -\frac{\nu}{8}(\nabla \ln \rho)^2 - \frac{\mathbf{v}^2}{2}, \tag{79}$$

where ν is a constant scalar. The corresponding reversible pressure is

$$\mathbf{P}^{\mathrm{r}} = -\frac{\nu}{8}\left(\Delta\rho\mathbf{I} + \nabla^2\rho - \frac{2\nabla\rho \circ \nabla\rho}{\rho}\right), \tag{80}$$

where $\circ$ denotes the tensorial/dyadic product, as mentioned before. The potential is

$$U_{SchM} = -\frac{\nu}{4\rho}\left(\Delta\rho - \frac{(\nabla\rho)^2}{2\rho}\right) = -\frac{\nu}{2}\frac{\Delta R}{R}, \tag{81}$$

where we introduced $R = \sqrt{\rho}$ to show that (81) is the quantum potential in the de Broglie-Bohm version of quantum mechanics (if $\nu = \hbar^2/m^2$) [4, 35].

The entropy flux of the Schrödinger-Madelung fluid is

$$\mathbf{J}_{SchM} = -\mathbf{v}\cdot\mathbf{P}^{\mathrm{r}} - \frac{\nu}{8}(\nabla\rho\nabla\cdot\mathbf{v} + \nabla\rho\cdot\nabla\mathbf{v}). \tag{82}$$

The results of this subsection may appear strange with a traditional interpretation of objectivity and frame independence. That question will be discussed from a more general point of view in the next section.

5.2.1 Discussion

An important property of the Schrödinger-Madelung fluid is that if the motion of the fluid is vorticity free, $\nabla\times\mathbf{v} = \mathbf{0}$, then the mass and momentum balances can be transformed into and united in the Schrödinger equation. Hence the balance of momentum (56) can be derived from a Bernoulli equation (in a given inertial reference frame). Defining a scalar valued phase (velocity potential) by

$$\mathbf{v} = \frac{\hbar}{m}\nabla S,$$

we obtain the Bernoulli equation observing that the second part of (78) is the gradient of

$$\frac{\hbar}{m}\frac{\partial S}{\partial t} + \frac{\mathbf{v}^2}{2} - U_{SchM} = 0. \tag{83}$$

Then, introducing a single complex-valued function $\psi := Re^{iS}$ that unifies $R = \sqrt{\rho}$ and S, it is easy to find that the sum of (55) multiplied by $i\hbar e^{iS}/(2R)$, and (83) multiplied by mRe^{iS}, form together the Schrödinger equation for free particles:

$$i\hbar\frac{\partial\psi}{\partial t} = -\frac{\hbar^2}{2m}\Delta\psi. \tag{84}$$

It is remarkable that the structure of quantum mechanics appears in a classical thermodynamic approach without any explicit distinctive assumption related to the microscopic quantum world. In this sense the basic assumptions of our derivation are rather weak and very general.

6 Summary and outlook

In this paper we have shown that the Second Law of Thermodynamics provides a general, uniform, rigorous and constructive method to investigate weakly nonlocal extensions of classical and nonclassical non-equilibrium thermodynamics. We have seen how one can generate evolution equations for internal variables, to understand the constitutive structure of Classical Irreversible Thermodynamics and to restrict the pressure tensor of ideal and viscous Korteweg fluids.

In the suggested approach the choice of the constitutive state space is a physical question. The number of the necessary constraints for a given constitutive state space was introduced intuitively, according to our calculational experience, always looking forward to an explicit solution of the Liu equations. Cimmelli in [9] introduces a different philosophy, requiring that the number of the constraints must equal the number of constitutive state variables. Additional gradients of the constraints are to be introduced in case of necessity. In general, the method given in [90] introduces fewer additional constraints, therefore the results are more easily applicable. On the other hand, the method proposed in [9] is more cumbersome and the interpretation of the consequences of the Second Law can be less straightforward. However, it is based on a precise rule.

One of the interesting observations was that we were able to derive some restrictions to the reversible part of the evolution equations. There we have encountered equations of Euler-Lagrange type in case of zero entropy production in the corresponding interaction (static Ginzburg-Landau, internal degrees of freedom, Bohm-potential for Korteweg pressure). In a sense we were able to derive Hamiltonian variational principles, as we have proved their existence from the Second Law of Thermodynamics.

There are several other classical and nonclassical weakly nonlocal equations of physics that could be investigated in this general frame. Weakly nonlocal extensions of the heat conduction equations (both Fourier and Cattaneo-Vernotte) were analyzed, too. In this respect we have obtained that special kinds of internal variables, the so-called *current multipliers*, can represent some kind of nonlocal effects in the theory. They can lead to theoretical structures like the Guyer-Krumhansl equation of heat conduction [85]. This concept originated in general extensions of the entropy flux [73, 102]. These possibilities have been investigated together with higher order weakly nonlocal extensions in the case of generalized heat conduction and generalized Ginzburg-Landau equations in [8, 91]. The flux hierarchy of extended irreversible thermodynamics and the origin of balance form evolution equations for internal variables was studied in [12, 13].

On the other hand, we have investigated phase separated multicomponent fluids and got a structure with natural instabilities (a kind of nonlocal phase boundaries) similar to the Goodman-Cowin pressure [26, 88, 89].

However, why should we restrict ourselves to weakly nonlocal extensions of the constitutive functions? Why do we not consider time derivatives in exploiting the Second Law? The answer to this question leads to one of the most fundamental problems of physics, the question of objectivity, in particular, to the question of material frame indifference. In this paper we restricted ourselves to evolution equations and physical quantities given in an inertial reference frame. However, we know well that the laws of physics are independent of an external observer. In particular, the constitutive functions can depend only on the material (e.g. on the motion of the continua) but not on the motion of an external observer. There are strong arguments that the usual formulations of the material frame indifference and the usual concept of objectivity are wrong [23, 53, 54, 56]. These investigations show that the core of the problem is in the usual intuitive concept of the nonrelativistic spacetime [55]. Therefore, we need more rigorous analysis of the kinematics of classical continua from this point of view, and an absolute, frame independent formulation of the exploitation method of the Second Law. As regards the new foundations of finite deformation kinematics, the investigations of Fülöp seem to settle on a good starting point [24].

The results of those investigations are exploited in two key points in this paper. First of all, from the objectivity point of view, considering mathematical models of nonrelativistic spacetime, the successes of gradient extensions can be well understood. The gradient of a physical quantity is a spacelike component of a four-covector (spacetime quantity) and therefore it is independent of an observer [52]: for inertial observers three-vectors are transforming by Galilei transformation, but three-covectors are invariant. Therefore *gradients are objective*. On the other hand, for fluids we have started our investigations with an explicitly velocity dependent constitutive state space. That is excluded by the old concept of objectivity, because the relative three-velocities are frame dependent. However, it is not excluded according to the new concept of objectivity, because the four-velocities are independent of the frame. Our treatment of a one component fluid here is not a true objective treatment, from more than one point of view. However, we have done the first steps toward the development of objective exploitation methods of the Second Law, too. As the problem of objectivity is in a sense more easily understandable from the point of view of a relativistic spacetime model, we have investigated the nonequilibrium thermodynamics of special relativistic fluids [92]. Moreover, we have started to investigate some possible practical consequences of incorporating our generalization of the objective time derivatives of rheology into a thermodynamic framework [93, 94].

Finally let us emphasize how surprising our result is regarding fluid mechanics. The original observation of Madelung was that the Schrödinger equation can be transformed into an interesting fluid mechanical form [49]. The observation of Bohm was that the same equation can be transformed into a Newtonian form [4]. In both cases the Schrödinger equation is postulated, coming out of the blue. Here we have *derived* the fluid mechanical equivalent of the Schrödinger equation in a very general framework, from a surprisingly minimal set of assumptions, from the basic balances, the Second Law of Thermodynamics and nonlocality of the interactions. The existence of such a derivation is a disturbing fact that is hardly understandable from the traditional point of view regarding the relation of the irreversible macro- and reversible microworld.

Acknowledgements This research was supported by OTKA T048489. The author is grateful to T. Matolcsi and C. Papenfuss for discussions and corrections and for the valuable remarks of the referees.

7 Appendix – Farkas's lemma and some of its consequences

Lemma 7.1 *(Farkas) Let* $\mathbf{a}_i \neq \mathbf{0}$ *be vectors in a finite dimensional vector space* $\mathbb{V}$, $i = 1...n$, *and* $S = \{\mathbf{p} \in \mathbb{V}^* | \mathbf{p} \cdot \mathbf{a}_i \geq 0, i = 1...n\}$. *The following statements are equivalent for all* $\mathbf{b} \in \mathbb{V}$:

(i) $\mathbf{p} \cdot \mathbf{b} \geq 0$, *for all* $\mathbf{p} \in S$.

(ii) There are nonnegative real numbers $\lambda_1, ..., \lambda_n$ *such that* $\mathbf{b} = \sum_{i=1}^n \lambda_i \mathbf{a}_i$.

Proof:

$(ii) \Rightarrow (i)$ $\mathbf{p} \cdot \sum_{i=1}^n \lambda_i \mathbf{a}_i = \sum_{i=1}^n \lambda_i \mathbf{p} \cdot \mathbf{a}_i \geq 0$ if $\mathbf{p} \in S$.

$(i) \Rightarrow (ii)$ Let us consider a maximal, linearly independent subset $\mathbf{a}_1, ..., \mathbf{a}_l$ of S. Let $S_0 = \{\mathbf{y} \in \mathbb{V}^* | \mathbf{y} \cdot \mathbf{a}_i = 0, i = 1...l\}$. Clearly $\emptyset \neq S_0 \subset S$.

If $\mathbf{y} \in S_0$ then $-\mathbf{y}$ is also in S_0, therefore $\mathbf{y} \cdot \mathbf{b} \geq 0$ and $-\mathbf{y} \cdot \mathbf{b} \geq 0$ together. Therefore for all $\mathbf{y} \in S_0$ it is true that $\mathbf{y} \cdot \mathbf{b} = 0$. As a consequence $\mathbf{b}$ is in the linear subspace generated by $\{\mathbf{a}_i\}$, that is there are real numbers $\lambda_1, ..., \lambda_l$ such that $\mathbf{b} = \sum_{i=1}^l \lambda_i \mathbf{a}_i$.

Moreover, for all $k \in \{1, ..., l\}$, there is a $\mathbf{p}_k \in \mathbb{V}^*$ such that $\mathbf{p}_k \cdot \mathbf{a}_k = 1$ and $\mathbf{p}_k \cdot \mathbf{a}_i = 0$ if $i \neq k$. Evidently, $\mathbf{p}_k \in S$ for all k, therefore $0 \leq \mathbf{p}_k \cdot \mathbf{b} = \mathbf{p}_k \cdot \sum_{i=1}^l \lambda_i \mathbf{a}_i = \sum_{i=1}^l \lambda_i \mathbf{p}_k \cdot \mathbf{a}_i = \lambda_k$ is valid for all k. Lastly, we can choose zero multipliers for the vectors that are not independent. ■

Remark 7.1. In the following the elements of $\mathbb{V}^*$ are called *independent variables* and $\mathbb{V}^*$ itself is called the *space of independent variables*. The inequality in the first statement of the lemma is called *objective inequality* and the nonnegative numbers in the second statement are called *Lagrange-Farkas multipliers*. The inequalities determining S are the *constraints*.

In the calculations an excellent reminder is to use Lagrange- Farkas multipliers similarly to the Lagrange multipliers in the case of conditional extremum problems:

$$\mathbf{p} \cdot \mathbf{b} - \sum_{i=1}^{n} \lambda_i \mathbf{p} \cdot \mathbf{a}_i = \mathbf{p} \cdot (\mathbf{b} - \sum_{i=1}^{n} \lambda_i \cdot \mathbf{a}_i) \geq 0, \quad \forall \mathbf{p} \in \mathbb{V}^*.$$

From this form we can read out the second statement of the lemma.

Remark 7.2. The geometric interpretation of the theorem is important and graphic: if the vector $\mathbf{b}$ does not belong to the cone generated by the vectors $\mathbf{a}_i$, there exists a hyperplane separating $\mathbf{b}$ from the cone.

7.1 Affine Farkas's lemma

This generalization of the previous lemma was first published simultaneously by A. Haar and J. Farkas as subsequent papers in the same journal, with different proofs [22, 32]. Later it was rediscovered independently by others several times (e.g. [72, 81]).

Theorem 7.2 *(Affine Farkas) Let $\mathbf{a}_i \neq \mathbf{0}$ be vectors in a finite dimensional vector space $\mathbb{V}$ and α_i real numbers, $i = 1...n$ and $S_A = \{\mathbf{p} \in \mathbb{V}^* | \mathbf{p} \cdot \mathbf{a}_i \geq \alpha_i, i = 1...n\}$. The following statements are equivalent for a $\mathbf{b} \in \mathbb{V}$ and a real number β:*

(i) $\mathbf{p} \cdot \mathbf{b} \geq \beta$, for all $\mathbf{p} \in S_A$.

(ii) There are nonnegative real numbers $\lambda_1, ..., \lambda_n$ such that $\mathbf{b} = \sum_{i=1}^n \lambda_i \mathbf{a}_i$ and $\beta \leq \sum_{i=1}^n \lambda_i \alpha_i$.

Proof:

$(ii) \Rightarrow (i)$ $\mathbf{p} \cdot \mathbf{b} = \mathbf{p} \cdot \sum_{i=1}^n \lambda_i \mathbf{a}_i = \sum_{i=1}^n \lambda_i \mathbf{p} \cdot \mathbf{a}_i \geq \sum_{i=1}^n \lambda_i \alpha_i \geq \beta$.

$(i) \Rightarrow (ii)$ First we will show indirectly that the first condition of Farkas's lemma is a consequence of the first condition here, that is if (i) is true then $\mathbf{p} \cdot \mathbf{b} \geq 0$, for all $\mathbf{p} \in S$.

Thus let us assume the contrary, hence there is $\mathbf{p}' \in S$, for which $\mathbf{p}' \cdot \mathbf{b} < 0$. Take an arbitrary $\mathbf{p} \in S_A$, then $\mathbf{p} + k\mathbf{p}' \in S_A$ for all real nonnegative numbers k. But now $(\mathbf{p} + k\mathbf{p}') \cdot \mathbf{b} = \mathbf{p} \cdot \mathbf{b} + k\mathbf{p}' \cdot \mathbf{b} < \beta$, if $k > (\mathbf{p} \cdot \mathbf{b} - \beta)(-\mathbf{p}' \cdot \mathbf{b})$. That is a contradiction.

Therefore, according to Farkas's lemma exist nonnegative Lagrange-Farkas multipliers $\lambda_1, ..., \lambda_n$ such that $\mathbf{b} = \sum_{i=1}^n \lambda_i \mathbf{a}_i$. Hence $\beta \leq inf_{p \in S_A} \{\mathbf{p} \cdot \sum_{i=1}^n \lambda_i \mathbf{a}_i\} = inf_{p \in S_A} \{\sum_{i=1}^m \lambda_i \mathbf{p} \cdot \mathbf{a}_i\} = \sum_{i=1}^n \lambda_i \alpha_i$. ■

Remark 7.3. The multiplier form is a good reminder again

$$(\mathbf{p} \cdot \mathbf{b} - \beta) - \sum_{i=1}^m \lambda_i (\mathbf{p} \cdot \mathbf{a}_i - \alpha_i) = \mathbf{p} \cdot (\mathbf{b} - \sum_{i=1}^m \lambda_i \cdot \mathbf{a}_i) - \beta + \sum_{i=1}^m \lambda_i \alpha_i \geq 0, \quad \forall \mathbf{p} \in \mathbb{V}^*.$$

Remark 7.4. The geometric interpretation is similar to the previous one, but with affine objects. If the line (one dimensional affine hyperplane) $(\mathbf{b}, \beta)$ does not belong to the (affine) cone generated by $(\mathbf{a}_i, \alpha_i)$, there exists an affine hyperplane separating $\mathbf{b}$ from the cone.

7.2 Liu's theorem

Here the constraints are equalities instead of inequalities, therefore the multipliers are not necessarily positive.

Theorem 7.3 *(Liu) Let $\mathbf{a}_i \neq \mathbf{0}$ be vectors in a finite dimensional vector space $\mathbb{V}$ and α_i real numbers, $i = 1...n$ and $S_L = \{\mathbf{p} \in \mathbb{V}^* | \mathbf{p} \cdot \mathbf{a}_i = \alpha_i, i = 1...n\}$. The following statements are equivalent for a $\mathbf{b} \in \mathbb{V}$ and a real number β:*

(i) $\mathbf{p}\cdot\mathbf{b} \geq \beta$, *for all* $\mathbf{p} \in S_L$,
(ii) There are real numbers $\lambda_1, ..., \lambda_n$ *such that*

$$\mathbf{b} = \sum_{i=1}^{n} \lambda_i \mathbf{a}_i, \tag{85}$$

and

$$\beta \leq \sum_{i=1}^{n} \lambda_i \alpha_i. \tag{86}$$

Proof:

A straightforward consequence of the previous affine form of Farkas's lemma because S_L can be given in a form S_A with the vectors $\mathbf{a}_i$ and $-\mathbf{a}_i$, $i = 1, ..., n$: $S_L = \{\mathbf{p} \in \mathbb{V}^* | \mathbf{p}\cdot\mathbf{a}_i \geq \alpha_i, \mathbf{p}\cdot(-\mathbf{a}_i) \geq \alpha_i, i = 1...n\}$.

Therefore there are nonnegative real numbers $\lambda_1^+, ..., \lambda_n^+$ and $\lambda_1^-, ..., \lambda_n^-$ such, that $\mathbf{b} = \sum_{i=1}^n (\lambda_i^+ \mathbf{a}_i - \lambda_i^- \mathbf{a}_i) = \sum_{i=1}^n (\lambda_i^+ - \lambda_i^-)\mathbf{a}_i = \sum_{i=1}^n \lambda_i \mathbf{a}_i$ and $\beta \leq \sum_{i=1}^n (\lambda_i^+ \alpha_i - \lambda_i^- \alpha_i)$. ■

Remark 7.5. The multiplier form is again helpful in the applications

$$0 \leq (\mathbf{p}\cdot\mathbf{b} - \beta) - \sum_{i=1}^{n} \lambda_i (\mathbf{p}\cdot\mathbf{a}_i - \alpha_i) = \mathbf{p}\cdot(\mathbf{b} - \sum_{i=1}^{n} \lambda_i \cdot \mathbf{a}_i) - \beta + \sum_{i=1}^{n} \lambda_i \alpha_i, \quad \forall \mathbf{p} \in \mathbb{V}^*.$$

Remark 7.6. In the theorem with Lagrange multipliers for local conditional extremum of differentiable function we apply exactly the above theorem of linear algebra after a linearization of the corresponding functions at the extremum point.

Considering the requirements of the applications we generalize Liu's theorem to take into account vectorial constraints:

Theorem 7.4 *(vector Liu) Let* $\mathbf{A} \neq \mathbf{0}$ *in a tensor product* $\mathbb{V} \otimes \mathbb{U}$ *of finite dimensional vector spaces* $\mathbb{V}$ *and* $\mathbb{U}$. *Let* $\alpha \in \mathbb{U}$ *and* $S_L = \{\mathbf{p} \in \mathbb{V}^* | \mathbf{p}\cdot\mathbf{A} = \alpha\}$. *The following statements are equivalent for a* $\mathbf{b} \in \mathbb{V}$ *and a real number* β:
(i) $\mathbf{p}\cdot\mathbf{b} \geq \beta$, *for all* $\mathbf{p} \in S_L$.
(ii) There is a vector λ *in the dual of* $\mathbb{U}$ *such that*

$$\mathbf{b} = \mathbf{A}\cdot\lambda, \tag{87}$$

and

$$\beta \leq \lambda\cdot\alpha. \tag{88}$$

Proof:

Let us observe that we can get back the previous form of the theorem by introducing a linear bijection $\mathbf{K} : \mathbb{U} \rightarrow \mathbb{R}^n$, a *coordinatization* in $\mathbb{U}$. Then by the vectors $\mathbf{a}_i := \mathbf{A}\cdot\mathbf{K}^*\cdot\mathbf{e}_i$, where $\mathbf{e}_i$ is the standard i-th unit vector in $\mathbb{R}^n$, and the numbers $\alpha_i := \mathbf{K}\cdot\alpha$, we obtain that $\mathbf{b} = \sum_{i=1}^n \lambda_i \mathbf{a}_i = \mathbf{A}\sum_{i=1}^n \lambda_i \mathbf{K}^* \mathbf{e}_i = \mathbf{A}\lambda$ where $\lambda := \sum_{i=1}^n \lambda_i \mathbf{K}^* \mathbf{e}_i$. ■

The previously excluded degenerate case of $\mathbf{A} = \mathbf{0}$ deserves a special attention. Then, of course, $\alpha = 0$ and $S_L = \mathbb{V}^*$; this is, in fact, a degenerate case of all the previous theorems and evidently the following statements are equivalent for a $\mathbf{b} \in \mathbb{V}$ and a real number β:

(i) $\mathbf{p} \cdot \mathbf{b} \geq \beta$ for all $\mathbf{p} \in \mathbb{V}^*$,

(ii) $\mathbf{b} = \mathbf{0}$ and $\beta \leq 0$.

In this case the assignment of the vector space of $\mathbb{V}^*$ is based on the inequality (i). We encountered various degeneracies in the calculations of the previous sections.

Remark 7.7. In continuum physics and thermodynamics the above algebraic theorems are applied to differential equations and inequalities. There the constraints are differential equations and therefore $\mathbb{V}$ is generated by derivatives of some constitutive functions in the differential equations. $\mathbb{V}^*$ is spanned by the *process directions*, the derivatives of the fields in the constitutive state space that are not already there. The corresponding form of (87) and (88) are called *Liu equation(s)* and the *dissipation inequality*, respectively.

References

1. Aranson, I.S., Kramer, L.: The world of the complex Ginzburg-Landau equation. Reviews of Modern Physics **74**(1), 99–143 (2002)
2. Béda, P.-B.: Dynamic systems, rate and gradient effects in material instability. International Journal of Mechanical Sciences **42**, 2101–2114 (2000)
3. Bedeaux, D., Johannessen, E., Rojorde, A.: A nonequilibrium van der Waals square gradient model (I): the model and its numerical solution. Physica A **330**, 329–353 (2003)
4. Bohm, D.: Quantum Theory. Prentice-Hall, New York (1951)
5. Brokate, M., Sprekels, J.: Hysteresis and Phase Transitions, Applied Mathematical Sciences Volume 121. Springer, New York (1996)
6. Capriz, G., Mariano, P.M.: Multifield theories: an introduction. International Journal of Solids and Structures **38**, 939–941 (2001)
7. Casimir, H.G.B.: On Onsager's principle of microscopic reversibility. Reviews of Modern Physics **17**, 343–350 (1945)
8. Ciancio, V., Cimmelli, V.A., Ván, P.: On the evolution of higher order fluxes in non-equilibrium thermodynamics. Mathematical and Computer Modelling **45**, 126–136 (2007) cond-mat/0407530.
9. Cimmelli, V.A.: An extension of Liu procedure in weakly nonlocal thermodynamics. Journal of Mathematical Physics **48**, 113510 (2007)
10. Cimmelli, V.A., Frischmuth, K.: Gradient generalization to the extended thermodynamic approach and diffusive-hyperbolic heat conduction. Physica B - Condensed matter **400**(1–2), 257–265 (2007)
11. Cimmelli, V.A., Kosiński, W.: Gradient generalization to internal state variables and a theory of superfluidity. Journal of Theoretical and Applied Mechanics **35**(4), 763–779 (1997)
12. Cimmelli, V.A., Ván, P.: The effects of nonlocality on the evolution of higher order fluxes in non-equilibrium thermodynamics. Journal of Mathematical Physics **46**(11), 112901–15 (2005) cond-mat/0409254.
13. Cimmelli, V.A., Ván, P.: First order weak nonlocality in extended thermodynamics of rigid heat conductors. In: Monaco, R., Mulone, G., Rionero, S., Ruggeri, T. (eds) Proceedings of WASCOM'05, pp. 122–127. World Scientific, Singapore (2006)

14. Coleman, B.D., Gurtin, M.E.: Thermodynamics with internal state variables. The Journal of Chemical Physics **47**(2), 597–613 (1967)
15. Coleman, B.D., Mizel, V.J.: Existence of caloric equation of state in thermodynamics. Journal of Chemical Physics **40**, 1116–1125 (1964)
16. Drouot, R., Maugin, G.A.: Application of diffusive internal variables in complex fluids. Journal of Non-Newtonian Mechanics **96**, 31–43 (2001)
17. Dunn, J.E., Serrin, J.: On the thermomechanics of interstitial working. Archive of Rational Mechanics and Analysis **88**, 95–133 (1985)
18. Eckart, C.: The thermodynamics of irreversible processes III .Relativistic theory of the simple fluid. Physical Review **58**, 919–924 (1940)
19. Engelbrecht, J.: Nonlinear Wave Mechanics. Kluwer, Dordrecht (1997)
20. Engelbrecht, J., Vendelin, M.: Microstructure described by hierarchical internal variables. Rendiconti del Seminario Matematico, Universitá e Politecnico di Torino **58**(1), 93–98 (2000)
21. Farkas, G.: A Fourier-féle mechanikai elv alkalmazásai. Mathematikai és Természettudományi Értesítő **12**, 457–472 (1894) (in Hungarian)
22. Farkas, G.: Egyenlőtlenség alkalmazásának új módjai. Mathematikai és Természettudományi Értesítő **36**, 297–308 (1918) (in Hungarian)
23. Frewer, M.: More clarity on the concept of material frame-indifference on classical continuum mechanics. Acta Mechanica **202**, 213–246 (2009)
24. Fülöp, T. A new interpretation of the kinematics of continua. In: Fülöp, T. (ed.) New Results in Continuum Physics, Notes on Engineering Geology and Rock Mechanics Vol.8, chapter 3, pp. 55–99. BME Publisher, Budapest (2008) (in Hungarian)
25. Glansdorff, P., Prigogine, I.: Thermodynamic Theory of Structure, Stability and Fluctuations. Wiley-Interscience, London (1971)
26. Goodman, M.A., Cowin, S.C.: A continuum theory for granular materials. Archive of Rational Mechanics and Analysis **44**, 249–266 (1972)
27. Grmela, M.: Weakly nonlocal hydrodynamics. Physical Review E **47**(1), 351–365 (1993)
28. Grmela, M., and Öttinger, H.C.: Dynamics and thermodynamics of complex fluids I. Development of a general formalism. Physical Review E **56**(6), 6620–6632 (1997)
29. Groot, de S.R., Mazur, P.: Non-equilibrium Thermodynamics. North-Holland, Amsterdam (1962)
30. Gurtin, M.G.: Generalized Ginzburg-Landau and Cahn-Hilliard equations based on a microforce balance. Physica D **92**, 178–192 (1996)
31. Gyarmati, I.: Non-equilibrium Thermodynamics /Field Theory and Variational Principles/. Springer, Berlin (1970)
32. Haar, A.: A lineáris egyenlőtlenségekről. Mathematikai és Természettudományi Értesítő **36**, 279–296 (1918) (in Hungarian)
33. Hauser, R.A., Kirchner, N.P.: A historical note on the entropy principle of Müller and Liu. Continuum Mechanics and Thermodynamics **14**, 223–226 (2002)
34. Hohenberg, P.C., Halperin, B.I.: Theory of dynamic critical phenomena. Reviews of Modern Physics **49**(3), 435–479 (1977)
35. Holland, P.R.: The Quantum Theory of Motion. Cambridge University Press (1993)
36. Hutter, K.: The foundations of thermodynamics, its basic postulates and implications. A review of modern thermodynamics. Acta Mechanica **27**, 1–54 (1977)
37. Ireman, P., Nguyen, Q.-S.: Using the gradients of temperature and internal parameters in Continuum Thermodynamics. Comptes Rendus Mécanique **332**, 249–255 (2004)
38. Johannessen, E., Bedeaux, D.: A nonequilibrium van der Waals square gradient model (II): local equilibrium of the Gibbs surface. Physica A **330**, 3354–372 (2003)
39. Johannessen, E., Bedeaux, D.: A nonequilibrium van der Waals square gradient model (III): heat and mass transfer coefficients. Physica A **336**, 252–270 (2004)
40. Kestin, J.: Internal variables in the local-equilibrium approximation. Journal of Non-Equilibrium Thermodynamics **18**, 360–379 (1993)

41. Kirchner, N.-P., Hutter, K.: Elasto-plastic behaviour of a granular material with and additional scalar degree of freedom. In: Ehlers, W., Blum, J. (eds.) Porous Media: Theory, Experiments and Numerical Applications, pp. 147–168. Springer, New York (2002)
42. Kjelstup, S., Bedeaux, D.: Non-equilibrium Thermodynamics of Heterogeneous Systems. World Scientific, Singapore (2008)
43. Kosiński, W., Wojno, W.: A gradient generalization to internal state variable approach. Archive of Mechanics **47**(3), 523–536 (1995)
44. Kosiński, W.: A modified hyperbolic framework for thermoelastic materials with damage. In: Kosiński, W., Boer, de R., Gross, D. (eds.) Problems of Environmental and Damage Mechanics, pp. 157–172. IPPT-PAN, Warszawa (1997)
45. Landau, L.D., Khalatnikov, I.M.: Ob anomal'nom pogloshehenii zvuka vblizi tochek fazovo perekhoda vtorovo roda. Dokladu Akademii Nauk SSSR **96**, 469–472 (1954). English translation: On the anomalous absorption of sound near a second order transition point. In: ter Haar, D. (ed.) Collected papers of L.D. Landau, pp. 626–633. Pergamon, Oxford (1965)
46. Lebon, G., Grmela, M.: Weakly nonlocal heat conduction in rigid solids. Physics Letters A **214**, 184–188 (1996)
47. Liu, I.: Method of Lagrange multipliers for exploitation of the entropy principle. Archive of Rational Mechanics and Analysis **46**, 131–148 (1972)
48. Machlup, S., Onsager, L.: Fluctuations and irreversible processes II. Systems with kinetic energy. Physical Review **91**(6), 1512–1515 (1953)
49. Madelung, E.: Quantentheorie in hydrodynamischer Form. Zeitschrift für Physik **40**, 322–326 (1926) in German
50. Mariano, P.M.: Multifield theories in mechanics of solids. Advances in Applied Mechanics **38**, 1–94 (2002)
51. Márkus, F., Gambár, K.: A variational principle in thermodynamics. Journal of Non-Equilibrium Thermodynamics **16**(1), 27–31 (1991)
52. Matolcsi, T.: Spacetime Without Reference Frames. Akadémiai Kiadó Publishing House of the Hungarian Academy of Sciences, Budapest (1993)
53. Matolcsi, T., Gruber, T.: Spacetime without reference frames: An application to the kinetic theory. International Journal of Theoretical Physics **35**(7), 1523–1539 (1996)
54. Matolcsi, T., Ván, P.: Can material time derivative be objective? Physics Letters A **353**, 109–112 (2006)
55. Matolcsi, T., Ván, P.: Absolute time derivatives. Journal of Mathematical Physics **48**, 053507–19 (2007)
56. Matolcsi, T., Ván, P.: On the objectivity of time derivatives. Atti dell'Accademia Peloritana dei Pericolanti, Classe di Scienze Fisiche, Matematiche e Naturali, Suppl. I. **86**, 1–13 (2008)
57. Maugin, G.A.: Nonlocal theories or gradient-type theories: a matter of convenience? Archives of Mechanics (Stosowanej) **31**(1), 15–26 (1979)
58. Maugin, G.A.: The principle of virtual power in continuum mechanics. Application to coupled fields. Acta Mechanica **35**, 1–70 (1980)
59. Maugin, G.A.: Internal variables and dissipative structures. Journal of Non-Equilibrium Thermodynamics **15**, 173–192 (1990)
60. Maugin, G.A.: The Thermomechanics of Nonlinear Irreversible Behaviors (An Introduction). World Scientific, Singapore (1999)
61. Maugin, G.A.: On the thermomechanics of continuous media with diffusion and/or weak nonlocality. Archive of Applied Mechanics **75**, 723–738 (2006)
62. Maugin, G.A., Drouot, R.: Internal variables and the thermodynamics of macromolecule solutions. International Journal of Engineering Science **21**(7), 705–724 (1983)
63. Maugin, G.A., Muschik, W.: Thermodynamics with internal variables. Part I. General concepts. Journal of Non-Equilibrium Thermodynamics **19**, 217–249 (1994)
64. Maugin, G.A., Muschik, W.: Thermodynamics with internal variables. Part II. Applications. Journal of Non-Equilibrium Thermodynamics **19**, 250–289 (1994)
65. Minkowski, H.: Geometrie der Zahlen. Teubner, Leipzig (1896)
66. Morro, A.: A phase-field approach to non-isothermal transitions. Mathematical and Computer Modelling **48**(3–4), 621–633 (2008)

67. Muschik, W.: Internal variables in non-equilibrium thermodynamics. Journal of Non-Equilibrium Thermodynamics **15**, 127–137 (1990)
68. Muschik, W., Ehrentraut, H.: An amendment to the Second Law. Journal of Non-Equilibrium Thermodynamics **21**, 175–192 (1996)
69. Muschik, W., Papenfuss, C., Ehrentraut, H.: A sketch of continuum thermodynamics. Journal of Non-Newtonian Fluid Mechanics **96**, 255–290 (2001)
70. Müller, I.: A thermodynamic theory of mixtures of fluids. Archive of Rational Mechanics and Analysis **28**, 1–39 (1968)
71. Müller, I.: On the frame dependence of stress and heat flux. Archive of Rational Mechanics and Analysis **45**, 241–250 (1972)
72. von Neumann, J.: Discussion of a maximum principle. In: Taub, A.H. (ed) Collected Works, volume VI, pp. 89–95. Pergamon Press, New York (1961)
73. Nyíri, B.: On the entropy current. Journal of Non-Equilibrium Thermodynamics **16**, 179–186 (1991)
74. Onsager, L.: Reciprocal relations of irreversible processes I. Physics Reviews **37**, 405–426 (1931)
75. Onsager, L.: Reciprocal relations of irreversible processes II. Physics Reviews **38**, 2265–2279 (1931)
76. Onsager, L., Machlup, S.: Fluctuations and irreversible processes. Physical Review **91**(6), 1505–1512 (1953)
77. Öttinger, H.C.: Beyond Equilibrium Thermodynamics. Wiley-Interscience, New York (2005)
78. Öttinger, H.-C., Grmela, M.: Dynamics and thermodynamics of complex fluids. II. Illustrations of a general formalism. Physical Review E **56**(6), 6633–6655 (1997)
79. Papenfuss, C., Forest, S.: Thermodynamical frameworks for higher grade material theories with internal variables or additional degrees of freedom. Journal of Non-Equilibrium Thermodynamics **31**(4), 319–353 (2006)
80. Penrose, O., Fife, P.C.: Thermodynamically consistent models of phase-field type for the kinetics of phase transitions. Physica D **43**, 44–62 (1990)
81. Schriver, A.: Theory of Linear and Integer Programming. Wiley and Sons, New York (1998)
82. Triani, V., Papenfuss, C., Cimmelli, V.A., Muschik, M.: Exploitation of the Second Law: Coleman-Noll and Liu procedure in comparison. Journal of Non-Equilibrium Thermodynamics **33**, 47–60 (2008)
83. Valanis, K.C.: A gradient theory of internal variables. Acta Mechanica **116**, 1–14 (1996).
84. Valanis, K.C.: A gradient thermodynamic theory of self-organization. Acta Mechanica **127**, 1–23 (1998)
85. Ván, P.: Internal thermodynamic variables and the failure of microcracked materials. Journal of Non-Equilibrium Thermodynamics **26**(2), 167–189 (2001)
86. Ván, P.: Weakly nonlocal irreversible thermodynamics - the Guyer-Krumhansl and the Cahn-Hilliard equations. Physics Letters A **290**(1–2), 88–92 (2001)
87. Ván, P.: Weakly nonlocal irreversible thermodynamics. Annalen der Physik (Leipzig) **12**(3), 146–173 (2003)
88. Ván, P.: One and two component fluids: restrictions from the Second Law. Physica A **340**(1–3), 418–426 (2004)
89. Ván, P.: Weakly nonlocal continuum theories of granular media: restrictions from the Second Law. International Journal of Solids and Structures **41**(21), 5921–5927 (2004)
90. Ván, P.: Exploiting the Second Law in weakly nonlocal continuum physics. Periodica Polytechnica, Ser. Mechanical Engineering **49**(1), 9–94 (2005)
91. Ván, P.: The Ginzburg-Landau equation as a consequence of the Second Law. Continuum Mechanics and Thermodynamics **17**(2), 165–169 (2005)
92. Ván, P.: Internal energy in dissipative relativistic fluids. Journal of Mechanics of Materials and Structures **3**(6), 1161–1169 (2008)
93. Ván, P.: Objective time derivatives in non-equilibrium thermodynamics. Proceedings of Estonian Academy of Sciences **57**(3), 127–131 (2008)

94. Ván, P., Asszonyi, Cs.: General principles of continuum mechanics. In: Asszonyi, Cs. (ed) Constitutive Functions of Isotropic Materials, Notes on Engineering Geology and Rock Mechanics, Vol. 3, chapter 2, pp. 25–87. Technical University Press, Budapest (2006) (in Hungarian)
95. Ván, P., Berezovski, A., Engelbrecht, J.: Internal variables and dynamic degrees of freedom. Journal of Non-Equilibrium Thermodynamics **33**(3), 235–254 (2008)
96. Ván, P., Fülöp, T.: Weakly nonlocal fluid mechanics - the Schrödinger equation. Proceedings of the Royal Society, London A **462**, 541–557 (2006)
97. Ván, P., Fülöp, T.: Stability of stationary solutions of the Schrödinger-Langevin equation. Physics Letters A **323**(5–6), 374–381 (2004)
98. Ván, P., Muschik, W.: The structure of variational principles in nonequilibrium thermodynamics. Physical Review E **52**(4), 3584–3590 (1995)
99. Ván, P., Nyíri, B.: Hamilton formalism and variational principle construction. Annalen der Physik (Leipzig) **8**, 331–354 (1999)
100. Van der Waals, J.D.: Thermodynamische Theorie der Kapillarität unter Voraussetzung stetiger Dichteänderung. Zeitschrift für Physikalische Chemie **13**, 657–725 (1894)
101. Vardoulakis, I., Aifantis, E.C.: On the role of microstructure in behaviour of solids: effects of higher order gradients and internal inertia. Mechanics of Materials **18**, 151–158 (1994)
102. Verhás, J.: On the entropy current. Journal of Non-Equilibrium Thermodynamics **8**, 201–206 (1983)
103. Verhás, J.: An application of Gyarmati's wave approach. Acta Physica Hungarica **66**(1–4), 95–97 (1989)
104. Verhás, J.: Thermodynamics and Rheology. Akadémiai Kiadó and Kluwer, Budapest (1997)
105. Verhás, J.: Non-equilibrium thermodynamics and the dynamics of particles. In: Gues, O., Iooss, G., Nouri, A. (eds) Trends in Applications of Mathematics to Mechanics, pp. 138–146. Chapman and Hall/CRC, Boca Raton (2000)

Part IV
Waves in Fluids

Overview

Tarmo Soomere

Surface waves at the cutting edge of research and applications

Waves in fluids became one of the central objects of study at the Institute of Cybernetics about one and a half decades ago, when CENS was still in the planning stage. Some pioneering research, initially in the context of mathematical questions of how to describe the phenomena occurring during interactions of a few shallow-water solitons, led not only to interesting results [5, 6], but also to further spirited discussions, to related PhD theses, and finally to establishing the topic as a research direction within CENS.

The scope of the research was widened explosively after the turn of the millennium, when it was realized that certain ships that frequently sail over Tallinn Bay systematically create long waves, which frequently evolve into almost perfect shallow-water solitons. A thorough analysis of the properties and specific features of interactions of long waves from high-speed ships led to the proof of existence of a new class of long-living freak or rogue waves in relatively shallow areas [3]. This class may play a role when, for example, the convergence of waves (introduced either by seabed features or by a spatio-temporal structure of surface currents) leads to the intersection of wave crests.

The original motivation for studying ship wakes in CENS was that they serve as an extremely complex and qualitatively new forcing factor of environmental processes in many semi-sheltered sea areas [7]. In addition to contributing to fundamental nonlinear science [9], the relevant studies may also give an insight, for example, into the reaction of beaches to abrupt changes to the local wave regime that may easily accompany relatively small changes in atmospheric forcing conditions in a changing climate. There is evidence that the reaction of the beach may be irreversible [4], a feature which might change our understanding of the predominantly cyclic nature of most of the processes on the seacoast. Last but not least, studies into ship waves serve as a clear example, in which the combination of competence from extremely different areas of science (such as classical surface wave theory, numeri-

E. Quak, T. Soomere (eds.), *Applied Wave Mathematics*,

DOI 10.1007/978-3-642-00585-5_11,

cal wave modelling experience, knowledge of the local wave climate, and elements of nonlinear analysis) is a precondition for progress.

The devastating tsunami of Boxing Day 2004 motivated scientists all over the world to undertake in-depth studies of similar processes and of additional threats connected with a specific shape or appearance of long waves attacking the coast. One of the largest threats stems from the potential for extreme run-up of certain wave classes. This potential depends not only on the basic properties of the wave, such as the wave height, length, or propagation direction, but also on the particular appearance of the wave. An important step towards a more adequate description of the related threats is the observation that the wave shape matters substantially [2].

While in the case of tsunamis the danger to people is explicit, it is equally present, albeit in somewhat modified form, for ship-induced waves. Such waves not only belong to the longest waves in many semi-sheltered sea areas but also may approach from unexpected directions. In many aspects, ship-induced wave groups can be used as a small model for the phenomena occurring during certain types of tsunamis. Although the similarity is not always perfect, a comparison of basic properties of, for example, landslide-induced tsunamis with those of ship waves is straightforward.

The following three contributions make a joint attempt to give insights into the complexity of wave-induced processes at the coastline. The contributions by T. Torsvik and T. Soomere provide a view onto a relatively new field of studies: wave phenomena occurring at the analogue of the speed of light in vacuum – the so-called critical velocity on the surface of relatively shallow sea areas, equal to the maximum phase velocity of long linear waves for the given depth. Although it has been well known for almost a century that at such a speed the classical ship wave system is substantially modified and already John Scott Russell knew that ships sailing in channels may produce "great waves of translation", called solitons today, the real foundations of this research area were only laid in the beginning of the 1980s. The most intriguing feature, according to [10], was that "a forcing disturbance moving steadily with a [...] critical velocity in shallow water can generate, continuously and periodically, a succession of solitary waves" [instead of reaching a steady wave field]. Today it is of course understood that disturbances moving at this speed may lead to the generation of a sequence of Korteweg-de Vries solitons. This phenomenon is a realization of the generic mechanism of resonant excitation of disturbances in situations where the nonlinear and dispersive effects are specifically balanced. It arises when the group velocity of the long waves radiated from the forcing area is close to the velocity of the disturbance [8].

The central goal of the presentation *Long Ship Waves in Shallow Water Bodies* by T. Soomere is to highlight the new, nonlinear features of wakes from fast ferries, and the basic consequences of their presence for the safety of people and the environment. It starts from the generic geometric features of the classical (Kelvin) ship wave system and provides the reader with a detailed analysis of what happens with this system in finite depth and shallow water, when the Froude number (the ratio of the ship's speed over the maximum phase speed $\sqrt{gH}$ of long waves for a given depth H) is approximately equal to or larger than unity. To some extent this approach complements the standard texts on ship waves that tend to focus on the deep water regime, while being rather brief on the finite depth aspects. For educational purposes the contribution explicitly addresses the related mathematical details and explains that only a few differences in the appearance of the wake are connected in a straightforward way with the increase in the ship's speed.

The majority of effects occurring at higher speeds reflect nonlinear processes of wave generation and propagation. At these speeds, qualitatively new structures such as groups of long, long-crested waves that resemble shallow-water solitons, and short monochromatic wave packets resembling envelope solitons may appear. These groups may remain compact for a long time and may considerably extend the area directly influenced by the ship traffic. The final sections of the paper present extensive evidence about methods and results concerning *in situ* measurements of ship-induced waves, about certain particular features of the impact of such waves in shallow areas, and about possible environmental aspects of the increased hydrodynamic activity.

The beginning of the contribution by T. Torsvik *Modeling of Ship Waves from High-speed Vessels* to some extent mirrors the preceding paper. It starts from the basic physics and generic equations that describe the problem and explains in some more detail how the pattern of ship waves undergoes alterations when the ship's speed becomes roughly equal to the critical speed, but soon it switches to the fascinating field of computational fluid dynamics (CFD) and ends with some very recent results in (ship) wave modelling. The line of thinking about the basic features of ship wave patterns deviates from the one used in several classical textbooks. The text presents a substantially different and interesting perspective on how to interpret and roughly estimate the properties of ship waves. The presentation should be very useful for those who think in physical rather than in geometrical or mathematical categories.

The central focus of the paper by T. Torsvik is, however, how to numerically reproduce the pattern of ship waves with reasonable costs and with an adequate accuracy. The reader learns a lot about the secrets usually hidden in the published papers on CFD. The mathematical equations that can be used to model ship waves, their basic properties, the numerical models based on these equations, and how these models can be used to simulate ship wakes together with several numerical tricks are described in a transparent manner. Perhaps the most intriguing part of the paper is where the author describes the results of modeling the spatial variability of ship wave patterns in realistic conditions, which apparently have great potential for sustainable planning of ship routes and speed regimes in order to maintain acceptable wash levels at the shore.

While the papers by T. Soomere and T. Torsvik focus on the generation, the propagation, and the properties of a specific class of (usually long) waves, the contribution *New Trends in the Analytical Theory of Long Sea Wave Runup* by I. Didenkulova concentrates on the recent highlights in the wave run-up theory and its applications for realistic coasts. She presents a modern review of the state of the art of the analytical theory of the long sea wave run-up on a plane beach, based on rigorous solutions of nonlinear shallow-water equations. These equations allow, in particular, a detailed analysis of the dynamics of the moving shoreline, or equivalently, the front of the inundation process. Here rigorous results can be achieved for long waves and very simple bottom geometry (plane beach), and most of the existing results are reproduced in the paper. Perhaps most importantly, the author demonstrates the link between a very important practical problem and the modern mathematical theory of hyperbolic equations.

Exploiting the line of thinking first introduced in [1] about half of a century ago, it is demonstrated that many extreme characteristics of the run-up process (such as run-up and rundown amplitudes, extreme values of on- and off-shore velocities, and critical amplitude of the breaking wave) can be determined using the solution of the linear shallow-water theory, whereas the description of the time series of the wave field requires the nonlinear theory. Perhaps the most striking feature is how strongly the run-up properties depend on the front-back asymmetry of the wave shape. The contribution also contains a number of results that can be directly applied in coastal operational oceanography, for example, simple parameterizations of basic formulas for extreme run-up characteristics for several wave shapes. These results show that for symmetric wave shapes the run-up properties only weakly depend on the initial wave shape. Finally, it is my pleasure to note that the joint efforts of several experts specializing in ship waves and wave run-up have already led to a successful continuation of the described studies during the preparation of this book.

References

1. Carrier, G.F., Greenspan, H.P.: Water waves of finite amplitude on a sloping beach. J. Fluid Mech. **4**, 97–109 (1958)
2. Didenkulova, I.I., Zahibo, N., Kurkin, A.A., Levin, B.V., Pelinovsky, E.N., Soomere, T.: Runup of nonlinearly deformed waves on a coast. Dokl. Earth Sci. **411**(8), 1241–1243 (2006)
3. Kharif, C., Pelinovsky, E.: Physical mechanism of the rogue wave phenomenon. Eur. J. Mech. B Fluids **22**, 603–634 (2003)
4. Parnell, K.E., McDonald, S.C., Burke, A.E.: Shoreline effects of vessel wakes, Marlborough Sounds, New Zealand. J. Coastal Res. Special Issue **50**, 502–506 (2007)
5. Peterson, P., van Groesen, E.: A direct and inverse problem for wave crests modelled by interactions of two solitons. Physica D **141**, 316–332 (2000)
6. Peterson, P., van Groesen, E.: Sensitivity of the inverse wave crest problem. Wave Motion **34**, 391–399 (2001)
7. Soomere, T.: Fast ferry traffic as a qualitatively new forcing factor of environmental processes in non-tidal sea areas: a case study in Tallinn Bay, Baltic Sea. Environ. Fluid Mech. **5**, 293–323 (2005)
8. Soomere, T.: Nonlinear components of ship wake waves. Appl. Mech. Rev. **60**(3), 120–138 (2007)
9. Soomere, T., Engelbrecht, J.: Weakly two-dimensional interaction of solitons in shallow water. European J. Mech. B Fluids **25**, 636–648 (2006)
10. Wu, T.Y.: Generation of upstream advancing solitons by moving disturbances. J. Fluid Mech. **184**, 75–99 (1987)

Long Ship Waves in Shallow Water Bodies

Tarmo Soomere

Abstract Wakes from large high-speed ships frequently reveal many interesting and important features that are not present in the classical Kelvin ship wave system. Only a few differences are connected with the increase in the ship's speed in a straightforward way. The majority of effects reflect nonlinear processes of wave generation and propagation. This overview concentrates on the recent results concerning the nature and consequences of these differences. The goal of the presentation is to highlight the new, nonlinear features of wakes from fast ferries, and the basic consequences of their presence for the safety of people and the environment in a comprehensive manner, but in terms understandable for non-experts. The starting point is the classical theory of the Kelvin wake and its modifications in shallow water. The pattern of ship waves undergoes major alterations when the ship's speed becomes roughly equal with the maximum phase speed of linear waves for a given depth $\sqrt{gH}$. At these speeds, qualitatively new structures such as groups of long, long-crested waves that resemble shallow-water solitons, and short monochromatic wave packets resembling envelope solitons may appear. These groups remain compact for a long time and considerably extend the area directly influenced by the ship traffic. Finally, we provide evidence of certain particular features of the impact of such waves in shallow areas and of possible ecological consequences of the increased hydrodynamic activity.

1 Introduction

The ever increasing density of marine transport combined with a limited number of possible fairways results in an extreme concentration of ship traffic in certain sea areas. The related concerns are traditionally associated with possible accidents (ship collisions or grounding, technical and navigational problems caused by severe

Centre for Nonlinear Studies, Institute of Cybernetics at Tallinn University of Technology, Akadeemia tee 21, 12618 Tallinn, Estonia, e-mail: soomere@cs.ioc.ee

E. Quak, T. Soomere (eds.), *Applied Wave Mathematics*,

DOI 10.1007/978-3-642-00585-5_12,

weather or human errors, etc.) that may lead to loss of lives or property, or to environmental pollution. The basic assertion in their management is that the risks of water surface transport are localized within a small area around the ship. A ship accident occurs indeed at a certain point and the possible contamination is transported relatively slowly owing to winds and currents.

The overall progress in the technology of shipping and in the detection of various ship-induced fields allowed the identification of several non-local agents during the latter decades. Apart from the growth of exhaust emissions [35] capable of creating substantial changes in the atmosphere at a height of many hundreds of meters above the sea surface [16], and an increase in external noise, also waves generated by large high-speed ships have become a key problem in certain coastal areas [65, 94].

The latter outcome was deeply surprising because ship waves form one of the most widely studied wave classes, and the relevant concepts and solutions were thought to be quite well understood. The classical theory of ship waves stems from the 19th century [25, 87] and has been comprehensively described in the international literature [46, 51, 82, 83]. The generic contribution of ship-induced water motion to the local hydrodynamic activity in certain water bodies was recognized a long time ago. Its importance is obvious in inland waterways, narrow straits, small and medium-sized lakes [34], and in the vicinity of vulnerable areas such as sheltered micro- or non-tidal areas, wetlands or low-energy coasts [6, 68], where the natural wave field is very weak. The list of potential problems is long and is connected with very different aspects of the marine management. Ship wakes and intense jets generated by the propulsion systems can essentially contribute to shoreline erosion [6, 26, 59, 66], may cause an extensive erosion and resuspension of bottom sediments [5, 84], and may easily trigger ecological disturbance and cause harm to the aquatic wildlife [2, 52], to mention just a few studies. An overview of navigation-caused threats to fish assemblages is given in [93]. Damage to structures and archaeological sites, and safety problems for navigation as well as for beach users may also arise [62]. Ecological aspects of intense ship traffic have been intensively studied for the archipelagos of the Baltic Sea [53].

A new aspect discovered in the 1990s is that waves excited by contemporary, strongly powered ships sailing at a high speed in shallow and moderate water depths acquire several qualitative differences compared to the classical ship wake patterns. One of the most intriguing features was that ship wakes demonstrated the potential for violent energy concentration not only in the vicinity of ship lanes but also in remote sea areas [31]. Such wakes not only affect the coastal environment, but also jeopardize the safety of people and their property [31, 62]. Early reports of dangerous ship wakes [31] contain descriptions of how holidaymakers were forced to "flee for their lives when enormous waves erupted from a millpond-smooth sea" or that waves (that caused a fatal accident near Harwich, a port on England's east coast, in July 1999) look like "the white cliffs of Dover."

Starting from the mid-1990s, the pioneering studies of the properties of this new type of ship wakes were performed more or less simultaneously in several countries [23, 41, 42, 43]. The major advance resulting from these studies consists in identifying the intriguing new features of the ship wave systems, depicting the major new threats, and formulating recommendations towards reasonable limitations for the ship traffic. The research was explicitly directed to understanding the role of new types of ships — so-called *High Speed Craft* — on routes near land or enclosed estuaries [41], and to adjusting the existing legislation so that the use of these ships would present no danger to people and the environment [65]. The research highlighted the particular features of waves created by large, strongly powered ships in relatively shallow areas. Such ships are now generally called *fast ferries*. The sailing speed, however, has, strictly speaking, direct relevance in wave matters only when the ship's speed and the water depth match each other in a certain manner. The basic parameter here is the relation of the ship's speed and the maximum phase speed (celerity) of linear waves for the given water depth.

The hydrodynamic impact of ship waves is usually negligible in coastal areas of large water bodies, where natural waves are the dominant driving force [52]. Yet for particular combinations of the existing hydrodynamic loads and the coastal environment, ship waves may become a major forcing factor for certain parts of coasts exposed to significant natural wave activity [80].

The goal of this paper is to give an overview of the major steps from the knowledge of the properties of ship waves as summarized in the classical texts towards understanding of the new, fascinating features of wakes from fast ferries. The presentation starts from a well-known exercise of how to calculate the basic properties of ship-induced waves in the framework of the linear theory. The main object of study in this framework is the so-called *Kelvin wave system*, or *Kelvin wake*. An instructive and useful feature of this material is the demonstration that simple underlying physics and mathematics may lead to conclusions that are very rich-in-content. While the mathematical treatment of deep water waves is fairly simple and transparent, many more efforts are needed to properly describe the features connected with finite-depth effects. Yet the qualitative aspects can be easily tracked with the use of a convenient scaling in terms of so-called *Froude numbers*. A fascinating feature of such a line of reasoning is the similarity of the effects occurring in a certain near-critical or supercritical range of speeds to well-known phenomena such as the Mach cone in supersonic aeronautics, or the Cherenkov radiation in classical physics.

In many cases, ship-induced waves exhibit certain properties of solitons. The classical nomenclature associates the term soliton with nonlinear, localised entities, which have a permanent form and retain their identity in interactions with other entities from the same class [15]. One part of the ship-induced disturbancies can be described in terms of solitons of a different kind with a high accuracy. Another major part of the waves from fast ferries resemble shallow-water solitons, but it is not clear whether they interact elastically with similar entitities; for that reason they are frequently called solitonic components. The deeply interesting topic of the generation, propagation and interactions of solitonic ship waves that can be identified as Korteweg-de Vries or Kadomtsev-Petviashvili solitons has been intentionally left out of this presentation. Comprehensive overviews of the relevant equations, mathematical methods, and observations are given in [71, 72]. Instead, this chapter focuses on the phenomenology of various more practical aspects connected with the wakes from fast ferries. The reason behind this is the fact that the explosive increase in the intensity of high-speed traffic during the last decade has initiated a number of studies of the properties of ship wakes in coastal areas.

One of the major scenes has been Tallinn Bay, a semi-enclosed area with dimensions of about 10×20 km. This water body has hosted extremely heavy fast ferry traffic since 1999. Various types of large, high-speed vessels crossed the bay up to 70 times per day during the high season at the turn of the millennium. The depth along the ship lane is 15–50 m. Several ships have occasionally entered into the transcritical regime [90]; however, none of the ships intentionally sailed in the clearly supercritical range of Froude numbers.

Tallinn Bay is a non-tidal area sheltered from open ocean swell and hosts relatively low intensity near-bottom currents. The unique feature of this region is that the impact of ship-induced hydrodynamic activity, in particular, of its long components, is comparable with or even exceeds the existing wave loads [79].

The method for understanding the role of this qualitatively new feature of hydrodynamic activity in this area involved a number of different approaches. It started from extensive field studies of single-point properties of ship waves in different sections of the coast that led to a striking estimate of the role of anthropogenic waves in the total wave activity [80]. In order to reach adequate estimates, the wind wave climate in the entire Baltic Sea and in sea areas adjacent to Tallinn Bay had to be established based both on historical wave measurements and on high-resolution wind wave simulations [7, 69]. As the knowledge of marine wind properties was relatively vague in this area, another focus of the relevant studies was how to reconstruct the open sea wind patterns from coastal wind data.

Later on, the presence of large variations in average properties of the wave fields was explained by numerical simulations of the spatial patterns of ship wakes. The specific role of nonlinear processes in the immediate vicinity of the coastline was estimated based on the cnoidal wave theory, and was found to be of large importance because of the excess length of a part of the ship waves. The direct impact of ship wakes on bottom sediments was quantified using *in situ* conditions with the use of combined measurements of wave properties, suspended matter, and optical properties of sea water.

2 Linear ship wakes

2.1 Kelvin wedge

Whenever a body (ship, bird, fish, insect) moves actively along the water surface, pressure variations at the water-air interface produce a series of waves. Since it does not matter whether the body moves or the fluid flows, one can always use a coordinate system attached to the ship. A steadily moving point source in deep water usually excites two sets of waves that move forward and out from the disturbance (*divergent waves*), and one set of waves that move in the direction of the disturbance (*transverse waves*, Fig. 1). The corresponding linear theory was developed by Lord Kelvin in 1887 [87], after whom this pattern is called Kelvin wave system, or *Kelvin wake*. Although wakes from fast ferries generally cannot be exactly described in the linear framework, conjectures inferred from the linear theory of the Kelvin wake are extremely helpful in explaining the nature of more complex phenomena attached to ship-induced waves.

In realistic conditions, different parts of the ship's hull usually produce several wave systems (Fig. 2). Generally only stern waves have a magnitude comparable with that of bow waves. The divergent bow and stern waves may travel independently if the ship is long enough, but the transverse bow and divergent stern waves are always superimposed (Fig. 1b). This feature opens a way towards decreasing the wave resistance by means of design, for which waves created by different parts of the hull meet in antiphase and annihilate each other.

The classical problem of kinematics of ship waves consists in determining the steady pattern of wave crests in the framework of the linear wave theory. The solution is described in many textbooks in hydrodynamics ([47], § 256; [51] § 3.10, or [83]). Note that in the linear framework the wave height does not directly depend on the other basic wave properties such as the length, period, or propagation direction, and must be treated separately. Note also that the full ship wave system may contain nonsteady components that are not accounted for in this theory.

The problem of the pattern of ship waves may arise for any type of waves (short, long, capillary, internal, etc.). Major progress towards the solution can be made provided the dispersion relation of the relevant wave class is known. The key to a solution consists in recognizing that (i) the constant phase curves (incl. the lines representing the wave crests and troughs) are always perpendicular to the wave vector, and that (ii) the local phase velocity (celerity) c_f of stationary waves is equal to the projection of the velocity of the ship onto the direction of the wave vector [97, 98].

Therefore the *wave crests* created by a ship sailing steadily with a speed V can only be stationary if the wave component travelling under an angle θ with respect to the sailing line has the *phase velocity*

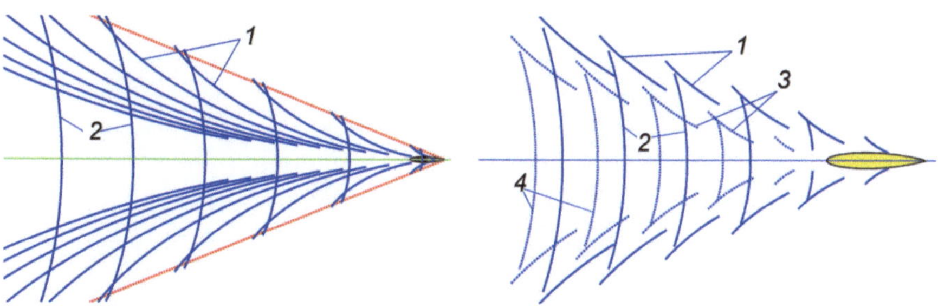

Fig. 1 Pattern of wave crests generated by a point pressure disturbance moving over deep water (left panel): 1-divergent waves, 2-transverse waves; realistic ship wave pattern after [4] (right panel): 1-divergent bow waves, 2-transverse bow waves, 3-divergent stern waves, 4-transverse stern waves.

$$c_f = \frac{\omega}{k} = V\cos\theta. \tag{1}$$

Let the ship sail from X to O (Fig. 3), crossing the distance Vt_g, within time t_g. The consequence of Eq. (1) is that during t_g steady waves only fill a certain circle. This circle is obviously symmetric with respect to the sailing line and goes through X.

If the phase and group velocity of ship-induced waves were equal, the pattern would be present in circle (1) of Fig. 3. Since *wave energy* generally travels with another velocity, namely the *group velocity* $c_g = \partial\omega/\partial k$, energy released at X actually reaches, e.g., in the sailing direction only until O^*. Therefore the energy of this wave crosses the distance $c_g t_g$. If $c_f > c_g$, which is the case in deep water and in water of finite depth, the waves we are looking for actually fill circle (2) that has a diameter $Vt_g c_g c_f^{-1}$ and is centred at $X_1 = X + \frac{1}{2}Vt_g c_g c_f^{-1} t_g$.

The direct consequence from this line of reasoning is that steady waves can only exist within a *Kelvin wedge* containing all circles (2) and limited by their common tangents going through O. As the diameter of the circle (2) is known, the relevant half-angle of the apex α of the wedge is defined from the purely geometrical condition

$$\sin\alpha = \frac{Vc_g c_f^{-1} t_g/2}{Vt_g - Vc_g c_f^{-1} t_g/2} = \frac{1}{2c_f c_g^{-1} - 1}. \tag{2}$$

Fig. 2 Pattern of ship waves in Geirangerfjord, Norway.

For deep water waves the linear dispersion relation is

$$\omega = \sqrt{gk}, \tag{3}$$

where g is the gravity acceleration, $k = 2\pi/L$ is the wave number (the length of the wave vector) and L is the wave length. It is easy to see that in this case $c_f = 2c_g$, $\sin\alpha = 1/3$, and $\alpha = \text{const} \approx 19°28'$. Therefore the geometry of the steady wave pattern does not explicitly depend on the sailing speed, a feature which seems to be counter-intuitive, but which still holds with very high accuracy in nature.

The relevant wave number k, wave length L, and period T for waves that have reached any point at circle (2) can be found from Eq. (1) as

$$k = V^{-2} g \cos^{-2}\theta, \tag{4}$$
$$L = 2\pi V^2 g^{-1} \cos^2\theta, \tag{5}$$
$$T = 2\pi V g^{-1} \cos\theta. \tag{6}$$

The situation is more complicated when the ship sails in water of finite depth. Then the dispersion relation

$$\omega = \sqrt{gk \tanh(kH)}, \tag{7}$$

wave properties and, most importantly, the ratio

$$\frac{c_f}{c_g} = \frac{2}{1 + 2kH \sinh^{-1}(2kH)} \tag{8}$$

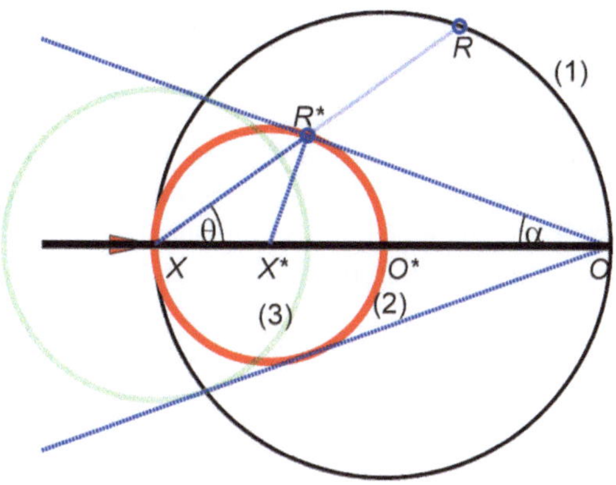

Fig. 3 Scheme of the Kelvin wedge for water waves. The wedge for deep water coincides with the supercritical wedge for $F_h = 3$.

additionally depend on the water depth H. Note that the shape of the Kelvin wake is still entirely defined by the ratio c_f/c_g, which is one of the decisive parameters of the linear ship wave theory in different environments [97, 98].

As there is no limitation for wave celerity in infinitely deep water, the ship wave pattern is theoretically valid for arbitrary ship speeds in the open ocean, where the effect of a finite depth is negligible.

The situation is different in shallow water, where wave celerity has a finite limit. It is obvious that the steady transverse waves propagating along the sailing line have celerity equal to the ship's speed. This can be easily verified, because for such waves $\theta = 0$ and Eq. (1) reduces to $V = c_f$. The intrinsic limitation of the classical Kelvin theory is that the celerity of linear waves (incl. steady ship waves) cannot exceed the maximum speed of long waves $\sqrt{gH}$ (also called *long wave speed*) for a particular depth.

This speed therefore plays a specific role in the formation of ship wave patterns in water bodies of finite depth. From above, it is clear that no steady transverse wave exists for speeds larger than $\sqrt{gH}$ and the steady wake of ships sailing even faster consists of divergent waves exclusively.

Since the ratio c_f/c_g decreases when the water depth H decreases, the Kelvin wedge widens in shallow water. The half-angle α can be found from Eq. (2) by substituting there the ratio c_f/c_g from Eq. (8):

$$\sin\alpha = \frac{1+2kH\sinh^{-1}(2kH)}{3-2kH\sinh^{-1}(2kH)} \text{ or } \cos^2\alpha = \frac{8[1-2kH\sinh^{-1}(2kH)]}{[3-2kH\sinh^{-1}(2kH)]^2}. \tag{9}$$

For finite speeds in very deep water ($H \to \infty$), expressions (9) give, as expected, $\sin\alpha \to 1/3$, because the wave vector k remains limited. If the water depth decreases, the wave length also decreases. As always $2kH < \sinh(2kH)$, Eqs. (9) always define a formal value of α; yet there exist certain external limitations for its physically meaningful values.

The properties of steady waves located at the border of the Kelvin wedge on circle (2) can still be easily found from Eq. (1) and the dispersion relation for divergent waves. The triangle OX^*R^* is obviously right-angled, and the angle OX^*R^* is $\pi/2-\alpha$. The triangle XR^*X^* is equilateral; therefore the angle R^*XX^* is $\theta = \frac{1}{2}[\pi/2-(\pi/4+\alpha)]$. Substituting this value in Eq. (1) gives

$$\omega^2 = k^2V^2\cos^2\theta = k^2V^2\frac{1+\cos(\pi/4-\alpha)}{2} = k^2V^2\frac{1+\sin\alpha}{2}. \tag{10}$$

Making use of Eq. (9) gives

$$\omega^2 = \frac{k^2V^2}{3-2kH\sinh^{-1}(2kH)}. \tag{11}$$

On the other hand, from Eq. (7) we have that $\omega^2 = gk\tanh kH$, which means that at the border of the Kelvin wedge the following relation between the wave number, water depth and ship's speed holds:

$$gk\tanh kH = \frac{2k^2V^2}{3-2kH\sinh^{-1}(2kH)}. \tag{12}$$

This relation can be interpreted as the following transcendental equation with respect to kH [82]:

$$\frac{\tanh kH}{kH}[3-2kH\sinh^{-1}(2kH)] = \left(\frac{V}{\sqrt{gH}}\right)^2. \tag{13}$$

2.2 Navigational speeds

An important consequence of Eq. (13) is that for a given water depth, the angle α and the properties of the ship waves at the border of the Kelvin wedge depend only on the ratio

$$F_h = \frac{V}{\sqrt{gH}} \tag{14}$$

of the ship's speed V and the maximum phase velocity of free surface waves $\sqrt{gH}$ for the given water depth. This convenient measure to characterize the relative speed of the ship with respect to the long wave speed is called (the depth-based) *Froude number* and is widely used in the contemporary theory of ship motion and ship waves.

Note that Eq. (13) implicitly sets a limit for the apex angle of the Kelvin wedge. Namely, as the left-hand side of Eq. (13) never exceeds 1, a real solution to this equation only exists if $F_h \leq 1$. Nonzero (that is, physically meaningful) values of kH only exist for $F_h < 1$. In the limit $F_h \to 1$ also $kH \to 0$ and $\alpha \to \pi/2$. In other words, if $V \to \sqrt{gH}$, the angle α reaches the maximum value $\alpha = \pi/2$ and, formally, the Kelvin wedge fills the entire half-plane aft of the ship.

Many authors write (perhaps after [32, 82]) that in this limit, the transverse and the divergent waves form a single large wave with its crest normal to the sailing line and that it travels at the same speed as the disturbance. Such a description should be interpreted as a figurative one, reflecting the fact that ship-generated wave heights increase considerably at $F_h \to 1$. It is, however, conceptually imprecise, because what exactly happens at speeds corresponding to $F_h \approx 1$ cannot be described by the linear theory. In particular, no linear steady waves exist at $F_h = 1$ [46]. Therefore, for near-critical speeds the linear approach is only conditionally applicable and must be used, if at all, with great care.

The range of the practical validity of this theory can be estimated as follows. Generally, shallow-water effects influencing, among other things, the apex angle of the Kelvin wedge, start to become important when the wavelength is twice as large as the water depth, or equivalently, when $kH \approx \pi$. The corresponding Froude number for divergent waves at the edge of the Kelvin wedge can be found from Eq. (3) as $F_{hd} \approx 0.687$. For somewhat longer transverse waves at the sailing line, this threshold is $F_{ht} \approx 0.56$ [82]. Therefore, at depth Froude numbers above 0.55–0.7, the ship-generated wave system should respond to the water depth (Fig. 4).

The threshold $F_h = 1$ serves as a natural basis of the classification of navigational speeds. Operating at speeds resulting in $F_h < 1$ is defined as *subcritical*, at $F_h > 1$ as *supercritical*, and at $F_h = 1$ as *critical*. There is a so-called *transcritical* speed range $0.84 < F_h < 1.15$ in realistic conditions, where no clear distinction between sub- and supercritical regimes is possible [36], and in which the linear theory generally fails.

Another important parameter of the sailing regime is the *depth based Froude number* $F_L = V/\sqrt{gL_W}$, where L_W is the length of the ship's waterline. A specific regime called *hump speed* occurs when the wavelength of the transverse waves propagating along the sailing line [at which $\cos\theta = 0$ in Eq. (4)] $L = 2\pi V^2 g^{-1}$ is about twice the ship's length $2L_W$, that is,

$$2\pi V^2 g^{-1} = 2L_W, \qquad (15)$$

from where we have that

$$F_{Lhump} = \frac{1}{\sqrt{\pi}} \approx 0.56. \qquad (16)$$

Sailing at this speed means that the ship is continuously moving "upwards" along the slope of her own wake. Wave resistance usually increases fast when $F_L \to \pi^{-1/2}$. The increase is particularly significant in shallow water conditions [83]. Approaching either the critical speed or the hump speed is commonly accompanied by a drastic increase in wave resistance, equivalently, by an increase in released wave energy. The highest waves eventually occur when the hump and the critical speed coincide [62, 65, 82].

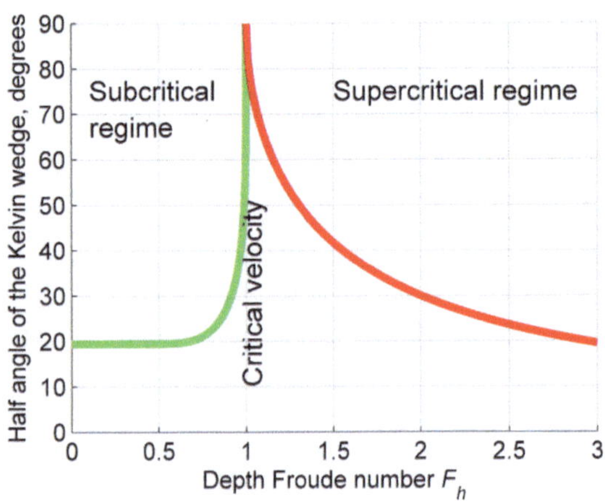

Fig. 4 Dependence of the half-angle of the Kelvin wedge on the Froude number.

2.3 Distribution of wave heights and periods

As mentioned above, in the linear theory the wave height is not directly connected with other properties of the wave and must be determined separately. It is straightforward to sketch the overall qualitative rule of the wave height changes with the distance r from the ship, provided wave dissipation is neglected. The width of the Kelvin wedge increases proportionally to r, whereas the rate of increase in the total energy of the ship wave system is constant. Therefore, the amount of wave energy behind the ship is constant along any line perpendicular to the sailing line and the wave height is expected to decrease as $r^{-1/2}$. Along the border of the wedge, however, a complex pattern of interfering wave components occurs. Wave amplitudes decay much more slowly there, as $r^{-1/3}$ [32, 46, 91]. The highest ship waves thus usually are found at the edges of the Kelvin wake [46, 82]. The decrease rate of wave heights with the distance from the ship is quite substantial also in this case; for this reason the waves from ships producing the classical Kelvin wave pattern usually cannot be distinguished from wind waves already at a distance of a few kilometres from the sailing line.

The real dependence of the ship wave heights on the ship parameters and the sailing regime is extremely complex. In deep water ($F_h \ll 1$), wave heights in the inner part of the Kelvin wedge are usually proportional to the ship's speed. Near the border of the Kelvin wedge they behave as $V^{2/3}$, obviously with a different proportionality coefficient. The maximum wave heights increase very rapidly if $F_h \to 1$, at a rate of F_h^p, with the power p at approximately 3–4 [82]. Beyond the critical speed, the wave amplitudes usually decrease somewhat with an increasing Froude number for a limited range of F_h, but still are higher than those at low Froude numbers.

Some other properties of the largest deep-water waves at the border of the Kelvin wedge can also be found in a straightforward manner. For such waves $\theta = \pi/4 - \alpha/2$ corresponds to the direction from X in Fig. 3, for which waves excited by the ship propagate through the common point of the circle (2) and the edge of the Kelvin wedge. As in deep water $\sin\alpha = 1/3$, one can easily show that for such waves $\cos\theta = \sqrt{2/3}$ and $T \approx 0.52V$, $L \approx 0.52V^2$. The units of periods and wave lengths are expressed in seconds and metres, respectively, provided the ship's speed is given in m/s.

These waves, however, are not the longest among steady ship waves. From Eqs. (4)–(6) it follows that the wave properties within the Kelvin wedge vary largely as the angle θ varies from $\theta = 0$ to $\theta = \pi/2$. The longest waves are the transverse waves propagating exactly along the sailing line, corresponding to $\theta = 0$. Their periods and lengths can be expressed as $T \approx 0.6405V$, and $L \approx 0.6405V^2$, respectively.

The basic consequence from this simple analysis is that in deep water the period of ship waves propagating at any angle with respect to the sailing line increases linearly with the increase in the ship's speed as shown by Eq. (6). The wave length is proportional to the ship's speed squared as indicated by Eq. (5). In particular, these conclusions are true for the largest and longest waves of the wake. This conjecture simply reflects the familiar fact that the faster a ship sails, the longer are the leading waves.

As the water depth in a real sea is always finite, the presented results are, strictly speaking, only valid asymptotically in the far field in deep water. They hold with a certain accuracy also for moderate Froude numbers, for which the effects of finite depth are still negligible [82]. For sea areas of a finite depth H, the period T can be calculated from the following transcendental equation, obtained from Eq. (1) by substituting the dispersion relation for surface waves and by expressing the wave number from Eqs. (4)–(6) as $k^{-1} = V^2 g^{-1} \cos^2\theta = (TV\cos\theta)/2\pi$:

$$\frac{gT}{2\pi} \tanh \frac{2\pi H}{TV\cos\theta} = V\cos\theta. \tag{17}$$

Similar to the case of deep water, the period of the longest transverse waves can be obtained from Eq. (17) when $\theta = 0$. The period of the largest divergent waves at the border of the Kelvin wedge occurs, as above, when $\theta = \pi/4 - \alpha/2$. In contrast to the deep water case, the angle α must be defined now from Eqs. (9) and (13). Some examples of the dependence of wave period on ship speed and water depth are shown in Fig. 5.

Although the appearance of the area filled by the waves from ships sailing at supercritical speeds is formally similar to the one in Fig. 3, the interpretation of the details of the wave propagation is quite different from that in deep water or at subcritical speeds. For this reason, a different line of thinking is used to describe the supercritical analogue of the Kelvin wedge. At such speeds, obviously no steady transverse waves exist. The wedge is therefore filled by divergent waves only. For the celerity of ship-generated waves holds $c_f \le \sqrt{gH}$. Therefore waves excited at X cannot propagate outside of the circle (3) that is centred at X and has a radius of $t_g\sqrt{gH}$, no matter what their propagation velocity and direction is. These circles fill a wedge with a half-angle given by

$$\sin\alpha = \frac{\max(c_g)t_g}{Vt_g} = \frac{\sqrt{gH}}{V} = \frac{1}{F_h}. \tag{18}$$

This wedge, formally similar to the Kelvin wedge, conceptually resembles the Mach cone of shock waves in supersonic flows. The entire process of wave radiation by a supercritical ship can be interpreted as an analogue to Cherenkov radiation. The apex angle of the resulting wedge is equivalent to the relevant Mach angle. The appearance of the wedge depends on the Froude number. Different to the subcritical case, its apex angle decreases when the ship's speed increases. For $F_h = 3$ this wedge exactly coincides with the wedge of ship waves in deep water.

The highest waves are again frequently found at the border of this wedge. As different from the subcritical case, they more closely resemble (shock) waves created by the ship at a point X (Fig. 6). Their front approximately coincides with the edge of the Kelvin wedge. Their propagation direction now makes the angle $\theta = \pi/2 - \alpha$ with the sailing line (Fig. 3). The approximate period of the leading wave implicitly depends on the Froude number. For example, for $F_h = 3$, it is $T = 2\pi V g^{-1} \sin\theta \approx 0.21V$. A comparison of the presented relationships suggests that the dependence of the period of the leading waves on the ship's speed is strongly nonlinear in the transcritical regime.

3 Patterns of wakes from fast ferries

3.1 Changes in the ship wave pattern

The above theory is basically based on linear approximation and fails to describe the spatial distribution of wave heights and the properties of the wave pattern at near-critical speeds. The major sources of relevant information are physical and numerical experiments. Apart from the increase in the apex angle of the Kelvin wedge, substantial changes of the entire ship wave system occur when the ship accelerates so that the depth Froude number approaches the transcritical range. As F_h increases, the leading divergent waves are accentuated at the expense of other waves. The lengths of the crests of the leading divergent waves gradually increase as well. At near-critical speeds, the major part of the energy is concentrated in a few long-crested waves, which have nearly straight crests [38, 82]. The details of these changes depend on many local features. For example, for a limited range of F_h, the amplitudes of transverse waves increase as well [82]. This increase, however,

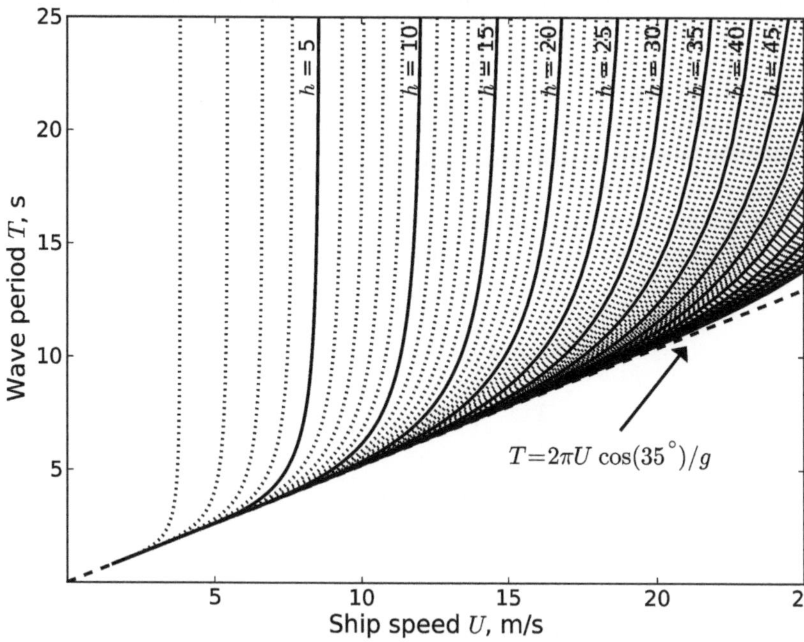

Fig. 5 Periods of the largest ship waves for subcritical speeds in the Kelvin wake. Dotted lines indicate results at different depths. Solid lines are drawn at 5 m intervals. Dashed line indicates wave period in deep water. The sudden steepening of the lines in the upper part of the figure does not indicate a dramatic increase in wave period, but rather that the assumption of the existence of a steady, linear Kelvin wake no longer applies [90]. Reprinted with permission from Estonian Academy Publishers.

will not continue indefinitely, because no transverse waves exist in the supercritical regime.

For very large (so-called high-speed supercritical) speeds, the Kelvin wedge is narrow. The energy of the ship waves is therefore concentrated in a small area [97]. Also, since the curvature of wave crests in this regime is relatively small, the extent of the area, where a particular wave crest is located in the vicinity of the border of the Kelvin wedge, is particularly large. Consequently, the high-speed supercritical waves have particularly long crests (Fig. 6).

The increase in the wave length when the ship's speed increases is accompanied by the gradual unification of the phase speed (celerity) of the largest and longest ship waves. This means that the impact of classical dispersion of such waves owing to differences in their frequency or length is quite small. As a result, a part of high-speed ship wakes in shallow areas consists of nearly non-dispersive long-crested waves.

Trans- and supercritical ship wave systems in the open sea possess an interesting feature that can be called geometric dispersion. The crest of each single wave among the leading ship wave group created at trans- and supercritical speeds has its own orientation and propagates in a different direction. Therefore the observed period of the leading waves increases not so much owing to frequency dispersion, but mostly owing to the process of purely geometric dispersion.

The viscous damping is negligible for long waves [82]. The other primary agent of spatial energy redistribution—the crest-lengthening process due to diffraction—is inefficient for such long-crested waves. Such waves only lose their energy owing to wave-bottom interactions. The packet of leading ship waves, accordingly, disperses particularly slowly and may remain compact for a long time. The combination of the listed features is responsible for substantial remote influence from ship traffic, the agents of which are severely high and compact wave groups [70].

Fig. 6 Waves from a small boat entering Nesna harbour (Norway, north of Mosjoen) at a supercritical speed.

Several other wake components may be created at large speeds (Fig. 7). A qualitative sketch of the full set of such components is presented in [73].

In the interior of the Kelvin wedge, a very specific wave packet may occasionally be generated at relatively high speeds [8, 65]. It makes an average angle with the sailing line about 10° [57]. There was an extensive debate about the nature of these waves until clear evidence was presented in [8] that this wave system was an autonomous part of the ship's wake. It consists of essentially plane waves that have long (but finite) crests. Such waves have been frequently observed in open sea conditions [78]. Some authors suggest that they are not stationary with respect to the ship [54]. This packet is frequently associated with trans- and supercritical wakes [41, 65]. In fact it becomes evident at low depth Froude numbers such as $F_h \approx 0.5$ [8]. The numerical models available before 1990s were not able to reproduce this feature [8], and this situation still largely continues today [73].

In many cases, this packet is a highly nonlinear spatially localized feature that resembles a bright envelope soliton solution of the nonlinear Schrödinger (NLS) equation. The (cubic) NLS equation has the nondimensional form $i\phi_t + p\phi_{xx} + q|\phi|\phi^2 = 0$ and is the universal equation describing the evolution of complex wave envelopes ϕ in a dispersive weakly nonlinear medium [15]. Its major applications are nonlinear optics, where its different versions describe the propagation and interactions of solitons of different kinds, and the theory of deep water waves in surface wave studies. The studies of its soliton solutions and their interactions date back to the 1960s. The classical soliton solution is called an envelope soliton. The name comes from the appearance of such structures that either have a shape of spatially localised relatively intense oscillations (a so-called bright soliton) or resemble an area of a decrease in the amplitude of oscillations with the carrier frequency (a so-called dark soliton) (Fig. 8).

Packets resembling bright envelope solitons are usually observed only in almost calm conditions [33]. A probable reason is that such packets have limited stability

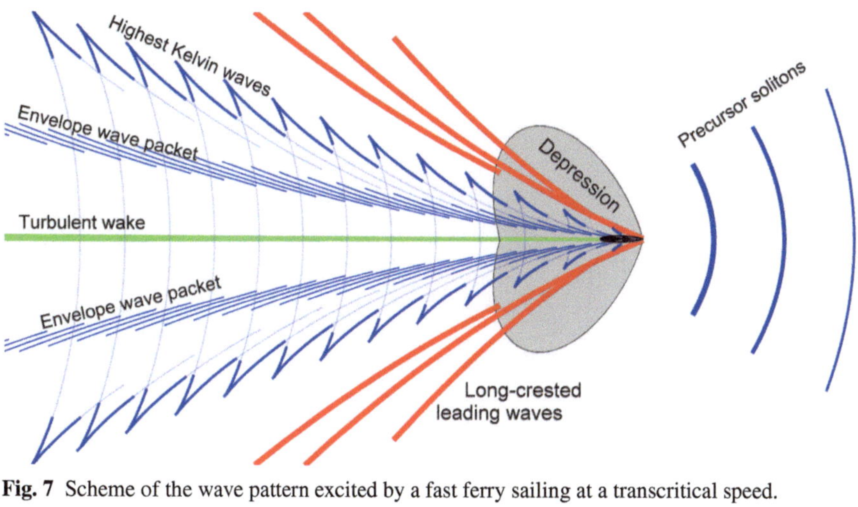

Fig. 7 Scheme of the wave pattern excited by a fast ferry sailing at a transcritical speed.

with respect to random disturbances [12]. They apparently are destroyed relatively fast unless there are no interacting harmonics with the sea state. There is still discussion as to which mechanism produces such wave systems [1, 30]. A better understanding of this phenomenon is particularly important, because this wave group is almost perfectly monochromatic, decays slowly in calm conditions, and frequently is the highest within the fast ferries' wakes in areas remote from the ship lane [78].

According to the linear theory, no steady waves exist ahead of the ship in deep water, where $F_h \ll 1$. This conclusion also holds for moderate speeds in shallow areas. For a certain range of speeds, however, nearly steady solitary waves can be generated ahead of the ship bow. John Scott Russell first documented this phenomenon as he watched, in 1834, a canal boat pulled by horses stopping suddenly (see his description reprinted, for example, in [15]). It has also been historically observed in towing tanks that a ship model advancing steadily can radiate waves upstream that move faster than the ship. These waves, called precursor solitons or *precursors* [95, 96], are perfect solitons. They not only propagate upstream keeping their shape and velocity constant, but also interact elastically and survive when they meet other similar entities. Their generation may start at Froude numbers as low as $F_h \approx 0.2$.

The solitons radiating ahead of a ship begin to break when $F_h \approx 1.1-1.2$ [20, 49] and can no longer be generated when F_h is greater than about 1.2. Both smooth and breaking solitons displace or carry a certain amount of water with them, as nonlinear entities usually do. Under specific conditions they may form a sort of bore ahead of the ship. The bore usually travels faster than the ship and should therefore disappear after some time. An important feature is that the bore-containing flow may become steady for a range of Froude numbers [27]. If the ship's speed increases further, a steady supercritical flow will result.

The radiated solitons as well as the supercritical bore rise almost entirely above the calm water level and displace a certain amount of mass. This mass is taken from the vicinity of the ship. A fascinating feature of the fast ferry traffic is that around the ship (usually immediately behind the moving disturbance, Fig. 9), there is an ever-longer region of depressed water of nearly uniform depth [20, 38, 49, 95]. The excess mass of the upstream-advancing solitons is found to come almost entirely from this region. The extent of the depression may be quite large. Its presence causes

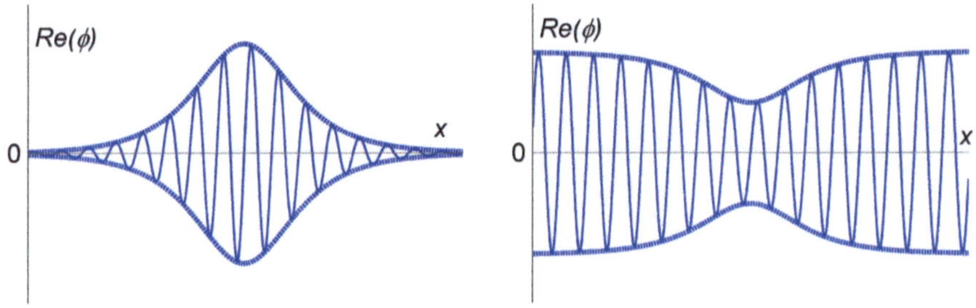

Fig. 8 The appearance of a bright (left) and a dark (right) envelope soliton.

the draw-down effect (called *squat* [56, 58]) of fast ferries. This phenomenon, a specific feature of shallow water navigation, may serve as an additional source of danger for fast ships [17, 56]. The depression may easily penetrate to bays and harbours located adjacent to a navigation channel. In some cases water dropdown in harbours adjacent to the ship lane may be unacceptable [24].

Solitonic precursors excited by contemporary high speed ships sailing at trans-critical speeds have been claimed to be responsible for some dangerous tsunami-like waves along shorelines [31, 41, 50]. As the precursors usually have small amplitudes, a more probable reason for danger is that the leading divergent waves become highly cnoidal or obtain solitonic nature when they propagate into shallow areas [62, 81]. As the distinguishing feature of long ship waves is their long crests, they frequently behave as almost perfect plane waves. Therefore, the concepts used in the analysis of weakly two-dimensional waves (that is, plane wave systems propagating under a small angle with respect to each other [75]) are frequently applicable. The ability to stay in nonlinear interaction with other waves of the same type for a long time is a decisive factor here. The most drastic result of nonlinear interactions of such waves (for example, when two solitonic ship wakes cross each other) is the four-fold amplification of the wave heights and up to eight-fold amplification of the surface slope. The potential influence of nonlinear effects upon ship-generated waves is described in detail in [71, 72, 75]. Since the height of solitonic ship waves in the open sea can reach 1–1.5 m [78], nonlinear amplification of either their height or their slope may result in acute danger.

3.2 Realistic spatial patterns of ship waves

The above description of the spatial pattern of ship waves is only correct for unlimited sea areas of a constant depth. Ship wakes in inland waterways and nar-

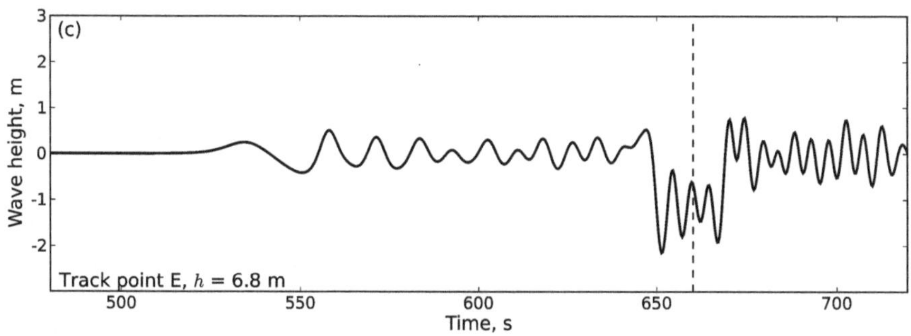

Fig. 9 Numerically simulated time series of the water surface at a selected point on the sailing line of a fast ferry sailing at a transcritical speed [90]. About 10 precursors are running ahead of the ship, the position of the centre of which is marked with the vertical dashed line. The effective length of the depression area is about three times as long as the ship's length. Reprinted with permission from Estonian Academy Publishers.

row straits can frequently be approximated by analogous model patterns excited by ships sailing steadily in uniform channels. However, the determination of such patterns and their spatial and temporal variability in semi-open sea areas with complex bathymetry and geometry is highly non-trivial. Additional effects arise when the ship's speed or the Froude number vary along the sailing line. Despite major advances in fluid mechanics and computational fluid dynamics, the numerical representation of ship wakes has still remained a major challenge [62, 65, 73].

The classical Kelvin theory of steady wakes is usually justified in cases when a ship sails steadily along a straight track over a sea area of a constant depth. In such cases, the major geometric properties of even highly nonlinear wave systems as well as the periods of the leading waves reasonably match those derived from the linear theory [73, 82, 90]. The classical wave propagation and transformation effects such as topographic refraction, lateral diffraction, or bottom-induced damping can be calculated with the use of standard wave models, provided the wake parameters are known in the neighbourhood of the track [41, 42, 43]. The relevant results together with semi-empirical relationships based on field measurements usually give good results in such cases for the far field [62]. Corrections reflecting the nonlinear (e.g. cnoidal [81]) nature of wave fields in shallow water are also straightforward. Several nonlinear effects occurring in the transcritical regime (such as generation and interactions of solitons) can be described, to a first approximation, within one- or two-dimensional weakly dispersive shallow-water frameworks represented by the Korteweg-de Vries and Kadomtsev-Petviashvili equations [48].

Theoretical studies of the effects occurring from ship wave systems when the ship has a variable speed and/or sails in water of non-uniform depth (in both cases the depth Froude number is not constant) are mostly limited to the 1D framework. If the ship is initially sailing at a supercritical speed, then decelerates to a subcritical speed (either because of a decrease in the ship's speed, or owing to sailing over a deeper area), and accelerates to the supercritical range again, a few solitons (two or three) are excited in the system during such a passage [67]. Solitary waves may be trapped for a certain range of constant accelerations. In such cases the waves are localised close to the forcing and are influenced by the forcing over a long time interval [28]. This mechanism may lead to the excitation of very high solitons provided the ship's speed perfectly matches the gradually increasing speed of the growing precursors.

Studies of two-dimensional wave patterns have mostly been performed numerically for relatively narrow channels, for example, in [39]. A typical problem is the study of the wave pattern occurring in a channel when a ship sailing at a supercritical speed passes a ramp to a much deeper area in which the same speed is clearly subcritical. The appearance of the pattern as well as the maximum wave amplitudes differ significantly, depending on whether the depth Froude number increases or decreases. The transition from super- to subcritical speeds is always accompanied by the generation of a solitary wave with a significant amplitude [89]. The amplitude of the solitary wave generated during acceleration is highly dependent on the transition time and may be fairly small for fast transitions. This feature could be used for reducing the wash effects through constructing a suitable bottom profile in order to achieve a fast "jump" from a sub- to supercritical regime [22].

The major lesson to be learned from these simulations is that several important properties of ship waves (such as the wave height, orientation of wave crests, and wave propagation direction) substantially depend on the sailing regime. In other words, the wave loads along the coasts of sea areas hosting high-speed traffic may vary considerably depending on wave generation conditions along the ship lane. Such variation becomes evident from field measurements of ship wave activity in Tallinn Bay: average wave properties vary more than 20% along a short coastal section with a length of about 0.5 km [78, 79].

Simulations performed for the inner part of Tallinn Bay for a randomly chosen fast ferry [90] demonstrate that the amplitudes of ship wakes reveal extremely large variability along the ship lane and impact mainly certain limited sections of the coast. Moreover, the wake pattern, different to the appearance of the Kelvin wedge, shows a striking asymmetry. As there are no evident topographical features that can create this asymmetry, the manoeuvres of the ship evidently have caused the wave focusing on one side of the track.

Another interesting and unexpected feature is that an extremely large wave is generated by the accelerating ship along a certain section of the sailing track. Earlier studies have shown that large waves may be generated during acceleration for $F_h \approx 1$ [67, 89], but in these studies little attention is paid to waves generated during acceleration in the range of speeds well below the critical one. This wave actually appears while the depth Froude number is $F_h \approx 0.7$. Although the large amplitude wave is most likely a highly localised and transient phenomenon, it may create a substantial danger for small vessels nearby and for the affected coastal sections.

3.3 Ship waves at a fixed point

There exist a variety of methods for tracing ship wakes at selected points. As direct measurements of the time series of the water surface position in such transient wave groups in the open sea are technically quite complex, usually certain indirect methods are used. The most widely used technique, in spite of its problems with a limited resolution for shorter waves, consists in the use of either bottom-mounted or sub-surface pressure sensors [78, 85]. This method is acceptable for the long components of ship wakes, the periods of which exceed a few seconds. This method is thus suitable for recording wakes from high-speed ships, where the periods of single waves typically vary from 2–3 s to 40 s [41, 42, 43]. In particular, these wakes contain a very limited amount of shorter waves in remote areas [78]. The technique represents long waves in shallow water well, where the pressure signal approximately follows the behaviour of the water surface. It adequately represents the periods of the ship waves and the wave energy spectrum. On the other hand, it tends to fail to reproduce the exact shape of waves and the water surface time series [78, 81] because of distortions introduced by the complicated procedure of restoring the surface elevation [65, 85].

Note that different concepts are used to describe the heights of wind waves and ship wakes. In contemporary physical oceanography, the surface wave fields are usually interpreted as stochastic fields and are described in terms of relevant average quantities [44]. The most widely used quantity to characterise the wind wave heights is the significant wave height, defined as $H_S = 4\sqrt{m_0} \approx H_{1/3}$, where m_0 is the zero-order moment of the wave spcctrum (the total variance of the water surface displacement). Its adequate estimate is the average height $H_{1/3}$ of 1/3 of the highest waves within a certain time interval. The latter definition has been extensively used in the past decades and is still a standard in many textbooks. Natural wave fields are usually also characterized by the peak or mean wave period [37].

None of these measures are applicable for ship wakes, in particular, for their long-wave part that consists of transient waves. Wave-by-wave analysis of a properly filtered water surface or pressure time series with the use of the zero-upcrossing, or the zero-downcrossing method [37] is usually recommended for their analysis [65, 78]. The wave height and period is thus estimated for each single wave crest. The cut-off frequency depends on the local conditions. For the long-wave components, the choice of 0.2 Hz gives good results [78], but for detecting the highly oscillating part of the wake the cut-off should be at about 0.5 Hz.

As expected, wakes from conventional ships reflect the features predicted by the theory of Kelvin wakes. The periods of the leading waves are around 3 s and the wave packet disperses relatively fast. It is detectable at a distance of about 3 km from the sailing line, but usually cannot be unambiguously identified at a distance in the order of 10 km. It is interesting to note that wakes from light hydrofoils (even if they sail at near-critical speeds) usually are indistinguishable from the natural background [78].

3.3.1 Group structure

A trans- or supercritical wake observed at a certain point remote from the sailing lane has a completely different appearance from those at the sailing line. Small-amplitude precursor solitons may arrive several minutes before the highest waves [60]. They are common in rivers and channels, but are also frequently observed in wide water bodies [41, 78]. The time interval between their crests may be a few tens of seconds; however, this cannot be interpreted in terms of wave periods.

The major part of the wake typically consists of two or three wave groups [41, 42, 43, 78]. The group structure becomes evident at a certain distance (typically about 500 m) from the ship track. At a medium range (about 2–3 km), the wake lasts about 10–20 min (Fig. 10).

The first group of transient waves contains the longest waves (periods 9–15 s; the long waves arriving first). The highest waves of a wake usually are found in the middle of this group. The group evidently consists of the leading, long, and long-crested waves in Fig. 7 that have undergone a certain transformation owing to geometrical defocusing and an increase in the group speed of the largest waves owing to nonlinear effects. These effects may lead to an increase in the periods and lengths of some waves.

Long waves higher than 0.4 m and with periods of approximately 10 s, exhibit nonlinear properties in the coastal area already at a depth of 4–5 m. In terms of cnoidal wave theory, the elliptic modulus k of the Jacobi elliptic functions is about 0.9–0.95 for about 0.5 m high waves, and reaches 0.99–0.999 for about 0.7–0.9 m high waves [81]. Therefore this group frequently evolves into a train of highly non-linear cnoidal waves. At some coasts, it may evolve into an ensemble of Korteweg-de Vries solitons.

The second group usually consists of waves with periods 7–9 s. It is usually interpreted as representing the classical Kelvin wave system. The typical wave periods in this group generally match the estimates for the linear Kelvin waves for a given ship speed.

The third wave group is present only occasionally. It has a typical duration of <1 min. It consists of relatively short, almost monochromatic waves with periods of 3–4 s. In favourable conditions for visual observations, it forms a clearly identifiable band on the sea surface with a width of about 100 m in shallow areas with a depth of about 3–5 m. Visual observations suggest that it exhibits properties of an envelope soliton. As mentioned above, it probably can be described in terms of solutions of the non-linear Schrödinger equation [29, 30].

At a moderate distance (<3 km) from the ship lane this group is usually smaller than the leading waves. Owing to its monochromatic nature and small wavelengths, it is non-dispersive and has much weaker interactions with the seabed at medium depths (10–30 m) than the leading waves. As a consequence, this group may remain compact for hours, may cross many kilometres, and may contain severely high (about 70 cm, at times the highest waves of the wake) waves at 8–12 km from the

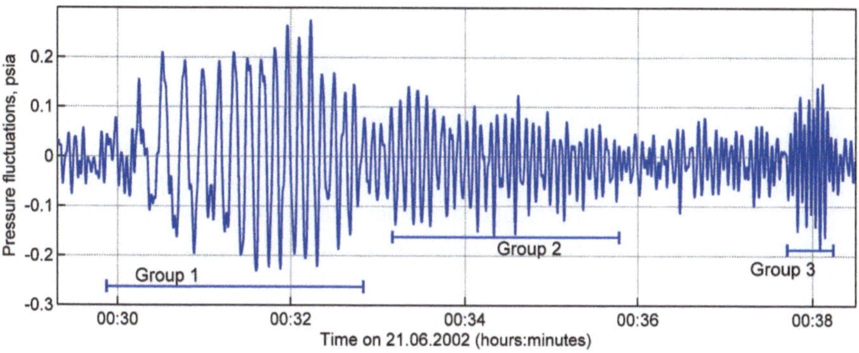

Fig. 10 Pressure fluctuations caused by the wake of a large high-speed ship near the western coast of the island of Aegna (Gulf of Finland, the Baltic Sea). The first wave group has a maximum height of 45 cm, the second group–of 25 cm. The highest is the third group (52 cm). The significant height of the natural wave background is about 30 cm. The pressure sensor was mounted at a depth of 3.5 m below the calm sea surface in an area with a total water depth of 6.7 m [78]. Reprinted with permission from Estonian Academy Publishers.

ship lane [78]. As a result of its small group speed, it arrives much later (more than one hour at a distance of 10 km from the sailing line) after the leading waves.

Although the maximum height of ship waves decreases and the duration of the wave groups gradually increases with the distance from the sailing line owing to both frequency and geometric dispersion, the bulk energy of the ship wakes weakly depends on this distance [78]. As the dispersion relatively weakly affects the propagation of long waves, the groups of long, leading waves excited by fast ferries frequently remain compact for a long time. Consequently, the area influenced by heavy ship traffic can be much larger than the immediate vicinity of the ship lane.

The energy of the waves from the first wave group usually considerably exceeds the energy carried by the rest of the wake. Such a clear prevalence is somewhat unexpected because the second group often is the longest [41, 65, 78]). The third group of short waves also has a very limited duration (usually below 1 minute) and carries a small fraction of the ship wave energy.

3.3.2 Remote impact

Owing to the rapid decay of the amplitudes of the waves from conventional ships ($\sim r^{1/2}$ within the Kelvin wedge and $\sim r^{1/3}$ along its borders [46, 47]), their effect on the remote shoreline is usually small compared to that of the wind waves [62]. At higher Froude numbers, the decay rate may be as low as $\sim r^{1/5}$ [13, 14, 65], and for the third wave group even about $\sim r^{1/6}$ [73, 78]. The reasons for such a low decay, described above, are as follows: (i) the limited wave dispersion in shallow water (that keeps the wave packets compact in the propagation direction), (ii) the relatively long crests of the leading ship waves (that diminish the intensity of the lateral energy redistribution), and (iii) the remarkably monochromatic nature of the third group.

The most important new aspect in this respect is the potential remote impact of the ship traffic in certain areas [70, 79, 80]. Two components of the wakes from fast ferries are able to carry massive amounts of energy far from the ship lane in a compact form: the long and high leading waves, and the group of monochromatic waves. Therefore it may happen that the wave energy is released far from the ship lane a long time after the ship has passed. As such waves usually appear unexpectedly, they create not only spectacular sights but also danger to lives [31]. Breaking of long waves may be responsible, e.g., for drastic thermal changes in shallow inlets several kilometres away from the fairway and even in the open sea [21]. The accompanying artificial upwelling may intensify the eutrophication effects and harmful algae blooms due to the transport of nutrients from sediments into the euphotic layer [52].

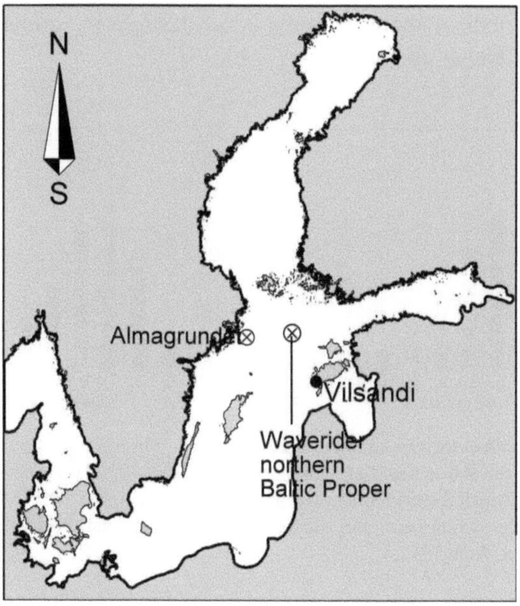

Fig. 11 Location of the wave measurement sites in the northern part of the Baltic Proper: Almagrundet near the Swedish coast [7] and the waverider operated by the Finnish Marine Research Institute in the open part of the Proper [40].

4 Contribution of ship wakes to local hydrodynamics

4.1 Parameters of wind waves in semi-enclosed seas

The typical period of waves excited by contemporary ships sailing at 25–35 knots may easily reach 10 s and occasionally may be even larger (Fig. 5). Although the period of open ocean swell and typical periods in very strong storms in the ocean conditions may be 15–20 s, such long-period waves are extremely uncommon in many semi-sheltered sea areas and smaller water bodies. While in the semi-sheltered bays the long-term components are frequently redirected and damped by extensive topographic refraction [69] or blocked by the specific geometry of the bay [9], wave periods remain intrinsically small in relatively small water bodies such as the Baltic Sea (Fig. 11).

Most frequently, waves with periods of 4–6 s dominate in the middle of the Baltic Proper, whereas in the coastal regions waves with periods of 3–4 s predominate (Fig. 12). Wave periods of about 5–6 s also occur with an appreciable frequency in the coastal areas. Periods up to 7 s are still common in the open sea (Fig. 12). This difference in periods apparently comes from a relatively large number of short-fetched wave conditions at sheltered measurement sites. Yet, wave conditions corresponding to periods of about 10 s only occur during very strong (and infrequent) storms or in very low swell conditions [7, 40, 74].

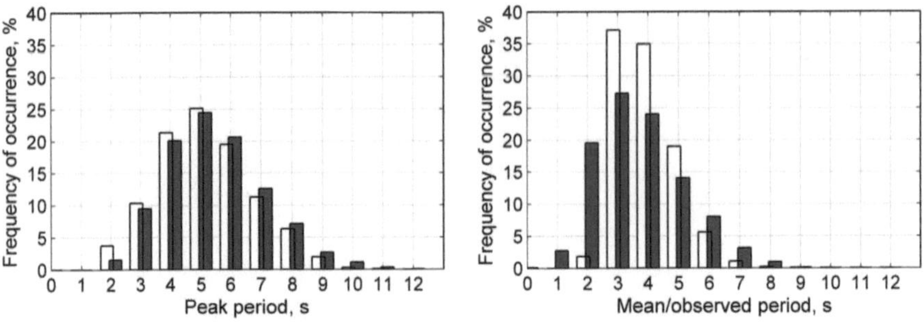

Fig. 12 Frequency of occurrence of wave periods at wave observation and measurement sites in the northern part of the Baltic Sea. Left panel: Bogskär 1982–1986 (waverider, white bars) and in the northern Baltic Proper 1996–2000 (waverider, filled bars); right panel: Almagrundet 1978–95 (inverted echosounder, white bars) and Vilsandi 1954–2005 (visual observations from the coast, filled bars). Reprinted from [74].

The joint distributions of wave heights and periods (Fig. 13) suggest that the proportion of relatively steep seas is quite large in the Baltic Sea. Periods of 2–3 s usually correspond to wave heights well below 1 m, whereas waves with periods of 4–5 s have a typical height of about 1 m. Periods 6–7 s usually correspond to wave heights of about 1.5–2 m. In coastal areas, peak periods are 7–8 s only when wave heights are about 3 m or higher (Fig. 13). Even longer waves are infrequent. Mean periods over 8 s (peak periods $\geq$10 s) dominate either in very rough seas (wave heights over 4 m) or in remote low swell conditions when the wave heights are well below 1 m. For example, at Almagrundet, the mean period never exceeded 9.5 s in very rough seas and was about 10 s in one case of rough seas, with the estimated significant wave height $H_{1/3} \sim 4$ m. Even in the final stage of an extreme storm in January 1984, when $H_{1/3} \sim 7$ m, the mean period was below 10 s [7].

Long and high waves occur somewhat more frequently in the open sea of the northern Baltic Proper, yet wave conditions with peak periods exceeding 10 s are seldom. For example, at the location of the waverider in Fig. 11 the peak period of about 12 s has been registered about twice a year (Fig. 13). The dominating wave periods in large semi-enclosed bays of the Baltic Sea usually are considerably smaller [69].

4.2 Ship wakes versus wind waves

The comparison of ship wakes and wind waves should account for all potential differences between the waves of different origin. The most important properties are the wave height, wave length (or period), and wave propagation direction. In such comparisons, one usually assumes that the shape of the waves of different origins are basically defined by these properties and by the water depth. These properties, however, characterise the relevant plane, albeit directional, wave fields. In many applications, multi-dimensional features of waves such as their typical crest length may play a role. Here, we only consider the above classical wave properties. The reader is referred to [71, 72] for an overview of some effects connected with essentially two-dimensional features. Finally, nonlinear properties of wave groups with certain parameters may add another aspect to the comparison in question.

High ship waves occurring several times each day may already be unsafe for low-energy coasts [63, 92]. The intensity of ship traffic is gradually increasing in many places, and certain coastal areas are hit by highly energetic wakes several dozen times each day. An attempt to quantify the influence of ship wakes on a high-energy coast affected by a specific wind wave regime led to a highly intriguing conclusion about the potentially very large role of ship waves even on the open sea coasts [79, 80].

The first step in a comparison of ship and wind waves usually consists in the comparison of the wave heights. As the number of single ship-induced waves is relatively small compared with the number of storm waves, this method usually shows that natural waves dominate in most of the environments in terms of extreme wave heights and total wave energy. The intensity of many fundamental coastal processes (e.g. wave-induced alongshore currents, sediment transport, wave setup, etc.) is proportional to the wave energy flux (frequently called wave power in coastal engineering applications) or radiation stress. These quantities also embrace the wave

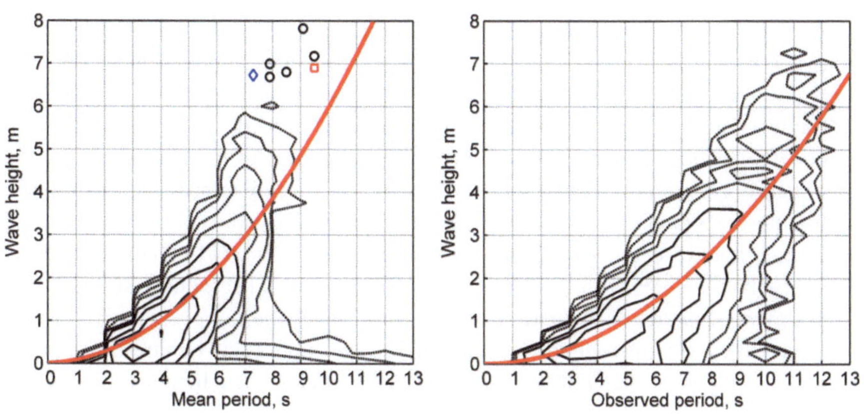

Fig. 13 Scatter diagram of wave heights and periods at Almagrundet 1978–95 (left, [7]) and in the northern Baltic Proper (right, [40]). The wave height step is 0.25 m and the period step is 1 s. The range of periods is shown on the horizontal axis: 2 s stands for $1.5 \leq T_p \leq 2.5$ s, 3 s stands for $2.5 \leq T_p \leq 3.5$ s, etc. Isolines for the probability of occurrence of 0.0033%, 0.01%, 0.033%, 0.1% (dashed lines), 0.33%, 1%, 3.3%, and 10% (solid lines) are plotted. Wave conditions with $H_{1/3} > 6.5$ m at Almagrundet are shown as follows: circles – the January 1984 storm, diamond – a storm in January 1988, square – a storm in August 1989. The bold line indicates saturated wave conditions with a Pierson-Moskovitz spectrum. Reprinted from [74].

period that may be an even more important parameter in such comparisons. The wave propagation direction plays a role in many aspects, such as the direction of wave-induced sediment transport, and also the damages and overtopping of waves approaching coastal structures from unusual directions.

4.2.1 Wave heights

The comparison of wave heights is frequently made in terms of the occurrence probability of waves of different height and origin in the coastal area of interest [79, 80]. The comparison is usually not straightforward because of insufficient resolution of the classical wave measurements. The networks of contemporary wave measurements are also quite scarce. For example, there is only one point of regular directional wave measurements in the Gulf of Finland [64]. The wave statistics obtained from such measurements cannot be directly extended to the coastal areas of interest that are usually located within semi-sheltered bays with complex bathymetry and geometry. Adequate comparisons thus can only be based on high-resolution modelling of the local wave climate.

For Tallinn Bay, these have been performed with the use of a high-resolution multi-nested version of the WAM wave model [44]. This is the first example of the third-generation spectral wave models that systematically include the main effects affecting the wave generation and dissipation as well as non-linear interactions between the wave harmonics. Its current version (cycle 4) takes into account the coastal line of the basin, topographic refraction, spatial and temporal variation of wind properties, wave propagation on the sea surface, quadruplet interactions between wave harmonics, whitecapping, shoaling and wave dissipation in shallow areas due to bottom friction, and interaction of waves and stationary currents. The application of the model for the Baltic Sea is discussed in [69].

The model used in [69] is forced with the marine wind data and takes the wind history into account to some extent. In this particular case of high latitudes, the sea is frequently covered by ice and such calculations are highly nontrivial even if high-quality marine wind fields are used. In calculations performed so far, the presence of sea ice is ignored. As the ice season, with a typical duration of 70–80 days annually [10], largely coincides with the most windy winter season [55], doing so leads to a certain overestimation of the parameters of wind waves [69].

The local wave climate in the test area of Tallinn Bay is relatively mild compared with the open part of the Gulf of Finland. The significant wave height exceeds 1.0–1.5 m, depending on the particular site, only with a probability of 1% (Fig. 14). An interesting feature is that the extreme wave heights in the bay are comparable with those in the open part of the Gulf of Finland. The significant wave height may reach 4 m in the central part of the bay during strong north-west storms (Fig. 14), and about 2–3 m at the 5 m isobath [69]. The listed properties of the local wave regime are caused by the interplay of the geometry and bathymetry of the bay, and the distinctive features of the wind climate. The wind regime in this area is strongly anisotropic. Although Tallinn Bay is well sheltered from the directions of the most frequent stormy winds (SW), particularly strong gales come from directions where winds generally are infrequent [77], and occasionally excite very high waves in certain parts.

The highest daily examples of ship waves, with a typical height of about 1 m, are mostly found in the range of the highest annual 1–5% of wind waves along the coasts of Tallinn Bay [79, 80]. Although the energy of a single wake is small compared with the energy of wind waves occurring during severe storms from unfavourable directions, such storms occur only seldomly. On the other hand, ship wakes are frequent enough so that the mean energy of ship-generated waves at the 5 m isobath of Tallinn Bay is about 5–8% of the annual mean wave energy. This share is considerably larger (6–12%) during the relatively calm spring and summer seasons [79, 80]. This share is actually not very unusual and much larger portions of ship-generated waves in the total wave activity have been identified in certain well-sheltered microtidal basins [68].

4.2.2 Contribution of wave periods

The comparison of the energy of ship and wind waves is equivalent to an evaluation of the squared wave heights and is only adequate provided all other properties of the waves coincide. Ship-generated waves frequently have a height of about 1 m and a period of about 10–15 s. From Fig. 13 it follows that such waves are extremely seldom even in the coastal area of the northern Baltic Proper. They almost never occur in natural conditions in many sections of the coasts of semi-sheltered subbasins of this water body. The introduction of fast ferry traffic in this area has therefore led to the unique situation, in which a qualitatively new hydrodynamic forcing factor with an intensity comparable to that of the existing one has abruptly emerged

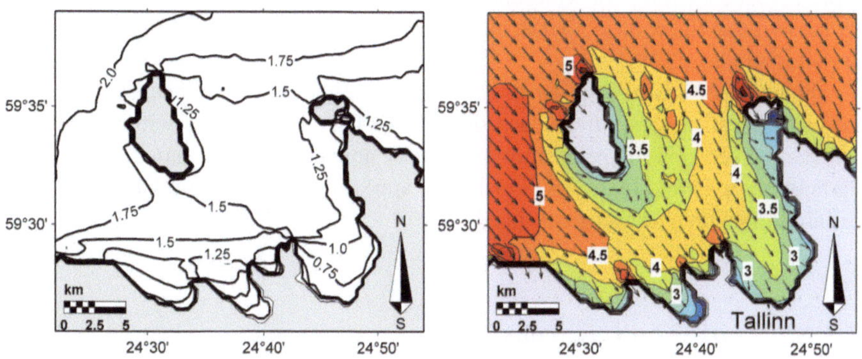

Fig. 14 Distributions of wave heights (m) occurring with the probability of 1% based on wind data from Kalbådagrund 1991–2000 (left); modelled significant wave height (m) and wave propagation direction (arrows) in Tallinn Bay during an extreme NNW storm when wind 23 m/s from the direction 330° blows for 6 hours (right). The wind data roughly correspond to the storm on November 15, 2001. Contour interval is 0.5 m. Reprinted from [69].

in open sea conditions [70]. The main reason is that many ship waves of modest height have considerably larger periods than even the longest natural waves in this area. This feature presents a rare possibility to study the effects of potential future changes of the local wave climate on the local marine ecosystem in this and similar semi-enclosed sea areas.

A convenient integral measure that implicitly accounts for the wave periods is the *wave energy flux*. Also commonly called *wave power*, it is defined as the product of the wave energy density and group speed, usually measured per unit of length of wave crests. It characterises the rate of energy supply to a particular sea area.

The annual mean power of ship waves constitutes 18–35% (27–54% during the summer season) of the annual wind wave power at a distance of 2–3 km from the ship lane. It is about 12% (17%) at sites located about 10 km from the ship lane in the coastal area of Tallinn Bay [70, 78]. As discussed above, the reason for the excessive content of ship waves in the total wave power, compared with the share of ship waves in the overall wave energy budget, is the moderate typical period of wind waves. This typical feature of semi-sheltered sea areas originates here from the interplay of the geometry of the basin and the specific angular structure of the dominant winds [69]. Even in cases when the 1-year return value of the significant wave height is reached, the dominant wave periods normally do not exceed 5–6 s [69]. In other basins of similar type, the blocking of long-period open sea waves may be the reason for an analogous effect [9].

The difference in the lengths of waves of different origin can be further quantified through a comparison of their spectral properties. Doing this is generally not justified for a single wake, which frequently has a very sharply peaked spectrum [78]. However, large ensembles of ship waves apparently can be described in terms of traditional wave energy spectra, e.g., as daily average wave energy spectra [70, 79]. The major part (up to 80%) of the wake energy is concentrated in waves with periods from 8 to 15 s along the coasts of Tallinn Bay. This energy lies outside the typical frequency range of windseas in many semi-sheltered areas. This is reflected by the fact that the wakes of large high-speed vessels are often distinguishable even at rough windseas [78].

The peak periods of ship waves may vary greatly at different measurement sites located a small distance from each other [78]. This is consistent with the perception that wakes of high-speed ships are usually non-stationary in natural conditions. Particularly long and high waves evidently appear only after a ship has sailed for a certain time with a transcritical speed [90].

4.3 Ship wakes and the coast

High, nonlinear waves excited by contemporary strongly powered ships sailing at moderate depths may result in large hydrodynamic loads not only in the immediate vicinity of ship lanes. As the groups of leading waves have very limited dispersion and diffraction, violent energy concentration may easily occur at a distance of several kilometres from the sailing line. Probably, the first documented evidence is presented in [63] about the significance of the remote influence of ship waves that usually becomes evident in the form of long period surge waves and causes limited beach erosion.

Under certain conditions the effects of nonlinear propagation and (basically linear) effects of refraction, shoaling, and diffraction may amplify each other. Considerable wave amplification in the downstream wave packet may occur in the shallow-water part of the water body owing to a nonlinear wave wave interaction [75, 88]. The nonlinear amplification combined with an extensive shoaling of long ship waves in shallow water and with the above-mentioned effect of trapping of precursors for gradually increasing Froude numbers is the probable reason of the first documented fatal event caused by waves from fast ships. This happened in the Gulf of Finland nearly a century ago in 1912. In this event, a boy was washed from a wharf and drowned. The wharf reached 2.7 m above water level and was located at a distance of about 10 km from the sailing line of the warship *Novik* [45].

The safety and environmental impacts associated with specific properties of wakes from high-speed vessels have attracted the attention of many scientists and experts. An early overview of the relevant studies, decision-making procedures and restrictions introduced in Denmark and in New Zealand can be found in [62]. Practical recommendations for managing wake wash have been formulated by the International Navigation Association [65].

One of the major agents for wave-induced effects is the orbital velocity created by ship waves. It is well known that the maximum near-bottom velocity linearly increases with the wave height, but shows much more complex dependence on the wave period (Fig. 15). In the nearshore, with a depth of 5–30 m, it has the highest variation when the wave period increases from 5 to 8 s. In areas with low natural hydrodynamic activity, the longest ship waves are able to excite quite large near-bottom velocities at considerable depths. The resulting impact of ship wakes on bottom sediments and aquatic wildlife at these depths may be comparable with or

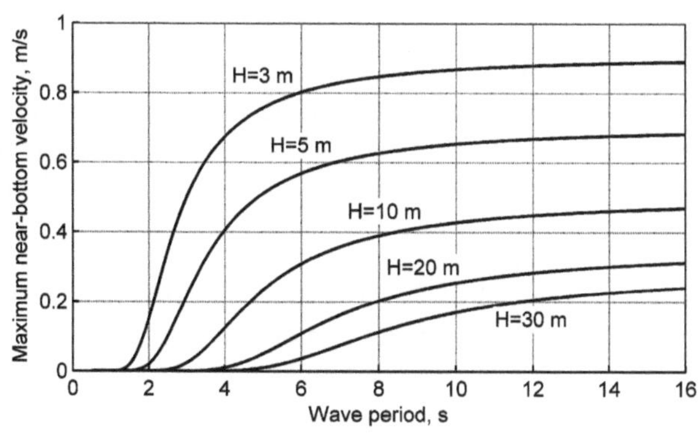

Fig. 15 The dependence of the maximum wave-induced near-bottom velocity on the wave period for the wave height 1 m and the water depth of 3–30 m.

even exceed the impact of wind waves occurring in the most violent storms in limited fetch conditions [79]. This effect is amplified by the nonlinear nature of a large part of long ship-generated waves at depths <10 m, for which the linear theory severely underestimates the near-bottom velocities in the leading ship waves [81]. Precursor solitons are usually of moderate height and cause no direct harm to the beach environment. In terms of the current velocity at a depth of a few meters, however, they are comparable with Kelvin waves [60].

The difference of prevailing periods of certain parts of ship wakes and the windseas is a generic vital issue in non-tidal areas sheltered from long ocean waves. In such seas, e.g. the Azov Sea and the Baltic Sea, which have either small size or limited fetch length for certain subbasins, the near-bottom hydrodynamic activity is governed by the short-period windseas and local currents [79, 80]. Many components of the marine ecosystem in sheltered non-tidal sea areas have been adjusted to low near-bottom velocities. The leading waves from high-speed ships may appear as a particularly hazardous new forcing component at certain depths in such areas [70]. Most importantly, they may cause unusually high near-bottom velocities.

4.4 Excessive near-bottom velocities and impulse loads

The increased hydrodynamic loads first become evident in re-suspension and transport processes of bottom sediments. Large vessel waves can create suspended sediment concentrations as large as 10–100 g/l in rivers [61]; yet they seem to have a tendency to maintain the existing shore profile rather than causing net sediment loss in the particular location.

An interesting form of the impact of ship wakes has been identified in Hillsborough Bay [68]. In this shallow, microtidal, subtropical estuary in West-Central Florida, long, solitonic wake waves generated by large vessels in a dredged ship channel frequently cause large water velocities and sediment resuspension. Consistent with the above description, such waves are extremely strong impulse loads that at times essentially exceed similar loads caused by natural factors. The estimated annual mass of sediment resuspended by these waves is by one order of magnitude greater than the annual mass of sediment resuspended by wind waves. A secondary impact of the long waves is that sediments that are resuspended and newly deposited are more susceptible to resuspension by natural currents than undisturbed bottom sediments.

Similar problems have been identified in the framework of recreational boating in very shallow environments [3]. The bottom of a marsh channel in an area with moderate tide activity (amplitude about 0.5 m) reacts readily to waves with heights of 10–15 cm from small personal watercraft, sailing with a speed of 16 km/h in a water depth of the order of 1 m at $F_h \approx 1.5$–2.

The specific contribution of ship waves to the wave climate of sheltered areas is comparable to that which would occur if open ocean swell had reached the areas. Even a small additional contribution of open ocean waves may lead to a great increase in sediment transport, in particular in conditions when bed stress, due to local wind waves or currents, is near a critical threshold for erosion or deposition [86]. Apart from this generic contribution, already relatively small levels of long-period energy in combination with wind waves can cause greater beach response than an equal amount of energy in only the windsea frequencies [11].

An extensive joint influence may occur when waves of different length (or origin) transport sediments at different water depths [76]. Usually, ship waves do not cause direct shoreline erosion of sea coasts exposed to dominant winds because such shores are often covered by coarse-grained deposits (gravel, pebbles, and cobbles) that protect the shore from further erosion. Long solitonic (ship) waves eventually resuspend unprotected sediments at deeper parts of the nearshore as compared to the natural waves. The potential sediment deficit will apparently be balanced by a more intense transport of material from the vicinity of the shoreline. In this way, ship waves may trigger a complicated pattern of sediment transport processes, possibly accompanied by a certain shoreline reduction [70, 76]. There is, however, very little experimental evidence about the intensity and direction of sediment transport induced by ship waves in such conditions.

The reaction of the bottom sediments was quantified with the use of optical methods in the Tallinn Bay conditions [18, 19]. The goal was to first adequately connect the wave-induced changes of the optical properties of the sea water and the underwater light regime in the coastal area of Tallinn Bay with the concentration of suspended matter. The changes of the latter quantity characterise the amount of sediments resuspended by a particular wave event. Based on this information, the total impact of waves of any origin can be roughly estimated by integrating the relevant optical data. In particular, the amount of material resuspended by a single ship wake, as well as the annual bulk sediment transport due to fast ferry traffic, can be estimated.

The analysis demonstrated that waves produced by fast ferries bring about significant changes in the optical parameters of sea water in an approximately 1 m thick near-bottom layer of the coastal areas of Tallinn Bay. The greatest of these changes occur at relatively small depths (2–4 m), but the duration of the influence increases with increasing depth. Detectable changes in the optical properties of sea water occurred down to the depths of about 14 m. Rough quantitative estimates suggest that the overall influence of fast ferry traffic in Tallinn Bay brings into motion, on average, some 10,000 kg of sediments per metre of coastline annually. This may result in an annual loss in the order of several hundred litres of fine sediments from each metre of the coastline. The additional coastline erosion owing to ship waves may be in the order of a few tens of centimetres annually.

5 Conclusions

Major progress in understanding the features of ship wakes has offered a handful of new and fascinating results, the importance of which extends far beyond the classical scope of navigational problems. The more exact identification of the rules governing the generation, evolution, and interaction of the different wake components has essentially enriched the general nonlinear dynamics. Apart from evident local applications, it has led to several inspired methods that make use of certain nonlinear properties of shallow-water waves for solving hot topics of contemporary physics and physical oceanography. In particular, progress in the understanding of the excitation and interactions of solitonic wave groups has essentially contributed to the freak wave theory by indicating a unique mechanism for the generation of long-living freak waves in shallow water [75].

Deeper knowledge of the properties of ship waves is obviously necessary for the sustainable development of ship traffic in shallow semi-enclosed sea areas. Such areas usually must be crossed in order to make a port call. Many fairways and ports are located in the vicinity of low-energy coasts that may be exceptionally vulnerable with respect to changes of the local wave activity. The major new understanding is that the most important decisive factor is not always the wave height, but in many cases it may be other properties such as the period or the shape of the largest waves.

The implicit influence of ship waves on the coasts and on the local ecosystem clearly requests further investigations. The extent and many properties of the environmental influence of intense ship traffic are still unclear and have to be addressed in further studies. The increase in wave activity in the vicinity of ship lanes has been recognised as inadmissible in many countries [62, 94]. There is an ongoing debate on whether the definition of pollution (that today commonly is interpreted as releasing certain substances or noise into the environment) should strictly include the releasing of energy in any form into the marine environment [85]. A wave wake is one of the central agents of energy released by a ship into the marine environment. Therefore, ship traffic may even be treated even as a potential source of energy pollution. The potential of a remote influence adds another dimension to this possibility [70, 79]. A comprehensive analysis of this feature may constitute a vital part of the environmental impact assessments of harbours and associated ship traffic in the neighbourhood of vulnerable areas in the future.

Acknowledgements Financial support from the Estonian Science Foundation (Grant 7413) and the Estonian block grant SF0140077s08 is gratefully acknowledged. A large part of the paper was written during a visit to the Centre of Mathematics for Applications in the framework of the Marie Curie Transfer of Knowledge project CENS-CMA (MC-TK-013909).

References

1. Akylas, T.R., Kung, T.J., Hall, R.E.: Nonlinear groups in ship wakes. In: 17th Symposium in Naval Hydrodynamics, The Hague, The Netherlands, pp. 477–482. National Academy Press, Washington DC (1988)
2. Ali, M.M., Murphy, K.J., Langendorff, J.: Interrelations of river ship traffic with aquatic plants in the River Nile, Upper Egypt. Hydrobiologia **415**, 93–100 (1999)
3. Anderson, F.E.: Effect of wave-wash from personal watercraft on salt marsh channels. J. Coast. Res. Special Issue **37**, 33–49 (2002)
4. Basin, A.M., Anfimov, V.N.: Ship Hydrodynamics. Retshnoj Transport, Leningrad (1961) (in Russian)
5. Bauer, B.O., Lorang, M.S., Sherman, D.J. Estimating boat-wake-induced levee erosion using sediment suspension measurements. J. Waterw. Port Coast. Ocean Eng. - ASCE **128**, 152–162 (2002)
6. Bourne, J.: Louisiana's Vanishing Wetlands: Going, Going.... Science **289**(5486), 1860–1863 (2000)
7. Broman, B., Hammarklint, T., Rannat, K., Soomere, T., Valdmann, A.: Trends and extremes of wave fields in the north–eastern part of the Baltic Proper. Oceanologia **48**(S), 165–184 (2006)
8. Brown, E.D., Buchsbaum, S.B., Hall, R.E., Penhune, J.P., Schmitt, K.F., Watson, K.M., Wyatt, D.C.: Observations of a nonlinear solitary wave packet in the Kelvin wake of a ship. J. Fluid Mech. **204**, 263–293 (1989)
9. Caliskan, H., Valle-Levinson, A.: Wind-wave transformations in an elongated bay. Continental Shelf Research **28**, 1702–1710 (2008)
10. Climatological Ice Atlas for the Baltic Sea, Kattegat, Skagerrak and Lake Vänern (1963–1979), Swedish Meteorological and Hydrological Institute, Norrkøping, Sweden, and Institute of Marine Research, Helsinki, Finland (1982)
11. Coates, T.T., Hawkes, P.J.: Beach recharge design and bi-modal wave spectra. In: Edge, B.L. (ed.) Coastal Engineering 1998: Proceedings of the 26th International Conference, Copenhagen, ASCE **3**, pp. 3036–3045 (1999)
12. Cohen, B.I., Watson, K.M., West, B.J.: Some properties of deep water solitons. Phys. Fluids **19**, 345–354 (1976)
13. Doctors, L.J., Phillips, S.J., Day, A.H.: Focussing the wave-wake system of a high-speed marine ferry. In: FAST 2001, 6th International Conference on Fast Sea Transportation, Southampton, **1**, pp. 97–106. The Royal Institute of Naval Architects, London (2001)
14. Doyle, R., Whittaker, T.J.T., Elsässer, B.: A study of fast ferry wash in shallow water. In: FAST 2001, 6th International Conference on Fast Sea Transportation, Southampton, **1**, pp. 107–116. The Royal Institute of Naval Architects, London (2001)
15. Drazin, P.G., Johnson, R.S.: Solitons: an Introduction. Cambridge Texts in Applied Mathematics. Cambridge University Press (1989)
16. Durkee, P.A., Chartier, R.E., Brown, A., Trehubenko, E.J., Rogerson, S.D., Skupniewicz, C., Nielsen, K.E.: Composite ship track characteristics. J. Atmos. Sci. **57**, 2542–2553 (2000)
17. El-Kader, F.A., El-Soud, M.S.A., El-Serafy, K., Hassan, E.A.: An integrated navigation system for Suez Canal (SCINS). J. Navigation **56**, 241–255 (2003)
18. Erm, A., Soomere, T.: Influence of fast ship waves on optical properties of sea water in Tallinn Bay, Baltic Sea. Proc. Estonian Acad. Sci. Biol. Ecol. **53**, 161–178 (2004)
19. Erm, A., Soomere, T.: The impact of fast ferry traffic on underwater optics and sediment resuspension. Oceanologia **48**(S), 283–301 (2006)
20. Ertekin, R.C., Webster, W.C., Wehausen, J.V.: Ship-generated solitons. In 15th Symposium on Naval Hydrodynamics, Hamburg, Germany, pp. 347–364. National Academy Press, Washington DC (1984)
21. Fagerholm, H.P., Rønnberg, O., Östman, M., Paavilainen, J.: Remote sensing assessing artificial disturbance of the thermocline by ships in archipelagos of the Baltic Sea with a note on some biological consequences. In: 11th Annual International Geoscience and Remote Sensing Symposium, Helsinki 1991, **2**, 377–380 (1991)

22. Feldtmann, M., Garner, J.: Seabed modifications to prevent wake wash from fast ferries. In: Proceedings of the RINA International Conference on Coastal Ships and Inland Waterways, London, Paper No 8. The Royal Institute of Naval Architects, London (1999)
23. Forsman, B.: High-Speed Ferries – Environmental Impact and Safety Assessment. PIANC Bulletin **96**, 23–24 (1997)
24. Forsman, B.: From bow to beach. SSPA Highlights **3**, 4–5 (2001)
25. Froude, W.: Experiments upon the effect produced on the wave-making resistance of ships by length of parallel middle body. Trans. Inst. Naval Architects **18**, 77–87 (1877)
26. Gaskin, S.J., Pieterse, J., Al Shafie, A., Lepage, S.: Erosion of undisturbed clay samples from the banks of the St. Lawrence River. Can. J. Civ. Eng. **30**, 585–595 (2003)
27. Gourlay, T.P.: The supercritical bore produced by a high-speed ship in a channel. J. Fluid Mech. **434**, 399–409 (2001)
28. Grimshaw, R., Pelinovsky, E., Sakov, P.: Interaction of a solitary wave with an external force moving with variable speed. Stud. Appl. Math. **97**, 235–276 (1996)
29. Grimshaw, R., Pelinovsky, D., Pelinovsky, E., Talipova, T.: Wave group dynamics in weakly nonlinear long-wave models. Physica D **159**, 35–57 (2001)
30. Hall, R., Buchsbaum, S. A model for the generation and evolution of an inner-angle soliton in a Kelvin wake. In: 18th Symposium on Naval Hydrodynamics, Ann Arbor, 1990, pp. 453–463. National Academy Press, Washington DC (1991)
31. Hamer, M.: Solitary killers. New Scientist **163**, No. 2201, 18–19 (1999).
32. Havelock, T.H.: The propagation of groups of waves in dispersive media, with application to waves on water produced by a travelling distance. Proc. R. Soc. London Ser. A - Math. Phys. Eng. Sci. **81**, 398–430 (1908)
33. Hennings, I., Romeiser, R., Alpers, W., Viola, A.: Radar imaging of Kelvin arms of ship wakes. Int. J. Remote Sens. **20**, 2519–2543 (1999)
34. Hofmann, H., Lorke, A., Peeters, F.: The relative importance of wind and ship waves in the littoral zone of a large lake. Limnol. Oceanogr. **53**, 368–380 (2008)
35. Hobbs, P.V., Garrett, T.J., Ferek, R.J., Strader, S.R., Hegg, D.A., Frick, G.M., Hoppel, W.A., Gasparovic, R.F., Russell, L.M., Johnson, D.W., O'Dowd, C., Durkee, P.A., Nielsen, K.E., Innis, G.: Emissions from ships with respect to their effects on clouds. J. Atmos. Sci. **57**, 2570–2590 (2000)
36. Hüsig, A., Linke, T., Zimmermann, C.: Effects from supercritical ship operation on inland canals. J. Waterw. Port Coast. Ocean Eng. – ASCE **126**, 130–135 (2000)
37. IAHR working group on wave generation and analysis: List of sea-state parameters. J. Waterw. Port Coast. Ocean Eng. - ASCE **115**, 793–808 (1989)
38. Jiang, T.: Ship Waves in Shallow Water, Fortschritt-Berichte VDI, Reihe 12, Nr. 466, VDI Verlag, Düsseldorf (2001)
39. Jiang, T., Henn, R., Sharma, S.D.: Wash wave generated by ships moving on fairways of varying topography. In: 24th Symposium on Naval Hydrodynamics, Fukuoka, Japan, 2002, pp. 441–457. National Academy Press, Washington DC (2003) web only: www.nap.edu
40. Kahma, K., Pettersson, H., Tuomi, L.: Scatter diagram wave statistics from the northern Baltic Sea. MERI – Report Series of the Finnish Institute of Marine Research **49**, 15–32 (2003)
41. Kirk McClure Morton: Investigation of high speed craft on routes near to land or enclosed estuaries. The Maritime and Coastguard Agency, UK, Research Report JR226 (1998)
42. Kofoed-Hansen, H., Kirkegaard, J.: Technical investigation of wake wave from fast ferries. Danish Hydraulic Institute, Report No. 5012 (1996)
43. Kofoed-Hansen, H., Mikkelsen, A.C.: Wake wave from fast ferries in Denmark. In: Proceedings of the 4th International Conference of Fast Sea Transportation FAST97, Sydney, 1997, **1**, pp. 471–478. Baird Publications, Hong Kong (1997)
44. Komen, G.J., Cavaleri, L., Donelan, M., Hasselmann, K., Hasselmann, S., Janssen, P.A.E.M.: Dynamics and Modelling of Ocean Waves. Cambridge University Press (1994)
45. Krylov, A.N.: On the wave resistance and large ship waves. In: My Memoirs, pp. 364–368. Politechnika, Sankt-Peterburg (2003) (in Russian)
46. Kuznetsov, N., Maz'ya, V., Vainberg, B.: Linear Water Waves. Cambridge University Press (2002)

47. Lamb, H.: Hydrodynamics. 6th edition Cambridge University Press (1997)
48. Lee, S.J., Grimshaw, R.H.J.: Upstream-advancing waves generated by 3-dimensional moving disturbances. Phys. Fluids A **2**, 194–201 (1990)
49. Lee, S.J., Yates, G.T., Wu, T.Y.: Experiments and analyses of upstream-advancing solitary waves generated by moving disturbances. J. Fluid Mech. **199**, 569–593 (1989)
50. Li, Y., Sclavounos, P.D.: Three-dimensional nonlinear solitary waves in shallow water generated by an advancing disturbance. J. Fluid Mech. **470**, 383–410 (2002)
51. Lighthill, J.: Waves in Fluids. Cambridge University Press (1978)
52. Lindholm, T., Svartström, M., Spoof, L., Meriluoto, J.: Effects of ship traffic on archipelago waters off the Långnäs harbour in Åland, SW Finland. Hydrobiologia **444**, 217–225 (2001)
53. Madekivi, O. (ed.): Alusten aiheuttamien aaltojen ja virtausten ympäristövaikutukset (The environmental effects of ship-induced waves and currents, in Finnish). Vesi ja Ympäristöhallinnon Julk. Sarja A **166**, 1–113 (1993)
54. Mei, C.C., Naciri, M.: Note on ship oscillations and wake solitons, Proc. R. Soc. London Ser. A - Math. Phys. Eng. Sci. **432**, 535–546 (1991)
55. Mietus, M. (co-ordinator): The Climate of the Baltic Sea Basin, Marine Meteorology and Related Oceanographic Activities, Report No. 41. World Meteorological Organisation, Geneva, (1998)
56. Millward, A.: A review of the prediction of squat in shallow water, J. Navigation **49**, 77–88 (1996)
57. Munk, W.H., Scully-Power, P., Zachariasen, F.: Ships from space. The Bakerian Lecture 1986. Proc. R. Soc. London Ser. A - Math. Phys. Eng. Sci. **412**, 231–254 (1987)
58. Naghdi, P.M., Rubin, M.B.: On the squat of a ship. J. Ship Res. **28**, 107–117 (1984)
59. Nanson, G.C., von Krusenstierna, A., Bryant, E.A.: Experimental measurements of river-bank erosion caused by boat-generated waves on the Gordon River, Tasmania. Regul. River. **9**, 1–14 (1994)
60. Neuman, D.G., Tapio, E., Haggard, D., Laws, K.E., Bland, R.W.: Observation of long waves generated by ferries. Can. J. Remote Sens. **27**, 361–370 (2001)
61. Osborne, P.D., Boak, E.H.: Sediment suspension and morphological response under vessel-generated wave groups: Torpedo Bay, Auckland, New Zealand. J. Coast. Res. **15**, 388–398 (1999)
62. Parnell, K.E., Kofoed-Hansen, H.: Wakes from large high-speed ferries in confined coastal waters: management approaches with examples from New Zealand and Denmark. Coastal Manage. **29**, 217–237 (2001)
63. Parnell, K.E., McDonald, S.C., Burke, A.: Shoreline effects of vessel waves, Marlborough Sounds, New Zealand. J. Coastal Res. Special Issue **50**, 502–506 (2007)
64. Pettersson, H.: Directional Wave Statistics from the Gulf of Finland. MERI 44, Finnish Institute of Marine Research (2001)
65. PIANC: Guidelines for Managing Wake Wash from High-speed Vessels. Report of the Working Group 41 of the Maritime Navigation Commission. International Navigation Association (PIANC), Brussels (2003)
66. Pickrill, R.A.: Beach changes on low energy lake shorelines, Lakes Manapouri and Te Anau, New Zealand. J. Coast. Res. **1**, 353–363 (1985)
67. Redekopp, L.G., You, Z.: Passage through resonance for the forced Korteweg-de Vries equation. Phys. Rev. Lett. **74**, 5158–5161 (1995)
68. Schoellhamer, D.H.: Anthropogenic sediment resuspension mechanisms in a shallow microtidal estuary. Estuar. Coast. Shelf Sci. **43**, 533–548 (1996)
69. Soomere, T.: Wind wave statistics in Tallinn Bay, Boreal Env. Res. **10**, 103–118 (2005)
70. Soomere, T.: Fast ferry traffic as a qualitatively new forcing factor of environmental processes in non-tidal sea areas: a case study in Tallinn Bay. Environ. Fluid Mech. **5**, 293–323 (2005)
71. Soomere, T.: Nonlinear ship wake waves as a model of rogue waves and a source of danger to the coastal environment: a review. Oceanologia **48**(S), 185–202 (2006)
72. Soomere, T.: Fast ferries as wavemakers in a natural laboratory of rogue waves. Rend. Sem. Mat. Univ. Pol. Torino **65**(2), 287–299 (2007)

73. Soomere, T.: Nonlinear components of ship wake waves. Appl. Mech. Rev. **60**, 120–138 (2007)
74. Soomere, T.: Extremes and decadal variations of the northern Baltic Sea wave conditions. In: Pelinovsky, E., Kharif, Ch. (eds.) Extreme Ocean Waves, pp. 139–157. Springer, Berlin (2008)
75. Soomere, T., Engelbrecht, J.: Weakly two-dimensional interaction of solitons in shallow water. Eur. J. Mech. B/Fluids **25**, 636–648 (2006)
76. Soomere, T., Kask, J.: A specific impact of waves of fast ferries on sediment transport processes of Tallinn Bay. Proc. Estonian Acad. Sci. Biol. Ecol. **52**, 319–331 (2003)
77. Soomere, T., Keevallik, S.: Directional and extreme wind properties in the Gulf of Finland. Proc. Estonian Acad. Sci. Eng. **9**, 73–90 (2003)
78. Soomere, T., Rannat, K.: An experimental study of wind waves and ship wakes in Tallinn Bay. Proc. Estonian Acad. Sci. Eng. **9**, 157–184 (2003)
79. Soomere, T., Elken, J., Kask, J., Keevallik, S., Kõuts, T., Metsaveer, J., Peterson, P.: Fast ferries as a new key forcing factor in Tallinn Bay. Proc. Estonian Acad. Sci. Eng. **9**, 220–242 (2003)
80. Soomere, T., Rannat, K., Elken, J., Myrberg, K.: Natural and anthropogenic wave forcing in Tallinn Bay, Baltic Sea. In: Brebbia, C.A., Almorza, D., López-Aguayo, F., (eds.) Coastal Engineering VI, pp. 273–282. WIT Press, Southampton (2003)
81. Soomere, T., Põder, R., Rannat, K., Kask, A.: Profiles of waves from high-speed ferries in the coastal area. Proc. Estonian Acad. Sci. Eng. **11**, 245–260 (2005)
82. Sorensen, R.M.: Ship-generated waves. Adv. Hydrosci. **9**, 49–83 (1973)
83. Sretenskii, L.N.: Theory of Wave Motions in Fluids. Nauka, Moscow (1977) (in Russian)
84. Stevens, R.L., Ekermo, S.: Sedimentation and erosion in connection with ship traffic, Gøteborg Harbour, Sweden. Environ. Geol. **43**, 466–475 (2003)
85. Stumbo, S., Fox, K., Dvorak, F., Elliot, L.: The prediction, measurement, and analysis of wake wash from marine vessels. Mar. Technol. SNAME News **36**, 248–260 (1999)
86. Talke, S.A., Stacey, M.T.: The influence of oceanic swell on flows over an estuarine intertidal mudflat in San Francisco Bay. Estuar. Coast. Shelf Sci. **58**, 541–554 (2003)
87. Thomson, W. (Lord Kelvin): On ship waves. Trans. Inst. Mech. Eng. **8**, 409–433 (1887)
88. Torsvik, T.: Long wave models with application to high speed vessels in shallow water. PhD Thesis, University of Bergen, Norway (2006)
89. Torsvik, T., Dysthe, K., Pedersen, G.: Influence of variable Froude number on waves generated by ships in shallow water. Phys. Fluids **18**, Paper 0621021 (2006)
90. Torsvik, T., Soomere, T.: Simulation of patterns of wakes from high-speed ferries in Tallinn Bay. Estonian J. Eng. **14**, 232–254 (2008)
91. Ursell, F.: On Kelvin's ship-wave pattern. J. Fluid Mech. **8**, 418–431 (1960)
92. Velegrakis, A.F., Vousdoukas, M.I., Vagenas, A.M., Karambas, Th., Dimou, K., Zarkadas, Th.: Field observations of waves generated by passing ships: a note. Coastal Eng. **54**, 369–375 (2007)
93. Wolter, C., Arlinghaus, R.: Navigation impacts on freshwater fish assemblages: the ecological relevance of swimming performance. Rev. Fish Biol. Fisher. **13**, 63–89 (2003)
94. Wood, W.A.: High-speed ferry issues for operators and designers. Mar. Technol. SNAME News **37**, 230–237 (2000)
95. Wu, D.M., Wu, T.Y.: Precursor solitons generated by three-dimensional disturbances moving in a channel. In: IUTAM Symposium on Non-linear Water Waves, Tokyo, Japan (1987)
96. Wu, T.Y.: Generation of upstream advancing solitons by moving disturbances. J. Fluid Mech. **184**, 75–99 (1987)
97. Yih, C.-S., Zhu, S.: Patterns of ship waves. Q. Appl. Math. **47**, 17–33 (1989)
98. Yih, C.-S., Zhu, S.: Patterns of ship waves II. Gravity-capillary waves. Q. Appl. Math. **47**, 35–44 (1989)

Modelling of Ship Waves from High-speed Vessels

Tomas Torsvik

Abstract Waves generated by high-speed vessels may differ considerably from the ship wakes generated by conventional ships. Wakes from high-speed vessels typically have long wave periods, and occasionally very large amplitudes. The wave fan is often much wider than for conventional ships, and the leading waves can be highly nonlinear. This is particularly an issue in shallow coastal waters, and for coastal regions not exposed to large sea swell or impact from large wind generated waves. As a steadily increasing number of high-speed crafts are put into service, particularly in passenger ferry traffic, more coastal regions are exposed to wash from high-speed vessel wakes, causing concern with respect to public safety, potential damage of coastal structures, and potential environmental impact. Speed reduction will often, but not always, mitigate the problem. Furthermore, such measures are in conflict with the interests of ferry operators and the general public concerning fast transportation at sea. A careful planning of ship routes and speed regimes is needed to optimize the ship operation, while maintaining acceptable wash levels at the shore. This review presents some of the basic properties of ship wakes, and show how numerical models can be used to model waves generated by high-speed vessels.

The presentation starts with a discussion of basic properties of ship wakes, and a demonstration of why high-speed vessel wakes are different from the wakes generated by conventional ships. Thereafter we discuss the mathematical equations that can be used to model such waves, and some basic properties of these equations. Finally we discuss numerical models based on these equations, and how these models can be used to simulate ship wakes.

Bergen Centre for Computational Science, UNIFOB, Thormøhlensgate 55, 5008 Bergen, Norway,
e-mail: tomas.torsvik@bccs.uib.no

E. Quak, T. Soomere (eds.), *Applied Wave Mathematics*,
DOI 10.1007/978-3-642-00585-5_13,

1 Waves generated by ships

Ship waves are generated by the displacement of water due to the moving hull. The study of ship waves has mainly been driven by the desire to operate the ship at its design speed with a minimum of wave resistance. Ship waves have not traditionally been considered a safety or environmental issue, exccpt in confined waterways such as rivers and canals. With the introduction of a new generation of large, high-speed ferries in the 1990s, ship operators and authorities started getting complaints about problems related to wake wash from high-speed vessels. Several studies were made, but much of the research is only available as unpublished reports for authorities and management agencies, and only a few articles have been published in scientific journals or conference proceedings [40, 43].

Problems related to high-speed vessel wash are mostly encountered in shallow (less than about 50 m), coastal waters. Safety concerns are associated with people on the shoreline who may be knocked down by unexpected large waves, swamping of small vessels and navigational problems for ships exposed to wake waves, and damage to moored vessels, docks and piers [33, 34, 35]. Environmental concerns include alteration of the coast due to changes in sediment transport, and biological impact due to large seabed velocities and turbidity due to wave breaking [40, 44, 46]. Since most coastal areas are also exposed to impact from wind generated waves, sea swell and tides, it can often be difficult to make an isolated assessment of the impact of wake waves in field studies [20].

In this section we consider basic properties of ship generated waves, including both deep water and shallow water wake characteristics, and discuss how the wakes from high-speed vessels differ from conventional vessel wakes.

1.1 Governing equations and results from linear wave theory

The basic equations governing water flow have been known for over a century [5, 6, 12], but the complexity of these equations is such that only a few analytical solutions are known, and these are limited to idealized case studies with simple boundary conditions. However, many of the basic features of ship generated waves can be described within the simplified framework of linear wave theory. The basic theory for water flow and the linear wave theory is described in several text books (see e.g. Kundu [23] or Lighthill [26]), and only a few of the most fundamental results will be presented here.

1.1.1 Basic equations

All equations are formulated using the coordinate system shown in Fig. 1. The equations for an inviscid, incompressible fluid consist of the continuity equation

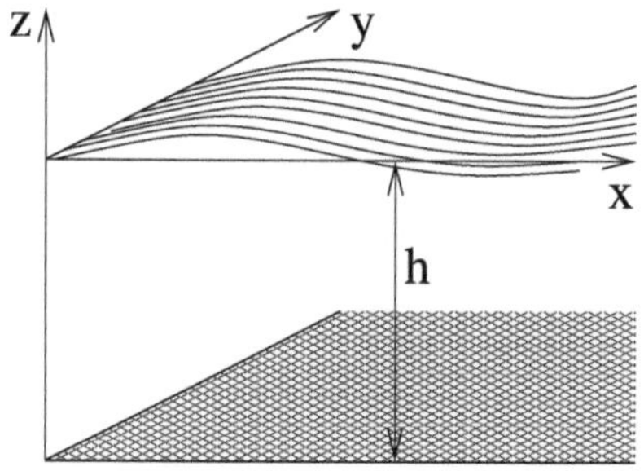

Fig. 1 Coordinate system for the fluid.

$$\nabla \cdot \mathbf{u} = \frac{\partial u}{\partial x} + \frac{\partial v}{\partial y} + \frac{\partial w}{\partial z} = 0, \tag{1}$$

and the momentum equations

$$\frac{\partial \mathbf{u}}{\partial t} + (\mathbf{u} \cdot \nabla)\mathbf{u} = -\frac{1}{\rho}\nabla p + \mathbf{g}, \tag{2}$$

where $\mathbf{u} = (u, v, w)$, and $\mathbf{g} = (0, 0, g)$. The momentum equations without viscous terms are also called Euler equations. In addition to these equations, we also need boundary conditions for the water flow at the bottom and at the free surface. For the bottom we need the kinematic boundary condition

$$w = u\frac{\partial h}{\partial x} + v\frac{\partial h}{\partial y}, \quad \text{at} \quad z = -h(x, y), \tag{3}$$

which states that a water particle at the boundary moves with the boundary interface. In particular, as there is no flux component directed normal to the boundary, there can be no flow through the boundary. Equation (3) is a sufficient condition, because we assume that the location of the bottom boundary is known, and remains fixed for all times. This is not the case for the free surface boundary between water and air, for which the surface location is a time dependent response to external and internal forces. The location of this surface is generally unknown, except for a specified initial condition, and the calculation of its position at different times is a part of the solution for the initial value problem. For the free surface boundary we need both the dynamic and kinematic boundary conditions,

$$p = p_a(x, y, t)\,, \quad \text{at} \quad z = \eta(x, y, t)\,, \tag{4}$$

$$w = \frac{\partial \eta}{\partial t} + u\frac{\partial \eta}{\partial x} + v\frac{\partial \eta}{\partial y}\,, \quad \text{at} \quad z = \eta(x, y, t)\,, \tag{5}$$

where p_a is the ambient pressure in the medium above the water. The dynamic boundary condition tells us that there should be no discontinuity in the pressure across the boundary. In many applications the ambient pressure p_a is just the air pressure, but a localized disturbance in the ambient pressure may also be used to represent a moving ship (see section 3.2).

1.1.2 Wave dispersion, speed and energy

A sinusoidal wave that propagates along the x-axis, and is uniform along the y-axis, is described by

$$\eta(x,t) = A\cos(kx - \omega t), \quad k = \frac{2\pi}{\lambda}, \quad \omega = \frac{2\pi}{T}, \tag{6}$$

where A is the wave amplitude, k is the wave number, ω is the angular frequency, λ is the wave length, and T is the wave period. Such a wave solution will satisfy the linearized version of the governing equations (1)–(5) if and only if the wave number and the angular frequency satisfy the dispersion relation [23, 26]

$$\omega^2 = gk\tanh kh\,. \tag{7}$$

The phase speed and group speed are defined as

$$c = \frac{\omega}{k}, \quad \text{and} \quad c_g = \frac{d\omega}{dk},$$

respectively. An approximation for deep water, valid for $h > \lambda/2$, is found by taking the limit $h \to \infty$ in Eq. (7), and a shallow water approximation, valid for $h < \lambda/20$, is found by taking the limit $h \to 0$. The expressions for the dispersion relation, phase speed, and group speed in deep and shallow water are listed in Table 1.

Table 1 Dispersion relation, phase speed, and group speed for linear waves.

	General Case	Deep Water	Shallow Water
Dispersion relation	$\omega^2 = gk\tanh(kh)$	$\omega^2 = gk$	$\omega^2 = ghk^2$
$c = \frac{\omega}{k}$	$\sqrt{\frac{g}{k}\tanh(kh)}$	$\sqrt{\frac{g}{k}}$	$\sqrt{gh}$
$c_g = \frac{d\omega}{dk}$	$\frac{1}{2}c\left(1+\frac{2kh}{\sinh(2kh)}\right)$	$\frac{1}{2}c$	c

The wave energy propagates with the wave group, not with individual waves, hence the energy flux over a volume will be associated with the group speed c_g. The wave energy E is the sum of the kinetic energy EK and the potential energy EP. The wave energy per unit horizontal area for the linear sinusoidal waves from Eq. (6), is found by integrating over the depth and averaging over the wave length [23, 26],

$$\bar{E} = \overline{EK} + \overline{EP} = \frac{1}{2}\rho g\overline{\eta^2} + \frac{1}{2}\rho g\overline{\eta^2} = \frac{1}{2}\rho g A^2, \tag{8}$$

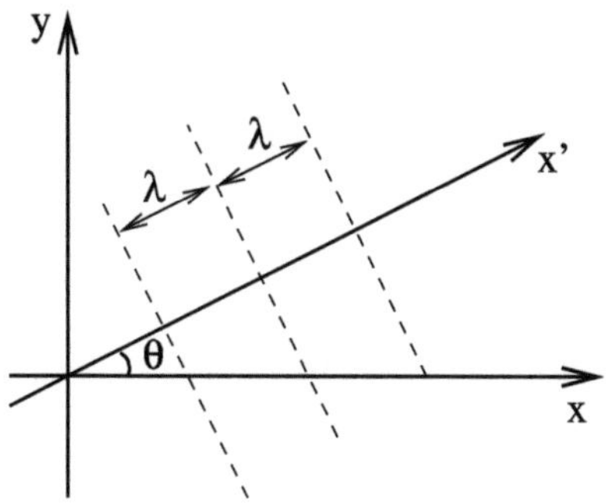

Fig. 2 Plane surface waves.

where

$$\overline{\eta^2} = \frac{1}{\lambda}\int_0^{\lambda} \eta^2 dx.$$

1.1.3 Superposition of plane waves

In realistic wave fields, waves are usually not confined to unidirectional motion. A wave propagating along an arbitrary direction $x' = x\cos\theta + y\sin\theta$, as shown in Fig. 2, is described by

$$\eta = A\cos(kx\cos\theta + ky\sin\theta - \omega t).$$

A reasonably large section of a water surface exposed to external forces will in most cases contain several different waves, each with a distinct wave number and direction of propagation. If the different wave components can be adequately described by linear theory, which is usually the case if the wave steepness is small, then there will be no exchange of wave energy between the different wave components. In this case the water surface is described by the superposition of the different wave modes, which can be written as

$$\eta(x,y,t) = \mathrm{Re}\int_0^{\infty}\int_{-\pi}^{\pi} A(\omega,\theta)\exp[-ik(\omega)(x\cos\theta + y\sin\theta) + i\omega t]\, d\theta\, d\omega, \qquad (9)$$

for an arbitrary number of simple sinusoidal wave solutions.

1.2 Steady ship wakes and ship wave parameters

The wake waves generated by a ship moving along a straight line at constant speed in water of constant depth, will be confined within a wedge shaped area with the ship located at the acute angle of the wedge. An example of such a wake is shown in

Fig. 3 Classical Kelvin ship wake pattern. Picture by Adrian Pingstone, taken at Avon Gorge, Bristol, England. Published on Wikipedia under the GNU Free Document License.

Fig. 3. Waves near the ship hull are typically unstationary, but at some distance away from the ship the stationary part of the wake is prominent, so that an observer on the ship will observe a steady pattern of wave crests and troughs in the wake. Although short crested, steep waves often occur near the ship hull, waves at a distance of a few ship lengths will usually have a small wave amplitude relative to the wave length, suggesting that the far field part of the wake should be appropriately described by linear wave theory.

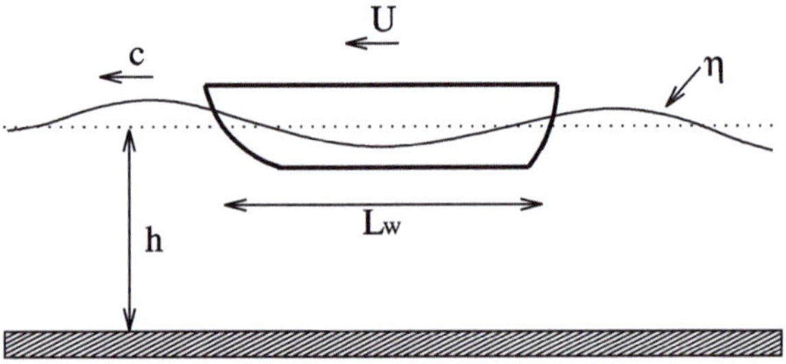

Fig. 4 Sketch of parameters for ship generated waves.

1.2.1 Ship wave parameters

Fig. 4 shows the main parameters that are important for the problem of ship wave generation, where the ship is moving with a speed U, and L_w is the length of the waterline of the ship. Wave generation can be characterized in terms of two non-dimensional parameters, the length Froude number F_L, and the depth Froude number F_h, defined by

$$F_L = \frac{U}{\sqrt{gL_w}}, \qquad F_h = \frac{U}{\sqrt{gh}},$$

respectively [11, 48]. The major part of wave generation occurs near the bow (front end) and stern (back end) of the ship, and the interactions between these wave groups are described by the length Froude number F_L. This is the dominating scale in deep water. Wake conditions in shallow water are restricted by the shallow water wave speed $\sqrt{gh}$, which is the maximum phase speed of linear waves in water with depth h. The presence of this threshold affects the long wave components of the wake in particular, but also waves with intermediate wave lengths $\lambda > 2h$ will be influenced by the finite depth (see Table 1). The wave making resistance for a high-speed vessel in coastal waters is typically a function of both F_L and F_h.

Table 2 Vessel classification

Vessel type	Conventional vessel	Semi-displacement vessel	Planing vessel
F_L range	$F_L < 0.4$	$0.4 - 0.5 < F_L < 1.0 - 1.2$	$F_L > 1.0 - 1.2$

Vessels are often classified according to the maximum achievable length Froude number. The ranges given in Table 2 are suggested by Faltinsen [11]. Wave making resistance is highly fluctuating in the range $0.4 < F_L < 0.6$, often called the "hump speed", and this speed range limits the operation of conventional vessels.

High-speed vessels are sometimes defined as vessels with a maximum speed or maximum length Froude number larger than a certain limit, e.g. $U > 30$–35 knots or $F_L > 0.4$. However, such definitions do not distinguish between small and large vessels. Since wake wave problems are mainly associated with fairly large vessels, a more sensible definition, which is used in the International Maritime Organization (IMO) High-Speed Craft Code, is to define a high-speed vessel as one that is capable of attaining a maximum speed

$$U = 3.7\nabla^{0.1667}\,\mathrm{m/s}\,, \qquad (10)$$

where ∇ is the numerical value of the vessel displacement measured in m^3 [43].

1.3 Wave making resistance to steady ship motion

Calculating the wave making resistance for a realistic hull is far from trivial, and will not be addressed here. However, it is possible to obtain some insight into the role of the length Froude number F_L by considering a few highly idealized cases. This discussion will be restricted to the deep water case, and 1D motion only.

The wake waves generated by a vessel moving at a constant speed along a straight line will be stationary relative to the vessel. This implies that

$$U = c = 2c_g\,, \tag{11}$$

along the direction of the vessel motion. Replacing either the phase or group speed by the deep water expression from Table 1, and the ship speed with the length Froude number gives

$$F_L\sqrt{gL_w} = \sqrt{\frac{g}{k}}\,,$$

and replacing the wave number k with the wave length λ from Eq. (6), and squaring both sides of the equation gives

$$\frac{\lambda}{L_w} = 2\pi F_L^2\,. \tag{12}$$

From this expression we find that $\lambda \geq L_w$ for $F_L \geq 0.4$, which is the lower limit of the “hump speed” region. Assuming this relation for the bow wave, which has a maximum near the bow of the ship, we see that the vessel will be “caught” between two wave crests at $F_L = 0.4$ (Fig. 5a), and will climb its own bow wave for larger values of F_L (Fig. 5b).

Fig. 5 Sketches of the 1D bow wave for different length Froude numbers, as defined by Eq. (12).

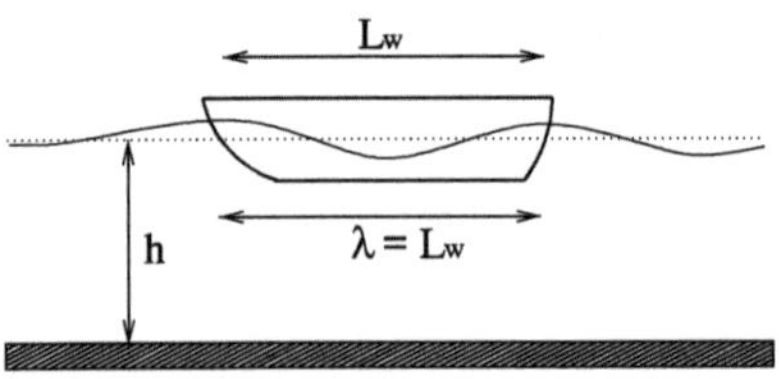

Fig. 5a Bow wave at $F_L = 0.4$

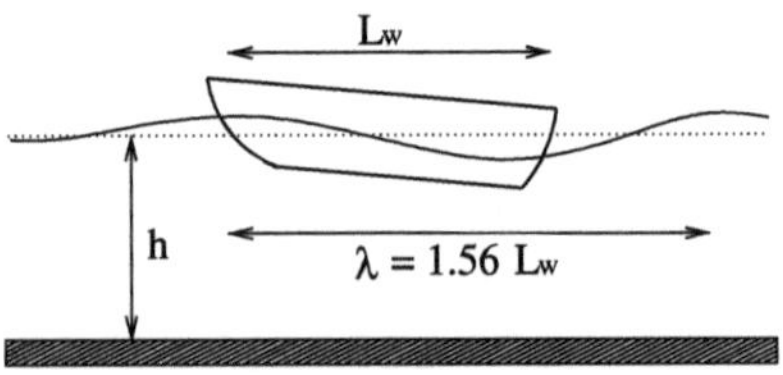

Fig. 5b Bow wave at $F_L = 0.5$

1.3.1 Wave making resistance in deep water

The wave making resistance experienced by the ship reflects a conversion of propulsion energy into wave energy. For a stationary wave pattern, the wave resistance will be independent of time. As a result, the energy conversion rate from ship propulsion to wave energy, a quantity called the wave drag, will have a constant value D that can be determined from

$$DU = \overline{\frac{dE}{dt}} = c_g\bar{E} = \frac{1}{2}\rho g\overline{\eta^2}U\,,$$

which can be simplified as

$$D = \frac{1}{2}\rho g\overline{\eta^2}\,. \tag{13}$$

In the following analysis, we assume that the waves originate at the bow and stern of the ship, as two point disturbances with equal magnitudes but opposing signs (see Newman [38])

$$\begin{aligned}\eta &= A\cos(kx-\omega t) - A\cos(k(x+L_w)-\omega t)\\ &= \mathrm{Re}\left\{Ae^{i(kx-\omega t)}(1-e^{ikL_w})\right\}.\end{aligned}$$

The mean square of the water elevation downstream of both disturbances will then be given by

$$\overline{\eta^2} = \frac{1}{\lambda}\int_0^{\lambda} A^2(1-e^{ikL_w})^2 e^{2i(kx-\omega t)}\,dx = 2A^2\sin^2(\frac{1}{2}kL_w)\,.$$

Inserting this expression into Eq. (13), we get

$$D = \rho g A^2\sin^2(\frac{1}{2}kL_w)\,,$$

as the expression for the wave drag. If we introduce the Froude length number F_L, and recall from Eq. (11) that $U = c = \sqrt{g/k}$, we can write the expression as

$$\frac{D}{\rho g A^2} = \sin^2\left(\frac{1}{2F_L^2}\right)\,. \tag{14}$$

Fig. 6 shows the non-dimensional wave drag as a function of the length Froude number F_L. The graph has a local maximum near $F_L = 0.5$, which is consistent with what we could expect from previous experience. The analysis presented here is of course very simplified compared to the problem of a two-dimensional wave field generated by a realistic hull shape, but nevertheless it gives us some indication of what the realistic wave drag will be. More extensive examples can be found in e.g. Baar and Price [1] and Sorensen [48].

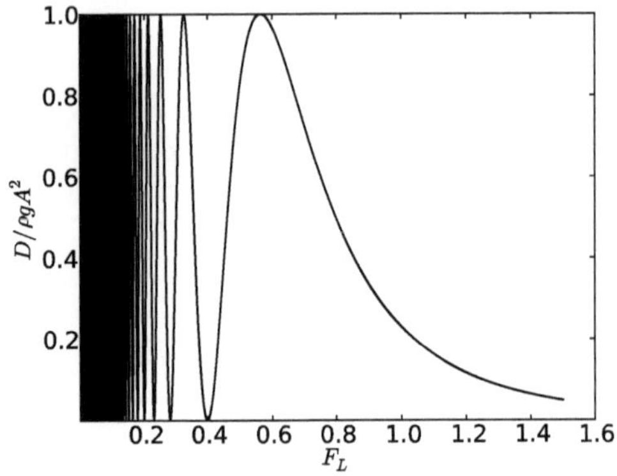

Fig. 6 Non-dimensional wave drag as function of F_L, according to the idealized model given in Eq. (14).

1.3.2 Wave making resistance in shallow water

The wave energy propagates with the group velocity c_g. In deep water we have $c_g = c/2$, which means that even if the wave pattern is stationary relative to the ship, the wave energy propagates at a slower speed and does not accumulate near the ship (Fig. 7a). In shallow water we have $c_g = c$, so the wave energy is transported with the individual waves (Fig. 7b). In this case a large portion of the wave energy is contained within a few long waves traveling in the vicinity of the ship. A study by Yang [59] suggested that wave resistance could increase significantly in shallow water if $h/L_w < 0.4$. According to linear wave theory, the wave amplitude should grow without bounds for this case. However, for larger amplitude waves nonlinear effects becomes significant and need to be taken into account, and linear wave theory is therefore not applicable for such waves.

Fig. 7 Wave energy propagation in deep and shallow water.

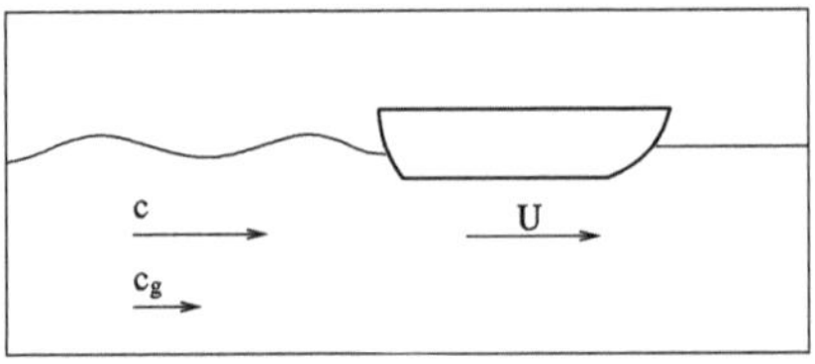

Fig. 7a Wave energy propagation for deep water.

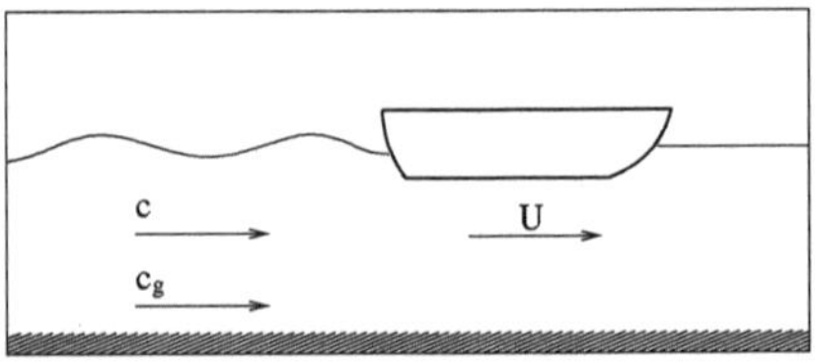

Fig. 7b Wave energy propagation for shallow water.

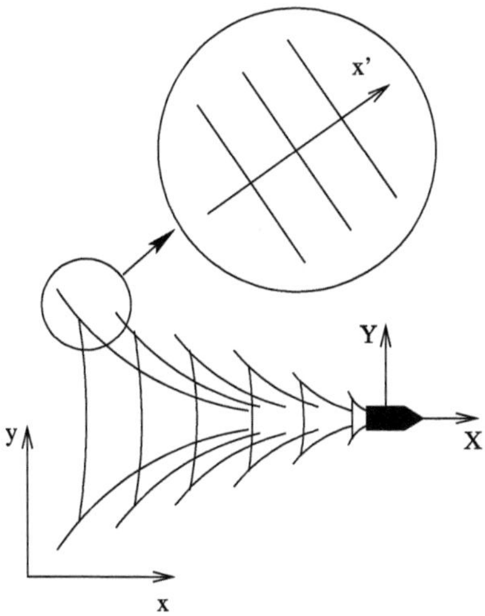

Fig. 8 Sketch of the steady state wake wedge.

1.4 Ship wake patterns in deep and shallow water

We define a fixed frame of reference (x,y) and a coordinate system (X,Y) which is moving with the ship, as shown in Fig. 8. The moving frame of reference is defined by

$$X = x - Ut\,, \quad Y = y\,,$$

and the general linear wave solution Eq. (9) in the moving frame of reference becomes

$$\eta(X,Y,t) = \mathrm{Re}\int_0^\infty \int_{-\pi}^{\pi} A(\omega,\theta)\exp[\,-ik(\omega)(X\cos\theta + Y\sin\theta) \\ +i(\omega - kU\cos\theta)t]\,d\theta\,d\omega\,. \qquad (15)$$

In the following analysis only the stationary part of the ship wake will be considered, which is the part that normally carries most of the wake energy. For this part of the wake there can be no time dependence in Eq. (15), which can only be achieved if

$$\omega - kU\cos\theta = 0\,.$$

This gives us the phase speed

$$c = \frac{\omega}{k} = U\cos\theta\,, \quad \cos\theta > 0 \qquad (16)$$

for the admissible wave components in the wake pattern.

1.4.1 The Kelvin wake pattern

The wake pattern generated by conventional ships in deep water is described by the classical Kelvin ship wave theory, named after the Irish mathematical physicist and engineer Lord Kelvin (William Thomson). The Kelvin wake pattern consists of two main families of waves, transverse waves which mostly propagate along the track of the vessel, and divergent waves that move at a relatively large angle to the track (see eg. Lighthill [26], pp 269–279).

Using the deep water dispersion relation ($k = \omega^2/g$), we can describe the waves in the moving frame of reference by

$$\eta(X,Y,t) = \mathrm{Re}\int_{-\frac{\pi}{2}}^{\frac{\pi}{2}} A(\theta)\exp[-ik(\theta)(X\cos\theta + Y\sin\theta)]\,d\theta$$

with wave numbers restricted by

$$k(\theta) = \frac{g}{U^2\cos^2\theta}\,. \tag{17}$$

Having found the description of the waves in the moving frame of reference, we now return to the fixed frame (x,y). An observer in the fixed frame, for instance on a beach some distance away from the track of the ship, will observe the ship generated waves as a wave system propagating in the x'-direction with the group velocity c_g. If the waves travel a distance x'_0 in the time t_0, we have

$$\frac{x'_0}{t_0} = c_g = \frac{d\omega}{dk}\,,$$

which can be written as

$$\frac{d}{dk}(kx'_0 - \omega t_0) = 0\,,$$

which, by integrating with respect to k, gives

$$kx'_0 - \omega t_0 = const\,.$$

In the reference frame of the moving ship, the location of x'_0 is determined by

$$x'_0 = x_0\cos\theta + y_0\sin\theta = X_0\cos\theta + Y_0\sin\theta + Ut_0\cos\theta\,,$$

where (X_0,Y_0) is the coordinate in the moving frame of reference of the wave generated at the time $t = 0$ propagating in the x'-direction. Combining the expressions above, using the condition for stationary waves $kUt_0\cos\theta - \omega t_0 = 0$, and replacing k with the expression from Eq. (17), gives us the following condition

$$\frac{g}{U^2\cos^2\theta}(X_0\cos\theta + Y_0\sin\theta) = const\,,$$

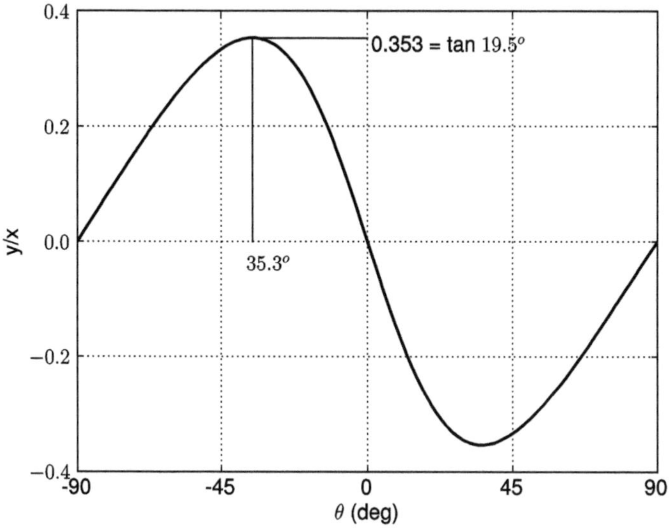

Fig. 9 The Kelvin angle for conventional ships in deep water.

which only depends on the direction of propagation θ. The ratio of Y_0/X_0 can be found by taking the derivative of the above expression with respect to θ:

$$\frac{d}{d\theta}\left[\frac{X_0\cos\theta + Y_0\sin\theta}{\cos^2\theta}\right] = 0, \quad \text{for} \quad -\frac{\pi}{2} < \theta < \frac{\pi}{2},$$

and solving this equation we find that

$$\frac{Y_0}{X_0} = -\frac{\cos\theta\sin\theta}{1+\sin^2\theta}.$$

The ratio Y_0/X_0 as a function of θ is shown in Fig. 9. The half value of the apex angle of the Kelvin wedge is defined by

$$\tan\alpha = \max\left(\frac{Y_0}{X_0}\right),$$

which gives a value of $\alpha \approx 19.5°$. Waves at the edge of the wedge will have an angle of $\theta \approx 35.3°$ relative to the track of the ship. The basic geometry of ship wakes in deep water is always the same, and does not depend on the size and speed of the ship. The derivation of the Kelvin wake properties can also be performed using a more geometrical approach (see e.g. Lighthill [26], or Soomere [45]), who derived the result $\sin\alpha = 1/3$, which is equivalent to the result presented here.

The period of the leading waves in the Kelvin wake can be derived from Eq. (16) and the deep water dispersion relation $k = \omega/g$, as

$$T = 2\pi \frac{U}{g} \cos\theta \, .$$

which shows that the period of the leading waves increases linearly with the ship speed U.

1.4.2 Ship wake in shallow water

Linear theory can also be used to calculate the half angle of the wake wedge for ships in shallow water. The stationary wave pattern in shallow water will have to satisfy

$$\omega - kU\cos\theta = 0, \qquad \omega = k\sqrt{gh},$$

which gives us

$$U\cos\theta = \sqrt{gh}, \qquad \text{or} \qquad \cos\theta = F_h^{-1}, \quad U > \sqrt{gh}.$$

Since the phase speed c and group speed c_g are equal in shallow water, the leading waves at the edge of the wake wedge will always be perpendicular to the wedge angle, i.e.

$$\theta = \frac{\pi}{2} - \alpha,$$

which gives the relation

$$\sin\alpha = F_h^{-1}, \quad U > \sqrt{gh},$$

between the depth Froude number and the wedge angle. In the critical limit we get

$$\lim_{F_h \to 1} \sin\alpha = 0,$$

which corresponds to a wedge apex half-angle of $\alpha = \pi/2$.

1.4.3 Wave patterns generated at different Froude numbers

The relation between the wedge angle α and the depth Froude number F_h is shown in Fig. 10. The wake angle approaches 90° at the critical value F_h, and decreases for super-critical values $F_h > 1$. This interpretation is somewhat misleading, since Fig. 10 is based on results from linear wave theory and does not include nonlinear effects. In particular, linear wave theory is not adequate to describe the long, nonlinear waves generated within the trans-critical speed regime $0.84 < F_h < 1.15$, where the nonlinear effect contributes to the wave celerity so that waves may propagate at a speed $c > \sqrt{gh}$.

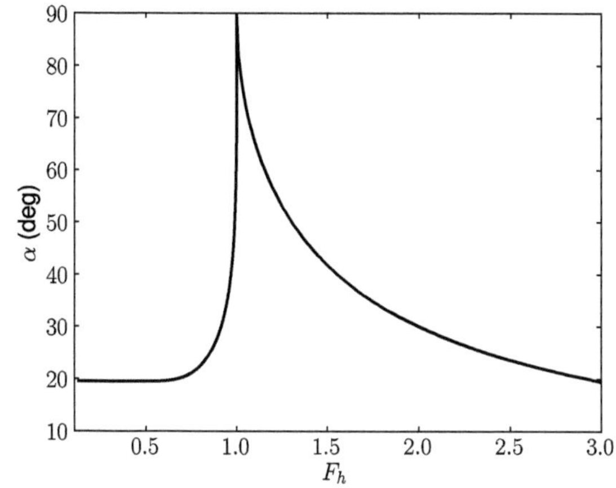

Fig. 10 Relation between wedge half angle α and depth Froude number F_h for linear wake waves.

Fig. 11 Wave pattern dependence on depth Froude number.

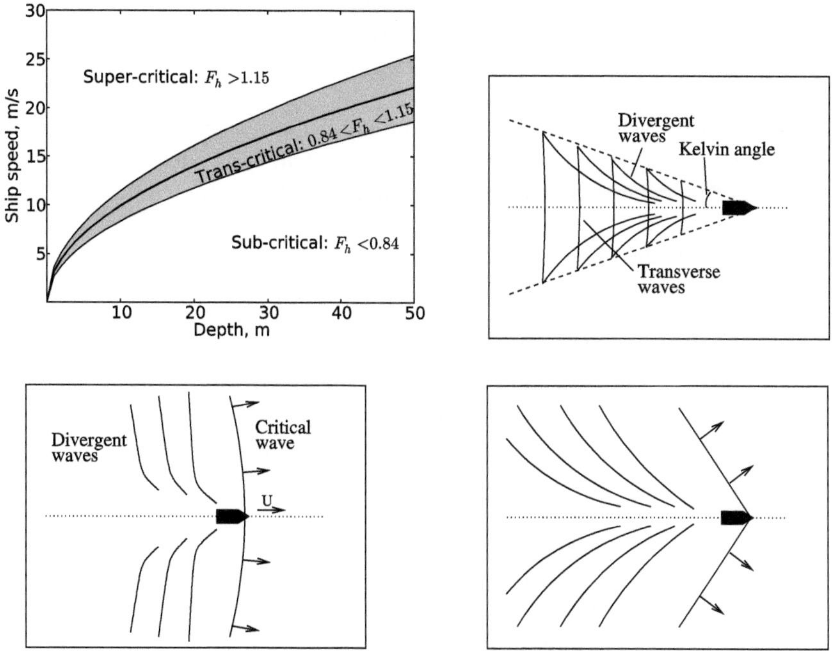

Fig. 11a Speed regimes at different depths

Fig. 11b Sub-critical wave pattern

Fig. 11c Near critical wave pattern

Fig. 11d Super-critical wave pattern

Fig. 11 shows how the ship wave pattern changes with different values of the depth Froude number F_h. Fig. 11a shows the different speed regimes in terms of the depth Froude number F_h, for various vessel speeds and water depths. Figs. 11b, 11c, and 11d show sketches of the classical Kelvin ship wake, a near-critical ship wake, and a supercritical ship wake, respectively.

Note that the Kelvin wave pattern has both divergent wave groups propagating away from the track of the ship, and transverse waves propagating along the direction of the ship. For the super-critical wave pattern Fig. 11d, no transverse wave is capable of moving with the speed of the ship, and only the divergent wave group can exist in the stationary state (see eg. [60, 61] for a more detailed study of ship waves).

1.4.4 Wave propagation and transformation

Wake waves decay as they propagate away from the vessel. The decay of waves within the classical Kelvin wake was analyzed by Havelock [14], who found that the decays of both transverse and divergent waves were proportional to ζ^n, where ζ is the distance from the vessel and n is a constant value. Waves inside the Kelvin wake were found to decay as $\zeta^{-1/2}$, whereas waves at the edge of the wedge decay as $\zeta^{-1/3}$.

The decay of the wash waves in intermediate and shallow water was discussed by Doyle et al. [8]. The wake waves were found to decay at different rates depending on the depth Froude number F_h and the water depth to ship length ratio h/L_w, ranging from $\zeta^{-0.2}$ to $\zeta^{-0.5}$. The decay rate is larger for wash generated in the near critical range $F_h \approx 1$, than in the super-critical regime $F_h > 1$, and decreases with decreasing values of h/L_w. It has, however, been shown that the decay rate of the leading wave system at $F_h \approx 1$ decreases with time. In the critical case, a significant portion of the wave energy is contained in a single wave group, which is moving with the vessel. Since the amplitude of these waves cannot grow indefinitely, an increasing amount of energy will have to be transferred in the lateral direction along the wave crest. It has been suggested that a state of equilibrium should eventually be attained, where the input of energy is equal to the energy transferred along the crest, but this has so far not been confirmed in experiments.

The geometrical divergence between the leading and following waves in the super-critical ship wake was studied by Whittaker et al. [57]. A divergence angle of up to 20° was found for $F_h = 1.1$ at a distance of less than 10 times the water depth from the ship track. Larger depth Froude numbers resulted in smaller values for the divergence angle, and the angle was found to decrease with the distance from the ship track. Due to this difference in the direction of propagation, the peak-to-peak distance increases with the distance from the ship track, and therefore the leading wave may arrive at the coast much earlier than the main wave group from the wake. In one case the time difference between the leading waves was measured to be more than 40 seconds at a distance of 2.7 km from the ship track [36].

1.5 Why is wake wash from high-speed vessels a problem?

As the preceding discussion has shown, there are significant differences between the wake created by conventional vessels, and wakes from high-speed vessels when travelling in coastal waters. Waves generated at trans-critical speeds are generally larger than normal wake waves, and waves generated at super-critical speeds decay less with the distance from the ship track, so waves from high-speed vessels will usually be larger than waves for conventional vessels, and influence coastal areas further away from the ship track. Whether or not this is a problem is highly dependent on local conditions of wind waves, sea swell and tides.

The long wave length and wave period may be a source of problems. The leading wave may arrive half a minute or more before the main part of the wake waves [36], depending on the distance from the ship track. The typical period for the long wave component of the wake will be case dependent, and measured wave periods vary from a range of 7–9 s [40] to a range of 10–15 s [47]. Although the amplitude of the leading wave may not be very large in intermediate water depth, it will increase in size near the coast due to wave shoaling. This may catch people on the shore by surprise, in particular on calm days without large wind waves [13]. Another effect of the large wave period is that it generates larger sea bed velocities at certain water depths than short period waves, altering the natural transport of sediments. Erosion or accretion due to high-speed vessel wakes may undermine coastal structures or alter fairways used for shipping, and may disturb marine habitats. Benthic habitats may also be directly affected by increased sea bed currents created by the long waves.

There may also be problems related to the increased wake wedge angle for near-critical wakes. With conventional ship wakes, people at exposed areas on the shore or in small boats near shallow water will see the vessel passing off shore long before the wake waves arrive, and will have time to take appropriate action. When the wedge angle is close to 90°, there will be little delay between the passing of the ship and the first wake waves hitting the shore.

Wave generation depends on both the length and depth Froude number, where the former is specific for each particular vessel, and the latter depends on the bathymetry along the ship track. A number of processes can transform the waves as they propagate towards the coast, such as wave refraction, diffraction, shoaling, and breaking, and the impact of the waves will usually depend on particular features of the coast. It is therefore important that local conditions are taken into account when formulating regulations of high-speed vessel wake wash. A simple measure such as speed reduction, may not have the desired effect, be excessively restrictive, or even aggravate the wake wave problem if the speed reduction causes the vessel to move at trans-critical speed, as seen in the case of Stavns Fjord in Denmark [40].

2 Properties of long wave model equations

Vessel wake waves can be adequately described by the Euler equations with appropriate boundary conditions, given by equations (1)–(5). However, only a few analytical solutions of these equations are known, and these are limited to simplified domains and idealized case studies. Simpler but less general model equations can be derived from the Euler equations (1)–(5) by assuming certain flow properties for the problem under investigation. From the discussion in the previous section, we know that the wake waves generated by high-speed vessels in shallow water have long wave lengths relative to the water depth, and waves with large amplitudes and high celerity that cannot be described properly by linear wave theory. The family of equations collectively called long wave equations are derived from the Euler equations by assuming that the waves are long relative to the depth. Examples of such equations include the linear and nonlinear shallow water equations, the Korteweg-de Vries (KdV) and Kadomtsev-Petviashvili (KP) equations, and Boussinesq-type equations, among others. Long wave equations are usually derived by writing the dependent variables $\mathbf{u}$ and η as series expansions in terms of the vertical coordinate z, and inserting these expressions into the Euler equations. Different formulations can be derived depending on how many terms are retained in the series expansion, and by including further assumptions such as weak nonlinearity and mild slopes for the bottom boundary.

Even with simplified model equations, analytical solutions are limited to idealized case studies. In most practical applications, where the water basin has a complex geometry, we need to solve the model equations numerically. Modern computers have reached a level of computational power where it is possible to run models that solve the Euler equations (1)–(5) without any modifications. However, the computational cost of running these models is very high, which makes them impractical for modeling flow on large computational domains. Long wave models are less computationally demanding, because the vertical flow profile is derived from the horizontal flow, and only equations for the 2D horizontal flow need to be solved.

2.1 Approximations for shallow water flow

The starting point for the derivation of long wave equations is a consideration of the relevant scales for surface gravity waves in the absence of external forces. These are summarized in Table 3, and the relevant length scales are shown in Fig. 12.

Table 3 Summary of physical scales for surface gravity waves.

Acceleration scale:	g - acceleration of gravity
Length scales:	λ - wave length, h - depth, A - amplitude/elevation
Time scale:	Derived from length and acceleration scales, e.g. $t = \lambda/\sqrt{gh}$

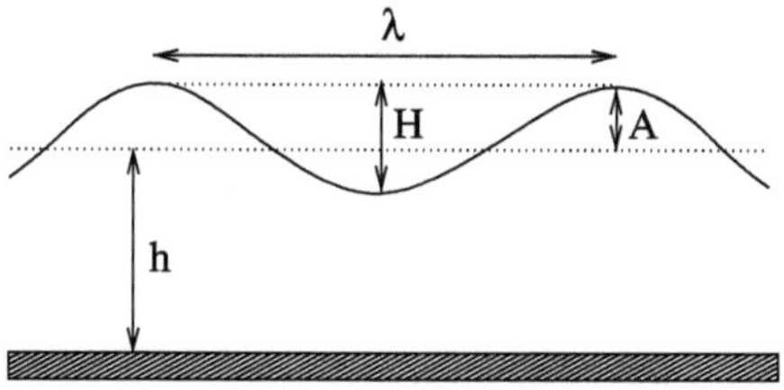

Fig. 12 Sketch showing the relevant length scales for surface gravity waves

The *shallow water* or *long wave* approximation states that the wave length λ is significantly larger than the water depth h, which can be formulated in terms of the non-dimensional parameter

$$\mu = \frac{h}{\lambda} \sim kh \ll 1 .$$

This parameter is related to wave dispersion, which can be seen if we carry out a Taylor expansion of the linear dispersion relation,

$$\omega = \sqrt{gk\tanh(kh)} = \sqrt{gk}\left\{kh - \frac{1}{3}(kh)^3 + \frac{2}{15}(kh)^5 - \dots\right\}^{\frac{1}{2}} . \tag{18}$$

For the longest waves, the first term in the Taylor expansion gives a reasonable approximation to the dispersion relation, but for larger values of μ, more terms must be included to obtain an adequate approximation of the dispersion relation.

For deep water waves, nonlinear wave properties depend only on the wave steepness

$$\sigma = \frac{A}{\lambda},$$

whereas nonlinear effects in shallow water will also depend on the ratio

$$\varepsilon = \frac{A}{h},$$

between wave amplitude and water depth. Ursell [54] showed that for long waves, the nonlinear effects depend both on the water depth and the dispersive properties of the wave, which is governed by the Ursell number

$$Ur = \frac{\varepsilon}{\mu^2} = \frac{A\lambda^2}{h^3} .$$

Waves can be adequately described by linear theory only if both $Ur \ll 1$ and $\sigma \ll 1$. If $Ur \gg 1$, the waves will be non-dispersive and highly nonlinear. In this case the front of the wave will gradually steepen until the wave eventually breaks. For intermediate values $Ur \approx 1$, wave dispersion is balanced by a weakly nonlinear effect. Even though the waves are nonlinear, they may propagate without changing wave shape. Solitary waves generated at $Ur \approx 1$ will not be limited by the shallow water wave speed $c=\sqrt{gh}$, but will propagate at an increased phase speed of $c=\sqrt{g(h+\eta)}$, which depends on the wave amplitude.

Consider a wave propagating along the x-axis. The flow must satisfy the continuity equation

$$\frac{\partial u}{\partial x}+\frac{\partial w}{\partial z}=0.$$

We now introduce U and W as characteristic horizontal and vertical velocities, respectively. The equation tells us that

$$\frac{W}{h}\sim\frac{U}{\lambda}.$$

Using the shallow water approximation, we see that the ratio between the vertical and horizontal velocities satisfies the following condition

$$\frac{W}{U}\sim\mu\ll 1.$$

2.2 Comparison between different model equations

Here we present three commonly used long wave equations, the KdV, KP and Boussinesq-type equations, and consider some of their properties.

2.2.1 Korteweg-de Vries equation

The Korteweg-de Vries equation (KdV)

$$\frac{\partial\eta}{\partial t}+\left(1+\frac{3}{2}\eta\right)\frac{\partial\eta}{\partial t}+\frac{1}{6}\frac{\partial^3\eta}{\partial x^3}=0,$$

is the simplest form of a long wave equation that includes both weakly nonlinear and weakly dispersive effects. It is instructive to compare the properties of this equation with properties for the corresponding nonlinear transport equation

$$\frac{\partial\eta}{\partial t}+\left(1+\frac{3}{2}\eta\right)\frac{\partial\eta}{\partial t}=0,$$

and the linear dispersive equation

$$\frac{\partial\eta}{\partial t}+\frac{\partial\eta}{\partial t}+\frac{1}{6}\frac{\partial^3\eta}{\partial x^3}=0.$$

Fig. 13 Comparison between the KdV equation and the corresponding nonlinear and dispersive transport equations.

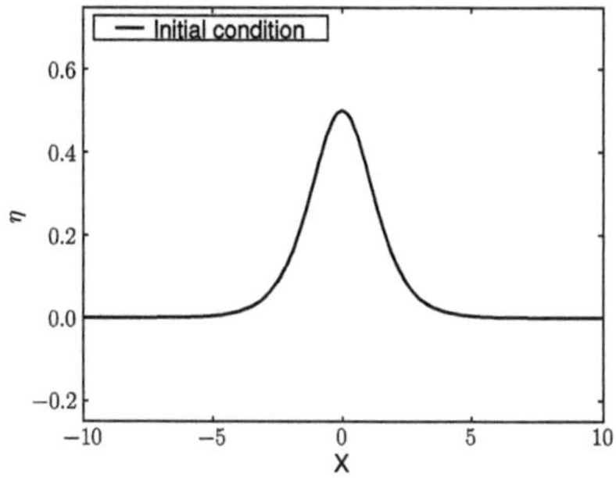

Fig. 13a Initial state

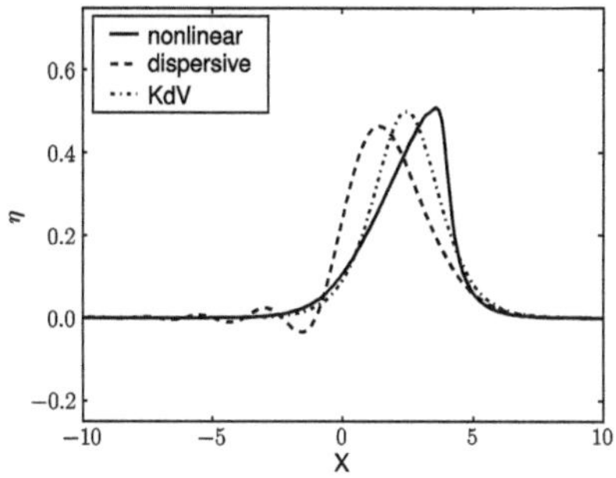

Fig. 13b End state

In this case we look at the evolution of an initial disturbance

$$\eta = A\cosh^{-2}\left(\frac{\sqrt{3A}}{2}x\right),$$

which is the *soliton* solution for the KdV equation.

The result is shown in Fig. 13. The linear dispersive equation breaks up the initial profile into a wave train of linear waves, while the nonlinear equation creates a wave with a steep wave front. In the KdV equation, these two effects are perfectly balanced for this particular initial profile, and the wave can propagate without any change in shape.

2.2.2 fKdV and KP equations

If we retain the ambient pressure p_a in the derivation of the equations, we can obtain the so-called forced KdV (fKdV) equation

$$\frac{\partial \eta}{\partial t} + \left(1 + \frac{3}{2}\eta\right)\frac{\partial \eta}{\partial t} + \frac{1}{6}\frac{\partial^3 \eta}{\partial x^3} = \frac{1}{2}\frac{\partial p_a}{\partial x}.$$

The ambient pressure does not influence the flow if it is homogeneous, but a localized disturbance in the pressure field can be used to represent a moving ship (see the discussion in section 3.2). The fKdV equation is only valid for waves propagating in one horizontal dimension. The forced Kadomtsev-Petviashvili (KP) equation

$$\frac{\partial}{\partial x}\left[\frac{\partial \eta}{\partial t} + \left(1 + \frac{3}{2}\eta\right)\frac{\partial \eta}{\partial t} + \frac{1}{6}\frac{\partial^3 \eta}{\partial x^3}\right] - \frac{1}{2}\frac{\partial^2 \eta}{\partial y^2} = \frac{1}{2}\frac{\partial^2 p_a}{\partial x^2},$$

is an extension of the fKdV equation that admits waves that propagate at a slight angle relative to the x-axis, which is assumed to be the main direction of propagation. The KdV and KP equations are interesting from an analytical point of view, because it is possible to find explicit, closed form solutions for these equations (see e.g. [9, 56] for a further discussion).

2.2.3 Boussinesq equations

Long waves can also be modelled by the so-called Boussinesq-type equations,

$$\frac{\partial \eta}{\partial t} + \nabla_H \cdot [(h+\eta)\bar{\mathbf{u}}] = 0, \tag{19}$$

$$\frac{\partial \bar{\mathbf{u}}}{\partial t} + (\bar{\mathbf{u}} \cdot \nabla_H)\bar{\mathbf{u}} = -\nabla_H \eta - \nabla_H p_a + \frac{1}{3}h^2 \nabla_H \nabla_H \cdot \left(\frac{\partial \bar{\mathbf{u}}}{\partial t}\right), \tag{20}$$

where $\bar{\mathbf{u}}$ is the depth averaged velocity and $\nabla_H = (\partial/\partial x, \partial/\partial y)$ is the horizontal gradient operator. Equation (19) is derived by integrating the continuity equation (1) over depth, and applying the kinematic boundary conditions for the bottom and free surface boundaries. Equation (20) is derived from the momentum equation (2), where the term

$$\frac{1}{3}h^2 \nabla_H \nabla_H \cdot \left(\frac{\partial \bar{\mathbf{u}}}{\partial t}\right),$$

corresponds to the third order term in the Taylor series Eq. (18) for the approximation of linear dispersion. Equations (19) and (20) are the classical formulations of the Boussinesq equations. Other Boussinesq-type equations may be written in terms of the surface velocity or the velocity at a specific depth, and may include higher order nonlinear and dispersive correction terms. Unlike the KdV and KP equations, there are no known closed form solutions for Boussinesq-type equations. However, Boussinesq-type equations can describe waves moving in any direction on the free surface, and do not assume that there is a balance between dispersive and nonlinear effects. For these reasons Boussinesq-type equations are often preferred over KdV or KP equations when doing numerical simulations.

2.2.4 Dispersion relation for long wave equations

Several properties may influence the choice of modelling equations. One such property is how the dispersion inherent in the equation relates to the dispersion relation for linear waves. This is shown in Fig. 14, where the phase velocity for the long wave equations is compared to the linear phase velocity as function of kh. The linear phase velocity is given by

$$c = \sqrt{(g/k)\tanh(kh)}.$$

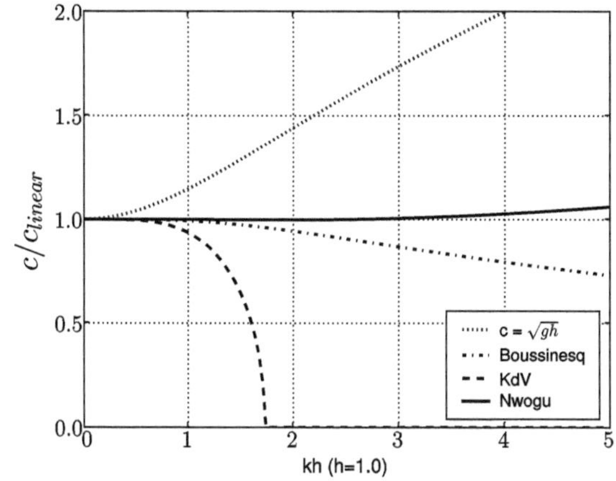

Fig. 14 Dispersion properties for shallow water equations.

The "Boussinesq" result refers to the classical formulation of the Boussinesq equations due to Peregrine [42]. Nwogu [39] modified the classical formulation by using variables at a reference depth $z_\alpha = -0.531\,h$, which optimizes the dispersive properties of the equations with respect to the linear dispersion relation. The modified Boussinesq equations are able to describe dispersion of waves up to $kh \approx \pi$, corresponding to $h \approx \lambda/2$, for which the error of the phase speed is

$$\left.\frac{c_{Nwogu}}{c_{linear}}\right|_{kh=\pi} = 1.00697\,.$$

The ability to more adequately represent the dispersion properties for waves with intermediate wave lengths is another important reason why Boussinesq equations are frequently preferred over the KdV and KP equations for numerical simulations. A thorough discussion on the properties of Boussinesq-type equations can be found in the review paper by Madsen and Schäffer [32]. Their analysis also includes 2nd order nonlinear properties. These properties generally display larger errors than the linear dispersive properties for intermediate values of kh, thus limiting the applicability of the Boussinesq equations for steep, large amplitude waves.

Improving the representation of flow properties for depth averaged equations generally requires a more detailed representation of the vertical flow structure. One approach is to include higher order nonlinear and dispersive terms in the formulation of the Boussinesq equations, as shown by Madsen and Schäffer [31, 32]. A different approach is to perform the depth integration in a piecewise manner over two or more vertical layers, as was done by Lynett and Liu [28, 29]. With this approach, separate equations are formulated for each layer, each with independent vertical velocity profiles, which need to be matched at the interfaces between the layers. The more advanced model equations (higher order or multi-layer) have been demonstrated to give reliable results for waves with $kh \approx 2\pi$, and may be extended further. However, each extension of the model equations adds to the computational cost of solving them, thereby reducing the advantage of using simplified model equations instead of the Euler equations.

3 Numerical modelling of ship waves

The ideal numerical model for studying ship generated waves would be a nonlinear, wave resolving model capable of calculating the unstationary flow and wave field around a vessel hull, the wave propagation and transformation over large distances in a non-homogeneous fluid, the wave dynamics in the surf zone including the maximum wave height before breaking, and the run-up of waves at the coast. These requirements are very difficult to satisfy within a single model, because the calculation of the non-stationary wave field near the vessel hull requires a high spatial and temporal resolution, which limits the size of the computational domain to a few ship-lengths. It is therefore customary to divide the numerical prediction of wake waves into two problems, (i) the prediction of the wave field near the vessel (near-field model), and (ii) the prediction of the propagation and transformation of wake waves far from the vessel (far-field model). A complete model of the wake waves from the point of generation to the surf zone and wave run-up at the coast can then be achieved by coupling a near-field and a far-field model [22].

Several computational fluid dynamics (CFD) codes for the simulation of ship generated waves have been developed in recent years, with capabilities of calculating details of the flow near a vessel hull [7, 16, 62]. Since the computational cost of these models is large, the thin ship theory based on the linear wave theory remains a valuable tool for simple calculations of the stationary wake pattern [37, 52, 53]. In more generic studies of wake waves, the vessel is often represented by a pressure disturbance at the free surface [10, 18, 41]. Although it is difficult to attribute a pressure disturbance to a specific hull shape [53], the far-field properties are qualitatively similar to results obtained using the slender body theory [25, 41]. The main difference between the two approaches is that waves generated by a pressure disturbance were found to be generally smoother than waves generated using the slender body theory. The main advantage of representing the ship by a pressure disturbance is that it can easily be combined with long wave equations for the modeling of far-field wave propagation. Boussinesq-type equations have been applied in studies of ship wakes in shallow channels [3, 17] and in coastal waters [51]. Spectral wave models, capable of predicting the propagation, growth, and decay of short-period waves have also been used to model the far-field wake waves in coastal waters [21, 22, 43]. These models are able to predict the average wave impact on the coast, but cannot be used to model the behavior of solitary or transient waves.

In the remainder of this section, the modelling of ship generated waves will be illustrated using a Boussinesq-type model, with waves generated by a moving pressure disturbance. This choice allows us to discuss the numerical modelling of ship generated waves without considering the coupling of different models. Such a model will normally be able to reproduce the correct wave periods for the ship generated waves, since this property mainly depends on the speed of the vessel, the water depth, and the distance from the ship track [43]. However, the amplitudes of the generated waves depend crucially on the design of the hull, and this property is difficult to represent by a pressure disturbance. The approach presented here is adequate to obtain a qualitative picture of the wake wave distribution, but conclusions about actual wave heights at a single point should not be drawn without comparing the results to data from measurements.

3.1 Numerical models based on Boussinesq-type equations

There is an extensive literature on the use of Boussinesq-type equations for the numerical modelling of waves in shallow water [19, 27, 30, 55]. Most of the existing models are formulated using finite differences, but there is also literature documenting the development of models using finite element [24, 58] and finite volume [4] methods. Models based on high order Boussinesq equations, such as FUNWAVE[1] and COULWAVE[2], are freely available for downloading from the World Wide Web. Most of the examples shown in this section have been obtained using a model based on COULWAVE, which has been slightly modified to include a moving pressure disturbance at the free surface.

3.1.1 Conditions for lateral boundaries

To make a working wave model, we need to define boundary conditions for the boundaries of the computational domain. For idealized case studies, periodic boundary conditions may sometimes be used. In cases where the domain is enclosed by a wall, such as in a wave tank, reflective boundary conditions are appropriate, where we prescribe

$$\mathbf{u} \cdot \mathbf{n} = 0, \quad \text{and} \quad \frac{\partial \eta}{\partial n} = 0,$$

on the boundary, where $\mathbf{n}$ is the unit normal vector of the reflective boundary. In many cases, in particular when dealing with realistic case studies, waves should be allowed to exit the computational domain without reflecting. Classical radiation conditions are difficult to use when waves with a wide range of wave numbers should be absorbed at the boundary. For this reason, many models use sponge layers to absorb waves approaching the boundary of the computational domain [15]. With this approach, damping terms with prescribed damping coefficients are added to the momentum equations to absorb the wave energy inside the sponge layer.

[1] `http://chinacat.coastal.udel.edu/programs/funwave/funwave.html`

[2] `http://ceprofs.tamu.edu/plynett/COULWAVE/`

3.1.2 Energy dissipation in numerical models

Numerical models for unsteady waves propagating in domains with complex geometry require some mechanism for energy dissipation in order to represent the waves in a realistic way. Dissipation of wave energy is particularly important when modelling waves in the shoaling and surf zones near the coast. For numerical reasons it is essential that the model includes adequate energy dissipation of short waves. The shortest admissible wave length in a model is defined by the resolution of the discretization grid (the difference Δx between grid points for finite difference methods, or the size of the finite volumes or elements). Accumulation of wave energy on the grid scale leads to a build up of numerical noise, which may eventually become the dominating feature in the simulation. The model equations presented in section 2 do not include any dissipative terms, so the numerical models based on such equations must include energy dissipation through other means.

A common way of dissipating wave energy is to include a term representing the bottom friction in the momentum equations. This term usually involves a quadratic dependence of the friction on the near-bottom velocity

$$\mathbf{R}_f = \frac{r}{h+\eta}\mathbf{u}|\mathbf{u}|,$$

where r is a bottom friction coefficient that needs to be set to an appropriate value. Lynett et al. [30] found an optimum value of $r = 5\times10^{-3}$ in their study of wave run-up using the COULWAVE model, but this value may be too small to avoid instabilities in large scale simulations [51]. The bottom friction term may not be sufficient to dissipate energy in the surf zone, where wave shoaling and breaking occur. A model for wave breaking should be applied locally to individual waves, which have a steepness larger than some threshold. An example of an eddy viscosity model for breaking waves was presented by Kennedy et al. [19]. The bottom friction and wave breaking models are examples of methods that try to represent a physical process responsible for energy dissipation. If these methods are not sufficient to remove the grid scale noise, explicit noise filtering may also be used, which smoothes out high frequency noise. Unfortunately, unless the spatial resolution is very high, filtering tends also to smooth the wave field for waves with intermediate lengths. For this reason filtering should be avoided if possible.

3.2 Ship representation by a moving pressure disturbance

As mentioned in the introduction to this section, it is difficult to represent a specific hull shape by a pressure disturbance. In studies that focus on the far-field wave properties, the ship is usually represented by a smooth, localized disturbance that bears only a superficial resemblance to a realistic ship hull [2, 10, 18, 25]. An example is presented by Ertekin et al. [10], who defined a pressure disturbance with a center point at (x^*, y^*) by

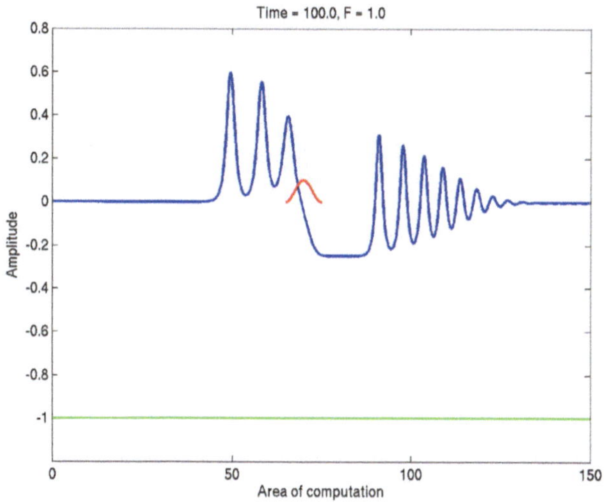

Fig. 15 Ship waves in one horizontal dimension.

$$p_a(\tilde{x},\tilde{y},t) = P\, f(\tilde{x},t)\, q(\tilde{y},t)\,,$$

$$f(\tilde{x},t) = \begin{cases} \cos^2\left(\frac{\pi(\tilde{x}-x^*(t))}{2\alpha L}\right), & \alpha L \le |\tilde{x}-x^*(t)| \le L\,, \\ 1 \quad , & |\tilde{x}-x^*(t)| \le \alpha L\,, \end{cases}$$

$$q(\tilde{y},t) = \begin{cases} \cos^2\left(\frac{\pi(\tilde{y}-y^*(t))}{2\beta R}\right), & \beta R \le |\tilde{y}-y^*(t)| \le R\,, \\ 1 \quad , & |\tilde{y}-y^*(t)| \le \beta R\,, \end{cases}$$

on the rectangle $-L \le \tilde{x} - x^*(t) \le L$, $-R \le \tilde{y} - y^*(t) \le R$, and zero outside this region. In the general case the coordinate system $(\tilde{x},\tilde{y})$ for the pressure disturbance may be rotated by some angle ϕ relative to the coordinate system (x,y) for the model equations, which requires the pressure patch to be transformed according to

$$\tilde{x} = x\cos\phi + y\sin\phi\,, \qquad \tilde{y} = -x\sin\phi + y\cos\phi\,,$$

to be correctly aligned within the model equations. A rough estimate of the displacement ∇ caused by the pressure disturbance can be derived from the Bernoulli equation, assuming a fluid at rest, which gives us

$$\nabla = \int_{-L}^{L}\int_{-R}^{R} p_a(\tilde{x},\tilde{y},0)\, d\tilde{y}\, d\tilde{x}\,.$$

An example of waves generated by a pressure disturbance is shown in Fig. 15, using the profile from Ertekin et al. [10] with $P = 0.1$, $L = 5$ and $\alpha = 0$. The shape and location of the pressure disturbance is marked by the inserted curve segment near $x \approx 65$. In this case the pressure disturbance is moving to the left with a speed

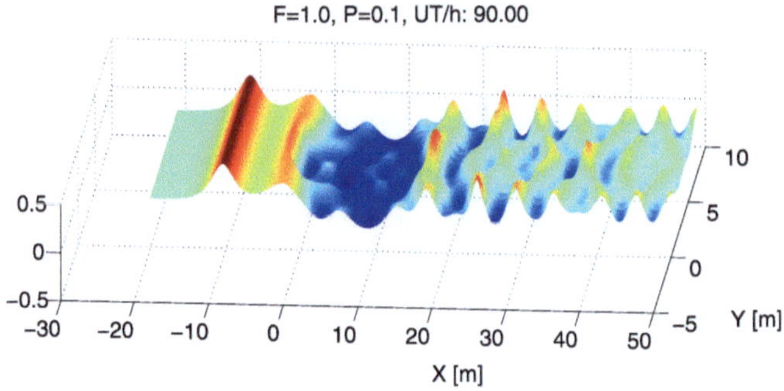

Fig. 16 Waves generated by a ship in a narrow channel.

corresponding to $F_h = 1$. At this speed the ship generates a train of almost soliton shaped waves that propagate upstream of the ship, and a depression of the water surface immediately downstream. A similar result is shown in Fig. 16, where the ship is moving in a narrow channel with a flat bottom and vertical side walls. The depression caused by the pressure disturbance can be seen near $x = 0$. The upstream waves are practically uniform in the cross-channel direction, whereas a complex wave pattern, conceptually similar to the Kelvin wake system, is found further downstream of the pressure disturbance. Upstream waves may be generated even if the bottom is not flat [17, 49, 50], but the waves will no longer be uniform in the cross channel direction.

Constructing the ship track for idealized studies such as the ones illustrated in Figs. 15 and 16 is trivial. Realistic ship tracks will usually involve maneuvering of the ship and sections of acceleration and deceleration. A typical record of a realistic ship track consists of coordinates recorded at fixed time intervals. The sampling rate of the recorded track will usually not coincide with the time step required for the numerical simulation. Standard cubic splines can be used to distribute the interpolated coordinates along smooth curves, but it can often be difficult to obtain a realistic velocity profile if the ship makes sudden turns and changes in speed. In addition to the position, we will also need to calculate the tangent of the track curve, in order to properly align the pressure disturbance with the track. Once the position and orientation of the pressure disturbance have been calculated, an interpolation routine can be used to include the effect of the pressure disturbance in the spatial grid used for solving the model equations.

3.3 The influence of the dispersive and nonlinear components

In practical applications we would like to follow the wake waves, or at least the leading wave group in the wake, from where they are generated by the ship until they shoal and break at the coast. If we allow up to a 5% error in the phase speed, a non-dispersive model can be used to model waves with $kh \leq 0.56$, or $h \leq 0.089\,\lambda$

(see Fig. 14). For a wave with wave length $\lambda = 100$ m, this threshold is satisfied until a depth of $h = 9$ m. In shallow sea areas, ship lanes are usually located along natural or artificially dredged trenches for the purpose of safe navigation. It is therefore rarely the case that wake waves can be considered non-dispersive throughout the computational domain.

The nonlinear contribution is strongest for large amplitude waves with steep wave fronts. Such waves can usually be found close to the wave generating disturbance, and near the shore where waves are shoaling. Fig. 17 shows a comparison between linear and nonlinear simulations of ship waves generated by fast ferries in Tallinn Bay [51]. The map in Fig. 17a shows a north-bound (solid line) and south-bound (dashed line) ship track, and the location of virtual wave gauges recording the wave profiles at different locations near the coast. The results shown in Figs. 17b and 17c are for a south-bound ship. Although the results are from the same simulation, there are significant differences in the wave profiles and periods, due to a high degree of spatial variability of the wake pattern, the details of which depend on ship speed and local bottom topography. Assuming the wave speed is close to the shallow water wave speed $c = \sqrt{gh}$, we can give a rough estimate for the Ursell number for the waves shown in Fig. 17, which is $Ur \approx 8$ for Fig. 17a and $Ur \approx 66$ for Fig. 17b. This difference in Ursell number is a combined effect of the difference in water depth and wave period. The wave records show that there is some difference in wave amplitude between the linear and nonlinear simulations, but that the wave phase and period are nearly the same. As long as the waves propagate in deep water, the nonlinear term does not contribute significantly to the phase speed of the waves. Nonlinear effects become important near the coast, when waves are shoaling and breaking. However, these effects are not prominent in the wave records shown in Figs. 17b and 17c, mainly because the wave gauges are located some distance from the shore, outside the main shoaling and surf zones.

The representation of large amplitude, steep waves is a challenge for models based on long wave equations. Such waves are often generated when simulating waves generated by ships operating in very shallow water. Fig. 18 shows results from a study of waves generated by a pressure disturbance moving along a channel with a variable cross channel topography [49]. The channel profile is uniform in the along-channel direction, and symmetric across the center line in the cross-channel direction, as shown in Fig. 18a. Fig. 18b shows the surface displacement for the upper half of the channel, at a near critical depth Froude number, simulated by a 1-layer Boussinesq model. The forcing disturbance is located at $\xi = 0$, and all axes units are scaled relative to the maximum water depth in the channel. In this case the wake waves include some large amplitude waves with intermediate wave lengths, and it is not clear whether these waves are represented correctly by the model. This is particularly of concern for regions where the reflected and incident waves interact to create unusually large amplitude waves, such as at $\xi \approx 12$ in Fig. 18b. To check the result we have run the same simulation using a 2-layer model, which has better nonlinear and dispersive properties than the standard model [29]. Fig. 18c shows a comparison between a 1-layer simulation and a 2-layer simulation, where the two simulated results are simply subtracted from each other in each grid point. The two

Fig. 17 Comparison between nonlinear and linear simulations. Reprinted from [51], with permission from *Estonian Journal of Engineering*.

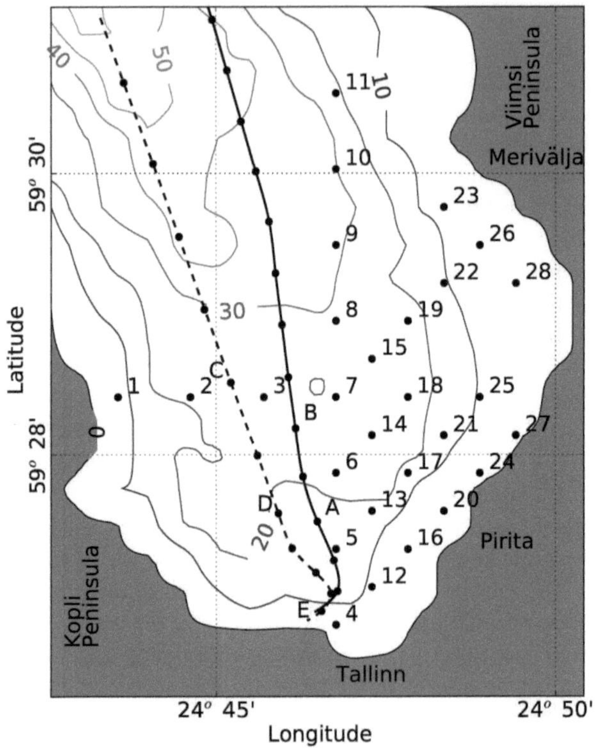

Fig. 17a Map of the interior of Tallinn Bay, including ship tracks and virtual wave gauge locations.

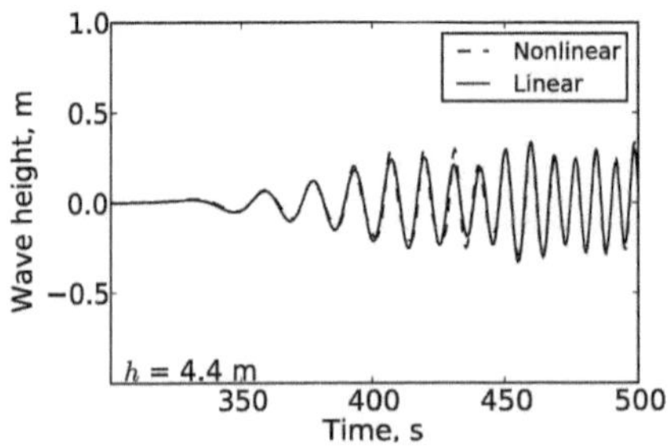

Fig. 17b Surface time series at wave gauge 1

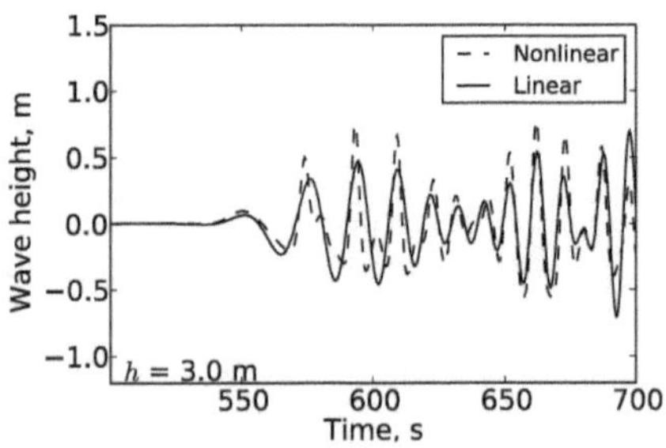

Fig. 17c Surface time series at wave gauge 16

Fig. 18 Ship waves in a channel with a variable bathymetry. Comparison between a 1-layer and a 2-layer model. Reprinted from Torsvik et al. [50], with permission from *Journal of Waterway, Port, Coastal, and Ocean Engineering*.

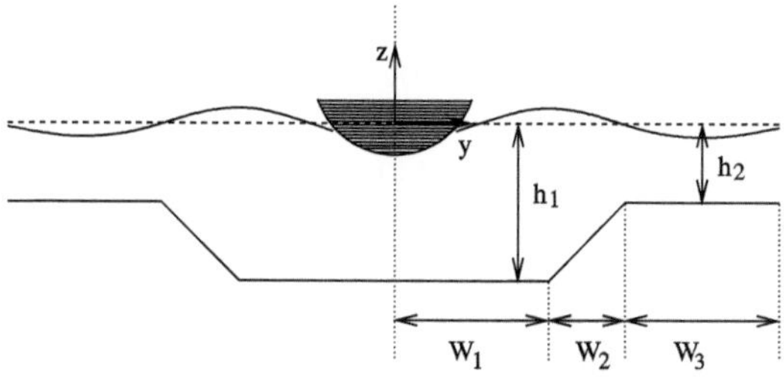

Fig. 18a Cross channel profile.

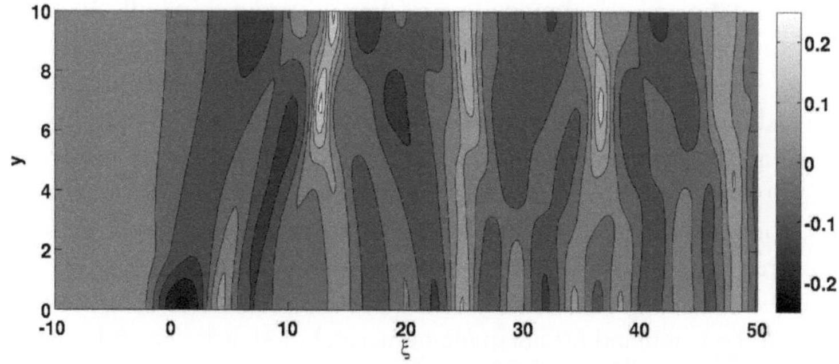

Fig. 18b Surface elevation shown for half of the channel width. The pressure disturbance is located at $\xi = 0$, and moves towards the left along the ξ-axis.

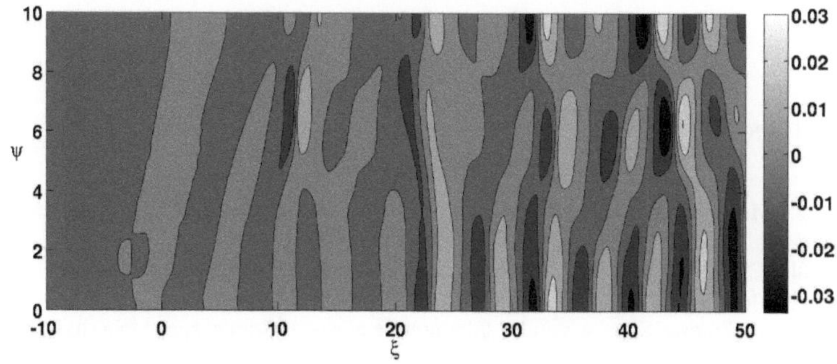

Fig. 18c Model comparison, showing the difference in surface elevation between the 1-layer and 2-layer model results.

models give similar results for wave amplitudes, but there is a distinct phase difference between the two results.

4 Concluding Remarks

In this paper we have presented some basic properties of ship wakes, considered mathematical equations that describe such waves, and numerical models to calculate them. There is ongoing research in developing appropriate mathematical equations describing the problem, for instance using higher order methods [31, 32], or including more vertical layers in the formulation [28, 29]. New numerical models are being developed using high order model formulations, and possibilities of using the finite element and finite volume approaches are being explored. There is also a development of proper models for energy dissipation, including advances in the representation of wave breaking and run-up [19, 30].

Related topics that have not been touched upon in this paper include issues about high-speed vessels, such as the design of high-speed vessels for minimized wake generation, and the numerical modelling of flow around a ship hull. We have partly touched upon the processes in the coastal zone, including wave transformation during shoaling and run-up, and the dynamics of breaking waves. This subject is closely related to sediment transport, and the resulting impact on coastal morphology.

There is a continued interest in putting high-speed vessels into operation, and a growing appreciation of the potential problems related to wake waves. In many countries, companies using high-speed vessels must assess the likely environmental impact before commencement of ship operations [40]. It is therefore likely that high-speed vessel wakes will continue to be an important problem for future studies, and that there will be a demand for adequate numerical models to simulate such waves.

Acknowledgements A large part of the work was written during a visit to the Centre for Nonlinear Studies in the framework of the Marie Curie Transfer of Knowledge project CENS-CMA (MC-TK-013909).

References

1. Baar, J.J.M., Price, W.G.: Developments in the calculation of the wavemaking resistance of ships. Proc. R. Soc. Lond. A **416**(1850), 115–147 (1988)
2. Casciola, C.M., Landrini, M.: Nonlinear long waves generated by a moving pressure disturbance. J. Fluid Mech. **325**, 399–418 (1996)
3. Chen, X.N., Sharma, S.D.: A slender ship moving at a near-critical speed in a shallow channel. J. Fluid Mech. **291**, 263–285 (1995)
4. Cienfuegos, R., Bathélemy, E., Bonneton, P.: A fourth-order compact fnite volume scheme for fully nonlinear and weakly dispersive boussinesq-type equations. Part I: model development and analysis. Int. J. Num. Meth. Fluids **51**, 1217–1253 (2006)
5. Craik, A.D.D.: The origins of water wave theory. Annu. Rev. Fluid Mech. **36**, 1–28 (2004)
6. Craik, A.D.D.: George Gabriel Stokes on water wave theory. Annu. Rev. Fluid Mech. **37**, 23–42 (2005)
7. Dommermuth, D.G., Sussman, M., Beck, R.F., O'Shea, T.T., Wyatt, D.C., Olson, K., MacNeice, P.: The numerical simulation of ship waves using cartesian grid methods with adaptive mesh refinement. In: The 25th Symposium on Naval Hydrodynamics. St. John's, Newfoundland and Labrador, Canada (2004)

8. Doyle, R., Whittaker, T., Elsäßer, B.: A study of fast ferry wash in shallow water. In: FAST2001, The 6th International Conference on Fast Sea Transportation, Southampton, UK (2001)
9. Drazin, P.G., Johnson, R.S.: Solitons: an Introduction. Cambridge University Press (1989).
10. Ertekin, R.C., Webster, W.C., Wehausen, J.V.: Waves caused by a moving disturbance in a shallow channel of finite width. J. Fluid Mech. **169**, 275–292 (1986)
11. Faltinsen, O.M.: Hydrodynamics of High-Speed Marine Vehicles. Cambridge University Press (2005).
12. Goldstein, S.: Fluid mechanics in the first half of this century. Annu. Rev. Fluid Mech. **1**, 1–29 (1969)
13. Hamer, M.: Solitary killers. New Scientist **163**(2201), 18–19 (1999)
14. Havelock, T.H.: The propagation of groups of waves in dispersive media, with application to waves on water produced by a travelling disturbance. Proc. Roy. Soc. London A **81**(549), 398–430 (1908). URL http://www.jstor.org/stable/93014
15. Hsu, T.W., Ou, S.H., Yang, B.D., Tseng, I.F.: On the damping coefficients of sponge layer in boussinesq equations. Wave Motion **41**, 45–57 (2005)
16. Idelsohn, S.R., Oñate, E., Sacco, C.: Finite element solution of free-surface ship-wave problems. Int. J. Num. Meth. Eng. **45**, 503–528 (1999)
17. Jiang, T., Henn, R., Sharma, S.D.: Wash wave generated by ships moving on fairways of varying topography. In: 24th Symposium on Naval Hydrodynamics, Fukuoka, Japan, 2002, pp. 441–457. National Academy Press, Washington DC (2003) web only: www.nap.edu
18. Katsis, C., Akylas, T.R.: On the exitation of long nonlinear water waves by a moving pressure distribution. Part 2. Three-dimensional effects. J. Fluid Mech. **177**, 49–65 (1987)
19. Kennedy, A.B., Chen, Q., Kirby, J.T., Dalrymple, R.A.: Boussinesq modeling of wave transformation, breaking, and runup. J. Waterw. Port Coast. Ocean Eng. **126**(1), 39–47 (2000)
20. Kirk McClure Morton: Investigation of high speed craft on routes near to land or enclosed estuaries. The Maritime and Coastguard Agency, Research Report JR226, UK (1998)
21. Kofoed-Hansen, H., Jensen, T., Kirkegaard, J., Fuchs, J.: Prediction of wake wash from high-speed craft in coastal areas In: Proceedings of the International Conference on Hydrodynamics of High Speed Craft. The Royal Institute of Naval Architects, London (1999)
22. Kofoed-Hansen, H., Jensen, T., Sørensen, O.R., Fuchs, J.: Wake wash risk assessment of high-speed ferry routes – a case study and suggestions for model improvements. In: Proceedings of the International Conference on Hydrodynamics of High Speed Craft, Wake Wash and Motion Control. The Royal Institute of Naval Architects, London (2000)
23. Kundu, P.K.: Fluid Mechanics. Academic Press (1990).
24. Langtangen, H.P., Pedersen, G.: Computational models for weakly dispersive nonlinear water waves. Comp. Meth. Appl. Mech. Eng. **160**, 337–358 (1998)
25. Li, Y., Sclavounos, P.D.: Three-dimensional nonlinear solitary waves in shallow water generated by an advancing disturbance. J. Fluid Mech. **470**, 383–410 (2002)
26. Lighthill, J.: Waves in Fluids. Cambridge University Press (1978)
27. Lynett, P., Liu, P.L.F.: A numerical study of submarine landslide generated waves and runup. Proc. Roy. Soc. London A **458**(2028), 2885–2910 (2002)
28. Lynett, P., Liu, P.L.F.: Linear analysis ofthe multi-layer model. Coast. Eng. **51**, 439–454 (2004)
29. Lynett, P., Liu, P.L.F.: A two-layer approach to wave modelling. Proc. R. Soc. Lond. A **460**(2049), 2637–2669 (2004)
30. Lynett, P., Wu, T.R., Liu, P.L.F.: Modeling wave runup with depth-integrated equations. Coast. Eng. **46**(2), 89–107 (2002)
31. Madsen, P.A., Schäffer, H.A.: Higher-order boussinesq-type equations for surface gravity waves: derivation and analysis. Phil. Trans. R. Soc. Lond. A **356**, 3123–3184 (1998)
32. Madsen, P.A., Schäffer, H.A. A review of boussinesq-type equations for surface gravity waves. In: Liu, P.L.F. (ed.) Advances in Coastal and Ocean Engineering, vol. 5, pp. 1–94. World Scientific, Singapore (1999)

33. MAIB: Report on the investigation of the man overboard fatality from angling boat Purdy at Shipwash Bank, off Harwich on 17 July 1999. Marine Accident Investigation Branch (2000). URL http://www.maib.gov.uk/cms_resources/purdy.pdf. Report No 17/2000
34. MAIB: Report on the investigation into a wash wave accident involving Portsmouth Express off East cowes on 18 July 2002. Marine Accident Investigation Branch (2003). URL http://www.maib.gov.uk/cms_resources/portsmouth-express.pdf. Report No 14/2003
35. MAIB: Report of investigation into swamping of unnamed cabin cruiser in Lady Bay on Loch Ryan, 3 September 2003, and associated wave generation issues. Marine Accident Investigation Branch (2004). URL http://www.maib.gov.uk/cms_resources/Loch Ryan-cabin cruiser.pdf. Report No 4/2004
36. MCA: Research project 457 - a physical study of fast ferry wash characteristics in shallow water. Final report. Tech. Rep., Maritime and Coastguard Agency (2001)
37. Molland, A.F., Wilson, P.A., Turnock, S.R., Taunton, D.J., Chandraprabha, S.: The prediction of the characteristics of ship generated near-field wash waves. In: FAST2001, The 6th international conference on fast sea transportation, Southampton, UK (2001)
38. Newman, J.N.: Marine Hydrodynamics. The MIT Press (1977).
39. Nwogu, O.: Alternative form of Boussinesq equations for nearshore wave propagation. J. Waterw. Port Coast. Ocean Eng. **119**(6), 618–638 (1993)
40. Parnell, K.E., Kofoed-Hansen, H.: Wakes from large high-speed ferries in confined coastal waters: management approaches with examples from New Zealand and Denmark. Coastal Manage. **29**, 217–237 (2001)
41. Pedersen, G.: Three-dimensional wave patterns generated by moving disturbances at transcritical speeds. J. Fluid Mech. **196**, 39–63 (1988)
42. Peregrine, D.H.: Long waves on a beach. J. Fluid Mech. **27**, 815–827 (1967)
43. PIANC: Guidelines for Managing Wake Wash from High-speed Vessels. Report of the Working Group 41 of the Maritime Navigation Commission. International Navigation Association (PIANC), Brussels (2003)
44. Soomere, T.: Fast ferry traffic as a qualitatively new forcing factor of environmental processes in non-tidal sea areas: a case study in Tallinn Bay, Baltic Sea. Environ. Fluid Mech. **5**(4), 293–323 (2005)
45. Soomere, T.: Long ship waves in shallow water bodies. In: Quak, E., Soomere, T. (eds.) Applied Wave Mathematics, pp. 193–228. Springer, Heidelberg (2009)
46. Soomere, T., Kask, J.: A specific impact of waves of fast ferries on sediment transport processes of Tallinn bay. Proc. Estonian Acad. Sci. Biol. Ecol. **52**, 319–331 (2003)
47. Soomere, T., Rannat, K.: An experimental study of wind waves and ship wakes in Tallinn Bay. Proc. Estonian Acad. Sci. Eng. **9**, 157–184 (2003)
48. Sorensen, R.M.: Ship-generated waves. Adv. Hydrosci. **9**, 49–83 (1973)
49. Torsvik, T., Pedersen, G., Dysthe, K.: Influence of cross channel depth variation on ship wave patterns. Tech. Rep. 2, Mechanics and Applied Mathematics, Dept. of Math., University of Oslo (2008). URL http://www.math.uio.no/eprint/appl_math/2008/02-08.pdf.
50. Torsvik, T., Pedersen, G., Dysthe, K.: Waves generated by a pressure disturbance moving in a channel with a variable cross-sectional topography. J. Waterw. Port Coast. Ocean Eng. **135** (2009).
51. Torsvik, T., Soomere, T.: Simulation of patterns of wakes from high-speed ferries in Tallinn Bay. Estonian J. Eng. **14**(3), 232–254 (2008).
52. Tuck, E.O.: Shallow-water flows past slender bodies. J. Fluid Mech. **26**(1), 81–95 (1966)
53. Tuck, E.O., Scullen, D.C., Lazauskas, L.: Wave patterns and minimum wave resistance for high-speed vessels. In: Proceedings of the 24th Symposium on Naval Hydrodynamics, Fukuoka, Japan (2002)
54. Ursell, F.: The long-wave paradox in the theory of gravity waves. Proc. Cambridge Philos. Soc. **49**, 685–694 (1953)
55. Wei, G., Kirby, J.T., Grilli, S.T., Subramanya, R.: A fully nonlinear boussinesq model for surface waves. Part 1. Highly nonlinear unsteady waves. J. Fluid Mech. **294**, 71–92 (1995)
56. Whitham, G.B.: Linear and Nonlinear Waves. John Wiley & Sons, New York (1970)

57. Whittaker, T., Doyle, R., Elsäßer, B.: An experimental investigation of the physical characteristics of fast ferry wash. In: 2nd International EuroConference on High-Performance Marine Vehicles HIPER'01 (2001)
58. Woo, S.B., Liu, P.L.F.: Finite-element model for modified Boussinesq equations. I: Model development. J. Waterw. Port Coast. Ocean Eng. **130**(1), 1–16 (2004)
59. Yang, Q.: Wash and wave resistance of ships in finite water depth. Ph.D. thesis, NTNU Trondheim (2002).
60. Yih, C.S., Zhu, S.: Patterns of ship waves. Q. Applied Math. **47**(1), 17–33 (1989)
61. Yih, C.S., Zhu, S.: Patterns of ship waves II: gravity-capillary waves. Q. Applied Math. **47**(1), 35–44 (1989)
62. Zhang, D., Chwang, A.T.: On nonlinear ship waves and wave resistance calculation. J. Mar. Sci. Technol. **4**, 7–15 (1999)

New Trends in the Analytical Theory of Long Sea Wave Runup

Ira Didenkulova

Abstract A modern view on the analytical theory of the long sea wave runup on a plane beach is presented. This theory is based on rigorous solutions of nonlinear shallow-water equations. The dynamics of the moving shoreline is studied in detail. It is demonstrated that extreme characteristics of the runup process (runup and rundown amplitudes, extreme values of on- and off-shore velocities, and critical amplitude of the breaking wave) can be found using the solution of the linear shallow-water theory, meanwhile the description of the time series of the wave field requires the nonlinear theory. The key and novel results presented here are: i) parameterization of basic formulas for extreme runup characteristics for bell-shape waves, showing that they weakly depend on the initial wave shape, which is usually unknown in real sea conditions; ii) runup analysis of periodic asymmetric waves with a steep front, as such waves are penetrating inland over larger distances and with greater velocities than symmetric waves.

1 Introduction

Giant sea waves approaching the coast can often lead to the destruction of coastal infrastructure and loss of lives. Such waves can have various origins: strong storms and cyclones, underwater earthquakes, and sub-aerial and sub-marine landslides. Recent such events were the catastrophic tsunami in the Indian Ocean (December 26, 2004) [18] and hurricane Katrina (August 28, 2005) in the Atlantic Ocean [17]. The huge storm in the Baltic Sea (January 9, 2005) should also be mentioned as an example of a regional marine natural hazard [30].

The prediction of the possible flooding and the properties of the water flow on the coast is an important practical task for coastal and port engineering. That is why

---

Institute of Cybernetics at Tallinn University of Technology, Tallinn, Estonia and Institute of Applied Physics, Nizhny Novgorod, Russia, e-mail: ira@cs.ioc.ee

E. Quak, T. Soomere (eds.), *Applied Wave Mathematics*,
DOI 10.1007/978-3-642-00585-5_14,

there are numerous empirical formulas describing runup characteristics in the engineering literature, see for example [19, 32]). Generally these formulas are specific to various geographic areas due to the characteristic features of the wave regimes in these areas (wind direction, coastal effects of wave refraction and diffraction).

An analytical approach to the wave runup problem has been developed after the pioneering work of Carrier and Greenspan [3], where the first rigorous solutions of the nonlinear shallow-water equations were found. Mathematically, they reduced the initial nonlinear shallow-water system describing waves in the domain with an unknown moving boundary, to the linear wave equation on the semi-axis, which can be solved by using methods of mathematical physics. This approach has been very popular during the last 25 years and this paper summarizes some of the results relating to the analytical theory of long wave runup over this period.

Various shapes of the periodic incident wave trains such as the sine wave [23], cnoidal wave [35] and nonlinear deformed periodic wave [11] have been analyzed in the literature. The relevant analysis has also been performed for a variety of solitary waves and single pulses such as a soliton [14, 26, 34], sine pulse [25], Lorentz pulse [28], Gaussian pulse [4, 15]), N-waves [36], "characterized tsunami waves" [38], and a random set of solitons [2]. However, as is often the case in nonlinear problems, reaching an analytical solution is seldom possible.

It is important to mention that many analytical formulas have been confirmed in laboratory tanks [20, 22], and are now actively used in the prediction of marine natural hazards; see for example [7, 12, 16].

The paper is organized as follows. The basic equations and main parameters are defined in section 2. An analytical model based on the nonlinear shallow-water equations and valid only for non-breaking waves is presented. A method of reducing the nonlinear shallow-water system to the linear wave equation, as suggested in the original paper of Carrier and Greenspan [3], is given in section 3. It can be rigorously applied only to the case of a plane beach with long waves approaching the coast from a perpendicular direction. The role of the linear approximation of the governing system is discussed in section 4. It is shown that extreme characteristics of the runup process (runup and rundown amplitudes, extreme values of on- and off-shore velocities, and critical amplitude of the breaking wave) can be found with a linear approximation. The "real" nonlinear dynamics of the moving shoreline is described in section 5. It is shown that with an increase of the amplitude the wave first breaks on the shoreline. In this case the velocity at the shoreline has the shape of a shock wave, and the first derivative of the water surface elevation is discontinuous in the trough. The runup of solitary waves of various shapes is analyzed in section 6. It is demonstrated that with the use of a definition of a "significant" wavelength, all formulas for extreme runup characteristics become universal and their dependence on the incident wave shape is very weak. These formulas can be used for engineering applications. The runup of periodic asymmetric waves with a steep front is discussed in section 7. It is shown that such waves penetrate inland over larger distances and with higher velocities than symmetric waves. All results are summarized in the conclusion.

2 Basic equations and parameters

The basic equations describing wave runup on a beach are the governing systems of fluid mechanics: the Euler or Navier-Stokes equations. In this paper the classical formulation of the wave runup problem within the ideal fluid model will be applied, with the basic equations being the Euler equations for incompressible fluids [1, 21],

$$\frac{\partial \mathbf{u}}{\partial t} + (\mathbf{u}\nabla)\mathbf{u} + w\frac{\partial \mathbf{u}}{\partial z} + \frac{1}{\rho}\nabla P = 0, \tag{1}$$

$$\frac{\partial w}{\partial t} + (\mathbf{u}\nabla)w + w\frac{\partial w}{\partial z} + \frac{1}{\rho}\frac{\partial P}{\partial z} = -g, \tag{2}$$

$$div\,\mathbf{u} + \frac{\partial w}{\partial z} = 0, \tag{3}$$

with boundary conditions on the bottom $z = -h(x,y)$ (water does not penetrate into the bottom)

$$w + \mathbf{u}\nabla h = 0, \tag{4}$$

and on the free surface $z = \eta\,(x,y,t)$

$$w = \frac{\partial \eta}{\partial t} + \mathbf{u}\nabla\eta, \quad P = P_{atm}. \tag{5}$$

Here $\mathbf{u} = \{u,v\}$ and w are the horizontal and vertical components of the flow velocity, ρ is the fluid density, P is the pressure and P_{atm} is the atmospheric pressure assumed to be constant, g is the gravity acceleration, z is the vertical axis and (x,y) are the horizontal coordinates, and $h(x,y)$ is the unperturbed fluid depth counted from the unperturbed sea surface $z = 0$. The differential operators ∇ and div act in the horizontal plane only.

The common geometry for the runup problem is the plane beach of a constant slope $h(x) = -\alpha x$ (Fig. 1). If the wave approaches the coast only in the perpendicular direction, the basic Euler equations reduce to 2D equations. The sea wave is characterized by its height R and the frequency ω (or wave duration ω^{-1}).

If we define non-dimensional variables

$$\tilde{\eta} = \eta/R, \quad \tilde{u} = \alpha u/\omega R, \quad \tilde{w} = w/\omega R,$$
$$\tilde{t} = \omega t, \quad \tilde{x} = \alpha x/R, \quad \tilde{z} = z/R,$$
$$P = P_{atm} + \rho g(\eta - z) + \rho\omega^2 R^2 \tilde{P},$$

the system (1)–(5) becomes (tilde is omitted)

$$\frac{\partial u}{\partial t} + u\frac{\partial u}{\partial x} + w\frac{\partial u}{\partial z} + \frac{1}{Br}\frac{\partial \eta}{\partial x} + \alpha^2\frac{\partial P}{\partial x} = 0, \tag{6}$$

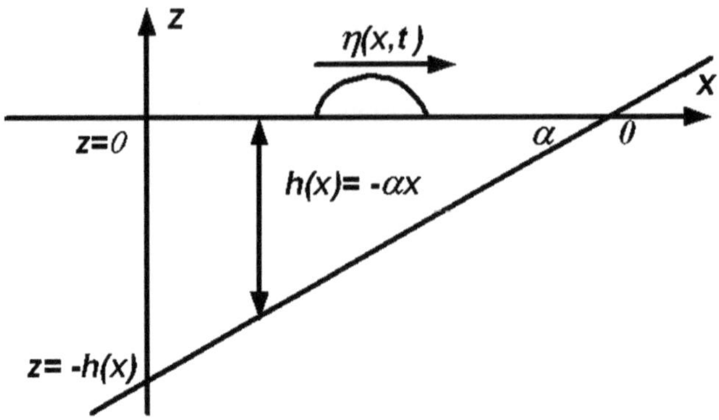

Fig. 1 The geometry applied to the wave runup problem.

$$\frac{\partial w}{\partial t}+u\frac{\partial w}{\partial x}+w\frac{\partial w}{\partial z}+\frac{\partial P}{\partial z}=0, \tag{7}$$

$$\frac{\partial u}{\partial x}+\frac{\partial w}{\partial z}=0 \tag{8}$$

with boundary conditions on the sea bottom

$$w=-u|_{z=-x}, \tag{9}$$

and sea surface

$$\frac{\partial \eta}{\partial t}+u\frac{\partial \eta}{\partial x}=w|_{z=\eta}, \quad P=0|_{z=\eta}. \tag{10}$$

These equations (6)–(10) depend on two parameters only: the beach slope α and the so-called breaking parameter (its name will become clear later)

$$Br=\frac{\omega^2 R}{g\alpha^2}. \tag{11}$$

Tsunami waves are long waves with characteristic lengths of 20–100 km, greatly exceeding oceanic water depths of 1–5 km. If the beach is relatively steep, the width of the coastal zone is much smaller than the tsunami wave height, and the coastal zone can be effectively replaced by a vertical wall. Results describing an interaction of the water wave with a vertical wall can be found, for instance, in the review paper [29].

In this paper we concentrate on the wave runup on a beach of a small slope. In this case the last term in Eq. (6) can be neglected. From Eq. (6) it follows that the horizontal velocity is uniform with depth. As a result, we obtain the first equation of the shallow-water theory

$$\frac{\partial u}{\partial t}+u\frac{\partial u}{\partial x}+\frac{1}{Br}\frac{\partial \eta}{\partial x}=0. \tag{12}$$

The second equation can be obtained from Eq. (8) by integration over the water depth with the use of the two boundary conditions (9) and (10):

$$\frac{\partial \eta}{\partial t} + \frac{\partial}{\partial x}[(-x+\eta)u] = 0. \tag{13}$$

The pressure in water is described by the hydrostatic formula (in dimensional form)

$$P = P_{atm} + \rho g(\eta - z), \tag{14}$$

and from Eq. (7) it follows that the vertical velocity (and also the vertical acceleration) is small in comparison to the horizontal velocity. The water particles follow elliptical trajectories, and the ellipse is elongated in a horizontal direction. That is why only the horizontal velocity (or depth-averaged velocity) appears in shallow-water theory.

It is important to mention that the shallow-water system contains only one parameter, Br. In the literature, wave processes in the coastal zone can be characterized in terms of the non-dimensional surf-similarity parameter (also known as the Iribarren number, see for instance [23])

$$\xi = \frac{\alpha}{\sqrt{H_0/\lambda}}, \tag{15}$$

where H_0 and λ are the height and length of the deep-water wave, respectively. Taking into account the dispersion relation for waves in deep water $\omega = \sqrt{gk}$, it is easy to find the relation between the surf-similarity and breaking parameters

$$\xi = \sqrt{\frac{2\pi}{Br}}. \tag{16}$$

Below we use mainly the parameter Br and demonstrate that it determines the wave breaking on a plane beach.

3 Method of solution: hodograph transformation

Taking into account the practical importance of the runup problem, we will use dimensional variables and re-write the shallow-water equations (12) and (13) in the form

$$\frac{\partial u}{\partial t} + u\frac{\partial u}{\partial x} + g\frac{\partial H}{\partial x} = -g\alpha, \tag{17}$$

$$\frac{\partial H}{\partial t} + \frac{\partial}{\partial x}(Hu) = 0, \tag{18}$$

where

$$H(x,t) = -\alpha x + \eta(x,t) \tag{19}$$

is the total water depth. Equations (17) and (18) represent a hyperbolic system of partial differential equations with constant coefficients. In this case, Riemann invariants can be determined as follows [33]:

$$I_{\pm} = u \pm 2\sqrt{gH} + \alpha g t, \tag{20}$$

and the system (17) and (18) can be re-written in new variables

$$\frac{\partial I_{\pm}}{\partial t} + c_{\pm}(I_{+}, I_{-}) \frac{\partial I_{\pm}}{\partial x} = 0, \tag{21}$$

where the characteristic velocities are

$$c_{\pm} = \frac{3}{4} I_{\pm} + \frac{1}{4} I_{\mp} - \alpha g t. \tag{22}$$

From Eq. (21) it follows that the Riemann invariants are conserved along the characteristics

$$\frac{dx}{dt} = c_{\pm}(I_{+}, I_{-}). \tag{23}$$

Nonlinear effects appear in the curvature of the characteristics due to the interaction between incident and reflected waves (in the linear theory the characteristics are straight lines).

For solving the hyperbolic system (21) the classical hodograph transformation (Legendre transformation) can be applied [3, 33]. The main advantage of such an approach is the transformation of a nonlinear system of equations to a linear system. This procedure is described below.

The system (21) can be represented in the Jacobian form

$$\frac{\partial(I_{\pm}, x)}{\partial(t, x)} + c_{\pm} \frac{\partial(t, I_{\pm})}{\partial(t, x)} = 0. \tag{24}$$

After multiplying Eqs. (24) by the non-zero Jacobian $\partial(t,x)/\partial(I_{+},I_{-}) \neq 0$ (this assumption is discussed later), the system (24) transforms to

$$\frac{\partial(I_{\pm}, x)}{\partial(I_{+}, I_{-})} + c_{\pm} \frac{\partial(t, I_{\pm})}{\partial(I_{+}, I_{-})} = 0, \tag{25}$$

or

$$\frac{\partial x}{\partial I_{\mp}} - c_{\pm} \frac{\partial t}{\partial I_{\mp}} = 0 \tag{26}$$

in differential form.

The system (26) remains nonlinear since the characteristic speeds (22) depend on time. Nevertheless it can be reduced to a second order linear equation by eliminating the coordinate x:

$$\frac{\partial^2 t}{\partial I_{+} \partial I_{-}} + \frac{3}{2(I_{+} - I_{-})} \left(\frac{\partial t}{\partial I_{-}} - \frac{\partial t}{\partial I_{+}} \right) = 0. \tag{27}$$

Equation (27) can be reduced to the Euler-Poisson-Darboux Equation [13]. Here we will follow [3]. Let us introduce new variables

$$\lambda = \frac{I_+ + I_-}{2} = u + \alpha g t, \tag{28}$$

$$\sigma = \frac{I_+ - I_-}{2} = 2\sqrt{gH}. \tag{29}$$

With these variables, Eq. (27) takes the form

$$\frac{\partial^2 t}{\partial \lambda^2} - \frac{\partial^2 t}{\partial \sigma^2} - \frac{3}{\sigma}\frac{\partial t}{\partial \sigma} = 0. \tag{30}$$

After substituting time t from Eq. (28) and introducing the new function Φ

$$u = \frac{1}{\sigma}\frac{\partial \Phi}{\partial \sigma}, \tag{31}$$

Eq. (30) can be presented in its final form

$$\frac{\partial^2 \Phi}{\partial \lambda^2} - \frac{\partial^2 \Phi}{\partial \sigma^2} - \frac{1}{\sigma}\frac{\partial \Phi}{\partial \sigma} = 0, \tag{32}$$

and all the variables are expressed through the wave function $\Phi(\lambda, \sigma)$:

$$t = \frac{1}{\alpha g}(\lambda - u), \tag{33}$$

$$x = \frac{1}{2\alpha g}\left(\frac{\partial \Phi}{\partial \lambda} - u^2 - \frac{\sigma^2}{2}\right), \tag{34}$$

$$\eta = \frac{1}{2g}\left(\frac{\partial \Phi}{\partial \lambda} - u^2\right). \tag{35}$$

Thus, the basic system of nonlinear shallow-water equations reduces to the linear wave equation (32), and all physical variables can be expressed through the wave function $\Phi(\lambda, \sigma)$.

The wave equation (32) is solved in the fixed semi-axis $\sigma \geq 0$ (the point $\sigma = 0$ corresponds to the moving shoreline, see Eq. (29)), in comparison to the initial shallow-water equations being solved in a domain with an unknown and unfixed boundary. The natural boundary condition at the point $\sigma = 0$ is the limitation of the physical variables η and u. It follows from Eq. (31) that

$$\frac{\partial \Phi}{\partial \sigma}(\lambda, \sigma = 0) = 0. \tag{36}$$

The boundary condition at infinity ($\sigma \to \infty$) will be discussed later.

The general initial conditions for η and u

$$\eta(x,0) = \eta_0(x) \quad \text{and} \quad u(x,0) = u_0(x) \tag{37}$$

should be transformed into initial conditions for $\Phi(\lambda,\sigma)$.

Thus, for example, for $u_0(x) = 0$ (a popular presentation of the tsunami source in the framework of the piston model) initial conditions are formulated for $\lambda = 0$ instead of for $t = 0$ as

$$\Phi(\sigma,0) = 0, \quad \frac{\partial \Phi}{\partial \lambda}(\sigma,0) = \Phi_1(\sigma), \tag{38}$$

where Φ_1 is parametrically defined by Eqs. (34) and (35).

The case of non-zero initial velocities is much more complicated from the mathematical point of view, since initial conditions for the wave equation (32) on the plane (λ,σ) are given along the curve $\sigma(\lambda)$, see Eq. (33). This case has been recently analyzed in [15].

The linear wave equation (32) defined on a semi-axis is well studied in mathematical physics [6] and the corresponding Green's function can be written in the integral form. The Green's function approach is also used in the runup problem [4, 15]. The transformation of the initial wave field $\eta_0(x)$ and $u_0(x)$ to the wave function $\Phi(\lambda,\sigma)$ and back to the water wave field $\eta(x,t)$ and $u(x,t)$ is described by the implicit formulas (31), (33)–(35), and there are only a few analytical examples when the solution of Eqs. (17)–(18) can be found explicitly [3, 26, 31]. Of course, modern computers easily perform this transformation.

The main goal of this paper is to calculate long wave runup characteristics, which are very important in practice. For this case, many rigorous mathematical procedures can be reduced to simple algorithms, convenient for engineers.

4 Linear approximation of nonlinear long wave runup

Waves usually approach the shore from the open sea where the depth is large and wave amplitudes are weak. It means that the incident wave is linear in most cases, with the exception of laboratory tanks, where the mechanically generated wave has an amplitude comparable with the depth. This means that the main formulas of the Carrier-Greenspan transformation (33)–(35) can be simplified far from the coast as

$$t = \frac{1}{\alpha g}\lambda, \tag{39}$$

$$x = -\frac{\sigma^2}{4\alpha g}, \tag{40}$$

$$\eta = \frac{1}{2g}\frac{\partial \Phi}{\partial \lambda}. \tag{41}$$

As a result, all formulas become explicit, and the transformation of the initial conditions (37) from the physical space (x,t) to the space (λ,σ) is trivial

$$\frac{\partial \Phi}{\partial \lambda}(\sigma,0) = 2g\eta_0[x(\sigma)], \tag{42}$$

$$\Phi(\sigma,0) = \int \sigma u_0[x(\sigma)]d\sigma. \tag{43}$$

Since the initial conditions for the wave equation (32) are fully determined for the general case of non-zero velocities, the general solution of the Cauchy problem can be obtained. This is the first advantage of the linear approximation to nonlinear long wave runup.

Moreover, if we consider the linear shallow-water system

$$\frac{\partial u}{\partial t} + g\frac{\partial \eta}{\partial x} = 0, \tag{44}$$

$$\frac{\partial \eta}{\partial t} + \frac{\partial}{\partial x}(-\alpha x u) = 0, \tag{45}$$

and apply the linearized Carrier-Greenspan transformation (31), (39)–(41), we reduce the system (44)–(45) to the wave equation (32), which should be solved for the same initial conditions (42)–(43) and the boundary condition (36) on the semi-axis. This means that the solutions of the basic nonlinear $\Phi(\lambda,\sigma)$ and basic linear $\Phi_l(\lambda_l,\sigma_l)$ systems are identical! Still the physical sense of the boundary point $\sigma = 0$ is different in the nonlinear and linear cases. In the nonlinear problem, the point $\sigma = 0$ corresponds to the moving shoreline with the total water depth (19) being equal to zero, while in the linear problem the same point corresponds to the unperturbed shoreline (still sea level) with an unperturbed water depth $h(x)$ equal to zero. The dynamics of the moving shoreline in the nonlinear problem is described by the function $\Phi(\lambda,\sigma=0)$, in particular the maximal runup height is achieved when $u = 0$ as

$$R = \max(\eta) = \frac{1}{2g}\max\frac{\partial \Phi}{\partial \lambda}(\lambda,\sigma=0), \tag{46}$$

see Eq. (35). The wave field on the shoreline ($x = 0$) in the linear problem is described by the function $\Phi_l(\lambda_l,\sigma_l)$, and the maximal wave height at this point is

$$R_l = \max(\eta_l) = \frac{1}{2g}\max\frac{\partial \Phi_l}{\partial \lambda_l}(\lambda_l,\sigma_l=0). \tag{47}$$

At the same time, as pointed out above, if initial conditions are given far from the shoreline where the wave is linear, the functions $\Phi(\lambda,\sigma)$ and $\Phi_l(\lambda_l,\sigma_l)$ are identical. This means that the maximal value of a runup height in the nonlinear theory is equal to the wave height on the unperturbed shoreline in the linear theory. Therefore, the maximal runup height can be found in the framework of the linear theory, this being very important for engineering applications. This non-trivial conclusion is rigorously proved, see for instance [28].

The same results can also be applied to the maximum value of rundown depth and the onshore and offshore velocities of the moving shoreline. Thus, all important extreme characteristics of the runup process can be found in the framework of linear shallow-water theory.

There is a significant advantage to using the linear approximation in the nonlinear problem of long wave runup. Instead of using the complicated Carrier – Greenspan transformation we solve the linear shallow-water system (44) and (45), which can be rewritten in the form of the variable-coefficient wave equation

$$\frac{\partial^2 \eta}{\partial t^2} - g\frac{\partial}{\partial x}\left[h(x)\frac{\partial \eta}{\partial x}\right] = 0, \tag{48}$$

where we insert the unperturbed depth $h(x)$ instead of $-\alpha x$. Equation (48) should be solved for the initial conditions (37), which can be written as

$$\eta(x,0) = \eta_0(x), \quad \frac{\partial \eta}{\partial t}(x,0) = -\frac{d}{dx}[h(x)u_0(x)]. \tag{49}$$

The solution of Eq. (48) in the "physical" space (x,t) is more obvious. The output of the solution of Eq. (48) should be extreme values of the water displacement $\eta\,(x=0,t)$ and horizontal velocity $u\,(x=0,t)$. These extremes obtained from the linear theory describe "real" nonlinear extremes of runup characteristics.

As an example, consider the runup of a sine wave with frequency ω on a plane beach. The well-known bounded solution of the linear wave equation (48) is expressed through Bessel functions

$$\eta(x,t) = R_0 J_0\left(\sqrt{\frac{4\omega^2\mid x\mid}{g\alpha}}\right)\cos(\omega t), \tag{50}$$

where $J_0\,(z)$ is the Bessel function of zero-th order. Far from the shoreline the wave field can be presented asymptotically as the superposition of two sine waves of equal amplitude propagating in opposite directions

$$\eta(x,t) = A(x)\left\{\sin\left[\omega(t-\tau)+\frac{\pi}{4}\right] + \sin\left[\omega(t+\tau)-\frac{\pi}{4}\right]\right\}, \tag{51}$$

where the instantaneous wave amplitude $A\,(x)$ is

$$A(x) = R_0\left(\frac{\alpha g}{16\pi^2\omega^2\mid x\mid}\right)^{1/4}, \tag{52}$$

and

$$\tau(x) = \int \frac{dx}{\sqrt{gh(x)}} \tag{53}$$

is the propagation time of this wave over the distance x in a fluid of variable depth.

The ratio of the maximal amplitude R_0 of the approaching wave to the wavelength λ_0 (determined from the shallow-water dispersion relation, namely $\omega = 2\pi\sqrt{gh(L)}/\lambda_0$) and an initial amplitude A_0 at the fixed point $|x| = L$ can be found from Eq. (52):

$$\frac{R_0}{A_0} = \left(\frac{16\pi^2\omega^2 L}{g\alpha}\right)^{1/4} = 2\pi\sqrt{\frac{2L}{\lambda_0}}. \tag{54}$$

We emphasize that the amplification factor R_0/A_0 in Eq. (54) represents the shoaling coefficient in the linear surface wave theory. As is pointed out above, it is the same in the nonlinear theory. This feature allows determining the extreme wave runup characteristics in both the linear and nonlinear cases as soon as the initial wave amplitude and wave length at a fixed point $|x| = L$ far offshore are known.

Formally, the solution of the wave equation (48) can be obtained for different depth profiles analytically or numerically. Still we should emphasize that the equivalence between nonlinear and linear theories in calculating extreme characteristics is proved only for the plane beach. This means that far offshore where the wave is linear, the bottom relief can have any shape, but in the vicinity of the shoreline it should almost be a plane beach for the applicability of rigorous nonlinear results. Such an example of joining a flat bottom and a plane beach is presented in Fig. 2.

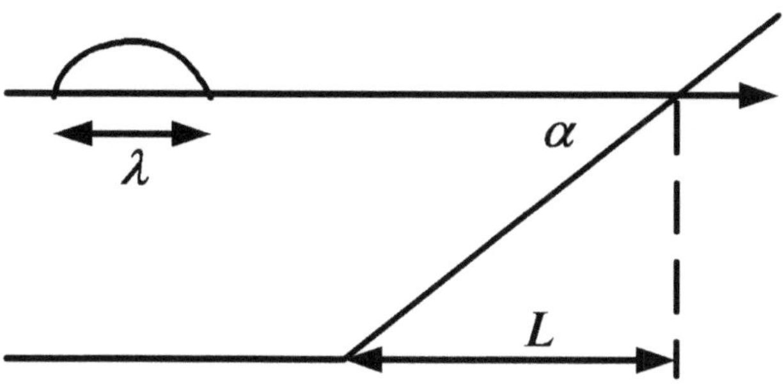

Fig. 2 Sketch of the geometry.

The wave field on a beach is described by Eq. (50). On a flat bottom, the solution of the wave equation (48) is

$$\eta(x,t) = A_0 \exp\left[i\omega(t - x/c)\right] + A_r \exp\left[i\omega(t + x/c)\right], \tag{55}$$

where A_0 and A_r are the amplitudes of the incident and reflected waves, and c is the long-wave speed on a flat bottom. The incident wave is known (the boundary condition at infinity, $x \to \infty$).

Matching the solutions (50) and (55) requires the continuity of water level and velocity at the junction of the flat bottom and the plane beach.

These conditions allow finding A_r and the runup height R, which is

$$\frac{R}{A_0} = \frac{2}{\sqrt{J_0^2(2kL) + J_1^2(2kL)}}, \tag{56}$$

where $k = \omega/c$ is the wave number of the incident wave. For large values of kL, Eqs. (54) and (56) coincide (Fig. 3). If the beach width tends to zero, the runup amplitude exceeds the incident wave amplitude by a factor of 2.

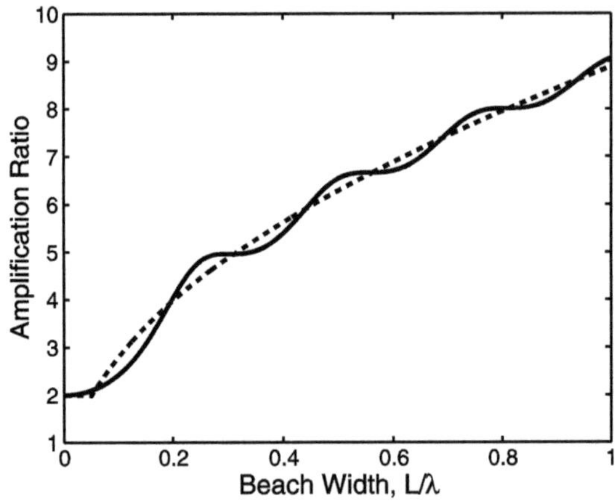

Fig. 3 Runup height versus beach width; the dashed line corresponds to Eq. (54), the solid line corresponds to Eq. (56).

5 The relation between linear and nonlinear runup properties

As was shown above, the linear theory predicts the extreme characteristics of long wave runup. Moreover, it can be used to simply calculations of "real" nonlinear dynamics of the moving shoreline. For instance, the velocity of the moving shoreline in the nonlinear theory can be expressed from Eq. (33) for $\sigma = 0$ as

$$u(\lambda) = \lambda - \alpha g t. \tag{57}$$

This equation implicitly determines the dependence of the velocity of the moving shoreline on time and can be written as

$$u(t) = U\left(t + \frac{u}{\alpha g}\right), \tag{58}$$

where the physical sense of the function $U(t)$ is clear; it is a "linear" velocity of the shoreline. The implicit formula (58) demonstrates that the "nonlinear" velocity of the moving shoreline can be obtained from the "linear" solution by deformation of the time axis (Riemann transformation of time). It is clear from Eq. (58) that the extreme values of the functions $u(t)$ and $U(t)$ coincide, which confirms the conclusion made in section 4. Thus, the nonlinearity influences only the deformation of the velocity time series.

It is simple to find the horizontal and vertical coordinates of the moving shoreline by knowing the "nonlinear" velocity of the moving shoreline:

$$x(t) = \int u(t)dt, \tag{59}$$

$$z(t) = \frac{x(t)}{\alpha}. \tag{60}$$

The vertical displacement of the water level at the shoreline in the linear theory $Z(t) = \eta_l(t, x = 0)$ can be calculated from the solution of the linear wave equation (48) with the use of traditional methods of mathematical physics or numerical modeling. It is related to the "linear" velocity

$$U(t) = \frac{1}{\alpha}\frac{dZ(t)}{dt}. \tag{61}$$

The "nonlinear" vertical displacement of the moving shoreline $r(t)$ can be obtained from Eqs. (35) and (58) as

$$r(t) = \eta(t, \sigma = 0) = Z\left(t + \frac{u}{\alpha g}\right) - \frac{u^2}{2g}. \tag{62}$$

The important conclusion from Eq. (62) is that the extremes of the vertical displacement (the runup and rundown heights) in the linear and nonlinear theories coincide, confirming the results of section 4. Therefore, the linear theory adequately describes the runup height, which is an extremely important characteristic of the long wave (tsunami, storm surge) action on the shore.

Therefore, the solution of the linear problem together with the Riemann transformation of time (two-step analysis) allows the calculation of the runup characteristics, which is much easier than the complicated Carrier – Greenspan transformation. This approach was first suggested in [28] and then applied for several cases in [9, 10, 11].

Another important outcome from the proposed approach is a simple definition of the first breaking condition of long waves on a beach. It is evident that long weak-amplitude waves do not break at all and result in a slow rise of the water level resembling a surge-like flooding. With an increase in the wave amplitude, breaking occurs seawards of the runup maximum and, depending on the wave amplitude and the bottom slope, may occur relatively far offshore ("plunging" breaking). The first breaking with an increase in wave amplitude should occur on the shoreline and it can be found with the two-step approach described above. The temporal derivative of the "nonlinear" velocity of the moving shoreline, calculated from Eq. (58),

$$\frac{du}{dt} = \frac{dU/dt}{1 - \frac{dU/dt}{\alpha g}}, \tag{63}$$

tends to infinity when the denominator of the right-hand side of Eq. (63) approaches zero. As follows from the theory of hyperbolic equations, this leads to the gradient catastrophe, which here can be related to the plunging breaking of the long water waves. In this case the water displacement contains a jump of its first derivative. This implies the condition of the first wave breaking as

$$Br = \frac{\max(dU/dt)}{\alpha g} = \frac{\max(d^2Z/dt^2)}{\alpha^2 g} = 1, \tag{64}$$

where the parameter Br has the same meaning as in section 2, as is evident from the dimensional analysis of Eqs. (64) and (11). The condition (64) has a simple physical interpretation: the wave breaks if the maximal acceleration of the shoreline $Z''\alpha^{-1}$ along the sloping beach exceeds the along-beach gravity component (αg). This interpretation is figurative but very convenient for illustration, because formally Z'' only presents the vertical acceleration of the shoreline in the linear theory and the "nonlinear" acceleration du/dt actually tends to infinity at the breaking moment. The criterion (64) can be re-written in another form with the use of the relation between dU/dt and $\partial\eta/\partial x$ from the linear equation (44) at the shoreline ($x = 0$):

$$Br = \frac{\max(\partial\eta/\partial x)}{\alpha} = 1. \tag{65}$$

This notation has the physical meaning that the wave steepness should be equal to the bottom slope for the case of the first breaking. In this case the curves of the water level and the bottom profile do not cross. This form of the criterion is popular in oceanography [24].

The breaking condition presented here is obtained from the assumption of the gradient catastrophe in the wave equation. It can be linked with the existence condition of the Carrier-Greenspan transformation. As was discussed in section 3, the Carrier-Greenspan transformation exists only for non-zero Jacobian

$$J = \frac{\partial\,(t,x)}{\partial\,(I_+, I_-)} \neq 0. \tag{66}$$

This Jacobian can be calculated using the formulas from section 3 as

$$J = -\frac{\sigma}{2}\frac{\partial t}{\partial I_+}\frac{\partial t}{\partial I_-} = \frac{\sigma}{8g^2\alpha^2}\left[\left(\frac{\partial u}{\partial \sigma}\right)^2 - \left(1 - \frac{\partial u}{\partial \lambda}\right)^2\right]. \tag{67}$$

The Jacobian is equal to zero ($J = 0$) at the point $\sigma = 0$. Yet the point $\sigma = 0$ is a singular point of the wave equation, and therefore $J = 0$ does not characterize the wave breaking in this point. When wave amplitudes are weak, the Jacobian

$$J = -\frac{\sigma}{8g^2\alpha^2} \tag{68}$$

is nonzero. With an increasing of the wave amplitude, the term in the second bracket of Eq. (67) tends to zero providing a zero value for the Jacobian at

$$\max\left(\frac{\partial u}{\partial \lambda}\right) = 1. \tag{69}$$

This condition in dimensional variables coincides with Eq. (64).

Thus, the nonlinear dynamics of the moving shoreline can be fully determined from Eqs. (58) and (62) by knowing the solution of the linear problem and by calculating the water displacement on the unperturbed shoreline $Z(t, x = 0)$. Moreover, the breaking condition can also be found in the same way.

Let us consider the runup of a sine wave on a beach. In the linear problem, the wave remains monochromatic in time for all distances. In this case it is convenient to use normalization of the maximum runup height R and a wave frequency ω. The breaking parameter in a simplified form is

$$Br = \frac{\omega^2 R}{\alpha^2 g}. \tag{70}$$

The maximum runup height of a non-breaking wave can be found from the breaking condition $Br = 1$ as

$$R_{\max} = \frac{g\alpha^2 T^2}{4\pi^2}, \tag{71}$$

where T is the wave period. It depends on the bottom slope and the wave period (Fig. 4).

Wind (short) waves have a characteristic period of 6 s and they break if their amplitudes exceed 10 cm ($\alpha \sim 0.1$). That is why the process of the wind wave breaking on a beach is observed very often. Tsunami waves have a characteristic period of 10 min and break if their heights exceed 10 m ($\alpha \sim 0.01$). As a result, we have the "paradoxical" conclusion that huge tsunami waves may climb the beach without breaking, while weak-amplitude wind waves always break. According to the observations [27] approximately 75% of tsunamis climb the shore without breaking.

The detailed analysis of the nonlinear dynamics of the moving shoreline for the case when a monochromatic wave approaches the beach can be done based on the Riemann transformation (58). Let us use the non-dimensional variables presented in section 2

$$u' = \alpha u/\omega R, \quad U' = \alpha U/\omega R, \quad t' = \omega t. \tag{72}$$

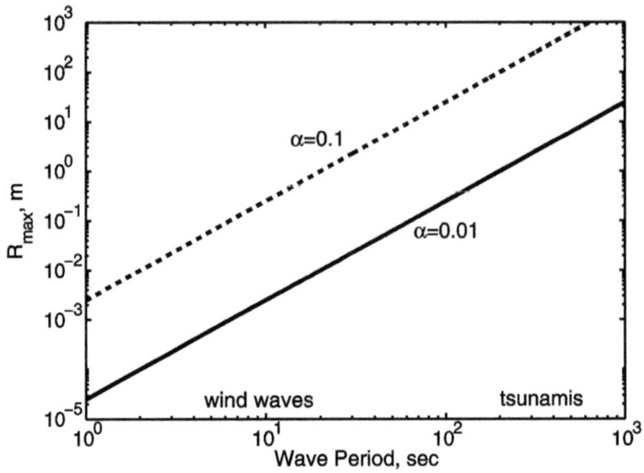

Fig. 4 The maximum runup height of a non-breaking wave.

In this case Eq. (58) becomes (primes are omitted)

$$u(t) = U(t + Bru), \tag{73}$$

where *Br* is determined by Eq. (70). The "linear" vertical displacement of water on the shoreline follows from Eq. (61) by introducing non-dimensional variables

$$Z(t) = \int U(t)dt. \tag{74}$$

The "real" nonlinear vertical displacement of the moving shoreline in non-dimensional variables follows from Eq. (35):

$$r(t) = Z(t + Bru) - \frac{Br}{2}u^2(t). \tag{75}$$

Eqs. (73) and (75) give a parametrical presentation of the moving shoreline for any shape of the incident wave. In particular, for a sine wave the parametric formulas are

$$t = \lambda - Br\cos\lambda, \quad r = \sin\lambda - \frac{Br}{2}\cos^2\lambda. \tag{76}$$

The dynamics of the moving shoreline when the sine wave approaches the coast can be computed from Eq. (76). It is presented in Fig. 5. If the wave amplitude is small enough ($Br << 1$), the shoreline changes in time as a sine function. If the wave amplitude increases and the parameter *Br* tends to 1, the water moves onshore faster and recedes to the sea more slowly. The first wave breaking with an increase of the amplitude occurs at the stage of the maximal rundown.

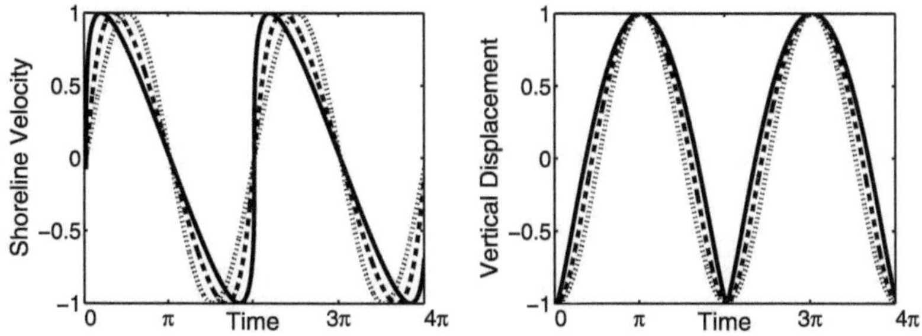

Fig. 5 Velocity and vertical displacement of the moving shoreline when the sine wave approaches the coast for $Br = 0.2$ (dotted line), 0.5 (dashed line), 1 (solid line).

6 Runup of solitary waves

The two-step procedure described above for calculating runup characteristics can be applied to all the types of incident waves approaching the shore from the open sea where such waves are linear. Due to the linearity of the "first-step" based on the solution of the linear wave equation for runup characteristics, its general solution can be presented as a Fourier superposition of elementary solutions (50) with various frequencies and spectral amplitudes. Fourier superposition is valid for an incident wave field (its elementary solution for the fixed point $x = -L$ is given by the left term in Eq. (51)) and for vertical oscillations of the water level in the point of the shoreline $x = 0$ (its elementary solution is given by Eq. (50)). An incident wave field far from the shoreline ($x = -L$) is

$$\eta(t) = \int_{-\infty}^{\infty} A(\omega)\exp(i\omega t)\,d\omega, \tag{77}$$

where the complex amplitude $A(-\omega) = A^*(\omega)$ provides that the function $\eta(t)$ is real. A similar formula can be written for the vertical oscillations of the water level on the shoreline

$$R(t) = \int_{-\infty}^{\infty} \sqrt{|\omega|}A(\omega)\exp\left[i\left(\omega t + \frac{\pi}{4}\operatorname{sign}(\omega)\right)\right]d\omega. \tag{78}$$

Here t should be replaced by $t - \tau$, where τ is the travel time to the coast, but this shift does not influence the maximum runup characteristics. For a monochromatic initial wave, Eq. (78) coincides with Eq. (54).

In the case when solitary waves approach the coast, an incident wave is characterized by two parameters: wave amplitude (height) H_0 and wave duration T_0.

Let us present an incident wave and runup displacement and velocity in a non-dimensional form

$$\eta(t) = H_0 f(t/T_0) = H_0 \int_{-\infty}^{\infty} B(\Omega)\exp(i\Omega\zeta)\,d\Omega, \tag{79}$$

$$R(t) = \sqrt{\frac{4\pi L}{\lambda_0}} H_0 \int_{-\infty}^{\infty} \sqrt{|\Omega|} B(\Omega)\exp\left[i\left(\Omega\zeta + \frac{\pi}{4}\operatorname{sign}(\Omega)\right)\right] d\Omega, \tag{80}$$

$$U(t) = \sqrt{\frac{4\pi L}{\lambda_0}} \frac{H_0}{\alpha T_0} \int_{-\infty}^{\infty} \sqrt{|\Omega|^3} B(\Omega)\exp\left[i\left(\Omega\zeta + \frac{\pi}{4}\operatorname{sign}(\Omega) + \frac{\pi}{2}\right)\right] d\Omega, \tag{81}$$

where

$$\zeta = t/T_0, \quad \Omega = \omega T_0, \quad \lambda_0 = \sqrt{gh_0}T_0, \quad B(\Omega) = \frac{1}{2\pi}\int_{-\infty}^{\infty} f(\zeta)\exp(-i\Omega\zeta)\,d\zeta. \tag{82}$$

As the breaking parameter is determined by the linear solution, it can also be presented by Fourier superposition as

$$Br = \frac{H_0 L}{\alpha\lambda_0^2}\sqrt{\frac{L}{\lambda_0}} \max\left\{\int_{-\infty}^{\infty} |\,\Omega\,|^{5/2} B(\Omega)\exp\left[i\left(\Omega\zeta + \frac{\pi}{4}\operatorname{sign}(\Omega) + \pi\right)\right] d\Omega\right\}. \tag{83}$$

The convergence of the integrals (79)–(83) provides a limitation on the smoothness of the incident wave shape. From Eq. (83) follows that the spectrum $B(\omega)$ should have roll-off faster than $\omega^{-7/2}$, providing a necessary criterion for non-breaking runup (the criterion requires the boundedness of the wave amplitude). This means that all the finite disturbances usually used in numerical simulations should have smoothed ends. In this case the runup of a sine shape crest cannot be considered, as its spectrum roll-off is proportional to ω^{-2}, and such a wave has to break for all values of the amplitude. Thus, only the third and higher derivatives of the pulse shape can have an abruption.

Formulas (79)–(83) can be re-written in a more convenient form

$$R_{\max} = R_0 p_R, \quad U_{\max} = \frac{R_0}{\alpha T_0} p_U, \quad Br = \frac{H_0 L}{\alpha\lambda_0^2}\sqrt{\frac{L}{\lambda_0}} p_b, \tag{84}$$

where

$$R_0 = \sqrt{\frac{4\pi L}{\lambda_0}} H_0, \quad \lambda_0 = \sqrt{gh_0}T_0, \tag{85}$$

λ_0 is the wavelength, h_0 is the water depth in the point $|x| = L$ and the numerical coefficients p_R, p_U and p_b characterize an incident wave shape.

In many cases it is not clear how to determine the wavelength λ_0 and the duration T_0 of a solitary pulse. In particular, most of the wave shapes (represented by analytical functions that are continuous for all derivatives) are nonzero everywhere for $-\infty < t < \infty$. There is obvious ambiguity in the definition of their wavelength (or duration) that can be interpreted as their width at any level of elevation, or by the value of an appropriate integral [10]. Even the incident wave has a finite duration, and its definition through two zero values does not characterize the wave shape properly. A convenient definition of the wavelength is the extension (spatial or temporal) of the wave profile elevation exceeding 2/3 of the maximum wave height. This choice is inspired by the definition of the significant wave height and length in physical oceanography and ocean engineering. For symmetric solitary waves, "significant" wave duration and "significant" wavelength are

$$T_s = 2T_0 f^{-1}\left(\frac{2}{3}\right), \quad \lambda_s = \sqrt{gh_0}T_s, \tag{86}$$

where f^{-1} is the inverse function of f. Thus the formulas (84) for the maximum displacement, the velocity of the moving shoreline, and the breaking parameter can be expressed as

$$R_{\max} = \mu_R^+ H_0 \sqrt{\frac{L}{\lambda_s}}, \quad U_{\max} = \mu_U^+ \frac{H_0 L}{\lambda_s}\sqrt{\frac{g}{\alpha\lambda_s}}, \quad Br = \mu_{Br}\frac{H_0 L}{\alpha\lambda_s^2}\sqrt{\frac{L}{\lambda_s}}, \tag{87}$$

where the coefficients μ_R^+, μ_U^+ and μ_{Br} (called form factors) depend on the wave shape:

$$\mu_R^+ = 2\sqrt{2\pi f^{-1}\left(\frac{2}{3}\right)}p_R, \quad \mu_U^+ = 4\sqrt{2\pi\left[f^{-1}\left(\frac{2}{3}\right)\right]^3}p_U,$$

$$\mu_{Br} = 8\sqrt{2\pi\left[f^{-1}\left(\frac{2}{3}\right)\right]^5}p_{Br}.$$

Analogous formulas can be written for rundown height and velocity also.

The analytical expressions for the maximum runup characteristics (runup and rundown heights, runup and rundown velocities and breaking parameter) become universal and only depend on the height and duration of the incoming onshore wave.

Let us first consider the runup of incident symmetrical positive waves (crests), having the shape of various "powers" of a sine pulse

$$f(\zeta) = \cos^n(\pi\zeta), \quad n = 3,4,5,... \tag{88}$$

which are defined on the segment [– 1/2,1/2]. Their shapes have a certain similarity, but their wave characteristics, such as mean water displacement, energy and wave duration on various levels, differ considerably (Fig. 6). The functions representing such impulses have different smoothness: their n-th order derivatives are discontinuous at their ends. The characteristic shapes of the incident crest and vertical displacement of the shoreline are presented in Fig. 7. Such shapes of oscillation are typical for runup of symmetric solitary waves on a beach of constant slope.

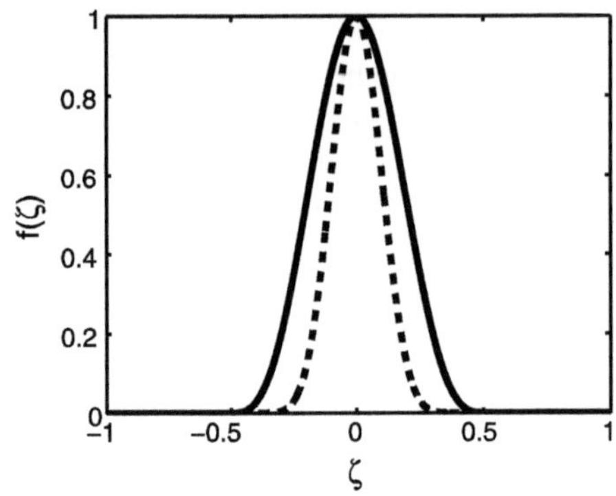

Fig. 6 Family of sine power pulses (88): solid line $n = 3$ and dashed line $n = 10$.

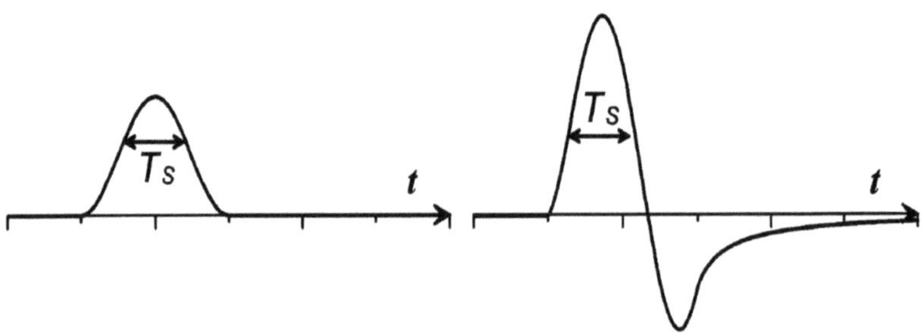

Fig. 7 The runup of a symmetric solitary wave on a beach of constant slope. The left figure shows water surface elevation at $x = -L$, the right one shows wave runup on shore. Notice that the duration T_s of the incident wave is defined at $x = -L$ and is not necessarily conserved during the runup and rundown process.

Form factors for runup μ_R^+ and rundown μ_R^- height, runup μ_U^+ and rundown μ_U^- velocity and breaking parameter μ_{Br} calculated for all sine power pulses (88) with the use of the definition of the characteristic wave length λ_s at $2/3$ of the maximum height, are presented in Figs. 9–11 at the end of this section. Calculating means and root-mean-square deviations, we obtain the values of the form factors for the maximum wave runup $\mu_R^+ = 3.61 \cdot (1 \pm 0.02)$ and rundown $\mu_R^- = 1.78 \cdot (1 \pm 0.28)$, these having fairly limited variation (Table 1), as also discussed at the end of this section.

First of all, it is significant that the runup height is higher than the rundown depth. This feature is observed for all sets of positive impulses. The form factor for the maximum wave runup is almost independent of the power n, showing that the influence of the initial wave shape on the extreme runup characteristics can be made fairly small by an appropriate choice of the characteristic wave length. The above choice of the (significant) wave length reduces the variation of the form factor for the sine power pulses to a remarkably small value, about 2%.

The deepest rundown is more affected by the wave shape, with the relevant form factor varying up to 28%. This feature can be explained by the presence of a complex field of motions in the rundown phase. A positive wave first executes runup and only later rundown (see Fig. 7). Therefore the runup process is predominantly governed by the incident wave dynamics, while the rundown phenomena occur under the influence of a set of distributed wave reflections from the slope, and consequently it is more sensitive to the wave shape variations.

A similar analysis can be applied to the maximum runup and rundown velocities of the moving shoreline. Calculated form factors for maximum runup μ_U^+ and rundown μ_U^- velocities are presented in Figs. 9 and 10. The maximal values for the rundown velocity are always greater than for the runup velocity for initial unidirectional impulses. The form factor for the rundown velocity $\mu_U^- = 6.98 \cdot (1 \pm 0.01)$ is almost constant for all values of n (root-mean-square deviation is 1%), whereas the runup velocity $\mu_U^+ = 4.65 \cdot (1 \pm 0.30)$ changes within a wider range (± 30%); see Table 1. The variations of the form factor for the breaking parameter are also weak (see Fig. 11, triangles). The relevant form factors $\mu_{Br} = 13.37 \cdot (1 \pm 0.10)$ can be considered a constant with a reasonable accuracy (Table 1).

Thus, the form factors for the most important parameters such as runup height and rundown velocity, and to some extent for the breaking parameter, are universal and do not depend on the particular shape of a sine power impulse. The variations of the form factors for rundown height and runup velocity are more significant (about 30%), but they can also be neglected for fast engineering estimates.

A similar analysis can be undertaken for the family of solitary waves, described by the following expression

$$f(\zeta) = \mathrm{sech}^{\mathrm{n}}(4\zeta), \quad \mathrm{n} = 1, 2, 3, \ldots \tag{89}$$

These impulses are unlimited in space with exponential decay of the elevation at their ends (see Fig. 8).

The case $n = 2$ corresponds to the well-known soliton solution of the Korteweg-de Vries (KdV) equation, which is frequently used as a generic example of shallow water solitary waves. The runup of the KdV solitons on a constant beach was studied previously by Synolakis [34], who presented both experimental and theoretical results. In our notation, the Synolakis formula [34] is

$$\frac{R_{\max}}{H_0} = 2.8312\sqrt{\frac{L}{h_0}}\left(\frac{H_0}{h_0}\right)^{1/4}. \tag{90}$$

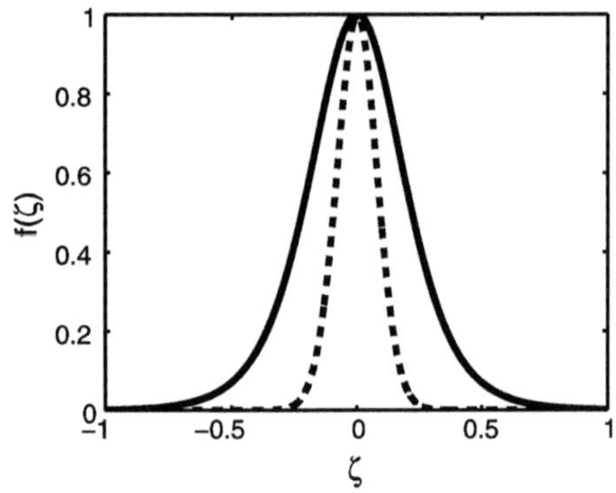

Fig. 8 Family of soliton-like impulses (89): solid line $n = 2$ and dashed line $n = 10$.

The "significant" wavelength of the soliton is easily calculated from the well-known analytical expression for a soliton in a constant-depth basin

$$\eta(x) = H_0 \mathrm{sech}^2\left(\sqrt{\frac{3\mathrm{H}_0}{4\mathrm{h}_0}}\frac{\mathrm{x}}{\mathrm{h}_0}\right) \tag{91}$$

and has the explicit form:

$$\lambda_s = 4\mathrm{sech}^{-1}\left(\sqrt{\frac{2}{3}}\right)\mathrm{h}_0\sqrt{\frac{\mathrm{h}_0}{3\mathrm{H}_0}}, \tag{92}$$

where $\mathrm{sech}^{-1}(\mathrm{z})$ is an inverse function of $\mathrm{sech}(\mathrm{z})$. Substituting this into the right-hand side of Eq. (90), we obtain

$$\frac{R_{\max}}{H_0} = 3.4913\sqrt{\frac{L}{\lambda_s}}. \tag{93}$$

Our numerical calculations lead to the same value of the form factor, $\mu_R^+ = 3.4913$ at $n = 2$.

Similar results are obtained for solitary bells of a Lorentz-like shape with algebraic decay

$$f(\zeta) = \frac{1}{\left[1 + (4\zeta)^2\right]^n}, \quad n = 1, 2, 3, \ldots \tag{94}$$

Results of computing all form factors are presented in Table 1 and Figs. 9–11. The maximum runup and rundown characteristics of a solitary wave on a beach virtually do not depend on the form of the incident wave if the wave duration is appropriately defined. It is especially evident that for the runup height and rundown velocity the variations for all given classes of wave shapes do not exceed 8%.

Table 1 Calculated form factors for different wave shapes

μ	Sine power	Soliton power	Lorentz pulse power
μ_R^+	$3.61 \cdot (1 \pm 0.02)$	$3.55 \cdot (1 \pm 0.05)$	$3.53 \cdot (1 \pm 0.08)$
μ_R^-	$1.78 \cdot (1 \pm 0.28)$	$1.56 \cdot (1 \pm 0.28)$	$1.51 \cdot (1 \pm 0.44)$
μ_U^+	$4.65 \cdot (1 \pm 0.30)$	$4.15 \cdot (1 \pm 0.22)$	$4.07 \cdot (1 \pm 0.26)$
μ_U^-	$6.98 \cdot (1 \pm 0.01)$	$6.98 \cdot (1 \pm 0.02)$	$6.99 \cdot (1 \pm 0.04)$
μ_{Br}	$13.37 \cdot (1 \pm 0.10)$	$12.90 \cdot (1 \pm 0.03)$	$12.99 \cdot (1 \pm 0.13)$

Based on Table 1 and Figs. 9–11, we recommend the following parameterized formulas for express estimations of the runup and rundown characteristics of long waves on a beach:

$$R_+ = 3.5 H_0 \sqrt{\frac{L}{\lambda_s}}, \quad R_- = 1.5 H_0 \sqrt{\frac{L}{\lambda_s}}, \tag{95}$$

$$U_+ = 4.5 \frac{H_0 L}{\lambda_s} \sqrt{\frac{g}{\alpha \lambda_s}}, \quad U_- = 7 \frac{H_0 L}{\lambda_s} \sqrt{\frac{g}{\alpha \lambda_s}}, \quad Br = 13 \frac{H_0 L}{\alpha \lambda_s^2} \sqrt{\frac{L}{\lambda_s}}, \tag{96}$$

where R_+ and R_- are the maximum values of the runup and rundown heights, U_+ and U_- are the maximum values of the runup and rundown velocities.

It is important that the obtained results hold only for symmetrical solitary waves. If the incident wave has the shape of an N-wave, also typical for the tsunami problem [36, 37, 38], the magnitude of the coefficients in the "runup" formulas will differ from that given in Eqs. (95)–(96); see for comparison the difference between runup heights of a soliton and its derivative [34, 36]. If the incident wave is an asymmetrical wave, with different steepness of the front and back slopes, the runup characteristics depend on the front-slope steepness [9]; this is also shown in section 7.

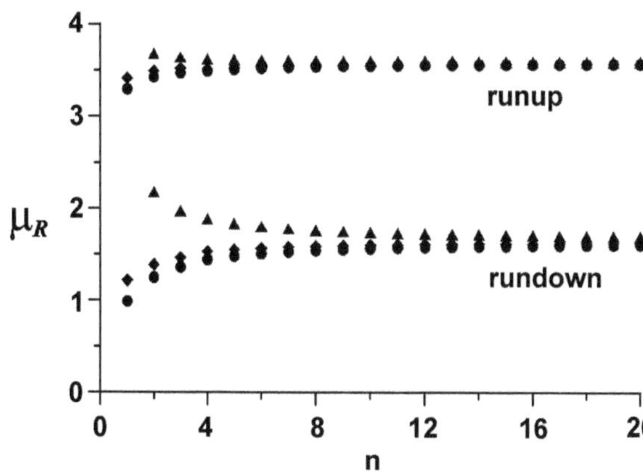

Fig. 9 Calculated form factors for the maximum runup μ_R^+ and rundown μ_R^- height for sine power pulses (triangles), soliton-like (diamonds) and Lorentz-like (circles) impulses.

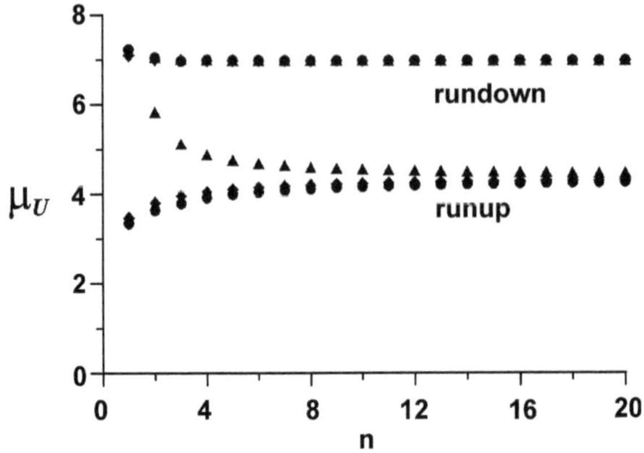

Fig. 10 Calculated form factors for the maximum runup μ_U^+ and rundown μ_U^- velocity for sine power pulses (triangles), soliton-like (diamonds) and Lorentz-like (circles) impulses.

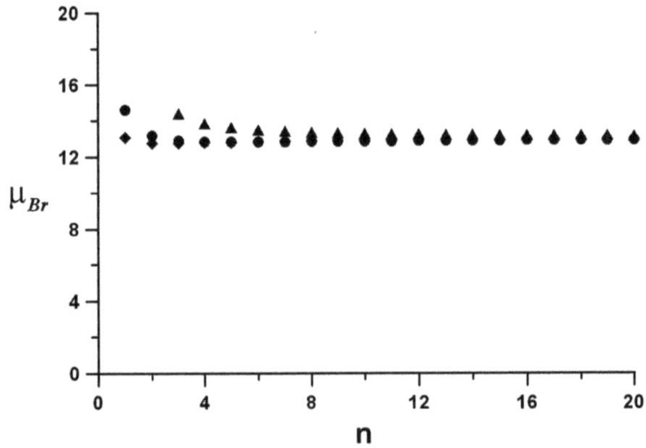

Fig. 11 Calculated form factors for the breaking parameter μ_{Br} for sine power pulses (triangles), soliton-like (diamonds) and Lorentz-like (circles) impulses.

As it was indicated above, extreme runup characteristics can be found from the linear theory. For calculating the real dynamics of the moving shoreline, the nonlinear theory based on the transformation (58) and (62) should be used. The computed shoreline velocity and vertical displacement for the incident KdV soliton are shown in Fig. 12. As for a sine wave, the first breaking occurs at the stage of maximum rundown, which occurs after the flooding induced by a long wave runup on the beach.

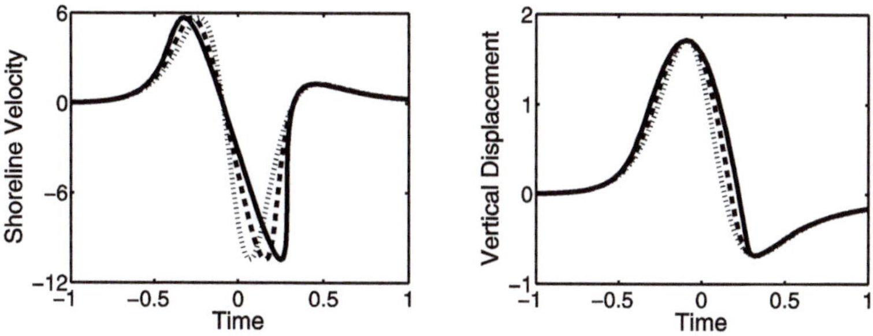

Fig. 12 Velocity and vertical displacement of the moving shoreline (dimensionless variables) for an incoming KdV soliton; with the breaking parameter $Br = 0$ (dotted line), 0.5 (dashed line) and 1 (solid line).

7 Runup of periodic waves

Another class of incident wave shapes of specific interest is the periodic asymmetric wave, the front slope steepness of which exceeds the back slope steepness. Such waves are often observed in the coastal zone (Fig. 13).

Fig. 13 Asymmetric wave in a coastal zone of the Island of Aegna, Estonia.

The theoretical model for a periodic wave runup on a beach is similar to a solitary wave runup, and the basic formulas are Fourier series, in comparison to Fourier integrals in Eqs. (79)–(83). In particular, the incident wave is presented as

$$\eta(t) = H_0 f(\omega t) = H_0 \mathrm{Re} \sum_{n=1} \mathrm{B_n} \exp(\mathrm{i} n \zeta), \tag{97}$$

where

$$\zeta = \omega t, \quad B_n = \frac{1}{2\pi}\int_0^{2\pi} f(\zeta)\exp(-in\zeta)\,d\zeta, \tag{98}$$

and Re denotes the real part of the complex sum in Eq. (97).

Similarly to section 6, the linear "runup" wave field can be written in non-dimensional form as

$$R(t) = \sqrt{\frac{4\pi L}{\lambda_0}} H_0 \mathrm{Re} \sum_{\mathrm{n}=1} \sqrt{\mathrm{n}} \mathrm{B}_\mathrm{n} \exp\left[\mathrm{i}\left(\mathrm{n}\zeta + \frac{\pi}{4}\right)\right], \tag{99}$$

$$U(t) = \sqrt{\frac{4\pi L}{\lambda_0}} \frac{H_0}{\alpha T_0} \mathrm{Re} \sum_{\mathrm{n}=1} \mathrm{n}^{3/2} \mathrm{B}_\mathrm{n} \exp\left[\mathrm{i}\left(\mathrm{n}\zeta + \frac{3\pi}{4}\right)\right]. \tag{100}$$

Here

$$\lambda_0 = \sqrt{gh_0}T_0 = 2\pi\sqrt{gh_0}/\omega \tag{101}$$

is the wavelength of the incident wave in the point $|x| = L$ with a water depth h_0. The breaking parameter can also be presented through Fourier series as

$$Br = \frac{H_0 L}{\alpha \lambda_0^2}\sqrt{\frac{L}{\lambda_0}} \max \mathrm{Re} \sum_{\mathrm{n}=1} \mathrm{n}^{5/2} \mathrm{B}_\mathrm{n} \exp\left[\mathrm{i}\left(\mathrm{n}\zeta + \frac{5\pi}{4}\right)\right]. \tag{102}$$

The spectral amplitudes B_n of the incident wave should decrease with increasing n faster than $n^{-7/2}$ for convergence of the series in Eq. (102). Therefore, the shape of the incident periodic wave should be smooth enough as has already been discussed for solitary waves.

There are several periodic incident wave shapes considered in the literature. Synolakis et al. [35] has considered the runup height of a cnoidal wave that is the steady-state solution of the Korteweg – de Vries equation

$$\eta(x,t) = H\left\{cn^2\left[2K\left(\frac{x}{\lambda} + \frac{t}{T}\right)|m\right] - < cn^2 >\right\}, \tag{103}$$

where cn is the elliptic Jacobi function, λ and T are the wavelength and the period of a periodic wave, K is the complete elliptic integral of the first kind, m is an elliptic parameter and $< cn^2 >$ is the mean value for the whole period. The cnoidal wave (103) is a monochromatic wave cosine for $m \to 0$, and it transforms to the set of solitary waves (KdV solitons) for $m \to 1$.

All extreme runup characteristics of a cnoidal wave can be computed by calculating its Fourier spectrum [35]. The runup height of the cnoidal wave is higher than for a sine wave, and therefore, the contribution of high harmonics (obertones) is important.

This problem is studied in [8] using the example of the bi-harmonic incident wave with an arbitrary relation between the phases of the harmonics in comparison to a cnoidal wave, where the phases of all harmonics are the same. Distributions of crest amplitudes in the incident wave and on the shoreline versus the ratio of harmonics amplitudes and phases (normalized by the amplitude of the first harmonic, its phase is equal to zero) are shown in Fig. 14. Crest amplitudes of the incident wave are minimal for a phase difference equal to π and maximal for zero. The distribution of the runup amplitude is also symmetrical with respect to the phase $5\pi/4$, where the runup amplitude is minimal and its maximal value is reached at $\pi/4$.

Computed time series of the incident wave (dashed line) and runup oscillations (solid line) are presented in Fig. 15 for a case of comparable amplitudes $B_2 = 0.8B_1$. This case corresponds to superimposed harmonics rather than a basic mode and a super-harmonic mode, since the amplitudes are comparable and the phases are unrelated.

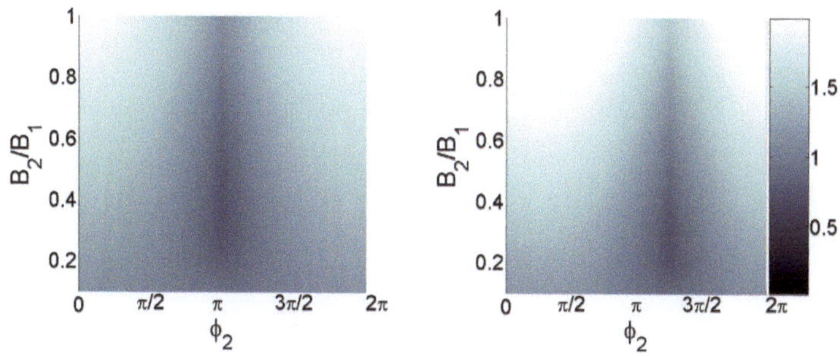

Fig. 14 Crest amplitude distribution: incident wave (left) and runup amplitude (right).

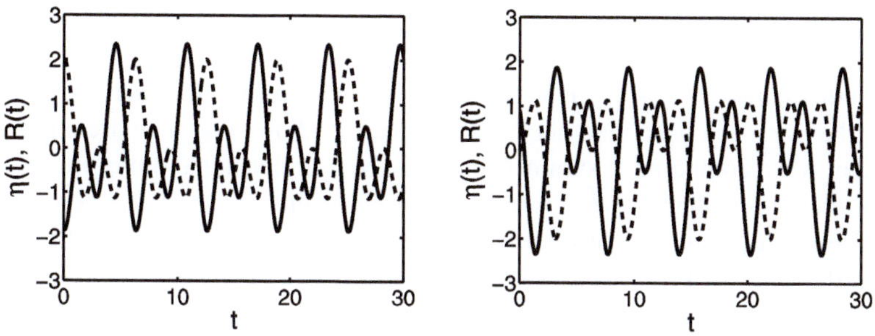

Fig. 15 Incident wave (dashed line) and water oscillations on the shoreline (solid line) for a phase difference 0 (left) and π (right).

The crests of the incident wave are more visible for a weak phase difference between harmonics. The runup extremes are phase shifted in comparison with the extremes of the incident wave.

The third example of periodic waves is the nonlinear deformed wave. The considered geometry is a flat bottom joined with a plane beach (Fig. 2). The sine wave, due to nonlinearity, becomes steeper, and this process can be described in the framework of the classical solution of the nonlinear shallow water theory in a basin of constant depth [9, 11, 39]. The wave shape gradually deforms due to the difference in speed of the crest and trough. The wave steepness increases when it propagates. The spectral amplitudes can be expressed through the face-slope steepness [9, 11, 39] as

$$B_n(s) = \frac{2B}{n(1-s_0/s)} J_n\left(n\left[1-\frac{s_0}{s}\right]\right), \tag{104}$$

where s_0 is the initial steepness (in our case, it is the steepness of a sine wave) and B is the non-dimensional amplitude of a sine wave in the open sea. This dependence is displayed in Fig. 16. It is clearly seen that the distribution of spectral amplitudes becomes universal for waves with steep fronts.

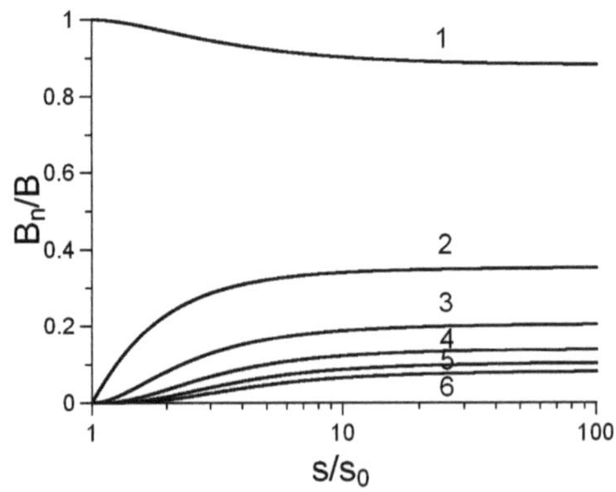

Fig. 16 Amplitudes of high harmonics (numbers on the figure) versus face-slope steepness.

Runup characteristics of such "deformed" waves on a plane beach are examined using the approach described above. This analysis shows that they are appreciably different when compared to symmetric waves. Non-dimensional runup and rundown amplitudes (normalized by R_0) turn out to be functions of the wave steepness (Fig. 17). The rundown amplitude weakly depends on the wave steepness, it differs from the rundown caused by a sine wave by no more than 30%, and Eq. (74) can still be used for the evaluation of its magnitude. However, the runup height significantly depends on the wave steepness. It tends to infinity for a shock wave within the model of non-breaking waves. The realistic runup height is limited by the wave breaking.

Therefore, wave asymmetry is the most important parameter in the runup process. An asymmetric wave with a steep front penetrates inland for a much larger distance than a symmetric wave of the same height and length. This result partially explains why tsunami waves with steep fronts (for example, the 2004 tsunami in the Indian Ocean) penetrate a long way inland.

A similar analysis for the extreme characteristics of the shoreline velocity shows that the runup velocities of asymmetric waves may considerably exceed the rundown velocities (Fig. 18).

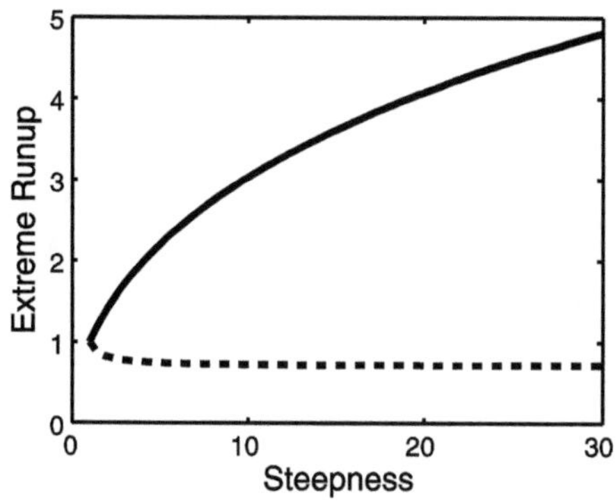

Fig. 17 Dependence of extreme runup (solid line) and rundown (dashed line) heights (dimensionless variables) on a wave steepness.

The "nonlinear" time history of the water level and velocity of the moving shoreline for different values of the breaking parameter Br shows that time records are substantially asymmetric even when the wave amplitude is small (Fig. 19). Runup height and velocity are higher than rundown height and velocity even for very low-amplitude, but asymmetric, waves. The runup process of a very asymmetric wave ($s = 10s_0$) causes extremely strong flow moving inland over a short time (Fig. 20). The runup height is considerably higher than the rundown height. Such intense flows can be identified on many photos of the 2004 tsunami in the Indian Ocean. Very steep waves, however, break relatively fast.

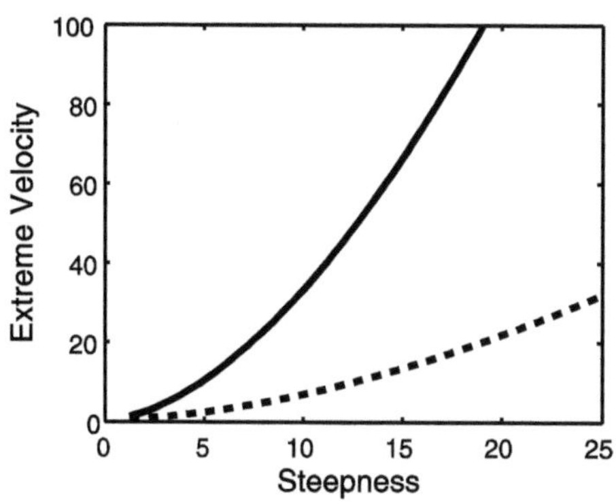

Fig. 18 Extreme runup (solid line) and rundown (dashed line) velocities (dimensionless variables) versus the wave steepness.

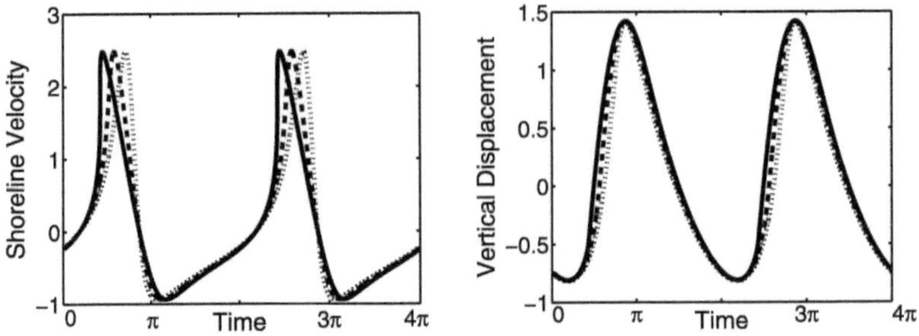

Fig. 19 Velocity and vertical displacement of the moving shoreline for nonlinear deformed waves ($s/s_0 = 2$) in non-dimensional variables; the breaking parameter $Br = 0$ (dotted line), 0.5 (dashed line) and 1 (solid line).

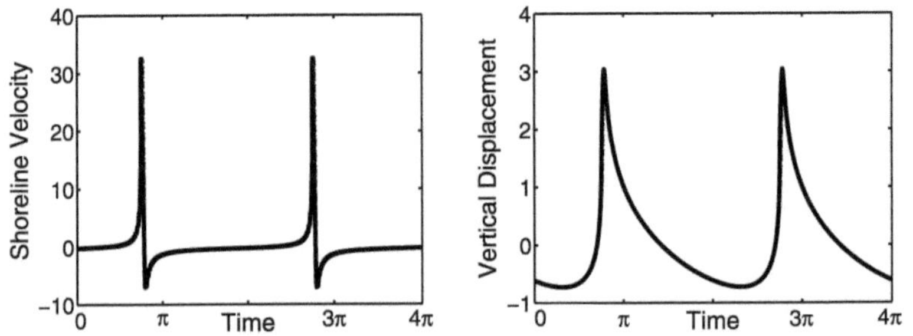

Fig. 20 Velocity and vertical displacement of the moving shoreline for nonlinear deformed waves ($s/s_0 = 10$) in non-dimensional variables; the breaking parameter $Br = 0$ (dotted line), 0.5 (dashed line) and 1 (solid line).

8 Conclusion

The long wave runup on a plane beach was described analytically in the framework of rigorous solutions of the nonlinear shallow-water theory. Such solutions are obtained with the use of the hodograph transformation of the nonlinear shallow-water system to the linear wave equation. This theory is limited by the wave breaking and the characteristics of the breaking wave, which were also investigated in this paper.

It was shown that the extreme characteristics of the runup process (runup and rundown amplitudes, extreme values of on- and off-shore velocities and the wave breaking condition) can be found within the linear approximation while the "real" dynamics of the moving shoreline must be described by the nonlinear theory. It has also been shown that with an increase in the amplitude the wave first breaks on the shoreline. In this case the velocity at the shoreline has the shape of a shock wave and the water displacement has a jump of the first derivative in the trough.

The runup of solitary waves of different shapes (various "powers" of a sine crest, soliton and Lorentz pulses) was analyzed using the same approach. It was demonstrated that the use of a wavelength definition as the "significant" wavelength can parameterize all formulas for extreme runup characteristics, and their dependence on the incident wave shape becomes very weak. Such an approach can be used for engineering estimates.

The runup process of periodic asymmetric waves was also analyzed here. It was shown that asymmetric waves penetrate inland over larger distances and with higher velocities than symmetric waves.

All analytical results described here were obtained for the 1D case of long wave runup on a plane beach. A few analytical results for a 2D nonlinear runup problem are discussed in [5, 40] but they are out of the scope of this paper.

Acknowledgements The author acknowledges Prof. Efim Pelinovsky for his long term valuable cooperation, discussion and collaboration, and Prof. Tarmo Soomere for his kind invitation to write this paper. This research is supported by grants from INTAS (06-1000014-6046), RFBR (08-05-00069, 08-02-00039), an EEA grant (EMP41), and the Marie Curie network SEAMOCS (MRTN-CT-2005-019374).

References

1. Batchelor, G.K.: An Introduction to Fluid Dynamics. Cambridge University Press (1967)
2. Brocchini, M., Gentile, R.: Modelling the run-up of significant wave groups. Cont. Shelf Res. **21**, 1533–1550 (2001)
3. Carrier, G.F., Greenspan, H.P.: Water waves of finite amplitude on a sloping beach. J. Fluid Mech. **4**, 97–109 (1958)
4. Carrier, G.F., Wu, T.T., Yeh, H.: Tsunami run-up and draw-down on a plane beach. J. Fluid Mech. **475**, 79–99 (2003)
5. Choi, B.H., Pelinovsky, E., Kim, D.C., Didenkulova, I., Woo, S.B.: Two- and three-dimensional computation of solitary wave runup on non-plane beach. Nonlin. Process. Geophys. **15**, 489–502 (2008)
6. Courant, R., Hilbert, D.: Methods of Mathematical Physics. Wiley & Sons, New York (1989)
7. Curtis, G.D., Pelinovsky, E.N.: Evaluation of tsunami risk for mitigation and warning. Science of Tsunami Hazards **17**, 187–192 (1999)
8. Didenkulova, I.I., Kharif, Ch.: Runup of biharmonic long waves on a beach. Izvestiya, Russian Academy of Engineering Science **14**, 91–97 (2005)
9. Didenkulova, I., Pelinovsky, E., Soomere, T., Zahibo, N.: Runup of nonlinear asymmetric waves on a plane beach. In: Kundu, A. (ed.) Tsunami and Nonlinear Waves, pp. 173–188. Springer, Berlin (2007)
10. Didenkulova, I.I., Pelinovsky, E.N.: Run up of long waves on a beach: the influence of the incident wave form. Oceanology **48**, 1–6 (2008)
11. Didenkulova, I.I., Zahibo, N., Kurkin, A.A., Levin, B.V., Pelinovsky, E.N., Soomere, T.: Runup of nonlinearly deformed waves on a coast. Dokl. Earth Sci. **411**, 1241–1243 (2006)
12. Dysthe, K., Krogstad, H.E., Muller, P.: Oceanic rogue waves. Annu. Rev. Fluid Mech. **4**, 287–310 (2008)
13. Jeffrey, A., Majorana, A.: Finite amplitude water waves above a sloping beach. Wave Motion **7**, 229–233 (1985)
14. Kânoğlu, U.: Nonlinear evolution and run-up-rundown of long waves over a sloping beach. J. Fluid Mech. **513**, 363–372 (2004)
15. Kânoğlu, U., Synolakis, C.: Initial value problem solution of nonlinear shallow water-wave equations. Phys. Rev. Lett. **97**, 148–501 (2006)

16. Kharif, C., Pelinovsky, E.: Physical mechanisms of the rogue wave phenomenon. Eur. J. Mech. B **22**, 603–634 (2003)
17. Kim, K.O., Lee, H.S., Yamashita, T., Choi, B.H.: Wave and storm surge simulations for Hurricane Katrina using coupled process based models. KSCE J. Civil Eng **12**, 1–8 (2008)
18. Lay, Th., Kanamori, H., Ammon, Ch.J., Nettles, M., Ward, S.N., Aster, R.C., Beck, S.L., Bilek, S.L., Brudzinski, M.R., Butler, Rh., DeShon, H.R., Ekström, G., Satake, K., Sipkin, S.: The great Sumatra-Andaman earthquake of 26 December 2004. Science **308**, 1127–1133 (2005)
19. Le Mehaute, B., Koh, R.C., Hwang, L.S.: A synthesis of wave run-up. J. Waterw. Port Coast. Ocean Eng. **94**, 77–92 (1968)
20. Li, Y., Raichlen, F.: Non-breaking and breaking solitary wave run-up. J. Fluid Mech. **456**, 295–318 (2002)
21. Lighthill, J.: Waves in Fluids. Cambridge University Press (2001)
22. Lin, P., Chang, K.-A., Liu, P.L.-F.: Runup and rundown of solitary waves on sloping beaches. J. Waterw. Port Coast. Ocean Eng. **125**, 247–255 (1999)
23. Madsen, P.A., Fuhrman, D.R.: Run-up of tsunamis and long waves in terms of surf-similarity. Coast. Eng. **55**, 209–223 (2008)
24. Massel, S.R.: Hydrodynamics of Coastal Zones. Elsevier, Amsterdam (1989)
25. Mazova, R.Kh., Osipenko, N.N., Pelinovsky, E.N.: Solitary wave climbing a beach without breaking. Rozprawy Hydrotechniczne **54**, 71–80 (1991)
26. Pedersen, G., Gjevik, B.: Run-up of solitary waves. J. Fluid Mech. **142**, 283–299 (1983)
27. Pelinovsky, E.: Nonlinear Dynamics of Tsunami Waves. Applied Physics Institute Press, Gorky (1982)
28. Pelinovsky, E., Mazova, R.: Exact analytical solutions of nonlinear problems of tsunami wave run-up on slopes with different profiles. Nat. Hazards **6**, 227–249 (1992)
29. Peregrine, D.H.: Water-wave impact on walls. Annu. Rev. Fluid Mech. **35**, 23–43 (2003)
30. Soomere, T.: Estonia got storm warning from newspaper. Scandinavian Shipping Gazette, **4**/2005, 26–29 (2005)
31. Spielfogel, L.O.: Runup of single waves on a sloping beach. J. Fluid Mech. **74**, 685–694 (1976)
32. Stockdon, H.F., Holman, R.A., Howd, P.A., Sallenger, A.H.: Empirical parameterization of setup, swash, and runup. Coast. Eng. **53**, 573–588 (2006)
33. Stoker, J.J.: Water Waves. The Mathematical Theory and Applications. Interscience, New York (1957)
34. Synolakis, C.E.: The runup of solitary waves. J. Fluid Mech. **185**, 523–545 (1987)
35. Synolakis, C.E., Deb, M.K., Skjelbreia, J.E.: The anomalous behavior of the run-up of cnoidal waves. Phys. Fluids **31**, 3–5 (1988)
36. Tadepalli, S., Synolakis, C.E.: The run-up of N-waves, Proc. Roy. Soc. London A **445**, 99–112 (1994)
37. Tadepalli, S., Synolakis, C.E.: Model for the leading waves of tsunamis, Phys. Rev. Lett. **77**, 2141–2145 (1996)
38. Tinti, S., Tonini, R.: Analytical evolution of tsunamis induced by near-shore earthquakes on a constant-slope ocean, J. Fluid Mech. **535**, 33–64 (2005)
39. Zahibo, N., Didenkulova, I., Kurkin, A., Pelinovsky, E.: Steepness and spectrum of nonlinear deformed shallow water wave. Ocean Eng. **35**, 47–52 (2008)
40. Zahibo, N., Pelinovsky, E., Golinko, V., Osipenko, N.: Tsunami wave runup on coasts of narrow bays. Int. J. Fluid Mech. Res. **33**, 106–118 (2006)

Part V
Mathematical Methods

Overview

Ewald Quak

Mathematics for Applications was the key concept for the collaboration of the CENS and CMA teams in the CENS-CMA project, yet it also illustrates the complementarity of these two centers. The research at CENS involves the application of a whole variety of available mathematical methods, often very sophisticated ones, to complex nonlinear problems in a lot of different fields. The CMA approach is to push the frontiers of mathematical research itself by extending existing mathematical methods and developing new ones.

In the previous parts of this book, *Waves in Solids* and *Waves in Fluids*, we have presented mathematics that is rooted in decades of research efforts at the Institute of Cybernetics. The parts *Mesoscopic Theory* and *Exploiting the Dissipation Inequality* represent subjects to which CENS members have been closely connected and for which the CENS-CMA project provided the opportunity to invite foreign collaborators for research stays of several months, in one case even leading to a prolonged stay through a still ongoing Feodor Lynen Fellowship of the Alexander von Humboldt Foundation.

In this last part of the book, we now will present three contributions on *Mathematical Methods*, seen from both the CENS and the CMA perspectives. True to the philosophy of this volume, also here the authors present overviews of their own experiences. Naturally these are only examples. The earlier introductions to the CENS and CMA teams by Jüri Engelbrecht and Ragnar Winther mention still a lot of other very interesting topics, which are dealt with by the teams but could not all be included in this book.

It has already become clear in previous parts of the book that solitary waves and solitons, arising in the solution of various nonlinear partial differential equations, are one of the central issues of CENS research. In the contribution *The Pseudospectral Method and Discrete Spectral Analysis*, Andrus Salupere now shares some of his long experience with the numerical treatment of an extensive list of model equations. The method of his choice in these investigations is the Discrete Fourier Transform in the framework of what is called the Pseudospectral Method. An important part of his exposition is also taken up by Discrete Spectral Analysis, showing how

E. Quak, T. Soomere (eds.), *Applied Wave Mathematics*,
DOI 10.1007/978-3-642-00585-5_15, © Springer-Verlag Berlin Heidelberg 2009

spectral characteristics can be used in the analysis of the properties of the numerical solutions. The application section of his paper provides a sample of results on solitons, some times hidden or multiple, which have been achieved.

The final two papers are provided by members of CMA, giving a survey of their research work related to wave equations. The first one, *Foundations of Finite Element Methods for Wave Equations of Maxwell Type* by Snorre Christiansen, illustrates the viewpoint of generalizing existing methods to more and more abstract settings, in this case by providing a description of finite element methods in a very general framework. After an overview of the theory of approximation of wave equations by Galerkin methods, the author proceeds to the construction of finite element spaces of differential forms on cellular complexes. Thus his approach leads us deep into homological algebra and tensor products for real vector spaces.

The contribution *An Introduction to the Theory of Scalar Conservation Laws with Spatially Discontinuous Flux Functions* by Nils Henrik Risebro provides a short but deep guide to conservation laws with a space-dependent flux function, restricting to one spatial dimension, for simplicity but also because the theory is more complete in this case. The Riemann and the Cauchy problem are studied, including various examples. The last part of the paper then deals with the uniqueness of entropy solutions.

The Pseudospectral Method and Discrete Spectral Analysis

Andrus Salupere

Abstract One of the focal points of the research at the Centre for Nonlinear Studies is related to the numerical simulation of the emergence, propagation and interaction of solitary waves and solitons in nonlinear dispersive media. Based on the discrete Fourier transform the pseudospectral method can be used for the numerical integration of the model equations, and the Fourier transform related discrete spectral characteristics for the analysis of the numerical results. The latter approach is called discrete spectral analysis. The main advantage of the pseudospectral method compared with the finite difference method is related to the computational costs. On the other hand, the Fourier transform related spectral characteristics carry additional information about the internal structure of the waves, which can be used for the analysis of the time-space behaviour of the numerical solutions. In the present paper several practical aspects of the application of the Fourier transform based pseudospectral method for the numerical integration of equations of different types are discussed and several examples of applications of the discrete spectral analysis are introduced.

1 Introduction

Wave phenomena occur in very different physical systems. One can say that waves surround us everywhere — sound waves, radio waves, water waves, deformation waves in solids, etc., etc. Waves in different media can interact, causing pleasant as well as dreadful effects. For example, standing waves in violin or guitar strings produce sound waves, which are audible as music. On the other hand, underwater earthquakes can cause tsunamis, which are able to destroy coastal settlements. See the textbook [4] for more examples.

Centre for Nonlinear Studies, Institute of Cybernetics at Tallinn University of Technology, Akadeemia tee 21, 12618, Tallinn, Estonia, e-mail: salupere@ioc.ee

E. Quak, T. Soomere (eds.), *Applied Wave Mathematics*,

DOI 10.1007/978-3-642-00585-5_16,

As a rule, the equations that govern wave propagation (in any media) are *partial differential equations* (PDEs). The simplest case here is the *wave equation*

$$u_{tt} - c^2 u_{xx} = 0 \qquad (1)$$

that describes the propagation of one-dimensional linear waves. Here t is time, x – a space coordinate, c – the phase velocity, and subscripts denote partial derivatives. The dependent variable u, for example, can be interpreted as displacement of a linear elastic medium. In more complex circumstances the governing equations must be able to describe various physical effects like nonlinearity, dispersion, etc. In order to model such wave processes numerical methods play as important a role as analytical ones. A typical mathematical problem, related to the numerical simulation of wave propagation, is the initial value problem. In the present paper we consider (i) how to apply the discrete Fourier transform (DFT) related *pseudospectral method* (PsM) for the numerical solution of the initial value problem, and (ii) how to use discrete spectral characteristics for the analysis of the numerical results, i.e., how to apply *discrete spectral analysis* (DSA).

By Fornberg and Sloan [26] the current popularity of the PsM dates back to the early 1970s and the works of Kreiss and Oliger [53] and Orszag [63]. The collocation approach was first used for PDEs with periodic solutions by Kreiss and Oliger [53], and it was referred to as the PsM in Orszag [63]. Due to its universality, high efficiency and accuracy, the PsM can be applied for the solution of very different problems. For example, in recent years, the PsM has been used for the numerical solution of the Kuramoto–Sivashinsky equation [25], the Schrödinger and Fokker–Planck equations [56, 90], the Klein–Gordon–Schrödinger equations [3], incompressible Navier–Stokes equations [51], the Hirota–Satsuma coupled KdV equation [12], the Ostrovsky–Hunter equation [7], the Korteweg–de Vries–Burgers equation [68], higher order-, perturbed- and extended KdV equations [33, 52, 67], delay partial differential equations [57], the two-dimensional hyperbolic heat transfer problem [87], and Fisher's equation [61]; for the numerical modelling of the KdV random wave field [64], for the numerical integration of stochastic models of surface growth [30], for simulations of nonlinear surface water waves over variable bathymetry [34], for seismic modelling [10], for the numerical simulation of the two-dimensional Navier–Stokes turbulence at high Reynolds numbers [91], for the free vibration analysis of cylindrical and non-cylindrical helical springs [54, 55], to validate the Nonlinear Elastic Wave Spectroscopy – Time Reversal (NEWS–TR) method for damage location [32]; for computing the ground and first excited states in Bose–Einstein condensates [1, 2]; etc.

The paper is organised as follows: a list of model equations, used in our workgroup for the modelling of wave propagation in nonlinear media, is presented in section 2; in section 3 the essence of the PsM is introduced and several aspects of the practical application of the method are discussed; section 4 is dedicated to the DSA and conclusions are drawn in section 5.

2 The model equations

One of the focal points of the research at the Centre for Nonlinear Studies (CENS) is related to the numerical simulation of the emergence, propagation and interaction of solitary waves and solitons in nonlinear dispersive media. Below a set of equations is listed in which we have been interested during the last years. In all these cases the PsM has been applied for the numerical integration, and in most cases discrete spectral characteristics are used for the analysis of the numerical results, i.e., the DSA is applied.

(i) The Korteweg–de Vries (KdV) equation

$$u_t + uu_x + du_{xxx} = 0, \tag{2}$$

where d is the dispersion parameter [22, 23, 71, 75, 76, 77, 81, 83]. Equation (2) is the simplest model equation for describing waves in nonlinear dispersive media. It includes the lowest order nonlinear and dispersive terms, uu_x and u_{xxx}, respectively.

(ii) The forced (perturbed) KdV (FKdV) equation

$$u_t + uu_x + du_{xxx} = F(u) \tag{3}$$

with two different force fields

$$F(u) = A \sin Bu \qquad \text{and} \qquad F(u) = \kappa\left[u(u-\upsilon_1)(u-\upsilon_2)\right]. \tag{4}$$

Here A, B, κ, υ_1 and υ_2 are constants and the function $F(u)$ represents the influence of an additional force (driven) field, which depends on the excitation u [20, 22, 66, 72, 80]. Equation (3) with cubic force field $(4)_2$ is derived by Engelbrecht and Khamidullin [18] and is found to describe deformation waves in a microstructured layer with energy influx like seismic waves in lithosphere [20].

(iii) The higher order KdV-like evolution equation

$$u_t + [P(u)]_x + du_{xxx} + bu_{xxxxx} = 0, \qquad P(u) = \frac{u^4}{4} - \frac{u^2}{2}, \tag{5}$$

where d and b are the third- and the fifth order dispersion parameters, respectively, and $P(u)$ is the quartic elastic potential [43, 44, 45, 46, 47, 73, 74, 78, 79, 82]. Because of the orders of the nonlinearity and dispersion, equation (5) is referred to as KdV435 below. The equation has been proposed by G. A. Maugin and is found to describe deformation waves in shape memory alloys.

(iv) The fifth order KdV (fKdV) equation

$$u_t + uu_x + du_{xxx} + bu_{xxxxx} = 0, \tag{6}$$

where d and b are the third- and the fifth order dispersion parameters, respectively [44].

(v) The hierarchical KdV (HKdV) equation

$$u_t + uu_x + \alpha_1 u_{xxx} + \beta\,(u_t + uu_x + \alpha_2 u_{xxx})_{xx} = 0 \tag{7}$$

as derived by P. Giovine and F. Oliveri is found to describe 1D wave propagation in dilatant granular materials [31]. Here, the variable u is the bulk density, x — the space coordinate, t — time; α_1 and α_2 are macro- and microlevel dispersion parameters, respectively, and β is a parameter involving the ratio of the grain size and the wavelength [36, 37, 38, 39, 40, 41, 42, 73].

(vi) The wave propagation in Mindlin-type microstructured continua [58] can be modelled by the equation

$$v_{tt} - bv_{xx} - \frac{\mu}{2}\left(v^2\right)_{xx} - \delta\,(\beta v_{tt} + \gamma v_{xx})_{xx} + \delta^{3/2}\frac{\lambda}{2}\left(v_x^2\right)_{xxx} = 0, \tag{8}$$

which has been derived by J. Engelbrecht, F. Pastrone and J. Janno in [16, 19, 48, 49] and analysed numerically in [17, 84, 85, 86]. Here v is the deformation, δ — a scale parameter, and $b, \mu, \beta, \gamma, \lambda$ are material parameters (see [48, 49] for details). This equation is referred to below as MEP equation.

(vii) The KdV-type evolution equation

$$w_\tau + qww_\xi + rw_{\xi\xi\xi} + s\left(w_\xi w_{\xi\xi}\right)_\xi = 0, \tag{9}$$

has been derived by M. Randrüüt from the MEP model. The one-wave equation (9) can be considered as the counterpart of the two-wave equation (8) [21, 84]. Here q and s are the nonlinearity parameters of the macrostructure and the microstructure, respectively, and r is the dispersion parameter.

(viii) The evolution equation

$$u_\tau + \sigma_1 uu_\xi + \sigma_2 u_{\xi\xi\tau} + \sigma_3(2u_\xi u_{\xi\xi} + uu_{\xi\xi\xi}) = 0, \tag{10}$$

derived by H.-H. Dai, describes the propagation of nonlinear axial-radial deformation waves in hyperelastic rods (HER) [11, 88]. Here the dimensionless independent variables ξ and τ correspond to the moving frame, u is a quantity measuring the dimensionless deformation, and σ_1, σ_2 and σ_3 are related to the material constants and the initial stretch of the rod. The term in parentheses presents the coupling effect of the material nonlinearity and the geometric size. This equation is referred to as the HER equation below.

(ix) The equation

$$w_{tx} + \mu(b_0 + b_1 w + b_2 w^2)w_x + b_{00} = 0 \tag{11}$$

presents a simple model that describes the propagation of a single nerve pulse; w is the excitation and b_0, b_1, b_2 and b_{00} are constants [89]. Equation (11) (referred to as the nerve pulse evolution (NPE) equation below) is based on the Nagumo model and has been derived by J. Engelbrecht (see [14] for details).

(x) the system of PDEs

$$\begin{aligned} u_t &= u(u-a-u^2+au) - v + I(t) + D_1 u_{xx}, \\ v_t &= \varepsilon(-v+bu) + D_2 v_{xx} \end{aligned} \qquad (12)$$

presents the classical Fitz-Hugh–Nagumo (FHN) model for the pulse transmission in nerve fibres (see [5, 89] for details). Here u represents the trans-membrane potential, v is the recovery current, D_1 and D_2 are the diffusion constants, $0 < \varepsilon \ll 1$ reflects the difference in the time scales of u and v, the parameters $0 < a < 1$, $b > 0$ characterise the nonlinearity and coupling of u and v, respectively, and $I(t)$ is an additional current. This system of PDEs is referred to as the FHN equations below.

3 The pseudospectral method

3.1 Approximation of space derivatives

In order to solve PDEs numerically one has to approximate the partial derivatives with respect to time as well as to the space coordinates. For the approximation of space derivatives two types of methods — global and local ones — are widely used.

3.1.1 Local methods

The derivative is a local property of a function, and therefore it is very reasonable to use the finite difference method (FDM) for the approximation of space derivatives. For example, in the simplest case the space derivative of a function $u(x)$

$$\frac{du(x)}{dx} \approx \frac{u(x+h)-u(x-h)}{2h}, \qquad (13)$$

where h is the space grid step. However, one needs to use a very high number of space grid points in order to get high accuracy results by FDM.

3.1.2 Global methods

In the case of global methods the function is approximated as a sum of very smooth basis functions $\Phi_k(x)$:

$$u(x) \approx \sum_{k=0}^{N} a_k \Phi_k(x). \qquad (14)$$

It is often useful to choose trigonometric functions for the basis $\Phi_k(x)$, because the *fast Fourier transform* (FFT) is an efficient and widely used algorithm (see [24, 26] for details). In other words, this means that e.g. we can choose the FFT-related cosine series

$$u(x) \approx \sum_{k=0}^{N} a_k \cos(kx + \phi_k) \tag{15}$$

in order to approximate a periodic function $u(x)$. The relationships between the FFT-related quantities and the coefficients a_k and ϕ_k will be discussed in section 4.1. It is clear that now derivatives or integrals of $u(x)$ can be calculated exactly (analytically) from the expression (15).

3.2 The discrete Fourier transform

Let a function $u(x,t)$ (periodic in space) be given on the interval $0 \leq x \leq 2\pi$ and the space grid be formed by N points with space step $\Delta x = 2\pi/N$. Now the *DFT* is defined by

$$U(k,t) = \mathrm{F}u = \sum_{j=0}^{N-1} u(j\Delta x, t) \exp\left(-\frac{2\pi ijk}{N}\right) \tag{16}$$

and the *inverse DFT* (IDFT) by

$$u(j\Delta x, t) = \mathrm{F}^{-1}U = \frac{1}{N}\sum_{k} U(k,t) \exp\left(-\frac{2\pi ijk}{N}\right). \tag{17}$$

Here F denotes the DFT and F^{-1} the IDFT, i is the imaginary unit and the wave numbers are

$$k = 0, \pm 1, \pm 2, \ldots, \pm(N/2-1), -N/2. \tag{18}$$

Note that the normalisation factors multiplying the DFT and IDFT (here 1 and $1/N$) and the signs of the exponents are merely conventions, and differ in some treatments (see [8, 62]). As a rule the FFT algorithm is used for the calculation of the DFT in scientific computing. In turn, the FFT algorithm requires that the number of grid points N must be a power of two, i.e., to consider reasonably large values, then $N = 64, 128, 256, \ldots$.

Making use of the properties of the DFT [8, 62], the space derivatives of the function $u(x,t)$ can be calculated as

$$\frac{\partial u(x,t)}{\partial x} = \mathrm{F}^{-1}\left[ik\mathrm{F}u\right],$$
$$\frac{\partial^2 u(x,t)}{\partial x^2} = -\mathrm{F}^{-1}\left[k^2\mathrm{F}u\right],$$
$$\frac{\partial^3 u(x,t)}{\partial x^3} = -\mathrm{F}^{-1}\left[ik^3\mathrm{F}u\right], \tag{19}$$
$$\dots$$
$$\frac{\partial^n u(x,t)}{\partial x^n} = \mathrm{F}^{-1}\left[(ik)^n\mathrm{F}u\right]$$

and the integral as

$$\int u(x,t)\mathrm{d}x = \mathrm{F}^{-1}\frac{\mathrm{F}u}{ik} + C. \tag{20}$$

Here C is the integration constant. If the length of the space period for $u(x,t)$ is not 2π, but $2m\pi$, then one must use the quantity k/m instead of k in formulae (19) and (20). In this case, for example,

$$\frac{\partial^n u(x,t)}{\partial x^n} = \mathrm{F}^{-1}\left[\left(\frac{ik}{m}\right)^n \mathrm{F}u\right]. \tag{21}$$

3.3 *The essence of the pseudospectral method*

3.3.1 The idea

In a nutshell the idea of the PsM is very simple. Let a PDE be given in a general form

$$u_t = \Phi\left(u, u_x, u_{xx}, \dots\right). \tag{22}$$

Making use of formulae (19) (which can be considered as numerical differential operators), one can formally reduce the original PDE (22) to an ordinary differential equation (ODE)

$$u_t = \Psi\left(u\right). \tag{23}$$

In the case of integro-differential equations, formula (20) must be used simultaneously in order to get equation (22).

The method is called pseudospectral because the integration with respect to time is carried out in the physical space, and the Fourier transform related quantities are used only for the calculation of space derivatives.

For the formulation of the initial value problem one needs to introduce the initial conditions

$$u(x,0) = u_0, \tag{24}$$

and to take into account that because of the usage of the DFT, the boundary conditions must be periodic, and the length of the space period must be $2m\pi$, where m is a positive integer, i.e.,

$$u(x+2m\pi,t) = u(x,t), \qquad m = 1,2,3,\ldots. \tag{25}$$

In order to solve the initial value problem, i.e., to integrate equation (23) with respect to time under the initial conditions (24) and the boundary conditions (25), standard ODE solvers can now be applied.

3.3.2 ODE solvers

Originally the PsM was proposed by Kreiss and Oliger [53] for the solution of the KdV equation. They used the finite difference *leap-frog scheme*

$$u(x,t+\Delta t) = u(x,t-\Delta t) + 2\Delta t u_t(x,t) \tag{26}$$

for the integration with respect to time. In this case, the KdV equation (2) leads to the following straightforward pseudospectral approximation

$$u(x,t+\Delta t) = u(x,t-\Delta t) - 2\Delta t u \mathrm{F}^{-1}(ik\mathrm{F}u) + 2d\Delta t \mathrm{F}^{-1}(ik^3\mathrm{F}u). \tag{27}$$

In the latter expression $u \equiv u(x,t)$ in the last two terms. The implementation of the leap-frog scheme is technically very convenient, but it has a very serious disadvantage — one has to use a very short time step in order to get a stable solution (see [24, 27, 69, 70]). For this reason, instead of the leap-frog scheme, many other ODE solvers (including adaptive, i.e. time step controlling solvers) are widely used nowadays (see e.g. [6, 9, 24, 26, 77, 85]).

In our workgroup in CENS we have used different Matlab ODE solvers for problems that need moderate computer resources. For problems that need more computer resources, the calculations are carried out with the package SciPy [50] using the FFTW library [28] for the DFT and the F2PY [65] generated Python interface to the ODEPACK Fortran code [35] for the ODE solver.

3.3.3 Advantages and disadvantages

The main advantage of the PsM compared to the FDM is the cost of computations, i.e., making use of the PsM the required accuracy can be obtained with a remarkably smaller number of space grid points [24, 26, 27]. In addition, when using the PsM, the DFT of the variable u, i.e. $U(k,t) = \mathrm{F}u$, is calculated at every discrete time-step. Making use of the quantities $U(k,t)$, different spectral characteristics (e.g. spectral amplitudes or the cumulative spectrum) can be obtained. These spectral characteristics carry additional information about the internal structure of the waves and can be applied for an analysis of the time-space behaviour of the numerical solutions, i.e., in DSA. This issue will be discussed in detail in section 4.

The main disadvantage of the PsM is related to the periodic boundary conditions. However, this drawback can be easily overcome because the length of the period is not limited, it must be just 2π-fold.

3.4 The pseudospectral method and different equation types

Now we return to the model equations listed in section 2 in order to consider how the PsM can be used for the solution of the initial value problem in particular cases. Roughly speaking, the PsM is directly applicable for the numerical integration of equations (2)–(5), (9) and (12), but equations (7)–(11) must be transformed significantly before the PsM can be applied.

3.4.1 Direct application of the PsM

Equations (2)–(5) and (9) can be expressed in the explicit form (22) with respect to the time derivative u_t, and therefore the PsM is directly applicable as it is described in Subsection 3.3.

FHN equations (12) are already given in an explicit form as

$$\begin{aligned} u_t &= \Phi_1(t, u, u_{xx}, v), \\ v_t &= \Phi_2(t, u, v, v_{xx}). \end{aligned} \tag{28}$$

The only difference compared to the equations (2)–(5) is that we now have a system of PDEs instead of a single PDE. As a rule, ODE solvers can be used for the numerical solution of systems of ODEs as well as of single ODEs. Therefore in order to apply the PsM one has to calculate the partial space derivatives u_{xx} and v_{xx}, making use of the formulae (19), to formulate initial conditions for u and v, and to use an ODE solver for the numerical solution of the initial value problem.

3.4.2 HKdV and HER equations

The HKdV equation (7) includes the mixed partial derivative term βu_{2xt}. Therefore the usual PsM algorithm cannot be applied directly, and we have to introduce a suitable change of variables. At first the HKdV equation (7) is rewritten in the form

$$(u+\beta u_{2x})_t + (u+3\beta u_{2x})\, u_x + (\alpha_1+\beta u)\, u_{3x} + \beta\alpha_2 u_{5x} = 0 \tag{29}$$

and a variable

$$\phi = u + \beta u_{2x} \tag{30}$$

is introduced. Making use of the DFT and its properties, the last expression can be rewritten in the form

$$\phi = \mathrm{F}^{-1}(\mathrm{F}u) + \beta\mathrm{F}^{-1}\left(-k^2\mathrm{F}u\right) = \mathrm{F}^{-1}\left[\left(1-\beta k^2\right)\mathrm{F}u\right]. \tag{31}$$

In turn, the variable u can be expressed from (31) in the form

$$u = \mathrm{F}^{-1}\left[\frac{\mathrm{F}\phi}{1-\beta k^2}\right]. \tag{32}$$

It is clear that in order to avoid division by zero, one can consider only such values for the parameter β, which result in $1-\beta k^2 \neq 0$. Now the space derivatives of u can be presented in terms of ϕ as

$$\frac{\partial^n u}{\partial x^n} = \mathrm{F}^{-1}\left[\frac{(ik)^n\mathrm{F}\phi}{1-\beta k^2}\right]. \tag{33}$$

Substituting expression (30) into equation (29) and expressing the time derivative ϕ_t results in the following equation

$$\phi_t = -(u+3\beta u_{2x})u_x - (\alpha_1+\beta u)\,u_{3x} - \alpha_2\beta u_{5x}. \tag{34}$$

In equation (34) the variable u and all its space derivatives could be expressed in terms of ϕ according to the expressions (32) and (33). Therefore equation (34) can be considered as an ODE with respect to the variable ϕ and could be integrated numerically making use of standard ODE solvers.

The HER equation (10) includes the mixed partial derivative term $\sigma_2 u_{\xi\xi\tau}$, and therefore a technique identical to that described above should be applied for the numerical solution of this equation by the PsM method.

3.4.3 The MEP equation

The MEP equation (8) contains a mixed partial derivative term $\delta\beta v_{xxtt}$ and therefore, as in the case of the HKdV equation, the PsM can be applied after a suitable change of variables. The procedure we follow now is very similar to that considered in the previous subsection. First we rewrite equation (8) in the form

$$(v-\delta\beta v_{xx})_{tt} - bv_{xx} - \frac{\mu}{2}\left(v^2\right)_{xx} + \delta\gamma v_{xxxx} + \delta^{3/2}\frac{\lambda}{2}\left(v_x^2\right)_{xxx} = 0, \tag{35}$$

and introduce a new variable

$$\phi = v - \delta\beta v_{xx}. \tag{36}$$

Applying the DFT and its properties, the previous expression can be rewritten in form

$$\phi = \mathrm{F}^{-1}(\mathrm{F}v) - \delta\beta\mathrm{F}^{-1}\left(-k^2\mathrm{F}v\right) = \mathrm{F}^{-1}[(1+\delta\beta k^2)\mathrm{F}v]. \tag{37}$$

In turn, from the latter variable v can be expressed as

$$v = \mathrm{F}^{-1}\left[\frac{\mathrm{F}\phi}{1+\delta\beta k^2}\right] \tag{38}$$

and its spatial derivatives as

$$\frac{\partial^n v}{\partial x^n} = \mathrm{F}^{-1}\left[\frac{(ik)^n \mathrm{F}\phi}{1+\delta\beta k^2}\right]. \tag{39}$$

Here we have similar limitations as in the case of the HKdV equation — the values of the parameters δ and β must be chosen to satisfy $1+\delta\beta k^2 \neq 0$. Finally we can rewrite equation (35) in the form

$$\phi_{tt} = \left[bv + \frac{\mu}{2}v^2 - \delta\gamma v_{xx} - \delta^{3/2}\frac{\lambda}{2}\left(v_x^2\right)_x\right]_{xx}. \tag{40}$$

Here the variable v and its space derivatives are expressed through a new variable ϕ making use of the expressions (38) and (39). After the introduction of a variable $\psi = \phi_t$, equation (40) can be easily reduced to a system of two first order differential equations:

$$\begin{aligned}\psi_t &= \left[bv + \frac{\mu}{2}v^2 - \delta\gamma v_{xx} - \delta^{3/2}\frac{\lambda}{2}\left(v_x^2\right)_x\right]_{xx},\\ \phi_t &= \psi.\end{aligned} \tag{41}$$

The system (41) can be considered as a system of ODEs (the variable v and its partial derivatives with respect to the space coordinate x are expressed in terms of the variable ϕ), and therefore it can be integrated with respect to time, making use of the methods known for solving systems of first order ODEs.

3.4.4 The NPE equation

As in the previous cases we have to introduce a suitable change of variables in order to apply the PsM for the numerical integration of the NPE equation (11). In the present case a new variable

$$\phi(x,t) = w_x(x,t) \tag{42}$$

is introduced. Application of the same approach as in the previous cases results now in

$$\phi = \mathrm{F}^{-1}(ik\mathrm{F}w) \qquad \text{and} \qquad w = \mathrm{F}^{-1}\left(\frac{\mathrm{F}\phi}{ik}\right). \tag{43}$$

Unfortunately, the wave numbers $k = 0, \pm 1, \pm 2, \ldots, \pm(N/2-1), -N/2$, and one cannot avoid division by zero in $(43)_2$ in this case. In order to remove this problem we define

$$\Phi(k,t) = \mathrm{F}\phi, \qquad \hat{w}(x,t) = \mathrm{F}^{-1}(\hat{k}\Phi), \qquad \hat{k} = \begin{cases} 0, & k = 0, \\ 1/ik, & k \neq 0. \end{cases} \tag{44}$$

According to expressions (42) and (44), the properties of the DFT, and the assumption that the functions w and ϕ are defined in the interval $0 \leq x \leq 2\pi$, we have

$$w(x,t) = \int_0^x \phi(x,t)dx = \hat{w}(x,t) - \hat{w}(0,t) + \frac{\Phi(0,t)}{N}x. \tag{45}$$

Substituting expression (42) into equation (11) and expressing the time derivative ϕ_t results in the equation

$$\phi_t = -\mu(b_0 + b_1 w + b_2 w^2)\phi - b_{00}, \tag{46}$$

which can be considered as an ODE with respect to the variable ϕ, because the variable w can be expressed in terms of ϕ according to (44) and (45). Therefore standard ODE solvers can be applied for solving equation (46).

3.5 Filtering and other practical tips

The issues considered in this subsection are related to the stability of the numerical technique. We start with a short historical review related to the KdV equation.

Fornberg and Whitham [27] investigated the stability of the numerical scheme (27) and found that the scheme is accurate for low enough wave numbers, but if one considers high wave numbers of u in (27), the dispersive term u_{xxx} dominates over the nonlinear term uu_x, and the scheme loses accuracy. They proposed to replace $\Delta t k^3$ by $\sin \Delta t k^3$ in the last term of expression (27), so the numerical scheme takes the form

$$u(x,t+\Delta t) = u(x,t-\Delta t) - 2\Delta t u \mathrm{F}^{-1}(ik\mathrm{F}u) + 2d\mathrm{F}^{-1}(i \sin \Delta t k^3 \mathrm{F}u). \tag{47}$$

From the spectral viewpoint, the considered replacement means that one uses a certain filter

$$f_\mathrm{s} = \frac{\sin \Delta t k^3}{\Delta t k^3} \tag{48}$$

in order to suppress the influence of higher harmonics in the last term of expression (27). The linear stability condition for the scheme (27) is

$$\frac{\Delta t}{\Delta x^3} < \frac{1}{\pi^3}, \tag{49}$$

but for scheme (47)

$$\frac{\Delta t}{\Delta x^3} < \frac{3}{2\pi^2}. \tag{50}$$

According to the linear stability conditions, the time step that can be used for scheme (47) is about 4.7 times bigger than for scheme (27). The modified PsM (47) is found to be sufficiently accurate, stable and fast by different authors (see [69] and the references therein). The accuracy and stability of scheme (47) for different values of the dispersion parameter d were discussed by the author of the present paper. Here we only point out the fact that the application of the linear stability condition for the numerical solution of the corresponding nonlinear PDE does not guarantee a stable solution [69, 70].

As was mentioned above, the third-order dispersion term u_{xxx} is the main source of instability in the case of the KdV equation because the dispersive term can dominate over that of the nonlinearity for higher wave numbers. However, the KdV equation (2) is the simplest model for nonlinear dispersive systems. If the model equation includes more complicated dispersive and nonlinear terms — e.g. equations (5)–(8) — one has additional sources of instability. On the other hand, instead of the simple but quite unstable leap-frog scheme, more stable ODE solvers that are able to use adaptive time steps are widely used nowadays. Unfortunately even these smart solvers are not able to always guarantee desirable results — sometimes a very short time step just slows down the computation process, and sometimes the solution still becomes unstable. Therefore it is recommended to apply a certain filtering technique in particular cases in order to get a stable solution or to optimise the computational time. In CENS, we have applied the simple filter

$$f(k,k_1,k_0) = \begin{cases} 1, & -k_1 \le k \le k_1, \\ 0, & k \ge k_0,\ k \le -k_0, \\ \sin^2 \dfrac{k+k_0-2k_1}{2(k_0-k_1)}\pi, & k_1 < k < k_0, \\ \sin^2 \dfrac{k-k_0+2k_1}{2(k_1-k_0)}\pi, & -k_1 > k > -k_0 \end{cases} \tag{51}$$

for this reason. If this filter is used, then wave numbers $|k| < k_1$ are passed, wave numbers $k_1 < |k| < k_0$ are partly suppressed, and the influence of wave numbers $|k| > k_0$ vanishes. In Fig. 1 the transfer functions of the filters (48) and (51) are plotted against the wave numbers k. For filter (48) the time step Δt corresponds to the linear stability condition (50) and for filter (51), $k_1 = n/4$ and $k_0 = 3n/8$.

In practice, the application of filtering means that instead of the wave numbers k given by expression (18), one has to use numbers $k_s = kf(k,k_1,k_0)$ in the formulae (19). In order to understand that the wave numbers $|k| > k_0$ are not significant, by means of an adequate description of the solution, one has (i) to fix the filter parameters $k_1 = \alpha_1 N < k_0 = \alpha_0 N < 0.5N$ (where α_0 and α_1 should be such constants that guarantee integer values for k_1 and k_0) and (ii) to solve the problem for a number of grid points $N = N^*$ and $N = 2N^*$, i.e., for $\Delta x = 2m\pi/N^*$ and $\Delta x = 2m\pi/(2N^*)$, where $2m\pi$ is the length of the space period. If these two results coincide (within accuracy ε) then one can use the filter with parameters $k_1 = \alpha_1 N$ and $k_0 = \alpha_0 N$ for $N = N^*$.

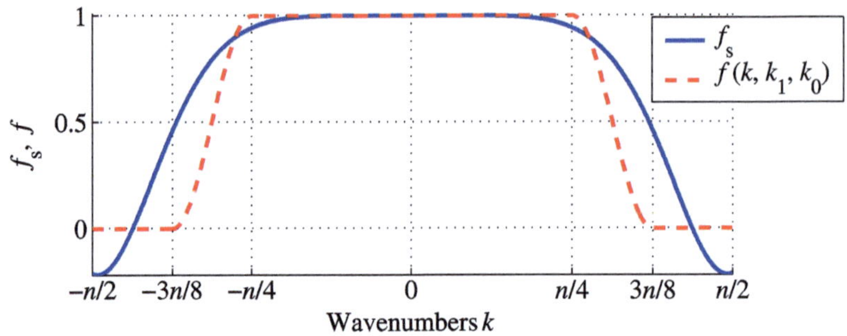

Fig. 1 The transfer functions of the filters f_s and $f(k,k_1,k_0)$.

If localised initial conditions are used and the space period $2m\pi$ is not long enough, then there exists another source for high wave number instabilities. In the case of localised initial conditions, the boundary conditions are considered to be asymptotic, i.e., u and its spatial derivatives u_x, u_{xx}, u_{xxx}, ... must tend to zero if $|x| \to \pm\infty$. However, for the PsM, the boundary conditions are periodic and therefore the length of the space period $2m\pi$ must be so long that $u(x,0)$ and its space derivatives are small enough at the boundaries $x = 2lm\pi$, $l = 0, \pm 1, \pm 2, \ldots$. If the space period is too short then the space derivatives at the boundaries cannot be considered to be continuous. The latter is reflected by the fact that in case of a "too short" space period, the Fourier components of the space derivatives $\hat{U}(k,t) = \mathrm{F}[u_{nx}(x,t)]$ have significantly higher values at $|k| \to N/2$ than in the case of a "long enough" space period. In the ideal case these Fourier components tend to zero if $|k| \to N/2$.

In conclusion, one has to point out that in order to get accurate and stable numerical results for a certain initial value problem, the number of grid points N, the length of the space period $2m\pi$ (in case of localised initial conditions) and the filter parameters k_1 and k_0 (if the filter is applied) must be chosen properly. The values of these four quantities depend on the character of the equation as well as on the character of the initial conditions. Probably the easiest way to get estimations whether or not these quantities are chosen properly, is to fix the initial conditions and calculate the Fourier components $\hat{U}(k,t)$ for all space derivative terms of the considered equation at $t = 0$. $\hat{U}(k,t)$ must tend to zero for $k \to k_0$ if filtering is applied and for $k \to N/2$ if filtering is not applied. If this is not the case, then the length of the space period is recommended to be increased. Of course, if the length of the period is increased significantly, the number of grid points is recommended to be increased simultaneously.

Last but not least, if the IDFT is applied in the expressions (17), (19)–(21), then the results are complex numbers. However, all terms in the equations (2)–(12) are realvalued. Therefore, if in computational algorithms these IDFT expressions are used, it is recommended to take the real part in order to speed up the computational process.

4 Discrete spectral analysis

4.1 Spectral amplitudes and spectral densities

In this subsection we return to the trigonometric series (15) and the DFT formula (16). At first we recall some elementary topics from complex numbers theory. By Euler's formula any complex number z can be represented in the form

$$z = |z|(\cos\phi + i\sin\phi) = |z|\exp i\phi. \tag{52}$$

Here $|z|$ and ϕ are called the *magnitude* and the *argument* of the complex number z, respectively. If $x = \mathrm{Re}\, z$ and $y = \mathrm{Im}\, z$ then the magnitude is

$$|z| = \sqrt{(x^2+y^2)}. \tag{53}$$

The argument ϕ can be calculated making use of the arctangent function. The latter cannot be used directly, because its results are limited to the interval $[-\pi/2, \pi/2]$. However, the values of the argument ϕ must be in the interval $[-\pi, \pi]$ (or equivalently in the interval $[0, 2\pi]$) because it must locate z in the complex plane. Therefore, besides the arctangent function the signs of the real part x and the imaginary part y must be taken into account in order to determine the specific quadrant in the complex plane. Matlab users can calculate the magnitude by the function 'abs' and the argument by 'angle'.

In case of an arbitrary waveform $u(x,t)$, the DFT related quantities $U(k,t)$ are complex numbers. Therefore one can calculate the magnitude $|U(k,t)|$ and the argument $\phi_k(t)$ for any quantity $U(k,t)$. Furthermore, if the DFT is defined by (16) (like e.g. in Matlab), then the multipliers of the cosine series are (15)

$$\begin{aligned} a_0(t) &= \frac{U(0,t)}{N}, \\ a_k(t) &= \frac{2|U(k,t)|}{N}, \qquad k = 1,2,\ldots,\frac{N}{2}-1, \\ a_k(t) &= \frac{|U(k,t)|}{N}, \qquad k = \frac{N}{2}. \end{aligned} \tag{54}$$

The phase shifts of the harmonic waves

$$\phi_k(t) = \mathrm{angle}\,[U(k,t)] \tag{55}$$

are just arguments of $U(k,t)$. Compared with expression (15), the argument t is added to a_k and ϕ_k in (54) and (55), because here $u(x,t)$ is considered instead of $u(x)$. Now we can "reconstruct" a wave $u(x,t)$ in terms of harmonics, i.e.,

$$u(x,t) = a_0(t) + \sum_{k=1}^{N/2} a_k(t)\cos\left[kx + \phi_k(t)\right]. \tag{56}$$

The quantities $a_k(t)$ are often referred to as the *spectral amplitudes* or just the *amplitude spectrum* and $\phi_k(t)$ as the *phase spectrum*. In order to understand the background of the formulae (54)–(56), one has to consider that (i) $U(-k,t)$ is the complex conjugate of $U(k,t)$; (ii) therefore $|U(-k,t)| = |U(k,t)|$ and angle $[U(-k,t)] = -$angle $[U(k,t)]$ and (iii) $\cos(-\alpha) = \cos\alpha$.

Next we introduce two important identities (see [8, 62]):

$$\sum_{j=0}^{N-1} u(j\Delta x,t) = U(0,t) \tag{57}$$

and Parseval's theorem

$$\sum_{j=0}^{N-1} [u(j\Delta x,t)]^2 = \frac{1}{N}\sum_{k}|U(k,t)|^2. \tag{58}$$

Expression (57) demonstrates that for a realvalued function u, the quantity $U(0,t)$ is always real.

It is well known that the KdV equation has an infinite number of conservation laws [13, 59, 60]. In discrete form the first two of them have the conserved densities $\sum_{j=0}^{N-1} u(j\Delta x,t)$ and $\sum_{j=0}^{N-1}[u(j\Delta x,t)]^2$, i.e., given by (57) and (58). The same conserved densities are applicable for some other equations (e.g., the KdV435). Therefore if $\sum_{j=0}^{N-1} u(j\Delta x,t)$ is conserved, then $U(0,t)$ has a constant value over the whole integration interval.

Parseval's theorem has an energy background — $\sum_{j=0}^{N-1}[u(j\Delta x,t)]^2$ is often interpreted as the total energy of $u(j\Delta x,t)$ at fixed t. In order to characterise how the energy is distributed between different harmonics (c.f. the harmonic series (56)) the *spectral densities*

$$S(k,t) = \begin{cases} \dfrac{2|U(k,t)|^2}{N} = \dfrac{Na_k^2}{2}, & k = 1;2,\ldots,\dfrac{N}{2}-1, \\ \dfrac{|U(k,t)|^2}{N} = Na_k^2, & k = \dfrac{N}{2}. \end{cases} \tag{59}$$

are introduced. In other words, the spectral densities (59) characterise the energy distribution over the wave numbers $k = 1,2,\ldots,N/2$. Parseval's theorem can now be rewritten in the form

$$\sum_{j=0}^{N-1} [u(j\Delta x,t)]^2 = \frac{[U(0,t)]^2}{N} + \sum_{k=1}^{N/2} S(k,t). \tag{60}$$

In the case of harmonic initial conditions, e.g. $u(x,0) = \sin x$, we have

$$\sum_{j=0}^{N-1} u(j\Delta x,0) = U(0,0) = 0 \tag{61}$$

and

$$\sum_{j=0}^{N-1} [u(j\Delta x,0)]^2 = \sum_{k=1}^{N/2} S(k,t) = N/2. \tag{62}$$

If these two quantities can be considered as conserved densities for a certain equation, then one can introduce normalised spectral densities (normalised energies for the kth harmonic [83]) in the form

$$\bar{S}(k,t) = \frac{S(k,t)}{\sum_{k=1}^{N/2} S(k,t)} = \frac{2S(k,t)}{N} = \begin{cases} \dfrac{4\,|U(k,t)|^2}{N^2} = a_k^2, & k = 1;2,\dots,\dfrac{N}{2}-1, \\ \dfrac{2\,|U(k,t)|^2}{N^2} = 2a_k^2, & k = \dfrac{N}{2}. \end{cases} \tag{63}$$

It is clear that $\bar{S}(k,t)$ and $S(k,t)$ differ only by the multiplier $N/2$ and therefore sometimes $\bar{S}(k,t)$ are considered by means of spectral densities even in cases of arbitrary initial conditions (i.e. $\sum_{j=0}^{N-1} [u(j\Delta x,t)]^2 \neq N/2$) and/or when $\sum_{j=0}^{N-1} [u(j\Delta x,t)]^2$ is not a conserved density.

4.2 *Cumulative spectrum and time averaged normalised spectral densities*

In the previous subsection we defined normalised spectral densities $\bar{S}(k,t)$ for a particular case. Here we consider the general case and we define *normalised spectral densities* in the form

$$S_{\text{norm}}(k,t) = \frac{S(k,t)}{S_{\text{sum}}(t)}, \qquad S_{\text{sum}}(t) = \sum_{k=1}^{n/2} S(k,t). \tag{64}$$

It is obvious that if $S_{\text{sum}}(t) = const$ and $U(0,t) = 0$ then $S_{\text{norm}}(k,t) = \bar{S}(k,t)$.

Now we can define the *cumulative spectrum* as

$$S_C(m,t) = \sum_{k=m}^{N/2} S_{\text{norm}}(k,t), \qquad m = 1,2,3,\dots,N/2. \tag{65}$$

It is clear that $S_C(1,t) = 1$ and $S_C(N/2,t) = S_{\text{norm}}(N/2,t)$. For an arbitrary value of $1 < m < N/2$, the quantity $S_C(m,t)$ reflects the amount of energy carried by the harmonics $m,\dots,N/2$. If e.g. $S_C(m^*,t) < 0.001$ in a certain time interval $0 < t < t_f$, then more than 99.9% of the energy is carried by the harmonics $1,\dots,m^*-1$ in this time interval. This quantity can be used for the estimation of the minimum number of harmonics one needs for an adequate description of waveforms.

Time averaged normalised spectral densities (TANSD) can be defined as

$$S_{\mathrm{a}}(k,t) = \frac{\int_0^t S_{\mathrm{norm}}(k,t)dt}{t}. \tag{66}$$

We are, however, analysing numerical solutions that are found at certain discrete time moments t_l. Therefore we have discrete values of the spectral densities S and S_{norm} at discrete time moments t_l ($l = 1,2,3,\ldots$), i.e. we have $S(k,t_l)$ and $S_{\mathrm{norm}}(k,t_l)$. Therefore at $t = t_l$

$$S_{\mathrm{a}}(k,t_l) = \frac{\sum_{m=1}^{l} S_{\mathrm{norm}}(k,t_m)}{l}. \tag{67}$$

TANSD (67) reflect the contribution of the k-th spectral density (or amplitude) over the time interval $[0,t_l]$. The idea of applying TANSD comes from the paper by Galgani et al. [29], where 'time average energies of single modes' are used in order to discuss the energy equipartition in systems of Fermi–Pasta–Ulam (FPU) type. Compared to spectral densities (or amplitudes), TANSD curves give a clearer understanding about the domination of certain harmonics [38, 39, 40, 41].

4.3 Applications

These spectral characteristics carry additional information about the internal structure of waves. This information is related to the existence of local and global minima and maxima in time dependence of the spectral curves, the relative and absolute values of the spectral quantities, the concavity and convexity of the spectral curves, etc. In this subsection several examples of applications of the DSA are presented.

4.3.1 The KdV equation, the number of emerging solitons and hidden solitons

The first example is related to the analysis of the numerical solution of the KdV equation (2) under harmonic initial conditions

$$u(x,0) = -\sin(x), \qquad 0 \le x < 2\pi. \tag{68}$$

The boundary conditions are periodic, the length of the space period is 2π, the dispersion parameter is $d = 10^{-1.9}$ and the length of the time interval is $t_f = 30$ (i.e. we consider $0 \le t \le t_f$). It is well known that the solution of this problem results in a train of interacting solitons (see e.g. [13, 92]). The specific solitonic behaviour of the solution can be traced in the timeslice plot (Fig. 2), in the pseudocolor plot (Fig. 3) and in the amplitude curves (Fig. 4). One can see that between interactions the emerged solitons propagate at constant speed; when two solitons interact then the amplitude of the higher soliton decreases and when they separate the amplitudes are restored; during interaction the trajectories of the solitons are phase shifted.

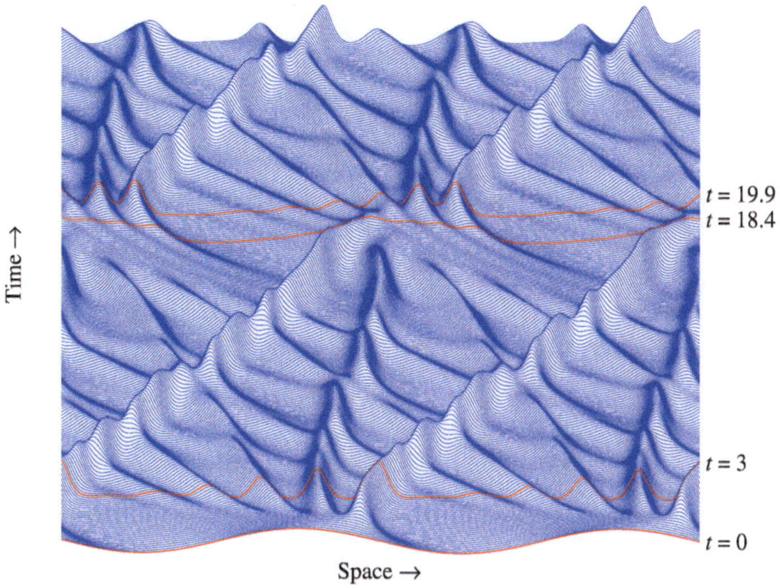

Fig. 2 Numerical solution of the KdV equation for $d = 10^{-1.9}$. Timeslice plot over two space periods. Four specific wave profiles are plotted in red (cf. Fig. 6).

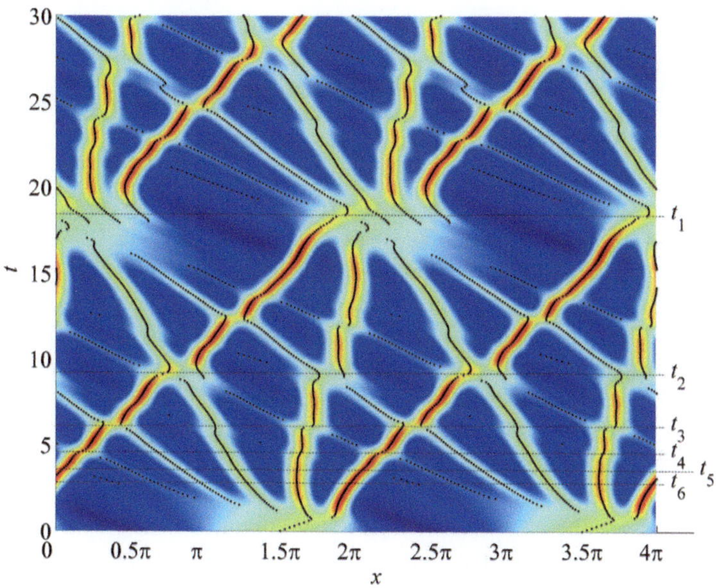

Fig. 3 Numerical solution of the KdV equation for $d = 10^{-1.9}$. Pseudocolor plot over two space periods. Black dots correspond to trajectories of wavecrests and time moments t_k to local maxima of spectral amplitudes (see Fig. 5).

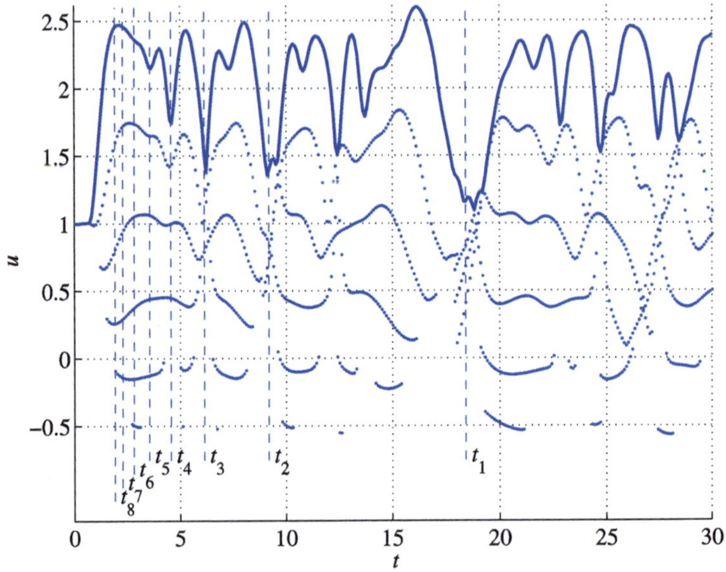

Fig. 4 Numerical solution of the KdV equation for $d = 10^{-1.9}$: wave profile maxima against time. Time moments t_k correspond to local maxima of spectral amplitudes (see Fig. 5).

In Fig. 5, the spectral amplitudes (54) are plotted against time. Making use of the spectral amplitude curves, the time moments $t_1 \dots t_8$ are introduced: t_1 corresponds to the time moment, when the initial spectral state is almost restored, i.e. $a_1 \to 1$ and the other $a_k \to 0$, $(k > 1)$; the time moments t_k $(k > 1)$ correspond to the maximum value of the k-th spectral amplitude a_k in time interval $0 < t < t_{k-1}$. For $d = 10^{-1.9}$, the time moments are $t_1 = 18.40$, $t_2 = 9.18$, $t_3 = 6.11$, $t_4 = 4.55$, $t_5 = 3.54$, $t_6 = 2.79$, $t_7 = 2.25$ and $t_8 = 1.88$. As a rule, these time moments, i.e. t_k for $k > 1$, correspond to global maxima of the k-th spectral amplitude in the time interval $0 < t < t_1$.

In Fig. 6, four specific wave profiles are plotted against the space coordinate x. One can see that at $t = 3$, a train of six solitons (i.e. six solitons per space period) is formed from the initial sine-wave. Therefore one can assume that the number of solitons is six for a value of the dispersion parameter of $d = 10^{-1.9}$. This assumption seems to be verified by Figs. 2–4, where one can trace six wave profile maxima (per one space period). However, the wave profile at $t = 19.9$ demonstrates that there exists a seventh soliton. Unfortunately this soliton is so small that it does not separate from the ensemble of interacting solitons to such an extent that the corresponding wave profile maximum could be detected from the wave profiles.

One question arises immediately: *what is the actual number of solitons in the present case*? For this reason we examine how the interaction process is reflected by spectral amplitudes (or spectral densities) in the time interval $0 \le t \le t_1$. It is clear from Fig. 3 that near t_1, the trajectories of all emerged solitons intersect and therefore the most complicated interaction takes place near t_1. Near t_2, interactions between odd number solitons and between even number solitons take place. Near

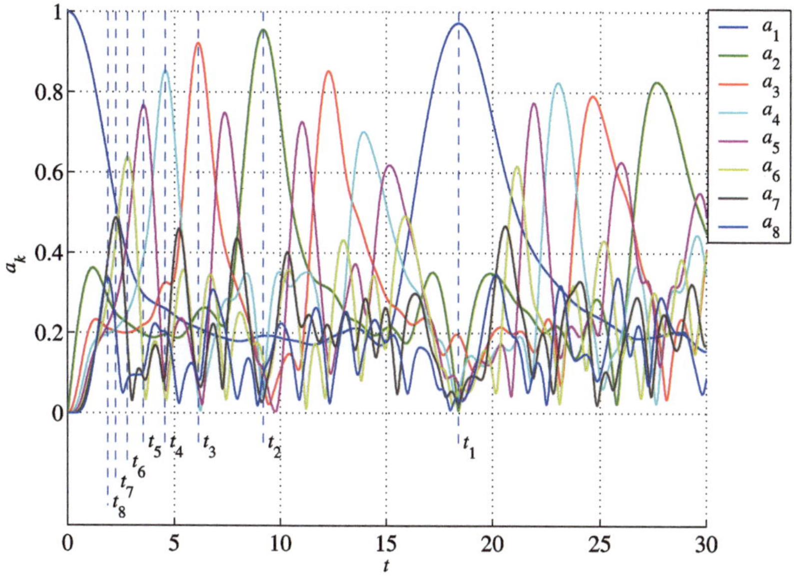

Fig. 5 Numerical solution of the KdV equation for $d = 10^{-1.9}$: spectral amplitudes $a_1 \ldots a_8$ against time.

t_3, the first soliton interacts with the fourth soliton, the second soliton with the fifth soliton, etc. Near t_4, the first soliton interacts with the fifth soliton, the second soliton with the sixth soliton, etc. In conclusion, if n solitons appear from the initial sine wave, then the first and the n-th soliton interact for the first time near the time moment t_{n-1}; the second and the n-th soliton interact for the first time near the time moment t_{n-2}; etc. Now we return to the amplitude curves in Fig. 4. It is characteristic for KdV solitons that their amplitudes are altered during interaction. The alteration process is more clearly reflected in the behaviour of the amplitude of higher solitons: if two solitons are interacting then the amplitude of the (higher) soliton decreases, and if the solitons are separating then the amplitude increases (see [75, 77] for a more detailed description of the interaction process). At the initial moment $t = 0$, all n solitons "are together" and the initial sine wave can be considered as a "peak" of ideal interaction between n solitons. For $t > 0$, the solitons start to separate: local maxima emerge, a train of solitons forms and amplitudes of (higher) solitons increase until new interactions start. In the amplitude curves (of higher solitons), each interaction is reflected by a local minimum or concave region. From Figs. 4 and 7, one can conclude that in the time interval $0 \leq t \leq 5$ the amplitude of the first soliton has a maximum between t_7 and t_8, the amplitude of the second soliton between t_6 and t_7, etc. The amplitude curve of the first soliton has a concave region near t_6 and local minimum near t_5, which correspond to its interactions with the seventh and the sixth soliton, respectively. The amplitude curve of the second soliton has a concave region near t_5 and a local minimum near t_4, which correspond to its interactions

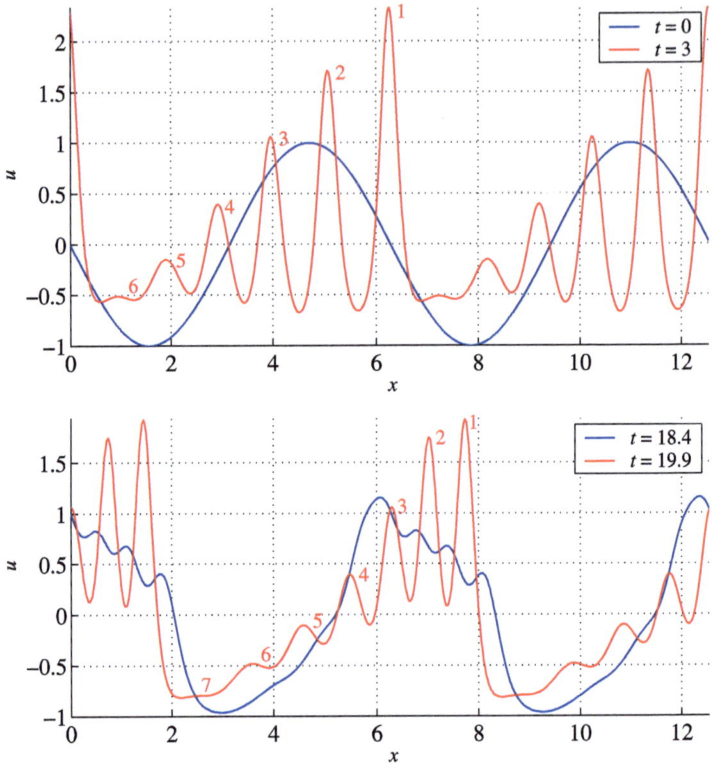

Fig. 6 Numerical solution of the KdV equation for $d = 10^{-1.9}$. Initial harmonic wave and train of six solitons at $t = 3$ (upper panel); wave profile at recurrence time $t = 18.4$ and train of seven solitons at $t = 19.9$ (lower panel) over two space periods.

with the seventh and the sixth soliton, respectively. Therefore we conclude that in the present case the actual number of solitons is seven (but neither six nor eight).

Based on the analysis above, one can distinguish *visible and hidden solitons* in a soliton train (see also our papers [22, 23, 71, 81, 83]). In our earlier papers [23, 71, 77, 83] we used the term 'virtual soliton' instead of 'hidden soliton'. In other words, besides clearly visible solitons, which interact with each other, there exist also solitons that either are visible only for a short time due to the fluctuation of the reference level or can be detected only by their influence on other solitons, i.e., by specific changes in the amplitude curves and in the soliton trajectories. The number of these hidden solitons over the large range of dispersion parameters is determined by making use of the time dependencies of soliton amplitudes, trajectories, and spectral amplitudes (or densities), as presented in [15, 81, 83]. The following concept has been used: (i) hidden solitons can emerge from harmonic excitations and have the same physical background as visible solitons; (ii) hidden solitons can be detected in wave profiles for a short-time interval only when several soliton interac-

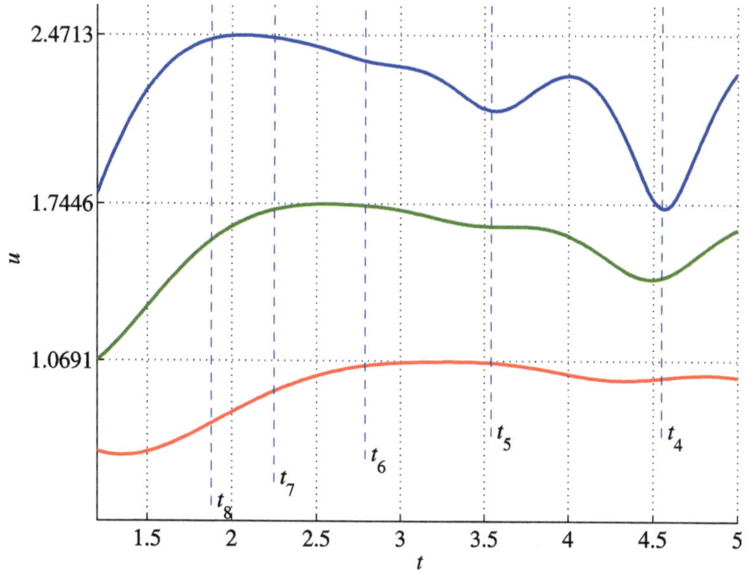

Fig. 7 Numerical solution of the KdV equation for $d = 10^{-1.9}$. First tree wave profile maxima against time for $1.2 \leq t \leq 5$.

tions have taken place and the reference state is fluctuating; and (iii) hidden solitons cause changes, specific to the soliton-type interaction, in the amplitudes and trajectories of the other solitons interacting with them. The existence of hidden (virtual) solitons is discussed in the papers [22, 23, 71, 72, 75, 76, 77, 80, 83] for the case of the KdV equation, and in the papers [43, 44] for the KdV435 equation.

4.3.2 Recurrence, super-recurrence, periodicity and spectrum

The concept of *recurrence* and *recurrence time* was first introduced by Zabusky and Kruskal in their celebrated paper [92], where the soliton concept was introduced. In the case of harmonic initial conditions recurrence means that solitons in an emerging train (propagating with their own speed and consequently interacting with each other) arrive at a certain time t_R in such a phase that they (almost) reconstruct the initial harmonic state.

In order to estimate the recurrence time one can use the spectral amplitudes (or spectral densities — their maxima and minima coincide): if $t \to t_R$, then $a_1(t) \to 1$ and $a_j(t) \to 0$, $j > 1$. It is clear that if the recurrence is repeated at $t_{R_k} \approx kt_R \equiv kt_{R_1}$, $k = 2, 3, \ldots$, then the solution can be called *(quasi)periodic*. The recurrence at t_{R_k} is called *super-recurrence* if $a_1(t_{R_k}) > a_1(t_{R_1})$. We have analysed the recurrence and the super-recurrence phenomena for the KdV equation (over a wide range of the dispersion parameter d and very long time intervals) in the papers [22, 76, 77] and for the KdV435 and the fKdV equations (over a wide range of dispersion parameters d and b) in [44].

In the KdV example considered in the previous section, the time moment $t_1 = 18.4$ corresponds to the recurrence time t_R (see Fig. 5). Corresponding wave profiles are presented in Figs. 2 and 6. It is clear that in the present case the (first) recurrence is incomplete. In [76] we have shown that for the considered value of the dispersion parameter d the next recurrences are even more incomplete and super-recurrence cannot be detected.

As a positive example (in the sense of super-recurrence), we demonstrate here the behaviour of the spectral amplitudes in the case of the fKdV equation (6). The initial value problem is solved under harmonic initial conditions for $d = 10^{-0.8}$, $b = 10^{-4.4}$, $0 \leq t \leq 1000$, and 206 recurrences were detected. In Fig. 8, the values of the first spectral amplitude at recurrence times t_{R_k} are plotted against the number of recurrences. One can see that in the present case the first recurrence belongs to the set of "the most incomplete recurrences" and in numerous cases the first spectral amplitude has values $0.999 < a_1(t_{R_k}) < 1$. In Fig. 9, the time intervals between succeeding recurrence times t_{R_i} and $t_{R_{i-1}}$ are plotted against the number of recurrences. The lengths of these time intervals Δ_R^i are not constant, but fluctuate between values of 4.76 and 4.94 in the present case. Nevertheless, one can say that the solution is (quasi)periodic in time — the maximal deviation of Δ_R^i from the mean value 4.8457 is less than 2%.

If the initial conditions are arbitrary, then the initial values of the spectral amplitudes are $a_i(0) \neq 0$, $i = 2,3,\ldots$, and the estimation of the recurrence time is not so easy as in the case of harmonic initial conditions. In order to determine the recurrence time in this case, one has to find the time moments when $a_i(t) \to a_i(0)$, $i = 1,2,3,\ldots$, simultaneously.

4.3.3 Multiple solitons and cumulative spectrum

In this subsection we present an example of the use of the cumulative spectrum. In the paper [82] numerical solutions of the KdV435 equation (5) under harmonic initial conditions are analysed for $1 \leq d \leq 10^{-2.4}$, $10^{-0.6} \leq b \leq 10^{-4.4}$. Four solution types are introduced: (i) a train of (positive) solitons, (ii) a train of negative solitons, (iii) a multiple soliton (multisoliton) solution, and (iv) a chaotic solution. A multiple soliton solution means that for this type an ensemble of n-solitons forms and all solitons in this ensemble are propagating with nearly the same speed. In Figs. 10 and 11, an example is presented where the number of solitons in the ensemble is five. In Fig. 12, the corresponding cumulative spectrum is plotted. One can see that there exists a jump over several orders of values between the cumulative spectrum measures $S_C(5,t)$ and $S_C(6,t)$. Our analysis has demonstrated that if such an ensemble of n-solitons forms, then one can detect a jump over several orders of values between the quantities $S_C(n,t)$ and $S_C(n+1,t)$. For other solution types such jumps do not exist. Therefore the cumulative spectrum can be used in order to determine

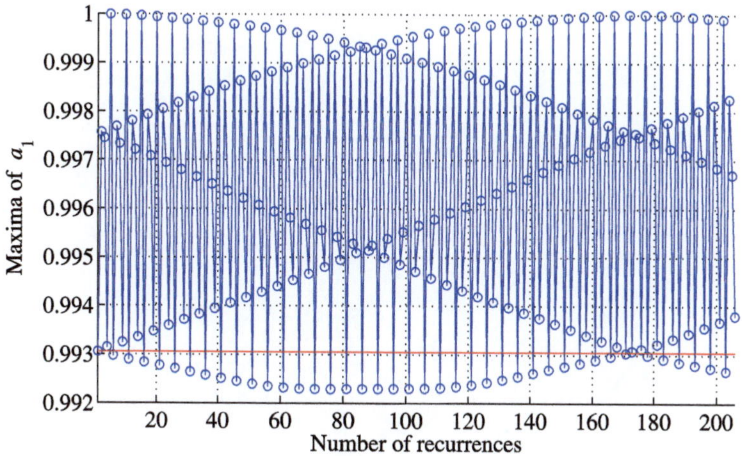

Fig. 8 Numerical solution of the fKdV equation for $d = 10^{-0.8}$ and $b = 10^{-4.4}$. Values of the first spectral amplitude a_1 at recurrence times t_{R_i} against number of recurrences. Red horizontal line corresponds to the value of a_1 at the first recurrence $t = t_{R_1}$.

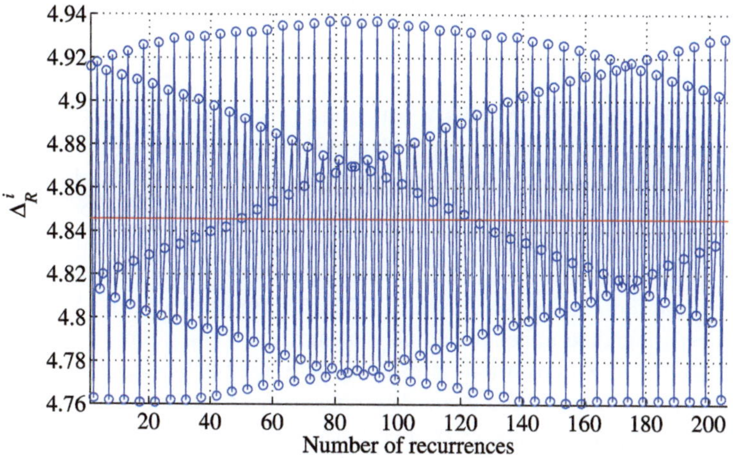

Fig. 9 Numerical solution of the fKdV equation for $d = 10^{-0.8}$ and $b = 10^{-4.4}$. Time intervals between succeeding recurrence times $\Delta_R^i = t_{R_i} - t_{R_{i-1}}$ against number of recurrences. Red horizontal line corresponds to the mean value of these time intervals over 206 recurrences.

the borders between the multiple soliton solution and the other three solution types in the dispersion parameter plane $d - b$.

4.3.4 Wave packets and time averaged spectral densities.

In this subsection TANSD (67) are applied for analysing the formation of wave packets. In order to study the propagation and interaction of solitary waves in di-

Fig. 10 Numerical solution of the KdV435 equation for $d = 10^{-0.8}$ and $b = 10^{-2.2}$. Timeslice plot over two space periods.

latant granular materials, the HKdV equation (7) is solved under sech^2-type initial conditions. One of the solution types that we detected for this problem is a 'solitary wave with tail and wave packets' (see [38, 39, 40, 41, 42] for details). In the timeslice plot in Fig. 13, all three components of the solution can be traced. The corresponding TANSD curves in Fig. 14 demonstrate that in this case the quantities $S_a(54,t),\ldots,S_a(58,t)$ dominate over those of $k < 54$ and $k > 58$. It is clear that wave packets are formed by the amplified harmonics, and the $S_a(k,t)$ having the highest value determines the number of maxima (oscillations) in a given wave profile. Similar situations are described in many textbooks (see e.g. [4]) in order to explain group velocity and dispersion phenomena — a sum of harmonic waves having nearly equal frequencies results in a wave packet. For other solution types such a phenomenon that some higher order harmonics dominate over those of lower order cannot be found [38, 39, 40, 41]. Vice versa, if wave packets are not formed, then the lower order harmonics are dominating.

5 Conclusions

During the last decades the PsM has been widely employed as a powerful computational tool in different areas, e.g. wave propagation, fluid dynamics and weather forecasting. The main advantage of the PsM is its high efficiency and accuracy, i.e., compared with the FDM, much less computer resource is required in order to attain the required accuracy of computations. In addition, if one needs to filter the data

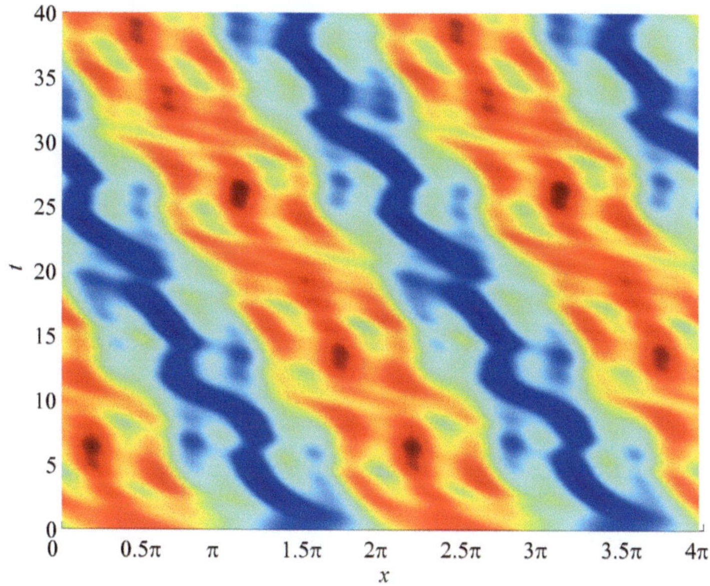

Fig. 11 Numerical solution of the KdV435 equation for $d = 10^{-0.8}$ and $b = 10^{-2.2}$. Pseudocolor plot over two space periods.

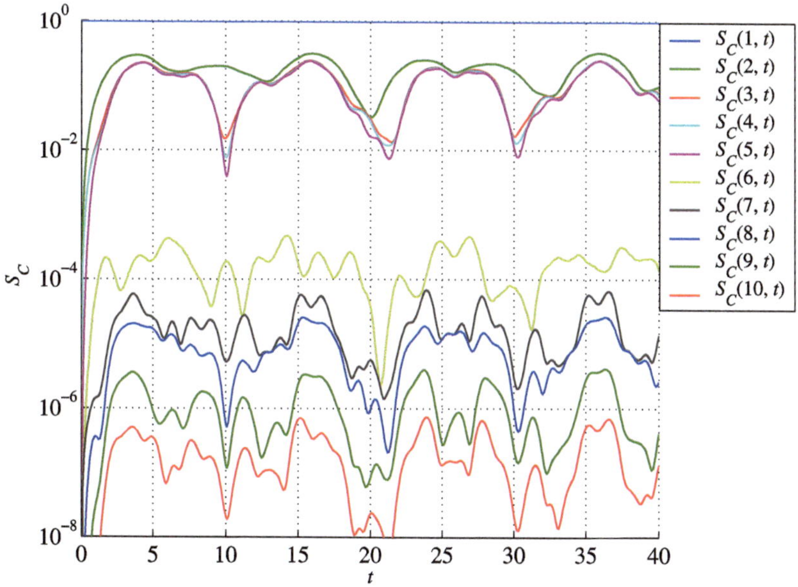

Fig. 12 Numerical solution of the KdV435 equation for $d = 10^{-0.8}$ and $b = 10^{-2.2}$. Cumulative spectrum $S_C(k,t)$, $k = 1, \ldots, 10$ against time.

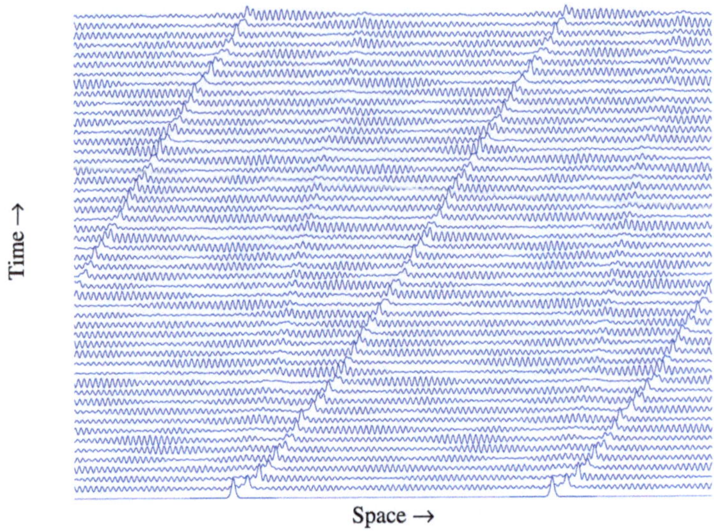

Fig. 13 Numerical solution of the HKdV equation for $\alpha_1 = 0.05$, $\alpha_2 = 0.11$, $\beta = 0.0111$. Timeslice plot over two space periods.

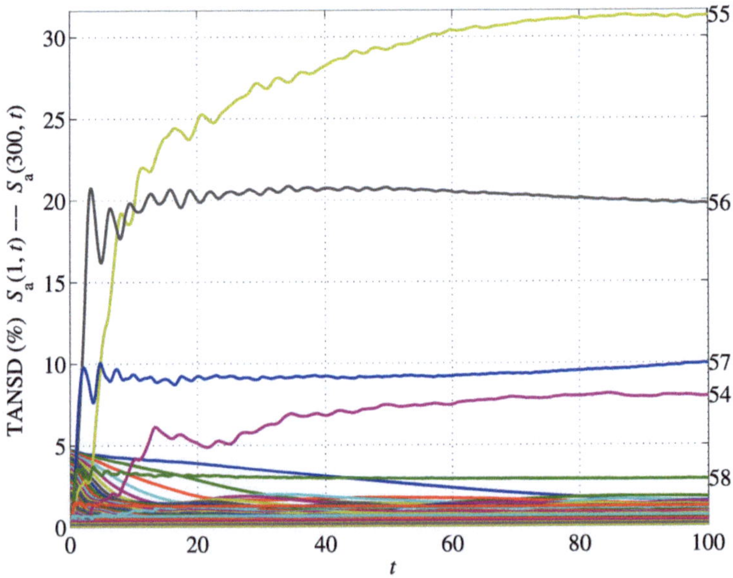

Fig. 14 Numerical solution of the HKdV equation for $\alpha_1 = 0.05$, $\alpha_2 = 0.11$, $\beta = 0.0111$. TANSD $S_a(k,t)$, $k = 1, \ldots 300$ against time.

during computations, then this is also very easily handled if the DFT-based PsM is used. In the present paper we demonstrated how the DFT-based PsM can be applied for the numerical solution of various nonlinear wave equations, as listed in section 2.

Compared to other numerical methods the DFT-based methods give additional information in the form of discrete spectra, which in turn can be used in DSA. DSA is a tool, which helps to understand the essence of the numerical solutions of nonlinear wave equations. In section 4.3 we demonstrated how the DSA can be applied for the determination of the actual number of solitons in the KdV soliton train, for finding the time moment when the KdV soliton train is formed, for distinguishing between different solution types in the case of the KdV435 equation, and for analysing the structure of wave packets for the HKdV equation and recurrence and super-recurrence phenomena. Additional examples can be found in [66], where the DSA is applied to determine the solution types of the FKdV equation; in [83], where it is demonstrated how the cumulative spectrum can be used for the estimation of the number of solitons, and in several other papers by the author.

In subsection 3.1 we stated that it is often useful to choose trigonometric functions as the basis $\Phi_k(x)$ in expression (14). However, in several cases another choice, e.g. Chebyshev polynomials, can be convenient. For corresponding examples one should consult the textbook by B. Fornberg [24].

Acknowledgements The research was supported by the FP6 EU Marie Curie Transfer of Knowledge project MTKD-CT-2004-013909 and by the Estonian Science Foundation Grant no. 7035. The author thanks Professors J. Engelbrecht and G.A. Maugin for fruitful discussions, Dr. P. Peterson for Python-related scientific software development and assistance, former and present students and colleagues Dr. O. Ilison, Dr. L. Ilison, MSc K. Tamm, MSc M. Randrüüt, MSc M. Sepp, MSc K. Veski, BSc M. Kukk and BSc M. Vallikivi for carrying out an enormous number of numerical experiments.

References

1. Bao, W., Chern, I.L., Lim, F.Y.: Efficient and spectrally accurate numerical methods for computing ground and first excited states in Bose–Einstein condensates. J. Comput. Phys. **219**(2), 836–854 (2006)
2. Bao, W., Lim, F.Y.: Computing ground states of spin-1 Bose–Einstein condensates by the normalized gradient flow. SIAM J. Sci. Comput. **30**(4), 1925–1948 (2008)
3. Bao, W., Yang, L.: Efficient and accurate numerical methods for the Klein–Gordon–Schrödinger equations. J. Comput. Phys. **225**(2), 1863–1893 (2007)
4. Billingham, J., King, A.C.: Wave Motion. Cambridge University Press (2000)
5. Bountis, T., Starmer, C.F., Bezerianos, A.: Stationary pulses and wave front formation in an excitable medium. Progr. Theoret. Phys. Suppl. **139**, 12–33 (2000)
6. Boyd, J.P.: Chebyshev and Fourier Spectral Methods. Dover, New York (2000)
7. Boyd, J.P.: Fourier pseudospectral method with Kepler mapping for travelling waves with discontinuous slope: Application to corner waves of the Ostrovsky–Hunter equation and equatorial Kelvin waves in the four-mode approximation. Appl. Math. Comput. **177**(1), 289–299 (2006)
8. Bracewell, R.N.: The Fast Fourier Transform and its Applications. McGraw-Hill, New York (1978)
9. Cabral, M., Rosa, R.: Chaos for a damped and forced KdV equation. Physica D **192**, 265–278 (2004)
10. Chen, J.B.: A multisymplectic pseudospectral method for seismic modeling. Appl. Math. Comput. **186**(2), 1612–1616 (2007)

11. Dai, H.H., Huo, Y.: Solitary shock waves and other travelling waves in a general compressible hyperelastic rod. Proc. R. Soc. Lond. A **456**, 331–363 (2000)
12. Darvishi, M.T.: Spectral collocation solution of a generalized Hirota–Satsuma coupled KdV equation. Int. J. Comput. Math. **84**(4), 541–551 (2007)
13. Drazin, P.G., Johnson, R.S.: Solitons: an Introduction. Cambridge University Press (1989)
14. Engelbrecht, J.: An Introduction to Asymmetric Solitary Waves. Longman, Harlow (1991)
15. Engelbrecht, J.: Nonlinear Wave Dynamics: Complexity and Simplicity. Kluwer, Dordrecht (1997)
16. Engelbrecht, J., Berezovski, A., Pastrone, F., Braun, M.: Waves in microstructured materials and dispersion. Phil. Mag. **85**, 4127–4141 (2005)
17. Engelbrecht, J., Berezovski, A., Salupere, A.: Nonlinear deformation waves in solids and dispersion. Wave Motion **44**(6), 493–500 (2007)
18. Engelbrecht, J., Khamidullin, Y.: On the possible amplification of nonlinear seismic waves. Phys. Earth Planet. Inter. **50**, 39–45 (1988)
19. Engelbrecht, J., Pastrone, F.: Waves in microstructured solids with strong nonlinearities in microscale. Proc. Estonian Acad. Sci. Phys. Math. **52**(1), 12–20 (2003)
20. Engelbrecht, J., Peipman, T.: Nonlinear waves in a layer with energy influx. Wave Motion **16**(2), 173–181 (1992)
21. Randrüüt, M., Salupere, A., Engelbrecht, J.: On modelling wave motion in microstructured solids. Proc. Estonian Acad. Sci. **58**(4), (2009)
22. Engelbrecht, J., Salupere, A.: On the problem of periodicity and hidden solitons for the KdV model. Chaos **15**, 015114 (2005)
23. Engelbrecht, J., Salupere, A., Kalda, J., Maugin, G.A.: Discrete spectral analysis for solitary waves. In: Guran, A., Maugin, G.A., Engelbrecht, J., Werby, M. (eds.) Acoustic Interactions with Submerged Elastic Structures. Part II: Propagation, Ocean Acoustics and Scattering, Series on Stability, Vibration and Control of Systems, Series B, Vol. 5, pp. 1–40. World Scientific, Singapore (2001)
24. Fornberg, B.: A Practical Guide to Pseudospectral Methods. Cambridge University Press (1998)
25. Fornberg, B.: A pseudospectral fictitious point method for high order initial-boundary value problems. SIAM J. Sci. Comput. **28**(5), 1716–1729 (2006)
26. Fornberg, B., Sloan, C.M.: A review of pseudospectral methods for solving partial differential equations. Acta Numerica **3**, 203–267 (1994)
27. Fornberg, B., Whitham, G.B.: A numerical and theoretical study of certain nonlinear wave phenomena. Phil. Trans. Roy. Soc. A **289**, 373–404 (1978)
28. Frigo, M., Johnson, S.G.: The design and implementation of FFTW3. Proc. IEEE **93**(2), 216–231 (2005)
29. Galgani, L., Giorgilli, A., Martinoli, A., Vanzini, S.: On the problem of energy equipartition for large systems of the Fermi–Pasta–Ulam type: analytical and numerical estimates. Physica D **59**, 334–348 (1992)
30. Gallego, R., Castro, M., López, J.M.: Pseudospectral versus finite-difference schemes in the numerical integration of stochastic models of surface growth. Phys. Rev. E **76**(5), 51121 (2007)
31. Giovine, P., Oliveri, F.: Dynamics and wave propagation in dilatant granular materials. Meccanica **30**(4), 341–357 (1995)
32. Goursolle, T., Callé, S., Dos Santos, S., Matar, O.B.: A two-dimensional pseudospectral model for time reversal and nonlinear elastic wave spectroscopy. J. Acoust. Soc. Am. **122**, 3220 (2007)
33. Guo, J., Taha, T.R.: Parallel implementation of the split-step and the pseudospectral methods for solving higher KdV equation. Math. Comput. Simulat. **62**(1-2), 41–51 (2003)
34. Guyenne, P., Nicholls, D.P.: A high-order spectral method for nonlinear water waves over moving bottom topography. SIAM J. Sci. Comput. **30**(1), 81–101 (2007)
35. Hindmarsh, A.C.: ODEPACK, a systematized collection of ODE solvers. In: Stepleman, R.S., et al. (eds.) Scientific Computing: Applications of Mathematics and Computing to the Physical Sciences, IMACS Trans. Sci. Comput., Vol. 1, pp. 55–64. North-Holland, Amsterdam (1983)

36. Ilison, L., Salupere, A.: On solitons in dilatant granular materials. In: Lund, E., Olhoff, N., Stegman, J. (eds.) Proceedings 15th Nordic Seminar on Computational Mechanics 2002, pp. 181–184. Institute of Mechanical Enginnering, Aalborg University (2002)
37. Ilison, L., Salupere, A.: Solitons in hierarchical Korteweg–de Vries type systems. Proc. Estonian Acad. Sci. Phys. Math. **52**(1), 135–144 (2003)
38. Ilison, L., Salupere, A.: Propagation of localised perturbations in granular materials. Research Report Mech 287/07, Institute of Cybernetics at Tallinn University of Technology (2007)
39. Ilison, L., Salupere, A.: Interactions of solitary waves in hierarchical KdV-type system. Research Report Mech 291/08, Institute of Cybernetics at Tallinn University of Technology (2008)
40. Ilison, L., Salupere, A.: Propagation of sech^2-type solitary waves in hierarchical KdV-type systems. Math. Comput. Simulat. **79**, 3314–3327 (2009)
41. Ilison, L., Salupere, A.: Numerical simulation of interaction of solitons and solitary waves in hierarchical KdV-type systems. Commun. Nonlinear Sci. Numer. Simulat. (2009) (submitted)
42. Ilison, L., Salupere, A., Peterson, P.: On the propagation of localised perturbations in media with microstructure. Proc. Estonian Acad. Sci. Phys. Math. **56**(2), 84–92 (2007)
43. Ilison, O., Salupere, A. On propagation of solitons in media with higher order dispersion. In: Lund, E., Olhoff, N., Stegman, J. (eds.) Proceedings 15th Nordic Seminar on Computational Mechanics 2002, pp. 177–180. Institute of Mechanical Enginnering, Aalborg University (2002)
44. Ilison, O., Salupere, A.: On the formation of solitons in media with higher-order dispersive effects. Proc. Estonian Acad. Sci. Phys. Math. **52**(1), 135–144 (2003)
45. Ilison, O., Salupere, A.: On the propagation of solitary waves in microstructured solids. In: Gutkowski, W., Kowalewski, T.A. (eds.) 21st International Congress of Theoretical and Applied Mechanics 2004, ICTAM04 CD-ROM Proceedings. IPPT PAN, Warszawa (2004)
46. Ilison, O., Salupere, A.: Propagation of sech^2-type solitary waves in higher order KdV-type systems. Chaos Soliton.Fract. **26**, 453–465 (2005)
47. Ilison, O., Salupere, A.: On the propagation of solitary pulses in microstructured materials. Chaos Soliton. Fract. **29**, 202–214 (2006)
48. Janno, J., Engelbrecht, J.: An inverse solitary wave problem related to microstructured materials. Inverse Problems **21**, 2019–2034 (2005)
49. Janno, J., Engelbrecht, J.: Solitary waves in nonlinear microstructured materials. J. Phys. A: Math. Gen. **38**, 5159–5172 (2005)
50. Jones, E., Oliphant, T., Peterson, P., et al.: SciPy: open source scientific tools for Python (2007). http://www.scipy.org
51. Keetels, G.H., D'Ortona, U., Kramer, W., Clercx, H.J.H., Schneider, K., Heijst, van G.J.F.: Fourier spectral and wavelet solvers for the incompressible Navier–Stokes equations with volume-penalization: Convergence of a dipole–wall collision. J. Comput. Phys. **227**(2), 919–945 (2007)
52. Kliakhandler, I.L., Porubov, A.V., Velarde, M.G.: Localized finite-amplitude disturbances and selection of solitary waves. Phys. Rev. E **62**(4), 4959–4962 (2000)
53. Kreiss, H.O., Oliger, J.: Comparison of accurate methods for the integration of hyperbolic equations. Tellus **24**, 199–215 (1972)
54. Lee, J.: Free vibration analysis of cylindrical helical springs by the pseudospectral method. J. Sound Vibration **302**(1-2), 185–196 (2007)
55. Lee, J.: Free vibration analysis of non-cylindrical helical springs by the pseudospectral method. J. Sound Vibration **305**(3), 543–551 (2007)
56. Lo, J., Shizgal, B.D.: Spectral convergence of the quadrature discretization method in the solution of the Schrödinger and Fokker–Planck equations: Comparison with sinc methods. J. Chem. Phys. **125**, 194108 (2006)
57. Mead, J., Zubik-Kowal, B.: An iterated pseudospectral method for delay partial differential equations. Appl. Numer. Math. **55**(2), 227–250 (2005)
58. Mindlin, R.D.: Micro-structure in linear elasticity. Arch. Rat. Mech. Anal. **16**, 51–78 (1964)
59. Miura, R.M.: The Korteweg-de Vries equation: a survey of results. SIAM Rev. **18**(3), 412–459 (1976)

60. Miura, R.M., Gardner, C.S., Kruskal, M.D.: Korteweg-de Vries equation and generalizations II. Existence of conservation laws and constants of motion. J. Math. Phys. **9**, 1204–1209 (1968)
61. Olmos, D., Shizgal, B.D.: A pseudospectral method of solution of Fisher's equation. J. Comput. Appl. Math. **193**(1), 219–242 (2006)
62. Oran Brigham, E.: The Fast Fourier Transform and Its Applications. Prentice Hall, London (1988)
63. Orszag, S.A.: Comparison of pseudospectral and spectral approximation. Stud. Appl. Math. **51**, 253–259 (1972)
64. Pelinovsky, E., Sergeeva, A.: Numerical modeling of the KdV random wave field. Eur. J. Mech. B Fluids **25**(4), 425–434 (2006)
65. Peterson, P.: F2PY: Fortran to Python interface generator (2005). http://cens.ioc.ee/projects/f2py2e/
66. Peterson, P., Salupere, A.: Solitons in a perturbed Korteweg-de Vries system. Proc. Estonian Acad. Sci. Phys. Math. **46**(1-2), 102–110 (1997)
67. Porubov, A.V., Maugin, G.A., Gursky, V.V., Krzhizhanovskaya, V.V.: On some localized waves described by the extended KdV equation. C.R. Mecanique **333**(7), 528–533 (2005)
68. Rashid, A.: Convergence analysis of three-level Fourier pseudospectral method for Korteweg–de Vries Burgers equation. Comput. Math. Appl. **52**(5), 769–778 (2006)
69. Salupere, A.: On the application of the pseudospectral method for solving the Korteweg-de Vries equation. Proc. Estonian Acad. Sci. Phys. Math. **44**(1), 73–87 (1995)
70. Salupere, A.: On the application of pseudospectral methods for solving nonlinear evolution equations, and discrete spectral analysis. In: Proceedings of 10th Nordic Seminar on Computational Mechanics, pp. 76–83. Tallinn Technical University, Tallinn (1997)
71. Salupere, A.: Technique for detection of solitons in numerical experiments and virtual soliton concept. In: Oñate, E., Bugeda, G., Suárez, G. (eds.) CD-ROM Proceedings of the European Congress on Computational Methods in Applied Sciences and Engineering – ECCOMAS 2000, Barcelona (2000).
72. Salupere, A., Engelbrecht, J. Hidden and driven solitons in microstructured media. In: Gutkowski, W. Kowalewski, T.A. (eds.) 21st International Congress of Theoretical and Applied Mechanics ICTAM04 CD-ROM Proceedings. IPPT PAN, Warszawa (2004)
73. Salupere, A., Engelbrecht, J., Ilison, O., Ilison, L.: On solitons in microstructured solids and granular materials. Math. Comput. Simulat. **69**, 502–513 (2005)
74. Salupere, A., Engelbrecht, J., Maugin, G.A.: Solitonic structures in KdV-based higher-order systems. Wave Motion **34**, 51–61 (2001)
75. Salupere, A., Engelbrecht, J., Peterson, P.: Long-time behaviour of soliton ensembles. Part I–Emergence of ensembles. Chaos Soliton. Fract. **14**, 1413–1424 (2002)
76. Salupere, A., Engelbrecht, J., Peterson, P.: Long-time behaviour of soliton ensembles. Part II–Periodical patterns of trajectories. Chaos Soliton. Fract. **15**, 29–40 (2003)
77. Salupere, A., Engelbrecht, J., Peterson, P.: On the long-time behaviour of soliton ensembles. Math. Comput. Simulat. **62**, 137–147 (2003)
78. Salupere, A., Ilison, O.: On the numerical determination of solitary waves for systems with quartic potential and higher order dispersion. In: Eriksson, A., Pacoste, C. (eds.) Proceedings of the NSCM-11: Nordic Seminar on Computational Mechanics. TRITA-BKN. Bulletin 39, pp. 106–109. KTH, Stockholm (1998)
79. Salupere, A., Ilison, O.: On solitonic structures in microstructured materials. In: Proceedings 13th Nordic Seminar on Computational Mechanics (NSCM-13) 2000, pp. 70–73. Oslo (2000)
80. Salupere, A., Kukk, M.: Periodically forced solitonic structures in dispersive media. Proc. Estonian Acad. Sci. Phys. Math. **52**(1), 145–156 (2003)
81. Salupere, A., Maugin, G.A., Engelbrecht, J.: Korteweg–de Vries soliton detection from a harmonic input. Phys. Lett. A **192**(1), 5–8 (1994)
82. Salupere, A., Maugin, G.A., Engelbrecht, J.: Solitons in systems with a quartic potential and higher-order dispersion. Proc. Estonian Acad. Sci. Phys. Math. **46**(1–2), 118–127 (1997)
83. Salupere, A., Maugin, G.A., Engelbrecht, J., Kalda, J.: On the KdV soliton formation and discrete spectral analysis. Wave Motion **23**(1), 49–66 (1996)

84. Salupere, A., Randrüüt, M., Tamm, K.: Emergence of soliton trains in microstructured materials. In: Denier, J., Finn, M., Mattner, T. (eds.) XXII International Congress of Theoretical and Applied Mechanics ICTAM 2008, CD-ROM Proceedings. The University of Adelaide, Adelaide (2008). Paper id 11311
85. Salupere, A., Tamm, K., Engelbrecht, J.: Numerical simulation of interaction of solitary deformation waves in microstructured solids. Int. J. Non-linear Mech. **43**, 201–208 (2008)
86. Salupere, A., Tamm, K., Engelbrecht, J., Peterson, P.: On interaction of deformation waves in microstructured solids. Proc. Estonian Acad. Sci. Phys. Math. **56**(2), 93–99 (2007)
87. Sarra, S.A.: A Pseudospectral Method with Edge Detection-Free Postprocessing for Two-Dimensional Hyperbolic Heat Transfer. Numer. Heat Transfer, Part B **54**(1), 52–61 (2008)
88. Vallikivi, M., Salupere, A., Dai, H.H.: Numerical simulation of propagation of solitary deformation waves in a compressible hyperelastic rod. Research Report Mech 293/08, Institute of Cybernetics at Tallinn University of Technology (2008)
89. Veski, K., Engelbrecht, J., Bountis, T.: Emergence and propagation of nerve pulses: parameter analysis and response to forcing. Chaos Soliton.Fract. (2008) (submitted)
90. Wang, T., Guo, B.: Composite generalized Laguerre–Legendre pseudospectral method for Fokker–Planck equation in an infinite channel. Appl. Numer. Math. **58**(10), 1448–1466 (2008)
91. Yin, Z., Clercx, H.J.H., Montgomery, D.C.: An easily implemented task-based parallel scheme for the Fourier pseudospectral solver applied to 2D Navier–Stokes turbulence. Comput. Fluids **33**(4), 509–520 (2004)
92. Zabusky, N.J., Kruskal, M.D.: Interaction of solitons in a collisionless plasma and the recurrence of initial states. Phys. Rev. Lett. **15**, 240–243 (1965)

Foundations of Finite Element Methods for Wave Equations of Maxwell Type

Snorre H. Christiansen

Abstract The first part of this paper is an overview of the theory of approximation of wave equations by Galerkin methods. It treats convergence theory for linear second order evolution equations and includes studies of consistency and eigenvalue approximation. We emphasize differential operators, such as the curl, which have large kernels, and use L^2-stable interpolators preserving them. The second part provides a setting for the construction of finite element spaces of differential forms on cellular complexes. Material on homological and tensor algebra as well as differential and discrete geometry is included. Whitney forms, their duals, their high order versions, their tensor products and their *hp*-versions all fit into this framework.

1 Introduction

The Yee scheme [62] is a very efficient finite difference scheme for simulating the initial/boundary value problem for Maxwell's equations. It is used in many industrial codes for problems ranging from antenna design to electromagnetic compatibility and medical imaging. However, it is only second order accurate, it treats boundary conditions in a rough way (stair-casing) and it is not clear how it should be formulated for anisotropic materials. All three problems can be addressed in the finite element framework. For electromagnetics, mixed finite elements [15, 49, 50, 53] have been found to give the best results.

Finite elements come with a natural way of deriving error estimates. The stability of the method is linked to energy conservation, which is almost automatic for variational Galerkin methods. Charge conservation is also ensured weakly by these methods. Through stability, the convergence of the method is reduced to the prob-

Centre of Mathematics for Applications, University of Oslo, PO Box 1053 Blindern, NO-0316 Oslo, Norway, e-mail: snorrec@math.uio.no

E. Quak, T. Soomere (eds.), *Applied Wave Mathematics*,
DOI 10.1007/978-3-642-00585-5_17,

lem of estimating errors of best approximation on the finite element space. These are usually obtained from interpolation operators.

The main drawback of the finite element method is that the mass matrix is not diagonal, leading to (linearly) implicit methods. In some circumstances a procedure known as mass-lumping solves this problem. In particular, one can recover the Yee scheme in this way.

One of the reasons for the success of the Yee scheme and mixed finite elements is that they respect the geometry of Maxwell's equations [12, 59]. The matrices of the differential operators *grad*, *curl* and *div* in the standard basis of lowest order mixed finite elements are the incidence matrices of algebraic topology. Mixed finite elements are equipped with interpolation operators, which behave well with respect to these differential operators: they form commuting diagrams linking continuous and discrete complexes of spaces.

The present paper is divided into two parts. The first part is devoted to the convergence analysis of the Galerkin method for waves, assuming just some basic properties of the discretization spaces. We start with the variational (weak) formulation of wave equations, adopting a general framework that covers scalar waves as well as electromagnetic waves. Convergence is proved using energy estimates, with an emphasis on rough data. Then we study consistency, which is the theoretical foundation for mass-lumping. Finally we treat the eigenvalue problem, basing the theory on the existence of L^2-stable interpolators vanishing on the kernel of the bilinear form.

The second part is devoted to the construction of finite element spaces. We first review some tools from algebraic topology, such as the long exact sequence and the five lemma, and tensor algebra including the Kunneth theorem. Then we introduce differential forms that enable one to treat the *grad*, *curl* and *div* operators in a uniform manner, as well as boundary conditions for various types of fields. Finally we develop a framework for the construction and study of finite element spaces of differential forms on cellular complexes as in [23]. The dual of a simplicial complex is not a simplicial complex, nor is the product of two simplicial complexes. However, duals and products of cellular complexes are again cellular complexes, so that we gain some flexibility from this level of generality. Tensor products of high order Whitney forms, which correspond to mixed finite elements on products of simplexes, fit the framework as well as the dual finite elements introduced in [18]. Even *hp*-finite elements [35] and projection-based interpolation are accommodated.

A number of surveys on these topics have been published in recent years, in particular [5, 40, 42, 48]. The algebra can be found in textbooks, but is most often mixed with considerations on homology (e.g., singular) and modules. The claim to originality of the present study is the emphasis on rough data through L^2-stable interpolation and the introduction of general constructions on cellular complexes.

2 Analysis of the finite element method for waves

2.1 Linear wave equations

When dealing with a time dependent quantity u, we denote its first and second time derivatives by $\dot{u}$ and $\ddot{u}$, respectively. Let U be an open bounded domain in $\mathbb{R}^n$. The linear wave equation for a function $u : \mathbb{R} \times U \to \mathbb{R}$, with forcing term $f : \mathbb{R} \times U \to \mathbb{R}$ is:

$$\ddot{u} = \Delta u + f, \tag{1}$$

where Δ is the Laplace operator on U. We will consider homogeneous Dirichlet boundary conditions, namely $u(t,x) = 0$ for all $t \in \mathbb{R}$ and $x \in \partial U$. Initial conditions u_0 and $\dot{u}_0$ are supplied for $u(0)$ and $\dot{u}(0)$. Our point of view is that u_0, $\dot{u}_0$ and f constitute the data of the problem, whereas u is the unknown function we want to compute a reasonable approximation of. It is necessary to interpret both the wave equation and initial-value/boundary conditions in a weak sense (not pointwise).

The L^2-scalar product is denoted $\langle \cdot, \cdot \rangle$, both on functions and vector fields on U (assuming the standard scalar product on vectors in $\mathbb{R}^n$). Recall that for smooth functions u, v satisfying the Dirichlet boundary condition we have:

$$\langle \Delta u, v \rangle = -\langle \operatorname{grad} u, \operatorname{grad} v \rangle.$$

The energy of u at time t is:

$$\mathcal{E}(u)(t) = \frac{1}{2}\langle \dot{u}(t), \dot{u}(t) \rangle + \frac{1}{2}\langle \operatorname{grad} u(t), \operatorname{grad} u(t) \rangle.$$

When the forcing term f is 0, energy is conserved for smooth enough solutions. We always deal with finite energy solutions, that is, functions for which $\mathcal{E}(u)$ is essentially bounded.

The weak formulation of (1) in space is to require that $u(t) \in \mathrm{H}^1_0(U)$ and for all test-functions u':

$$\langle \ddot{u}(t), u' \rangle = -\langle \operatorname{grad} u(t), \operatorname{grad} u' \rangle + \langle f(t), u' \rangle. \tag{2}$$

It remains to give a meaning to $\ddot{u}(t)$ and the initial conditions. Given $u_0 \in \mathrm{H}^1_0(U)$ and $\dot{u}_0 \in \mathrm{L}^2(U)$, we find:

$$u \in C([0,T]; \mathrm{H}^1_0(U)) \cap C^1([0,T]; \mathrm{L}^2(U)),$$

satisfying the initial conditions and such that (2) holds in some sense. For instance, we can require that for all $u' \in C^2_0([0,T[; \mathrm{H}^1_0(U))$:

$$\int \langle u(t), \ddot{u}'(t) \rangle \mathrm{d}t + \langle \dot{u}_0, u'(0) \rangle - \langle u_0, \dot{u}'(0) \rangle =$$
$$- \int \langle \operatorname{grad} u(t), \operatorname{grad} u'(t) \rangle \mathrm{d}t + \int \langle f(t), u'(t) \rangle \mathrm{d}t.$$

When $f \in \mathrm{L}^1([0,T];\mathrm{L}^2(U))$, the problem has a unique solution. This can be proven using the Galerkin method, to which we now turn our attention.

More generally we consider the following situation. Let H be a Hilbert space with scalar product $\langle\cdot,\cdot\rangle$ and norm $|\cdot|$. Suppose H contains a Hilbert space X, which is dense and such that the inclusion $X \to H$ is continuous. The wave equation (1) corresponds to the choices $H = \mathrm{L}^2(U)$ equipped with the standard L^2-product and $X = \mathrm{H}^1_0(U)$. Suppose a is a continuous symmetric bilinear form on X such that for all $u \in X$:

$$a(u,u) \geq 0, \tag{3}$$

and:

$$\langle\cdot,\cdot\rangle + a(\cdot,\cdot) \text{ is coercive on } X.$$

The scalar wave equation corresponds to the choice:

$$a(u,v) = \int \mathrm{grad}\, u \cdot \mathrm{grad}\, v.$$

Notice that we allow a to have a possibly infinite-dimensional kernel. This will be the case when we apply the abstract setting to Maxwell's equations. As a compatible norm on X we can take the one defined by:

$$\|u\|^2 = \langle u,u\rangle + a(u,u).$$

We are interested in the initial value problem for the second order evolution equation:

$$\forall u' \in X \quad \langle \ddot{u}, u'\rangle = -a(u,u') + \langle f, u'\rangle. \tag{4}$$

Given $u_0 \in X$ and $\dot{u}_0 \in H$, as well as $f \in \mathrm{L}^1([0,T];H)$, we look for $u \in C([0,T];X) \cap C^1([0,T];H)$ such that for all $u' \in C^2_0([0,T[;X)$:

$$\int \langle u(t), \ddot{u}'(t)\rangle \mathrm{d}t + \langle \dot{u}_0, u'(0)\rangle - \langle u_0, \dot{u}'(0)\rangle = \\ - \int a(u(t),u'(t))\mathrm{d}t + \int \langle f(t), u'(t)\rangle \mathrm{d}t.$$

We call this the abstract wave equation. The energy of a function u is defined by:

$$\mathcal{E}(u)(t) = \frac{1}{2}\langle \dot{u}(t), \dot{u}(t)\rangle + \frac{1}{2} a(u(t),u(t)).$$

This level of generality includes wave equations in media with variable, possibly discontinuous, coefficients and also Maxwell's equations, for which we provide a short sketch.

We work with an open domain U in $\mathbb{R}^3$, filled with vacuum. The unknowns are two time-dependent vectorfields E and B on U called the electric and magnetic field.

The first pair of Maxwell's equations is:

$$\dot{B} = -\operatorname{curl} E,$$
$$\operatorname{div} B = 0.$$

The second pair is:

$$\dot{E} = \operatorname{curl} B + J,$$
$$\operatorname{div} E = Q.$$

The right hand sides are a time-dependent scalar field Q called charge, and a time-dependent vector field J called current.

Notice that we must have conservation of charge:

$$\dot{Q} = \operatorname{div} J. \tag{5}$$

The first pair of equations guarantees the existence (at least on simply connected domains) of a magnetic potential A which is a time dependent vectorfield on U such that:

$$E = -\dot{A},$$
$$B = \operatorname{curl} A.$$

The first pair of equations is then automatically satisfied. The second pair of equations can be written in terms of A:

$$\ddot{A} = -\operatorname{curl}\operatorname{curl} A - J, \tag{6}$$
$$\operatorname{div} \dot{A} = -Q. \tag{7}$$

Notice that if the divergence constraint (7) is satisfied initially, then the evolution equation (6) and the conservation of charge (5) guarantee that the divergence constraint is satisfied at all times. The usual boundary condition for A is that its tangential component A_{T} on ∂U should vanish. Then E also has a zero tangential component on ∂U – this is the perfect conductor boundary condition.

The second order evolution equation (6) can be cast into the above framework. We take H to be the space of square integrable vector fields equipped with its usual scalar product. We define:

$$X = \{u \in H \ : \ \operatorname{curl} u \in H \text{ and } u_{\mathrm{T}} = 0\}.$$

It is a non-trivial fact that the vanishment of the tangential component u_{T} of u on the boundary makes sense in this topology [48]. The bilinear form a is defined on X by:

$$a(u,v) = \langle \operatorname{curl} u, \operatorname{curl} v \rangle.$$

2.2 Convergence theory for linear equations

We now study the construction of approximate solutions to the abstract wave equation using the Galerkin method. The literature on the topic includes [31, 46, 47, 63]. The author was particularly inspired by [42].

We suppose we are given a family (X_h) of finite dimensional subspaces of X with the property that for all $v \in X$:

$$\lim_{h\to 0} \inf_{v_h \in X_h} \|v - v_h\|_X = 0.$$

We now look for $u_h \in \mathrm{C}^1([0,T];X_h)$ such that $\dot{u}_h$ is absolutely continuous and for almost every $t \in [0,T]$ we have:

$$\forall u' \in X_h \quad \langle \ddot{u}_h(t), u' \rangle = -a(u_h(t), u') + \langle f(t), u' \rangle. \tag{8}$$

Initial conditions are approximated in X_h. This problem has a unique solution by ODE theory.

The first result is a stability estimate. We denote the time-derivative of a quantity $\mathcal{U}$, by $\mathcal{U}^\bullet$. We have:

$$\mathcal{E}(u_h)^\bullet(t) = \langle f(t), \dot{u}_h(t) \rangle \leq 2^{1/2} |f(t)| \mathcal{E}(u)(t)^{1/2}.$$

From this it follows that:

$$\mathcal{E}(u)(t)^{1/2} - \mathcal{E}(u)(0)^{1/2} \leq 2^{-1/2} \int_0^t |f(s)| \mathrm{d}s. \tag{9}$$

From the stability one can conclude that a subsequence converges weak-star in the space of bounded energy functions:

$$\{u \in \mathrm{L}^\infty([0,T];X) \; : \; \dot{u} \in \mathrm{L}^\infty([0,T];H)\}, \tag{10}$$

where the time derivative is defined a priori in the sense of X-valued distributions.

One can show that the weak limit is a solution. Proving uniqueness and continuity/differentiability of the solution requires extra work. Assuming now that these results for the continuous problem have been established, we can examine the strong convergence of u_h to u in the natural norm provided by the energy.

Set $e_h = u - u_h$. We have for any $e' \in X_h$:

$$\langle \ddot{e}_h(t), e' \rangle + a(e_h(t), e') = 0.$$

Let v_h be any smooth enough function $[0,T] \to X_h$. At almost every time t we have:

$$\begin{aligned}\mathcal{E}(u)^\bullet &= \langle \ddot{e}_h, \dot{u} - \dot{u}_h \rangle + a(e_h, \dot{u} - \dot{u}_h) \\ &= \langle \ddot{e}_h, \dot{u} - \dot{v}_h \rangle + a(e_h, \dot{u} - \dot{v}_h) \\ &= \langle \dot{e}_h, \dot{u} - \dot{v}_h \rangle^\bullet - \langle \dot{e}_h, \ddot{u} - \ddot{v}_h \rangle + a(e_h, \dot{u} - \dot{v}_h).\end{aligned} \tag{11}$$

The two last terms are bounded by:

$$|\dot{e}_h|\ |\ddot{u} - \ddot{v}_h| + a(e_h, e_h)^{1/2} a(\dot{u} - \dot{v}_h, \dot{u} - \dot{v}_h)^{1/2} \leq 2\mathcal{E}(e_h)^{1/2}\mathcal{E}(\dot{u} - \dot{v}_h)^{1/2}. \tag{12}$$

Inserting this in (11) and integrating gives:

$$\mathcal{E}(e_h)(t) - \mathcal{E}(e_h)(0) \leq \Big[\langle \dot{e}_h, \dot{u} - \dot{v}_h \rangle\Big]_0^t + 2\Big(\int_0^t \mathcal{E}(e_h)\Big)^{1/2}\Big(\int_0^t \mathcal{E}(\dot{u} - \dot{v}_h)\Big)^{1/2}. \tag{13}$$

The first term on the right hand side is bounded by:

$$2\mathcal{E}(e_h)(t)^{1/2}\mathcal{E}(u - v_h)(t)^{1/2} + 2\mathcal{E}(e_h)(0)^{1/2}\mathcal{E}(u - v_h)(0)^{1/2}.$$

Thus we get:

$$\begin{aligned}(1/2)\mathcal{E}(e_h)(t) \leq & 2\mathcal{E}(u - v_h)(t) + 2\mathcal{E}(e_h)(0) + \mathcal{E}(u - v_h)(0) + \\ & \int_0^t \mathcal{E}(e_h) + \int_0^t \mathcal{E}(\dot{u} - \dot{v}_h).\end{aligned}$$

Gronwall's lemma then shows that there is a C depending only on T such that:

$$\sup_{0 \leq t \leq T} \mathcal{E}(e_h)(t) \leq C\Big(\mathcal{E}(e_h)(0) + \sup_{0 \leq t \leq T} \mathcal{E}(u - v_h)(t) + \int_0^T \mathcal{E}(\dot{u} - \dot{v}_h)(s)\mathrm{d}s\Big). \tag{14}$$

For any p denote:

$$\mathcal{X}^p = \{u \in C^p([0,T];X)\ :\ u^{(p+1)} \in C([0,T];H)\}.$$

We also define:

$$\mathcal{X}^p_h = \{u \in C^{p+1}([0,T];X_h)\}.$$

One checks that for all p and for all $u \in \mathcal{X}^p$ one has:

$$\lim_{h \to 0} \inf_{v_h \in \mathcal{X}^p_h} \|u - v_h\|_{\mathcal{X}^p} = 0.$$

To get convergence in (14) it is therefore enough to have u in $\mathcal{X}^1$. This is a stronger condition than just to be in the energy space $\mathcal{X}^0$. It is guaranteed if the data is more regular than we assumed initially: $\dot{u}(0) \in X$, $\ddot{u}(0) \in H$ and $\dot{f} \in \mathrm{L}^1(0,T;H)$ will do, since then $\dot{u}$ satisfies an abstract wave equation, with finite energy initial data and with forcing term $\dot{f}$. Note also that if $u(0)$ is in the domain of a, that is $u(0) \in X$ and :

$$\sup_{u' \in X} |a(u(0), u')|/|u'| < +\infty,$$

then the abstract wave equation for u guarantees that the condition $\ddot{u}(0) \in H$ is satisfied.

Finally since we have stability in $\mathcal{X}^0$ for data of the type $u_0 \in X$, $\dot{u}_0 \in H$ and $f \in \mathrm{L}^1(0,T;H)$, and norm-convergence in $\mathcal{X}^0$ for a dense subset of these, we get norm-convergence in $\mathcal{X}^0$ in general.

Orders of convergence are usually deduced from (14) by assuming a certain amount of extra regularity on the data, deducing regularity of u in various Sobolev spaces, letting $v_h = \Pi_h u$ for a certain X_h-valued interpolation operator Π_h, and using known error estimates for $u - \Pi_h u$ given this Sobolev regularity of u.

We now turn to time discretization of (8). We will consider only the most popular scheme, namely the leap-frog scheme, also called the Störmer-Verlet scheme. The time-step is denoted Δt. Approximations $u_h^n \approx u_h(n\Delta t)$ are obtained by the recursion formula (for the time being we omit the subscript h):

$$\langle \frac{u^{n+1} - 2u^n + u^{n-1}}{(\Delta t)^2}, u' \rangle = -a(u^n, u') + \langle f^n, u' \rangle. \tag{15}$$

Here we put $f^n = f(n\Delta t)$. Inserting

$$u' = u^{n+1} - u^{n-1} = (u^{n+1} - u^n) + (u^n - u^{n-1})$$

into this formula gives:

$$|\frac{u^{n+1} - u^n}{\Delta t}|^2 - |\frac{u^n - u^{n-1}}{\Delta t}|^2 = -a(u^n, u^{n+1}) + a(u^n, u^{n-1}) + \tag{16}$$

$$\langle f^n, u^{n+1} - u^{n-1} \rangle. \tag{17}$$

We introduce the discrete energy of u^n:

$$\mathcal{E}^{n+1/2} = \frac{1}{2}|\frac{u^{n+1} - u^n}{\Delta t}|^2 + \frac{1}{2}a(u^n, u^{n+1}).$$

When f is zero, it is conserved. Unfortunately it need not even be positive due to the second term. We can write:

$$2a(u^n, u^{n+1}) = a(u^{n+1}, u^{n+1}) + a(u^n, u^n) - a(u^{n+1} - u^n, u^{n+1} - u^n).$$

We wish to control the third term on the right hand side. In finite element theory, if h denotes the mesh-width (the largest diameter of a cell of a given mesh), and a is the bilinear form associated with a second order operator (so is continuous on $\mathrm{H}^1(U)$), there exists a constant $C > 0$ such that for all h and all $u, v \in X_h$:

$$a(u, v) \leq Ch^{-2}|u|\ |v|.$$

Such bounds are called inverse estimates. Notice that the bound blows up as $h \to 0$, in accordance with the possible non-continuity of a on $\mathrm{L}^2(U)$. We henceforth suppose that such an estimate exists in our abstract setting. Suppose also that, with this constant C, Δt is chosen so small that:

$$1 - Ch^{-2}(\Delta t)^2/2 \geq \delta > 0.$$

In finite element theory the interpretation is that Δt must be smaller than something comparable to the mesh-width h. This is known as the CFL condition. Then we have:

$$\mathcal{E}^{n+1/2} \geq \frac{\delta}{2}|\frac{u^{n+1}-u^n}{\Delta t}|^2 + \frac{1}{4}(a(u^{n+1},u^{n+1}) + a(u^n,u^n)),$$

consisting of non-negative terms.

From (16) we deduce:

$$\begin{aligned} 2(\mathcal{E}^{n+1/2} - \mathcal{E}^{n-1/2}) &= \langle f^n, u^{n+1} - u^n\rangle + \langle f^n, u^n - u^{n-1}\rangle, \\ &\leq (2/\delta)^{1/2}\Delta t|f^n|((\mathcal{E}^{n+1/2})^{1/2} + (\mathcal{E}^{n-1/2})^{1/2}). \end{aligned}$$

Therefore:

$$(\mathcal{E}^{n+1/2})^{1/2} - (\mathcal{E}^{n-1/2})^{1/2} \leq (2\delta)^{-1/2}|f^n|\Delta t,$$

yielding:

$$(\mathcal{E}^{n+1/2})^{1/2} - (\mathcal{E}^{1/2})^{1/2} \leq (2\delta)^{-1/2}\sum_{k=1}^{n}|f^k|\Delta t.$$

This is a stability estimate comparable to (9), but we must require that f is Riemann integrable (with values in H) rather than merely Lebesgue integrable.

Next we wish to establish convergence. We estimate the distance from the semidiscrete $u_h(n\Delta t)$ to the fully discrete u_h^n. Set $e_h^n = u_h(n\Delta t) - u_h^n$. We find that e_h^n satisfies:

$$\langle\frac{e_h^{n+1} - 2e_h^n + e_h^{n-1}}{(\Delta t)^2}, e'\rangle = -a(e_h^n, u') + \langle\varepsilon_h^n, e'\rangle,$$

with:

$$\varepsilon_h^n = \frac{u_h((n+1)\Delta t) - 2u_h(n\Delta t) + u_h((n-1)\Delta t)}{(\Delta t)^2} - \ddot{u}_h(n\Delta t).$$

Put:

$$\hat{\mathcal{E}}_h^{n+1/2} = \frac{1}{2}|\frac{e_h^{n+1} - e_h^n}{\Delta t}|^2 + \frac{1}{2}a(e_h^n, e_h^{n+1}).$$

As before we get:

$$(\hat{\mathcal{E}}^{n+1/2})^{1/2} - (\hat{\mathcal{E}}^{1/2})^{1/2} \leq (2\delta)^{-1/2}\sum_{k=1}^{n}|\varepsilon_h^k|\Delta t.$$

We want to bound ε_h^k. One has the formula:

$$\varepsilon_h^k = \int_0^1 \Big(\ddot{u}_h((k+s)\Delta t) - 2\ddot{u}_h(k\Delta t) + \ddot{u}_h((k-s)\Delta t)\Big)(1-s)\mathrm{d}s. \qquad (18)$$

For any $T > 0$ if $\dot{u}(0) \in X$, $\ddot{u}(0) \in H$ and $\dot{f} \in \mathrm{L}^1([0,T];H)$, we get convergence of u_h to u in $\mathrm{C}^2([0,T];H)$. Actually we need to be careful with the way initial conditions are handled to guarantee this. We need to choose $u_h(0) \in X_h$ and $\dot{u}_h(0) \in X_h$ so that:

$$u_h(0) \to u(0) \text{ in } X, \tag{19}$$

$$\dot{u}_h(0) \to \dot{u}(0) \text{ in } H, \tag{20}$$

but in addition we need that whenever $\dot{u}(0) \in X$ and $\ddot{u}(0) \in H$ we have:

$$\dot{u}_h(0) \to \dot{u}(0) \text{ in } X, \tag{21}$$

$$\ddot{u}_h(0) \to \ddot{u}(0) \text{ in } H. \tag{22}$$

To guarantee (20) and (21), we can approximate $\dot{u}_h(0)$ using projections $H \to X_h$, which are stable not only in H but also in X. We will comment later on the construction of such projections for finite element spaces. To guarantee (19) and (22), we can use the scalar product $\langle\cdot,\cdot\rangle + a(\cdot,\cdot)$ to project $u(0)$ to X_h. Then we have for all $u' \in X_h$:

$$\begin{aligned}\langle \ddot{u}_h(0), u'\rangle &= -a(u_h(0),u') + \langle f(0),u'\rangle\\ &= -a(u(0),u') - \langle u(0),u'\rangle + \langle u_h(0),u'\rangle + \langle f(0),u'\rangle\\ &= \langle \ddot{u}(0),u'\rangle - \langle u(0)-u_h(0),u'\rangle,\end{aligned}$$

from which (22) follows.

Since now u_h converges to u in $\mathrm{C}^2([0,T];H)$ and $\ddot{u} : [0,T] \to H$ is uniformly continuous, we have:

$$\sup\{|\ddot{u}_h(t') - \ddot{u}_h(t)| \; : \; 0 \leq t,t' \leq T \text{ and } |t'-t| \leq \Delta t\} \to 0,$$

which thanks to (18) gives:

$$\sum_{k=1}^{\lfloor T/\Delta t \rfloor} |\varepsilon_h^k| \Delta t \to 0.$$

From this, one deduces convergence of the linear interpolant of the sequence (u_h^n) to u in the space of bounded energy functions defined in (10) under the CFL condition.

Finally, stability for all finite energy initial conditions and Riemann integrable f, together with convergence for a dense subset of such data, imply convergence for all such data.

2.3 Consistency

Usually the bilinear forms involved are not just restricted to the Galerkin space, they are also approximated. This is necessary when material parameters vary, for instance. This can also have advantages. Equation (15) requires solving a linear system involving the mass matrix (the matrix of $\langle \cdot, \cdot \rangle$ in the chosen basis) at each time-step. If the mass matrix can be approximated by a diagonal matrix, without loss of accuracy, a lot of work is saved. We investigate such approximations in general. We start with stationary problems.

Let X, Y be Banach spaces and a a continuous bilinear form $X \times Y \to \mathbb{R}$. It induces a map:

$$A : \begin{cases} X \to Y^\star, \\ u \mapsto a(u, \cdot). \end{cases}$$

If it is invertible the following number is positive:

$$\|A^{-1}\|^{-1} = \inf_{u \in X} \sup_{v \in Y} \frac{|a(u,v)|}{\|u\| \, \|v\|}.$$

Suppose (X_h) and (Y_h) are families of finite dimensional subspaces of X and Y respectively. We also suppose $\dim X_h = \dim Y_h$ and that (X_h) is approximating in the sense that:

$$\forall u \in X \quad \inf_{u_h \in X_h} \|u - u_h\| \to 0,$$

We make a similar hypothesis for (Y_h). Let a_h be bilinear forms $X_h \times Y_h \to \mathbb{R}$.

Given $l \in Y^\star$, we want to find approximations to the solution $u \in X$ of:

$$\forall v \in Y \quad a(u,v) = l(v).$$

We construct linear forms $l_h \in Y_h^\star$ which approximate the restriction of l. We find $u_h \in X_h$ such that:

$$\forall v \in Y_h \quad a_h(u_h, v) = l_h(v). \tag{23}$$

Two conditions play a crucial role in the analysis of the convergence of u_h: The inf-sup condition [7, 14], and consistency [58].

- *Inf-sup condition*. There is a $C > 0$ such that for all h:

$$\inf_{u \in X_h} \sup_{v \in Y_h} \frac{|a_h(u,v)|}{\|u\| \, \|v\|} \geq 1/C.$$

- *Consistency*. For any $u \in X$ there is a sequence $\tilde{u}_h \in X_h$ such that $\tilde{u}_h \to u$ in X as $h \to 0$ and:

$$\sup_{v \in Y_h} \frac{|a(u,v) - a_h(\tilde{u}_h, v)|}{\|v\|} \to 0 \text{ as } h \to 0. \tag{24}$$

Going back to the analysis of (23), we suppose that the inf-sup condition and consistency hold. Choosing $\tilde{u}_h \in X_h$ as in the definition of consistency, we have:

$$\begin{aligned}\|u_h - \tilde{u}_h\| &\leq C \sup_{v \in Y_h} \frac{|a_h(u_h - \tilde{u}_h, v)|}{\|v\|} \\ &\leq C \sup_{v \in Y_h} \frac{|l_h(v) - l(v) + a(u,v) - a_h(\tilde{u}_h, v)|}{\|v\|} \\ &\leq C\Big(\|l - l_h\|_{Y_h^\star} + \sup_{v \in Y_h} \frac{|a(u,v) - a_h(\tilde{u}_h, v)|}{\|v\|}\Big).\end{aligned}$$

We suppose $\|l - l_h\|_{Y_h^\star} \to 0$ and obtain $u_h \to u$.

A diagonal argument shows that it is enough to ensure the consistency requirement for all u in a dense subset X_0 of X. There usually are such dense subsets for which one obtains rather explicit rates of convergence in (24). When the solution to the continuous problem is in such a subset, the above computation gives convergence rates for $u_h \to u$.

Reciprocally, if convergence $u_h \to u$ is to hold for any $l \in Y^\star$ (with simply $l_h = l|_{Y_h}$) then the inf-sup condition and the consistency must hold. We prove this. First concerning the inf-sup condition we do the following. We define the mappings:

$$A_h : \begin{cases} X_h \to Y_h^\star, \\ u \mapsto a_h(u, \cdot). \end{cases}$$

They must be invertible. Let $r_h : Y^\star \to Y_h^\star$ be the restriction map. We suppose that for all $l \in Y^\star$ we have:

$$A_h^{-1} r_h l \to A^{-1} l.$$

By the uniform boundedness principle there must be C such that for all h:

$$\|A_h^{-1} r_h\|_{Y^\star \to X} \leq C.$$

By the Hahn-Banach theorem it follows that:

$$\|A_h^{-1}\|_{Y_h^\star \to X_h} \leq C.$$

This can be rephrased as:

$$\inf_{u \in X_h} \sup_{v \in Y_h} \frac{|a_h(u,v)|}{\|u\| \, \|v\|} \geq 1/C.$$

So we get the inf-sup condition. That consistency must hold is clear: For a given u simply let $\tilde{u}_h = u_h$ be the discrete solution:

$$\forall v \in Y_h \quad a_h(u_h, v) = a(u, v).$$

Since the inf-sup condition is a stability condition, we get an instance of the Lax equivalence principle: convergence is equivalent to stability and consistency.

Usually there is a $C > 0$ such that for all h:

$$\forall u \in X_h \, \forall v \in Y_h \quad |a_h(u,v)| \leq C\|u\| \, \|v\|. \tag{25}$$

If the method is consistent and this equiboundedness condition holds, we have that for any $u \in X$ and *any* sequence $\tilde{u}_h \in X_h$ such that $\tilde{u}_h \to u$ in X as $h \to 0$:

$$\sup_{v \in Y_h} \frac{|a(u,v) - a_h(\tilde{u}_h, v)|}{\|v\|} \to 0 \text{ as } h \to 0. \tag{26}$$

Reciprocally, if this last condition holds then (25) holds and the method is consistent. Indeed suppose (25) does not hold. Then there are subsequences $u_h \in X_h$, $v_h \in Y_h$, such that $|a_h(u_h, v_h)| = 1$, $\|u_h\| \to 0$ and $\|v_h\| = 1$. This contradicts (26).

We now return to the wave equation, in its abstract formulation. We suppose we have constructed bilinear forms:

$$\langle \cdot, \cdot \rangle_h,\ a_h(\cdot, \cdot) : X_h \times X_h \to \mathbb{R},$$

with the properties that for some C we have for all h:

$$\forall u \in X_h \quad \frac{1}{C}\langle u, u \rangle \leq \langle u, u \rangle_h \leq C \langle u, u \rangle, \tag{27}$$

and:

$$\forall u \in X_h \quad \frac{1}{C} a(u,u) \leq a_h(u,u) \leq C a(u,u). \tag{28}$$

We require $\langle \cdot, \cdot \rangle_h$ to be a consistent discretization of $\langle \cdot, \cdot \rangle$ with respect to the H norm and $a_h(\cdot, \cdot)$ to be a consistent discretization of $a(\cdot, \cdot)$ with respect to the X norm.

The new semidiscrete equation we solve is:

$$\langle \ddot{u}_h(t), u' \rangle_h = -a_h(u_h(t), u') + \langle f(t), u' \rangle. \tag{29}$$

Notice that we do not do anything with the f term, though this is possible.

Let $\mathcal{E}_h$ be the corresponding energy:

$$\mathcal{E}_h(u)(t) = \frac{1}{2}\langle \dot{u}(t), \dot{u}(t) \rangle_h + \frac{1}{2} a_h(u(t), u(t)).$$

We get stability for this energy as before. Stability for the true energy follows from the comparison estimates (27) and (28).

To get convergence we first make the usual extra assumptions that $u(0) \in X$ is in the domain of a, $\dot{u}(0) \in X$ and $\dot{f} \in \mathrm{L}^1([0,T]; H)$. They guarantee that $u \in \mathcal{X}^1$. We compare u_h with some function $v_h : [0,T] \to X_h$. We have:

$$\begin{aligned}\mathcal{E}_h(u_h - v_h)^\bullet = &\langle \ddot{u}_h - \ddot{v}_h, \dot{u}_h - \dot{v}_h \rangle_h + a_h(u_h - v_h, \dot{u}_h - \dot{v}_h) \\ = &\langle \ddot{u}, \dot{u}_h - \dot{v}_h \rangle - \langle \ddot{v}_h, \dot{u}_h - \dot{v}_h \rangle_h + \\ &a(u, \dot{u}_h - \dot{v}_h) - a_h(v_h, \dot{u}_h - \dot{v}_h).\end{aligned}$$

We can ensure convergence of v_h to u in $\mathcal{X}^1$. We also have boundedness of (u_h) in the space $\mathcal{X}^1$ by the stability argument applied to the time-differentiated equation. Consistency augmented by a compactness argument then shows:

$$\sup_{0 \le t \le T} \mathcal{E}_h(u_h - v_h)^\bullet(t) \to 0.$$

This gives convergence in $\mathcal{X}^0$. As usual stability for general data and convergence on a dense subset give convergence in general.

The treatment of the time-discretized case can be carried out along the path sketched for the case of exact bilinear forms.

Finally we include a proof that the procedure known as mass lumping gives, on scalar degree one continuous elements, a stable and consistent discretization of the L^2-scalar product on functions. The techniques involved are detailed for instance in [30]. For the extension to higher order finite element methods, see [32], for the case of vectorial finite elements in relation to the Yee scheme, see [47].

The computational domain $U \subseteq \mathbb{R}^n$ is equipped with a regular simplicial mesh $\mathcal{T}_h$. The space X_h consists of the functions that are continuous and piecewise affine with respect to $\mathcal{T}_h$. The parameter h is identified with the largest diameter of a cell of $\mathcal{T}_h$. We denote by I_h the nodal interpolator, which is defined on continuous functions and is a projection onto X_h. The mass-lumped L^2 product is defined on continuous functions by:

$$\langle u, v \rangle_h = \int I_h(uv).$$

Notice that in the standard basis of X_h, the matrix of this bilinear form is diagonal. A rather explicit formula is:

$$\langle u, v \rangle_h = \sum_{T \in \mathcal{T}_h} \frac{|T|}{n+1} \sum_{x \in T} u(x)v(x).$$

Here the points x in the sum are the $n+1$ vertices of T.

Consider a reference element $\hat{T}$. By equivalence of norms in finite dimensions and unisolvence of the vertex degrees of freedom on affine functions, there are constants C, C' such that for any affine function u on T:

$$\frac{1}{C'} \int_{\hat{T}} |u|^2 \le \frac{|\hat{T}|}{n+1} \sum_{x \in \hat{T}} |u(x)|^2 \le C \int_{\hat{T}} |u|^2.$$

Transporting this to all elements T of $\mathcal{T}_h$ and adding gives, for all h and all $u \in X_h$:

$$\frac{1}{C'} \langle u, u \rangle \le \langle u, u \rangle_h \le C \langle u, u \rangle,$$

which ensures stability and equiboundedness.

Let $l > n/2$, which ensures that $\mathrm{H}^l(U)$ is continuously injected into the space of continuous functions on $\overline{U}$. Standard Sobolev norms will be denoted $\|\cdot\|$ whereas seminorms are denoted as $|\cdot|$. We shall prove that there is a $C > 0$ such that for all $u \in \mathrm{H}^l(U)$, all h and all $v \in X_h$:

$$|\langle u, v \rangle - \langle u, v \rangle_h| \le Ch \|u\|_{\mathrm{H}^l(U)} \|v\|_{\mathrm{L}^2(U)}. \tag{30}$$

We want to estimate the errors $E_T(uv)$ with E_T defined by:

$$E_T(w) = \int_T w - \frac{|T|}{n+1} \sum_{x \in T} w(x).$$

The above integration rule is exact for polynomials of degree at most 1. Working on the reference element, a Bramble-Hilbert lemma gives an estimate of the form, for all $w \in \mathrm{H}^l(\hat{T})$:

$$|E_{\hat{T}}(w)| \leq C(|w|^2_{\mathrm{H}^2(\hat{T})} + |w|^2_{\mathrm{H}^l(\hat{T})})^{1/2}.$$

We apply a Leibniz rule for the differentiation of products. For $u \in \mathrm{H}^l(\hat{T})$ and v confined to the finite-dimensional space of affine functions on $\hat{T}$, we deduce:

$$|E_{\hat{T}}(uv)| \leq C(|u|^2_{\mathrm{H}^1(\hat{T})} + \cdots + |u|^2_{\mathrm{H}^l(\hat{T})})^{1/2} \|v\|_{\mathrm{L}^2(\hat{T})}.$$

Transporting this estimate to a simplex T of diameter h gives:

$$|E_T(uv)| \leq Ch \|u\|_{\mathrm{H}^l(T)} \, \|v\|_{\mathrm{L}^2(T)}.$$

Summing these estimates and applying a Cauchy-Schwartz inequality gives (30).

The same arguments also show:

$$|\langle u,v \rangle - \langle u,v \rangle_h| \leq Ch^2 \|u\|_{\mathrm{H}^l(U)} \|v\|_{\mathrm{H}^1(U)}.$$

2.4 Eigenvalue approximation

Linear evolution problems are closely linked to eigenvalue problems. For instance, the solution of the wave equation (1) can be expressed quite simply in terms of the eigenvalues and eigenvectors of the Laplace operator. Similarly, the discrete solutions of the wave equation obtained by the Galerkin method can be expressed in terms of discrete eigenvalues and eigenvectors. In this section we comment on the relation between continuous and discrete eigenproblems. The emphasis is on problems of Maxwell type, for which the bilinear form a will have a large kernel. References on the subject include [9, 10, 19, 43]. For p-version finite elements there are some new developments [33, 41].

We introduce the following notations, which construct a decomposition of X generalizing the Helmholtz decomposition of vectorfields (into a gradient and a curl) to our abstract setting. We let W be the kernel of a defined by:

$$W = \{u \in X \ : \ \forall u' \in X \quad a(u,u') = 0\}.$$

Notice that by (3) we have, for all $u \in X$:

$$u \in W \iff a(u,u) = 0.$$

We also let V be the orthogonal of W in X with respect to $\langle\cdot,\cdot\rangle$:

$$V = \{u \in X \ : \ \forall w \in W \quad \langle u,w \rangle = 0\}.$$

Thus V and W are closed subspaces of X and:

$$X = V \oplus W.$$

We make the additional assumption that the injection $V \to H$ is compact (when V is equipped with the norm inherited from X). In particular, we get a Poincaré-Friedrichs inequality. There is a $C > 0$ such that:

$$\forall v \in V \quad a(v,v) \geq \frac{1}{C}|v|^2. \tag{31}$$

Coercivity of a on V follows.

Notice that W is a closed subspace of H. Let $\overline{V}$ be the orthogonal of W in H with respect to $\langle\cdot,\cdot\rangle$:

$$\overline{V} = \{u \in H \ : \ \forall w \in W \quad \langle u,w \rangle = 0\}.$$

The notation is justified by the fact that $\overline{V}$ is the closure of V in H. We have the decomposition:

$$H = \overline{V} \oplus W.$$

In H, let P denote the projector with range $\overline{V}$ and kernel W. It is orthogonal. Notice that if $u \in X$, then $Pu \in X$ since $u - Pu \in W \subseteq X$. On X, P is the projector with range V and kernel W. It is characterized by the property that for any $u \in X$, Pu is the element of V solving:

$$\forall u' \in V \quad a(Pu,u') = a(u,u').$$

This equation is well posed by (31). When it holds, it actually holds for all $u' \in X$. Thus we get:

$$\forall u,u' \in X \quad a(Pu,Pu') = a(u,u').$$

In this setting look for pairs $(\lambda,u) \in \mathbb{R} \times X$ with $u \neq 0$ such that:

$$\forall u' \in X \quad a(u,u') = \lambda \langle u,u' \rangle.$$

For intance $(0,w)$ is an eigenpair of a for any non-zero $w \in W$; 0 is an eigenvalue with associated eigenspace W (when $W \neq 0$).

We introduce the operator $T : H \to V$ defined as follows. For any $u \in H$, Tu is the element of V satisfying:

$$\forall u' \in V \quad a(Tu,u') = \langle u,u' \rangle.$$

That T is well defined and continuous follows from the coercivity on V of a, compare with (31). If we consider T to be an operator $H \to H$, it is immediate that T is compact and symmetric. Thus H has an orthonormal basis consisting of eigenvectors in X. The eigenspace of 0 is W (when $W \neq 0$). The non-zero eigenvalues of T correspond to elements of V and can be ordered in a decreasing sequence converging to 0. Moreover λ is a non-zero eigenvalue of a iff $1/\lambda$ is a non-zero eigenvalue of T and the corresponding eigenspaces are the same.

Discrete eigenvalues are defined as follows. For each h we look for pairs $(\lambda, u) \in \mathbb{R} \times X_h$ with $u \neq 0$ such that:

$$\forall u' \in X_h \quad a(u, u') = \lambda \langle u, u' \rangle.$$

In general this eigenvalue problem is not so well-behaved: there are cases where discrete eigenvalues cluster as $h \to 0$ at values, which are not eigenvalues of the continuous problem (this cannot happen if W is finite dimensional). The approximation of eigenvalues requires more properties of the Galerkin spaces than does the approximation of the evolution problem. We make the following assumption:

- There are projectors $\Pi_h : H \to H$ with range X_h, which are uniformly bounded in H and have the property that W is mapped to W.

We define:

$$W_h = \{u \in X_h \ : \ \forall u' \in X_h \quad a(u, u') = 0\},$$

and:

$$V_h = \{u \in X_h \ : \ \forall w \in W_h \quad \langle u, w \rangle = 0\}.$$

Thus we have decompositions:

$$X_h = V_h \oplus W_h$$

and $W_h = X_h \cap W$. A crucial point is that V_h need not be a subspace of V. Using our assumption we shall prove that V_h is close to V, in a precise sense.

We have by the approximation property of (X_h) – which holds in X and therefore in H by density – and the uniform boundedness of the projectors Π_h:

$$\forall u \in H \quad \Pi_h u \to u \text{ in } H.$$

Since the injection $V \to H$ is compact, there is a sequence ε_h converging to 0 as $h \to 0$ such that:

$$\forall u \in V \ \forall h \quad |u - \Pi_h u| \leq \varepsilon_h \|u\|.$$

Choose v_h in V_h. We aim to prove that $v_h - P v_h$ is small and write:

$$|v_h - P v_h| \leq |v_h - \Pi_h P v_h| + |P v_h - \Pi_h P v_h|.$$

We have:

$$|P v_h - \Pi_h P v_h| \leq \varepsilon_h \|P v_h\| \leq \varepsilon_h \|v_h\|.$$

Note that $v_h - P v_h \in W$ so, by our hypothesis on Π_h, we have:

$$v_h - \Pi_h P v_h = \Pi_h (v_h - P v_h) \in W_h.$$

Therefore v_h and Pv_h are both orthogonal to $v_h - \Pi_h P v_h$:

$$|v_h - \Pi_h P v_h|^2 = \langle v_h - \Pi_h P v_h, P v_h - \Pi_h P v_h \rangle$$
$$\leq |v_h - \Pi_h P v_h| \; |P v_h - \Pi_h P v_h|,$$

so that:

$$|v_h - \Pi_h P v_h| \leq |P v_h - \Pi_h P v_h|.$$

Thus we get:

$$|v_h - P v_h| \leq 2|P v_h - \Pi_h P v_h| \leq 2\varepsilon_h \|v_h\|.$$

Hence:

$$\|v_h - P v_h\| \leq 2\varepsilon_h \|v_h\|.$$

This property can be rephrased in terms of gaps between subspaces of X:

$$\delta(V_h, V) \leq 2\varepsilon_h.$$

Since a is coercive on V, it follows that there is a C such that for all h:

$$\forall v_h \in V_h \quad a(v_h, v_h) \geq \frac{1}{C}\|v_h\|^2, \tag{32}$$

More precisely we can write, for all small enough h and all $v_h \in V_h$:

$$a(v_h, v_h) = a(P v_h, P v_h)$$
$$\geq \frac{1}{C}\|P v_h\|^2$$
$$\geq \frac{1}{C}(1 - 2\varepsilon_h)^2 \|v_h\|^2.$$

Estimate (32) follows for small h. For a finite number of large h, one uses the non-negativity of a.

We introduce the discrete analogue of T, which is the map $T_h : H \to V_h$ taking any $u \in H$ to the element $T_h u \in V_h$ such that:

$$\forall u' \in V_h \quad a(T_h u, u') = \langle u, u' \rangle.$$

The operator T_h has finite-dimensional range and is symmetric. Note that λ is a non-zero discrete eigenvalue (of a on X_h) iff $1/\lambda$ is a non-zero eigenvalue of T_h. The corresponding eigenspaces are the same. In order to relate the discrete eigenvalues of a to the continuous ones, we aim to prove:

$$\|T - T_h\|_{H \to H} \to 0. \tag{33}$$

Let $P_h : X \to V_h$ be the projector defined by the property that for all $u \in X$, $P_h u$ is the element of V_h such that:

$$\forall u' \in V_h \quad a(P_h u, u') = a(u, u').$$

This equation then holds for all $u' \in X_h$. Notice that we have, for all $u \in H$, all h and all $u' \in V_h$:

$$a(T_h u, u') = \langle u, u' \rangle = a(Tu, u') = a(P_h T u, u').$$

Therefore:

$$T_h = P_h T.$$

Now we establish that $T : H \to V$ is compact. Indeed, if (u_n) converges weakly to 0 in H, then we have already stated that (Tu_n) converges strongly to 0 in H (compactness of $T : H \to H$). In addition:

$$a(Tu_n, Tu_n) = \langle u_n, Tu_n \rangle \to 0.$$

Therefore (Tu_n) converges strongly to 0 in X. This shows that $T : H \to V$ is compact.

Choose $u \in V$. We shall prove that $P_h u \to u$ in X. Let u_h be elements of X_h converging to u in X. The following is a Céa type argument.

$$\begin{aligned}
\|u - PP_h u\|^2 &\leq Ca(u - PP_h u, u - PP_h u) \\
&\leq Ca(u - P_h u, u - P_h u) \\
&\leq Ca(u - P_h u, u - u_h) \\
&\leq Ca(u - PP_h u, u - u_h) \\
&\leq C\|u - PP_h u\| \, \|u - u_h\|.
\end{aligned}$$

Hence:

$$\|u - PP_h u\| \leq C\|u - u_h\| \to 0.$$

We also have stability of P_h by (32). Hence:

$$\|P_h u - PP_h u\| \leq C\varepsilon_h \|u\| \to 0.$$

Combining the above estimates gives:

$$\|u - P_h u\| \to 0.$$

Compactness of $T : H \to V$ and pointwise convergence of $P_h : V \to X$ combine to prove that:

$$\|T - T_h\|_{H \to X} \to 0,$$

which is slightly stronger than we need. Whenever (33) holds, we have (see [8]) that for any $\delta > 0$ not in the spectrum of T, the eigenvalues of T_h, which are above δ can be split into sets, each set converging to an eigenvalue of T (in the sense of the Hausdorff distance). If one such set converges to say λ, the sum of the corresponding discrete eigenspaces converges to the continuous eigenspace of λ (in the sense of gaps between subspaces of H).

3 Construction of finite element spaces

We expressed Maxwell's equations in terms of vector fields. The relevant operators on scalar and vectorfields are the gradient, the curl and the divergence. For any of these three operators op one defines:

$$\mathrm{H}_{\mathrm{op}} = \{u \in \mathrm{L}^2 \ : \ \mathrm{op}\, u \in \mathrm{L}^2\}.$$

We then have a diagram of spaces linked by operators:

$$\mathrm{H}_{\mathrm{grad}} \xrightarrow{\mathrm{grad}} \mathrm{H}_{\mathrm{curl}} \xrightarrow{\mathrm{curl}} \mathrm{H}_{\mathrm{div}} \xrightarrow{\mathrm{div}} \mathrm{H}\ .$$

An important property is that the composition of any two consecutive operators is 0. One says that the spaces form a complex.

To obtain good discretizations of Maxwell's equations one now tends not only to construct subspaces X_h^1 of $\mathrm{H}_{\mathrm{curl}}$, but also subspaces X_h^0 of $\mathrm{H}_{\mathrm{grad}}$, X_h^2 of $\mathrm{H}_{\mathrm{div}}$ and X_h^3 of H, also forming, for each h, a complex for the same operators. Moreover, one wishes to relate the two complexes by projections onto these subspaces, forming a commuting diagram:

$$\begin{array}{ccccccc}
\mathrm{H}_{\mathrm{grad}} & \xrightarrow{\mathrm{grad}} & \mathrm{H}_{\mathrm{curl}} & \xrightarrow{\mathrm{curl}} & \mathrm{H}_{\mathrm{div}} & \xrightarrow{\mathrm{div}} & \mathrm{H} \\
\downarrow \Pi_h^0 & & \downarrow \Pi_h^1 & & \downarrow \Pi_h^2 & & \downarrow \Pi_h^3 \\
X_h^0 & \xrightarrow{\mathrm{grad}} & X_h^1 & \xrightarrow{\mathrm{curl}} & X_h^2 & \xrightarrow{\mathrm{div}} & X_h^3
\end{array}$$

Commutativity of the diagram means that one can follow the arrows along any path between two spaces and still obtain the same operator between these spaces.

We first give some results concerning such diagrams of complexes and tensor product constructions. Then we recall the definition and basic properties of differential forms on manifolds. They provide a more flexible framework than vector fields. We also define cochains associated with decompositions of a space into cells, which are at the basis of discretizations. Finally we apply these concepts to the construction of spaces X_h^k as above.

3.1 Algebra

3.1.1 Homological algebra

We now recall some definitions and facts from homological algebra, and refer the reader to Lang [45] and Gelfand-Manin [36] for thorough expositions. In this paper by a *complex* we mean a sequence $A^\bullet$ of vector spaces equipped with linear operators $d^k : A^k \to A^{k+1}$, called *differentials* and satisfying, for each k, $d^{k+1}d^k = 0$. The *cohomology group* $\mathrm{H}^k A^\bullet$ is (the vectorspace) defined by:

$$\mathrm{H}^k A^\bullet = (\ker d^k : A^k \to A^{k+1})/(\operatorname{im} d^{k-1} : A^{k-1} \to A^k)$$

Most often the index k in d^k is dropped and the complex is represented by a diagram:

$$\cdots \to A^{k-1} \to A^k \to A^{k+1} \to \cdots,$$

where it is implicit that the arrows represent instances of the differential d.

A complex $A^\bullet$ is said to be *exact* at an index k if $\mathrm{H}^k A^\bullet = 0$, or equivalently:

$$\ker d^k : A^k \to A^{k+1} = \operatorname{im} d^{k-1} : A^{k-1} \to A^k.$$

It is said to be exact if it is exact at each index where the cohomology group is defined.

Example 3.1. The complex $A^0 \to A^1 \to 0$ is exact iff the first arrow is surjective.
The complex $0 \to A^0 \to A^1$ is exact iff the second arrow is injective.
If $B \subseteq A$, we have an exact sequence $0 \to B \to A \to A/B \to 0$.
If $C = A \oplus B$, we also have an exact sequence of the form $0 \to A \to C \to B \to 0$.

Remark 3.1. If we have an exact sequence of finite dimensional spaces $0 \to A \to B \to C \to 0$ we have $\dim A - \dim B + \dim C = 0$ and any two terms of the sequence determine the third term up to isomorphism. A similar result holds for longer exact sequences.

If $A^\bullet$ and $B^\bullet$ are two complexes, a *morphism of complexes* $f^\bullet : A^\bullet \to B^\bullet$, is a sequence of linear operators $f^k : A^k \to B^k$ such that the following diagrams commute:

$$\begin{array}{ccc} A^k & \longrightarrow & A^{k+1} \\ \downarrow f^k & & \downarrow f^{k+1} \\ B^k & \longrightarrow & B^{k+1} \end{array}$$

Proposition 3.1. *A morphism of complexes $f^\bullet : A^\bullet \to B^\bullet$ induces a sequence of linear maps $\mathrm{H}^k f^\bullet : \mathrm{H}^k A^\bullet \to \mathrm{H}^k B^\bullet$, namely to the class of $u \in \ker d^k : A^k \to A^{k+1}$, we associate the class of $f^k u$.*

Proof. This is based on two facts:
If $u \in A^k$ satisfies $d^k u = 0$, then $d^k f^k u = f^{k+1} d^k u = 0$.
Also, if $u = d^{k-1} v$ for some $v \in A^{k-1}$, then $f^k u = f^k d^{k-1} v = d^{k-1} f^{k-1} v$ with $f^{k-1} v \in B^{k-1}$, hence the class of u determines the class of $f^k u$.

We have $\mathrm{H}^k(f^\bullet g^\bullet) = (\mathrm{H}^k f^\bullet)(\mathrm{H}^k g^\bullet)$ and $\mathrm{H}^k(\mathrm{id}) = \mathrm{id}$.

As an illustration of the above concepts we notice:

Remark 3.2. Suppose $A^\bullet$ is a complex and, for each k, B^k is a subspace of A^k such that the differential $A^k \to A^{k+1}$ induces differentials $B^k \to B^{k+1}$ by restriction. Suppose furthermore that we have projections $p^k : A^k \to B^k$ commuting with the differential. Then $p^\bullet$ induces surjections in cohomology.

Proof. If $u \in B^k$ satisfies $du = 0$, then $u \in A^k$ satisfies $du = 0$ and $pu = u$.

Given two morphisms of complexes $f^\bullet, g^\bullet : A^\bullet \to B^\bullet$, a *homotopy operator* from $f^\bullet$ to $g^\bullet$ is a family $h^\bullet$ of linear operators $h^k : A^k \to B^{k-1}$ such that:

$$g^k - f^k = h^{k+1}d^k + d^{k-1}h^k.$$

Proposition 3.2. *If $f^\bullet$ and $g^\bullet$ are homotopic in the sense that a homotopy operator exists from one to the other, they induce the* same *maps* $\mathrm{H}^kA^\bullet \to \mathrm{H}^kB^\bullet$.

Proof. Pick $u \in A^k$ such that $du = 0$. Then $gu - fu = dhu$ so gu and fu determine the same class modulo $\mathrm{d}B^{k-1}$ in $\mathrm{H}^kB^\bullet$.

Recall that if $\alpha : A \to A'$ then $\operatorname{coker}\alpha = A'/(\alpha A)$.

We will use the *snake lemma*:

Lemma 3.1. *Suppose we have a commuting diagram:*

$$\begin{array}{ccccccc}
 & & A & \xrightarrow{f} & B & \xrightarrow{g} & C & \longrightarrow & 0 \\
 & & \downarrow{\scriptstyle \alpha} & & \downarrow{\scriptstyle \beta} & & \downarrow{\scriptstyle \gamma} & & \\
0 & \longrightarrow & A' & \xrightarrow{f'} & B' & \xrightarrow{g'} & C' & &
\end{array}$$

We suppose each row is exact. Then we have a well-defined morphism:

$$\delta : \ker\gamma \to \operatorname{coker}\alpha.$$

It is defined on any $w \in C$ such that $\gamma w = 0$, by taking first any *reciprocal v of w by g, then mapping v to $v' = \beta v$, then taking the reciprocal u' of v' by f', and finally considering the equivalence class of u' modulo αA.*

Proof. Pick $w \in C$ such that $dw = 0$ and choose $v \in B$ such that $gv = w$.
We have $g'\beta v = \alpha gv = dw = 0$, so there is $u' \in A'$ such that $f'u' = \beta v$.
If we had chosen $\hat{v} \in B$ instead of v, we would have obtained $\hat{u}' \in A'$.
We have $g(\hat{v} - v) = 0$, so $\hat{v} - v = fu$ for a $u \in A$.
Then $f'(\hat{u}' - u') = \beta(\hat{v} - v) = \beta fu = f'\alpha u$, so $\hat{u}' - u' = \alpha u$.

Theorem 3.1. *Suppose we are given three complexes $A^\bullet$, $B^\bullet$ and $C^\bullet$, and morphisms of complexes $f^\bullet : A^\bullet \to B^\bullet$ and $g^\bullet : B^\bullet \to C^\bullet$, providing for each k a* short exact sequence*:*

$$0 \to A^k \to B^k \to C^k \to 0.$$

Then one can construct a long exact sequence *linking the cohomology groups:*

$$\begin{array}{ccccc}
\mathrm{H}^kA^\bullet & \longrightarrow & \mathrm{H}^kB^\bullet & \longrightarrow & \mathrm{H}^kC^\bullet \\
 & \swarrow & & & \\
\mathrm{H}^{k+1}A^\bullet & \longrightarrow & \mathrm{H}^{k+1}B^\bullet & \longrightarrow & \mathrm{H}^{k+1}C^\bullet
\end{array}$$

Here the first, second, fourth and fifth arrows are the natural ones and the third, diagonal, one – called connecting morphism and usually denoted δ^k – is constructed by the snake lemma.

Proof. We denote by f^k and g^k the maps $f^k : A^k \to B^k$ and $g^k : B^k \to C^k$. The proof consists of a series of straightforward verifications, to the effect that we have a complex and that it is exact.

Exactness at $\mathrm{H}^k B^\bullet$:
Pick $v^k \in B^k$ such that $dv^k = 0$ and $g^k v^k = dw^{k-1}$ for some $w^{k-1} \in C^{k-1}$.
Choose $v^{k-1} \in B^{k-1}$ such that $g^{k-1} v^{k-1} = w^{k-1}$.
We have $g^k(v^k - dv^{k-1}) = dw^{k-1} - dg^{k-1}v^{k-1} = 0$.
Put $v^k - dv^{k-1} = f^k u^k$ for some $u^k \in A^k$.
We have $f^{k+1} du^k = df^k u^k = d(v^k - dv^{k-1}) = 0$, hence $du^k = 0$.

Construction of δ^k:
Recall that given $w^k \in C^k$ such that $dw^k = 0$, we choose any $v^k \in B^k$ such that $g^k v^k = w^k$. Then we take the $u^{k+1} \in A^{k+1}$ such that $f^{k+1}u^{k+1} = dv^k$, and consider the class of u^{k+1}.
If $w^k = dw^{k-1}$ for some $w^{k-1} \in C^{k-1}$, we have $w^{k-1} = g^{k-1}v^{k-1}$ with $v^{k-1} \in B^{k-1}$. Then $g^k dv^{k-1} = dg^{k-1}v^{k-1} = dw^{k-1} = w^k$, so we may suppose $v^k = dv^{k-1}$. Then $dv^k = ddv^{k-1} = 0$ so $u_{k+1} = 0$.
Remark also that $f^{k+2}du^{k+1} = df^{k+1}u^{k+1} = ddv^k = 0$, so $du^{k+1} = 0$.

Concerning $\delta^k \mathrm{H}^k g^\bullet$:
If $w^k = g^k v^k$ for an v^k such that $dv^k = 0$, then $f^{k+1}u^{k+1} = dv^k = 0$, so $u^{k+1} = 0$, hence the composition is zero.
If $u^{k+1} = du^k$ for some $u^k \in A^k$, then $g^k(v^k - f^k u^k) = w^k$ and $d(v^k - f^k u^k) = dv^k - f^{k+1}du^k = dv^k - f^{k+1}u^{k+1} = 0$.

Concerning $\mathrm{H}^{k+1} f^\bullet \delta^k$:
We have $f^{k+1}u^{k+1} = dv^k$, hence the composition is zero.
If $f^{k+1}u^{k+1} = dv^k$ for some $v^k \in B^k$, then $dg^k v^k = g^{k+1}dv^k = g^{k+1}f^{k+1}u^{k+1} = 0$ and $\delta^k g^k v^k = u^{k+1}$.

Such proofs are called *diagram chasing*.

Example 3.2. If we are given a morphism of complexes $g^\bullet : B^\bullet \to C^\bullet$ consisting of surjections, and define A^k to be the kernel of g^k, we remark that $A^\bullet$ is a subcomplex of $B^\bullet$ called the *kernel complex*, which has trivial cohomology iff $g^\bullet$ induces isomorphisms in cohomology $\mathrm{H}^k B^\bullet \to \mathrm{H}^k C^\bullet$.

Example 3.3. Suppose $A^\bullet$ is a complex. Define $Z^k, B^k \subseteq A^k$ by:

$$Z^k = \{u \in A^k \ : \ du = 0\},$$
$$B^k = \{du \ : \ u \in A^{k-1}\}.$$

These are both subcomplexes of $A^\bullet$ and their differentials are the 0 morphisms. We have short exact sequences:

$$0 \to Z^k \to A^k \to B^{k+1} \to 0.$$

Moreover the arrows here are morphisms of complexes. We thus get a long exact sequence:

$$\cdots B^k \to Z^k \to \mathrm{H}^k A^\bullet \to B^{k+1} \to Z^{k+1} \cdots$$

It turns out that the connecting morphism is the inclusion $B^k \subseteq Z^k$, which is injective, so we get exact sequences:

$$0 \to B^k \to Z^k \to \mathrm{H}^k A^\bullet \to 0,$$

which confirm, if need be, that:

$$\mathrm{H}^k A^\bullet \simeq Z^k / B^k.$$

Remark 3.3. Suppose we have an exact sequence:

$$A^1 \xrightarrow{\phi} A^2 \longrightarrow A^3 \longrightarrow A^4 \xrightarrow{\psi} A^5.$$

Then we have a short exact sequence:

$$0 \to \operatorname{coker} \phi \to A_3 \to \ker \psi \to 0.$$

In particular, full knowledge of the connecting morphism associated with a short exact sequence of complexes determines the cohomology of the middle complex.

The construction of the connecting morphism δ has the following property. Suppose that the above situation holds for complexes $A_0^\bullet$, $B_0^\bullet$ and $C_0^\bullet$ (with $f_0^\bullet$ and $g_0^\bullet$) and also for $A_1^\bullet$, $B_1^\bullet$ and $C_1^\bullet$ (with $f_1^\bullet$ and $g_1^\bullet$). Suppose furthermore that we have morphisms of complexes $\alpha^\bullet : A_0^\bullet \to A_1^\bullet$, $\beta^\bullet : B_0^\bullet \to B_1^\bullet$ and $\gamma^\bullet : C_0^\bullet \to C_1^\bullet$, intertwining $f_1^\bullet$ with $f_0^\bullet$ and $g_1^\bullet$ with $g_0^\bullet$. Explicitly we have commuting diagrams:

$$\begin{array}{ccccccccc}
0 & \longrightarrow & A_0^k & \xrightarrow{f_0^k} & B_0^k & \xrightarrow{g_0^k} & C_0^k & \longrightarrow & 0 \\
 & & \downarrow{\scriptstyle \alpha^k} & & \downarrow{\scriptstyle \beta^k} & & \downarrow{\scriptstyle \gamma^k} & & \\
0 & \longrightarrow & A_1^k & \xrightarrow{f_1^k} & B_1^k & \xrightarrow{g_1^k} & C_1^k & \longrightarrow & 0
\end{array}$$

Then the following diagram commutes:

$$\begin{array}{ccc}
\mathrm{H}^k C_0^\bullet & \xrightarrow{\delta_0^k} & \mathrm{H}^{k+1} A_0^\bullet \\
\downarrow{\scriptstyle \mathrm{H}^k \gamma^\bullet} & & \downarrow{\scriptstyle \mathrm{H}^{k+1} \alpha^\bullet} \\
\mathrm{H}^k C_1^\bullet & \xrightarrow{\delta_1^k} & \mathrm{H}^{k+1} A_1^\bullet
\end{array}$$

(This property is a functoriality property of δ.)

The following result is known as the *five lemma*.

Lemma 3.2. *Suppose we have a commuting diagram:*

$$\begin{array}{ccccccccc} A^1 & \longrightarrow & A^2 & \longrightarrow & A^3 & \longrightarrow & A^4 & \longrightarrow & A^5 \\ \downarrow{\scriptstyle f^1} & & \downarrow{\scriptstyle f^2} & & \downarrow{\scriptstyle f^3} & & \downarrow{\scriptstyle f^4} & & \downarrow{\scriptstyle f^5} \\ B^1 & \longrightarrow & B^2 & \longrightarrow & B^3 & \longrightarrow & B^4 & \longrightarrow & B^5 \end{array}$$

where each row is exact. If f^1, f^2, f^4 and f^5 are isomorphisms, then f^3 is also an isomorphism.

Proof. We let u^i denote elements of A^i and v^i elements of B^i. All horizontal arrows are denoted d. We prove:

If f^1 is surjective, and f^2 and f^4 are injective, then f^3 is injective:
Suppose $f^3u^3 = 0$. Then $f^4du^3 = df^3u^3 = 0$, hence $du^3 = 0$. Put $u^3 = du^2$.
We have $df^2u^2 = f^3du^2 = f^3u^3 = 0$. Put $f^2u^2 = dv^1$.
We have $f^2du^1 = df^1u^1 = dy^1 = f^2u^2$. Hence $du^1 = u^2$.
We have $u^3 = ddu^1 = 0$.

If f^5 is injective, and f^2 and f^4 are surjective, then f^3 is surjective:
Pick v^3. Write $dv^3 = f^4u^4$.
We have $f^5du^4 = df^4u^4 = ddv^3 = 0$, hence $du^4 = 0$. Put $u^4 = du^3$.
We have $dv^3 = f^4u^4 = f^4du^3 = df^3u^3$, hence $v^3 - f^3u^3 = dv^2$ and $v^2 = f^2u^2$.
We have $v^3 = f^3u^3 - dv^2 = f^3u^3 - df^2u^2 = f^3(u^3 - du^2)$.

3.1.2 Multilinear algebra and tensor products

Differential forms (which will be defined later) involve multilinear maps on the tangent spaces of a manifold. The tensor product of multilinear maps provides a convenient tool for their study. Moreover differential forms on a product manifold have a special structure best described in terms of tensor products of spaces.

If U is a vector space, $U^\star$ denotes its dual. If U and V are vector spaces, $\mathrm{L}(U,V)$ is the space of linear maps $U \to V$. If U, V and W are vector spaces, $\mathrm{B}(U,V;W)$ is the space of bilinear maps $U \times V \to W$. Recall that if $b : U \times V \to W$ is bilinear, $b(U,V)$ designates the subspace of W generated by elements of the form $b(u,v)$, with $u \in U$ and $v \in V$.

Let U, V be vector spaces. We say that $(W, \otimes)$ is a *tensor product* of U and V if W is a vector space, $\otimes : U \times V \to W$ is bilinear, and for any W' and any $b : U \times V \to W'$ there is a unique $\beta : W \to W'$ such that for all $u \in U$ and all $v \in V$:

$$b(u,v) = \beta(u \otimes v).$$

Thus tensor products, when they exist, are unique up to canonical isomorphism.

Remark 3.4. $(W, \otimes)$ is a tensor product of U and V iff the following two conditions hold. First, for any free families $(u_i)_{i\in I}$ in U and $(v_j)_{j\in J}$ in V, the family $(u_i \otimes$

$v_j)_{(i,j)\in I\times J}$ is free in W. Second, the vector space W is generated by elements of the form $u\otimes v$ for $u\in U$ and $v\in V$.

Let U, V be vector spaces. Put $W = \mathrm{B}(U,V)^\star$ and define, for $u\in U$, $v\in V$ and $b\in \mathrm{B}(U,V)$:

$$(u\otimes v)(b) = b(u,v).$$

Then $\otimes : U\times V \to W$ is bilinear and $(U\otimes V, \otimes)$ is a tensor product of U and V. Thus tensor products exist. Unless otherwise specified, this is the tensorproduct we will use.

When U, V and W are vector spaces and a tensor product of U and V is chosen, we have a canonical isomorphism:

$$\mathrm{L}(U\otimes V, W) \to \mathrm{B}(U,V;W).$$

Let V be a vectorspace. $\mathrm{L}^k(V)$ denotes the space of k-linear maps $V^k \to \mathbb{R}$. In particular, $\mathrm{L}^1(V)$ is the dual of V, also denoted $V^\star$.

If $u\in \mathrm{L}^k(V)$ and $v\in \mathrm{L}^l(V)$, then $u\otimes v\in \mathrm{L}^{k+l}(V)$ is defined by:

$$u\otimes v(\xi_1,\cdots,\xi_{k+l}) = u(\xi_1,\cdots,\xi_k)v(\xi_{k+1},\cdots,\xi_{k+l}).$$

Lemma 3.3. *If $(u_i)_{i\in I}$ is free in $\mathrm{L}^k(V)$ and $(v_j)_{j\in J}$ is free in $\mathrm{L}^l(V)$, then the family $(u_i\otimes v_j)_{(i,j)\in I\times J}$ is free in $\mathrm{L}^{k+l}(V)$.*

Comparing dimensions we find that if V is finite-dimensional, $(\mathrm{L}^{k+l}(V),\otimes)$ is a tensor product of $\mathrm{L}^k(V)$ and $\mathrm{L}^l(V)$.

Suppose $f_i : U_i \to V_i$ for $i = 0,1$. Define $f_0\otimes f_1 : U_0\otimes U_1 \to V_0\otimes V_1$ by:

$$(f_0\otimes f_1)(u_0\otimes u_1) = (f_0u_0)\otimes(f_1u_1).$$

This is called the *Kronecker product* of f_0 and f_1. When all spaces are finite dimensional, $\mathrm{L}(U_0\otimes U_1, V_0\otimes V_1)$ equipped with the Kronecker product is indeed a tensor product of $\mathrm{L}(U_0,V_0)$ and $\mathrm{L}(U_1,V_1)$.

Remark 3.5. If $b : U\times V \to Z$ is bilinear, any $w\in b(U,V)$ can be written as $w = \sum_{ij}\lambda_{ij} b(u_i,v_j)$, with (u_i) and (v_j) free.

The following canonical isomorphisms are useful. If $U\subseteq V$, we have for any W:

$$(V\otimes W)/(U\otimes W) \simeq (V/U)\otimes W.$$

If $U_i \subseteq V_i$ for $i = 0,1$, we have:

$$(V_0\oplus V_1)/(U_0\oplus U_1) \simeq (V_0/U_0)\oplus(V_1/U_1).$$

An element $u\in \mathrm{L}^k(V)$ is alternating iff for any permutation $\sigma\in\mathfrak{S}_k$:

$$u(\xi_{\sigma_1},\cdots,\xi_{\sigma_k}) = \varepsilon(\sigma)u(\xi_1,\cdots,\xi_k).$$

The subspace of $\mathrm{L}^k(V)$ consisting of alternating forms is denoted $\mathrm{L}^k_a(V)$. The canonical projection in $\mathrm{L}^k(V)$ onto $\mathrm{L}^k_a(V)$ is denoted α_k.

$$(\alpha_k u)(\xi_1, \cdots, \xi_k) = \frac{1}{k!} \sum_{\sigma} \varepsilon(\sigma) u(\xi_{\sigma_1}, \cdots, \xi_{\sigma_k}),$$

where the sum extends over the set of permutations of $\{1, \cdots, k\}$.

The wedge product of $u \in \mathrm{L}^k_a(V)$ and $v \in \mathrm{L}^l_a(V)$ is $u \wedge v \in \mathrm{L}^{k+l}_a(V)$ defined by:

$$u \wedge v = \frac{(k+l)!}{k!\,l!} \alpha_{k+l}(u \otimes v).$$

The wedge product is bilinear and associative. We have, for $u_i \in V^\star$:

$$(u_1 \wedge \cdots \wedge u_k)(\xi_1, \cdots, \xi_k) = \sum_{\sigma} \varepsilon(\sigma) u_1(\xi_{\sigma_1}) \cdots u_k(\xi_{\sigma_k}).$$

Moreover, if $u \in \mathrm{L}^k_a(V)$ and $v \in \mathrm{L}^l_a(V)$:

$$v \wedge u = (-1)^{kl} u \wedge v.$$

This property is known as graded commutativity.

Suppose V is n-dimensional. If $(e_i)_{1 \le i \le n}$ is a basis of $V^\star$, then the following family is a basis of $A_k(V)$:

$$e_{i_1} \wedge \cdots \wedge e_{i_k} \text{ for } 1 \le i_1 < \cdots < i_k \le n.$$

It may be regarded as being indexed by the set of subsets of $\{1, \cdots, n\}$ with cardinality k. In particular:

$$\dim \mathrm{L}^k_a(V) = \binom{n}{k}.$$

If $F : V_0 \to V_1$ is linear and $u \in \mathrm{L}^k(V_1)$, then $F^\star u \in \mathrm{L}^k(V_0)$ is defined by:

$$F^\star u(\xi_1, \cdots, \xi_k) = u(F\xi_1, \cdots, F\xi_k).$$

If $u \in \mathrm{L}^k_a(V_1)$, then $F^\star u \in \mathrm{L}^k_a(V_0)$.

Suppose $V = V_0 \oplus V_1$. Let P_0 be the projection onto V_0 with kernel V_1, and P_1 be the projection onto V_1 with kernel V_0. Then the spaces $P_0^\star \mathrm{L}^{k_0}_a(V_0) \otimes P_1^\star \mathrm{L}^{k_1}_a(V_1)$ form a direct sum.

Proposition 3.3. *With a slight abuse of notations, the following map is an isomorphism:*

$$\alpha_k : \bigoplus_{k_0+k_1=k} \mathrm{L}^{k_0}_a(V_0) \otimes \mathrm{L}^{k_1}_a(V_1) \to \mathrm{L}^k_a(V).$$

Proof. Prove injectivity and note that, by the formula:

$$\sum_{k_0+k_1=k} \binom{n_0}{k_0}\binom{n_1}{k_1} = \binom{n}{k},$$

where $n = n_0 + n_1$, the dimensions are equal.

A scalar product on V is a $g \in \mathrm{L}^2(V)$, which is symmetric and positive definite. It induces a scalar product on $V^\star$ by requiring that $u \mapsto g(u,\cdot)$ is an isometry. In other words, if $l, l' \in V^\star$ are represented as $l = g(u,\cdot)$ and $l' = g(u',\cdot)$, then we define $g(l,l') = g(u,u')$. There is a unique scalar product on multilinear forms such that:

$$g(u_1 \otimes \cdots \otimes u_k, v_1 \otimes \cdots \otimes v_k) = g(u_1,v_1)\cdots g(u_k,v_k).$$

If we restrict this scalar product to alternating forms, we notice:

$$g(u_1 \wedge \cdots \wedge u_k, v_1 \wedge \cdots \wedge v_k) = k! \sum_\sigma \varepsilon(\sigma) g(u_1, v_{\sigma_1}) \cdots g(u_k, v_{\sigma_k}).$$

Actually, one scales this scalar product and defines:

$$g(u_1 \wedge \cdots \wedge u_k, v_1 \wedge \cdots \wedge v_k) = \det g(u_i, v_j).$$

With this scaling, if $(e_i)_{1\le i\le n}$ is an orthonormal basis of $V^\star$, then the family:

$$e_{i_1} \wedge \cdots \wedge e_{i_k} \text{ for } 1 \le i_1 < \cdots < i_k \le n,$$

indexed by subsets of $\{1,\cdots,n\}$ of cardinality k, is orthonormal in $\mathrm{L}^k_a(V)$.

A scalar product and an orientation of V give rise to a volume form ω, namely the determinant in any orthonormal oriented basis. Suppose $\dim(V) = n$. The *Hodge star* is the map $\star : \mathrm{L}^k_a(V) \to \mathrm{L}^{n-k}_a(V)$ uniquely determined by:

$$u \wedge \star v = g(u,v)\omega.$$

If $(e_i)_{1\le i\le n}$ is an orthonormal basis of $V^\star$, then for any set $i_1 < \cdots < i_k$ with complement $j_1 < \cdots < j_{n-k}$, we have:

$$\star(e_{i_1} \wedge \cdots \wedge e_{i_k}) = \varepsilon(I,J) e_{j_1} \wedge \cdots \wedge e_{j_{n-k}},$$

where $\varepsilon(I,J)$ is the sign of the permutation $(i_1,\cdots,i_k,j_1,\cdots,j_{n-k})$.

One checks that: $\star\star = \pm\mathrm{id}$ and $\star' = \pm\star$.

We now turn to the interplay between tensor products and homological algebra.

A *graded* vector space is a sequence $A^\bullet = (A^k)_{k\in\mathbb{Z}}$ of vector spaces, identified with the space:

$$A = \bigoplus_k A^k.$$

Thus a complex is a particular kind of graded vector space, as well as the family of its cohomology groups. The map H, taking a complex to its cohomology sequence, is a functor from complexes to graded vector spaces.

If F and G are graded vector spaces, their tensor product is by definition the graded vector space defined by:

$$(F \otimes G)^l = \bigoplus_k F^k \otimes G^{l-k},$$

and

$$(u \otimes v)^l = \sum_k u^k \otimes v^{l-k}.$$

If F and G are both complexes (i.e., they are equipped with differentials), then $F \otimes G$ is a complex when equipped with the differential defined on $F^k \otimes G^{l-k}$ by:

$$d(u \otimes v) = (du) \otimes v + (-1)^k u \otimes (dv).$$

If A is graded, we introduce the map $\sigma : A \to A$ defined on A^k by $\sigma u = (-1)^k u$. Thus $\sigma\sigma = \mathrm{id}_A$, and if A is equipped with a differential d, we notice that $d\sigma = -\sigma d$.

With this definition we can write for any $u \in F$ and $v \in G$:

$$d(u \otimes v) = (du) \otimes v + (\sigma u) \otimes (dv).$$

This gives a proof that this is indeed a differential:

$$\begin{aligned} dd(u \otimes v) &= d((du) \otimes v + (\sigma u) \otimes (dv)) \\ &= ddu \otimes v + (\sigma du + d\sigma u) \otimes (dv) + (\sigma\sigma u) \otimes (ddv) \\ &= 0. \end{aligned}$$

Remark 3.6. Suppose F_i and G_i are complexes for $i = 0,1$ and $f_i : F_i \to G_i$ are morphisms of complexes. Then the Kronecker product $f_0 \otimes f_1$ is a morphism of complexes $F_0 \otimes F_1 \to G_0 \otimes G_1$.

The following is the algebraic Kunneth theorem.

Theorem 3.2. *We have a canonical isomorphism:*

$$\mathrm{H}(F) \otimes \mathrm{H}(G) \simeq \mathrm{H}(F \otimes G).$$

Proof. We will use the following remark. Suppose that A is a graded vectorspace equipped with the 0 differential. Then $A \otimes G$ has the differential:

$$d : A^k \otimes G^{l-k} \to A^k \otimes G^{l+1-k} \text{ with } u \otimes v \mapsto (-1)^k u \otimes dv.$$

Thus we get:

$$\mathrm{H}^l(A \otimes G) \simeq (A \otimes \mathrm{H}(G))^l.$$

Return now to the case of $F \otimes G$. Recalling the construction of Example 3.3, we denote by $B^k, Z^k \subseteq F^k$ the image and kernel of the differential of F. We have short exact sequences:

$$0 \to Z^k \to F^k \to B^{k+1} \to 0.$$

Tensoring with the identity of G^{l-k}, we get short exact sequences, for each l:

$$0 \to \bigoplus_k Z^k \otimes G^{l-k} \to \bigoplus_k F^k \otimes G^{l-k} \to \bigoplus_k B^{k+1} \otimes G^{l-k} \to 0.$$

Thus we have a short exact sequence of complexes:

$$0 \to (Z \otimes G)^\bullet \to (F \otimes G)^\bullet \to (B \otimes G)^{\bullet+1} \to 0.$$

This gives a long exact sequence:

$$\mathrm{H}^l(B \otimes G) \to \mathrm{H}^l(Z \otimes G) \to \mathrm{H}^l(F \otimes G) \to \mathrm{H}^{l+1}(B \otimes G) \to \mathrm{H}^{l+1}(Z \otimes G).$$

Using the above remark we get:

$$(B \otimes \mathrm{H}(G))^l \to (Z \otimes \mathrm{H}(G))^l \to \mathrm{H}^l(F \otimes G) \to (B \otimes \mathrm{H}(G))^{l+1} \to (Z \otimes \mathrm{H}(G))^{l+1}.$$

The connecting morphism, appearing here as the first and last arrow, is the Kronecker product of the inclusion $B^k \subseteq Z^k$ with the identity of $\mathrm{H}(G)$. It is injective. Thus we get short exact sequences:

$$0 \to (B \otimes \mathrm{H}(G))^l \to (Z \otimes \mathrm{H}(G))^l \to \mathrm{H}^l(F \otimes G) \to 0.$$

Yet we have:

$$(Z \otimes \mathrm{H}(G))^l / (B \otimes \mathrm{H}(G))^l \simeq \bigoplus_k (Z^k / B^k) \otimes \mathrm{H}^{l-k}(G).$$

Thus we get:

$$\mathrm{H}^l(F \otimes G) \simeq (\mathrm{H}(F) \otimes \mathrm{H}(G))^l,$$

which is what we want.

3.2 Differential geometry

In this section we outline continuous geometry (differential forms), discrete geometry (cellular complexes) and the relation between the two. For details see [13, 44, 54, 61].

3.2.1 Continuous geometry

Let M be a topological space. A *chart* on M is a continuous map $F : U \to M$, where U is open in some normed vector (half–) space, such that $F(U)$ is open in M and the induced map $U \to F(U)$ is a homeomorphism. An *atlas* on M is a family of charts $F_i : U_i \to M$ such that the sets $F_i(U_i)$ cover M.

A *manifold* is a topological space equipped with an atlas. Suppose $(F_i : U_i \to M)$ is an atlas. Put $V_i = F(U_i)$ and $V_{ij} = V_i \cap V_j$. The atlas is said to be differentiable if the maps:

$$F_{ji} = F_j^{-1} F_i : F_i^{-1}(V_{ij}) \to F_j^{-1}(V_{ij})$$

are differentiable. These maps are called *transition maps*.

Put $V_{ijk} = V_i \cap V_j \cap V_k$. Then, on $F_i^{-1}(V_{ijk})$, we have well defined maps satisfying:

$$F_{ki} = F_{kj} F_{ji}.$$

Notice also that:

$$F_{ii} = \mathrm{id}.$$

The manifold M can be identified with the disjoint union $\cup_{i \in I}\{i\} \times U_i$, quotiented by the finest relation such that:

$$(i,x) \sim (j, F_{ji}(x)),$$

whenever F_{ji} is defined on x. Explicitly an identification is well-defined by:

$$(i,x) \mapsto F_i(x).$$

The tangent manifold is the disjoint union $\cup_{i \in I}\{i\} \times U_i \times V$ quotiented by the relation:

$$(i,x,\xi) \sim (j, F_{ji}(x), \mathrm{D}F_{ji}(x)\xi).$$

The canonical projection $\pi : \mathrm{T}M \to M$ is well defined by:

$$\pi(i,x,\xi) = (i,x).$$

The tangent space of M at x is $\mathrm{T}_x M = \pi^{-1}(x)$. It is a vector space.

The space of smooth functions $M \to \mathbb{R}$ is denoted $\mathcal{F}(M)$.

A vector field on M is a smooth map $X : M \to \mathrm{T}M$ such that $\pi X = \mathrm{id}_M$. In other words, it is an object, which to any $x \in M$ associates an element $X_x \in \mathrm{T}_x M$. The space of vector fields on M is denoted $\mathcal{V}(M)$.

A differential form of degree k on a manifold M is an object, which to every point x associates an element $u_x \in \mathrm{L}_a^k(\mathrm{T}_x M)$. Such an object is also called a k-form. The space of smooth k-forms is denoted by $\Omega^k(M)$. If $u \in \Omega^k(M)$ and $v \in \Omega^l(M)$, then their wedge product is $u \wedge v \in \Omega^{k+l}(M)$, defined pointwise. Explicitly:

$$(u \wedge v)_x(\xi_1, \cdots, \xi_{k+l}) = \frac{1}{k!l!} \sum_\sigma \varepsilon(\sigma) u_x(\xi_{\sigma_1}, \cdots, \xi_{\sigma_k}) v_x(\xi_{\sigma_{k+1}}, \cdots, \xi_{\sigma_{k+l}}),$$

where the sum extends over all permutations.

If $F : M \to N$ and $u \in \Omega^k(N)$, the pull-back of u to N is denoted $F^\star u$ and defined by:

$$(F^\star u)_x(\xi_1, \cdots, \xi_k) = u_{F(x)}(\mathrm{D}F(x)\xi_1, \cdots, \mathrm{D}F(x)\xi_k).$$

In other words:

$$(F^\star u)_x = \mathrm{D}F(x)^\star u_{F(x)}.$$

If $F_0 : M_0 \to M_1$ and $F_1 : M_1 \to M_2$, then $(F_1 F_0)^\star = F_0^\star F_1^\star$.

Let U be open in a vector space V. An element $u \in \Omega^k(U)$ can be considered as a map $U \to \mathrm{L}^k_a(V)$. Its total (or Fréchet) derivative is a map:

$$\mathrm{D}u : U \to \mathrm{L}(V, \mathrm{L}^k_a(V)) \subseteq \mathrm{L}^{k+1}(V).$$

The exterior derivative $\mathrm{d} : \Omega^k(U) \to \Omega^{k+1}(U)$ is defined by:

$$(\mathrm{d}u)_x(\xi_0, \cdots, \xi_k) = \sum_{i=0}^{k} (-1)^i (\mathrm{D}u)_x(\xi_i, \xi_0, \cdots, \hat{\xi}_i, \cdots, \xi_k).$$

In other words: $\mathrm{d}u = (k+1)\alpha_{k+1}(\mathrm{D}u)$, where α_k is the canonical antisymmetrizing projector on k-linear forms.

Lemma 3.4. *If $F : U_0 \to U_1$ is a map:*

$$\mathrm{d}F^\star u = F^\star \mathrm{d}u.$$

Proof. We have:

$$(F^\star u)_x(\xi_1, \cdots, \xi_k) = u_{F(x)}(\mathrm{D}F(x)\xi_1, \cdots, \mathrm{D}F(x)\xi_k).$$

Hence:

$$\begin{aligned} &\mathrm{D}(F^\star u)_x(\xi_0, \xi_1, \cdots, \xi_k) \\ &= (\mathrm{D}u)_{F(x)}(\mathrm{D}F(x)\xi_0, \mathrm{D}F(x)\xi_1, \cdots, \mathrm{D}F(x)\xi_k) + \\ &\sum_{i=1}^{k} u_{F(x)}(\mathrm{D}F(x)\xi_1, \cdots, \mathrm{D}^2F(x)(\xi_0, \xi_i), \cdots, \mathrm{D}F(x)\xi_k). \end{aligned}$$

The sum disappears upon anti-symmetrization, and the first term corresponds to the pull-back of $F^\star \mathrm{d}u$.

From this it follows that the exterior derivative can be defined on manifolds, yielding a map $\mathrm{d} : \Omega^k(M) \to \Omega^{k+1}(M)$. It has the following properties:

Proposition 3.4. *For any map $F : M_0 \to M_1$ and any $u \in \Omega^k(M_1)$, we have:*

$$\mathrm{d}F^\star u = F^\star \mathrm{d}u.$$

For any $u \in \Omega^k(M)$ and $v \in \Omega^l(M)$ (Leibniz formula):

$$\mathrm{d}(u \wedge v) = (\mathrm{d}u) \wedge v + (-1)^k u \wedge (\mathrm{d}v). \tag{34}$$

Moreover, $\mathrm{d}^2 = 0$. In other words we have a complex:

$$\cdots \to \Omega^k(M) \to \Omega^{k+1}(M) \to \cdots \tag{35}$$

Proof. The first property follows from Lemma 3.4. We likewise prove the two others in a chart over $U \subseteq V$.

We have:

$$\begin{aligned} \mathrm{d}(u \wedge v) &= (k+l+1)\alpha_{k+l+1}\mathrm{D}(u \wedge v) \\ &= \frac{(k+l+1)!}{k!\,l!}\alpha_{k+l+1}\mathrm{D}(u \otimes v) \\ &= \frac{(k+l+1)!}{k!\,l!}\alpha_{k+l+1}((\mathrm{D}u) \otimes v + \tau(u \otimes \mathrm{D}v)), \end{aligned}$$

where τ is the cyclic permutation of the variables represented by:

$$(0,\cdots,k+l) \to (1,\cdots,k,0,k+1,\cdots,k+l).$$

The sign of this permutation is $(-1)^k$. This gives the second identity.

We have:

$$\begin{aligned} \mathrm{dd}u &= \alpha_{k+2}\mathrm{D}(\alpha_{k+1}\mathrm{D}u) \\ &= \alpha_{k+2}\mathrm{D}^2 u. \end{aligned}$$

Since $\mathrm{D}^2 u$ is symmetric in its first two arguments, this gives the third identity.

Remark 3.7. The exterior derivative is uniquely determined by properties (34) and (35), together with its action on $\Omega^0(M)$. Indeed elements of the form $u_0 \mathrm{d}u_1 \wedge \cdots \wedge \mathrm{d}u_k$ with $u_i \in \Omega^0(M)$ generate $\Omega^k(M)$.

If u is a differential k-form and X a vectorfield, the contraction of u by X is $i_X u$, defined by:

$$(i_X u)_x(\xi_2,\cdots,\xi_k) = u_x(X(x),\xi_2,\cdots\xi_k).$$

We shall define integration of n-forms on n-dimensional oriented manifolds with boundary. A non-degenerate smooth n-form is called a volume form. The local model of such a manifold is:

$$\mathbb{R}^n_+ = \{(x_1,\cdots,x_n) \in \mathbb{R}^n \ : \ x_1 \leq 0\}.$$

Let ω denote the canonical volume form on $\mathbb{R}^n$. Let U_0 and U_1 be two open subsets of $\mathbb{R}^n_+$ and $F : U_0 \to U_1$ an orientation preserving diffeomorphism. The determinant of $\mathrm{D}F(x)$ in orthonormal oriented bases (the Jacobian) is denoted $\mathrm{Jac}\,F(x)$. For $i = 0,1$ let $u_i = f_i\omega$ be a differential n-form on U_i, with f_i a function on U_i. We suppose that $u_0 = F^\star u_1$. Then $f_0 = f_1 \circ F\, \mathrm{Jac}\,F$, hence f_0 is integrable iff f_1 is integrable and in this case:

$$\int_{U_1} f_1 = \int_{U_0} f_0.$$

An atlas is said to be orientation-preserving if all transition maps are orientation-preserving. A manifold is orientable if it has an orientation-preserving atlas. Two oriented atlases are orientation-equivalent if the union is oriented. If a manifold with N connected components is orientable, this equivalence relation on oriented atlases has 2^N equivalence classes. An orientation of the manifold is the choice of such an equivalence class, and an oriented atlas is an atlas in this equivalence class.

Let M be a compact manifold of dimension n, which is oriented. Let $(F_i : U_i \to M)$ be a countable oriented atlas, and (ϕ_i) be a subordinated partition of unity. Let $u \in \Omega^n(M)$. Put $u_i = F_i^\star u$ and let $u_i = f_i \omega$. We say that u is integrable if the f_i are measurable and:

$$\sum_i \int_{U_i} (\phi_i \circ F_i)|f_i| < +\infty.$$

If u is integrable we put:

$$\int u = \sum_i \int_{U_i} (\phi_i \circ F_i) f_i.$$

One checks that the above notions of integrability and integral are independent of the choice of the (countable oriented) atlas and partition of unity. We notice that for any chart $F : U \to M$ and any $u \in \Omega^n(M)$, if we define f by $F^\star u = f\omega$ we have:

$$\int_{F(U)} u = \int_U f.$$

The charts $F_i : U_i \to M$ are such that:

$$\forall x \in U_i \quad F_i(x) \in \partial M \iff x_1 = 0.$$

The maps $(F_i|_{x_1=0})$ constitute an orientation-preserving atlas of ∂M. It determines the induced orientation of ∂M.

Let M be a manifold with boundary ∂M oriented outwardly by the orientation of M. If ω is a positively oriented volume form on M, and X is a field on ∂M of vectors in $\mathrm{T}M$ (a section of the injection $\partial M \to \mathrm{T}M$) pointing out of M, then $i_X \omega$ is a positively oriented volume form on ∂M.

The following result is known as *Stokes' theorem.*

Theorem 3.3. *For any compact manifold M, any $u \in \Omega^{n-1}(M)$, which is continuous up to the boundary, and whose exterior derivative is integrable:*

$$\int_M \mathrm{d}u = \int_{\partial M} u.$$

The following result concerning the cohomology of contractible domains is known as *Poincaré's lemma* . Recall that a domain is contractible if the identity map is homotopic to a constant map. Using Lie derivatives and Cartan's formula, one shows that a homotopy $t \mapsto f_t$ between maps f_0 and f_1 gives a homotopy operator (in the algebraic sense) between the pullbacks $f_0^\star$ and $f_1^\star$.

Lemma 3.5. *If M is a contractible domain then $\mathrm{H}^k(M) = 0$ for $k \geq 1$ (and $\mathrm{H}^0(M)$ consists of the constant functions).*

Suppose M is a three dimensional manifold equipped with a Riemannian metric g. Given a vector field $u \in \mathcal{V}(M)$, one can associate with it the two forms

$\Psi^1 u = g(u,\cdot) \in \Omega^1(X)$ and $\Psi^2 u = \omega(u,\cdots) \in \Omega^2(X)$. A function $u \in \mathcal{F}(M)$ can be associated with itself, considered as an element $\Psi^0 u = u \in \Omega^0(M)$, and with the form $\Psi^3 u = u\omega \in \Omega^3(M)$. Then one has the commuting diagram:

$$\begin{array}{ccccccc}
\mathcal{F}(M) & \xrightarrow{\text{grad}} & \mathcal{V}(M) & \xrightarrow{\text{curl}} & \mathcal{V}(M) & \xrightarrow{\text{div}} & \mathcal{F}(M) \\
\downarrow{\scriptstyle \Psi^0} & & \downarrow{\scriptstyle \Psi^1} & & \downarrow{\scriptstyle \Psi^2} & & \downarrow{\scriptstyle \Psi^3} \\
\Omega^0(M) & \xrightarrow{\mathrm{d}} & \Omega^1(M) & \xrightarrow{\mathrm{d}} & \Omega^2(M) & \xrightarrow{\mathrm{d}} & \Omega^3(M)
\end{array}$$

where the vertical arrows are isomorphisms. Through this correspondence, problems expressed in terms of vector fields can be translated into problems expressed with differential forms.

3.2.2 Discrete geometry

A k-simplex is a set with $k+1$ elements ($k \geq 0$); its elements are called vertices. The standard simplex is $[k] = \{0,\cdots,k\}$. Its geometric realization is Δ_k defined by:

$$\Delta_k = \{(x_0,\cdots,x_k) \in \mathbb{R}^{k+1} \ : \ x_i \geq 0 \text{ and } x_0 + \cdots + x_k = 1\}.$$

The vertices of Δ_k are the standard basis vectors $(e_i^k)_{i\in[k]}$ of $\mathbb{R}^{k+1}$. If $\sigma : [k'] \to [k]$ is a map, there is a unique affine map $\Delta_{k'} \to \Delta_k$ sending $e_i^{k'}$ to $e_{\sigma(i)}^k$.

A simplicial complex $\mathcal{T}$ is a set of finite non-empty sets (simplices), such that for any T in $\mathcal{T}$, any non-empty subset of T is also in $\mathcal{T}$. Such a subset of T is called a face. It is proper when it is different from T. A k-simplex is said to be k-dimensional. For each k we let $\mathcal{T}^k$ denote the subset of $\mathcal{T}$ consisting of k-dimensional simplices.

A simplicial complex determines a topological space by gluing together spaces Δ_k using the connectivity of $\mathcal{T}$. More precisely $|\mathcal{T}|$ denotes the set of maps $x : \mathcal{T}^0 \to [0,1]$ such that x has support in a $T \in \mathcal{T}$ (i.e., there exists a $T \in \mathcal{T}$ such that x is non-zero only on the vertices of T) and moreover:

$$\sum_{\alpha} x(\alpha) = 1.$$

For any simplex $T \in \mathcal{T}$ we denote by $|T|$ the set of $x \in |\mathcal{T}|$ with support in T. Notice that for any bijection $\alpha : [k] \to T$, we have a bijection $|\alpha| : \Delta_k \to |T|$. The topology of $|\mathcal{T}|$ is defined by the property that $F \subseteq |\mathcal{T}|$ is closed iff $F \cap |T|$ is closed for each T, when $|T|$ is equipped with the topology inherited from some Δ_k.

For simplices, orientation can be defined combinatorially. For a k-simplex T, we consider the equivalence relation on bijections $\alpha : [k] \to T$ defined by $\alpha \sim \alpha'$ iff $\alpha^{-1}\alpha'$ is an even permutation of $[k]$. This equivalence relation has two equivalence classes when $k \geq 1$, and a combinatorial orientation of T is a choice of one of the equivalence classes. The standard simplex $[k]$ has a canonical combinatorial orientation, namely the equivalence class of the identity map, the set of even permutations. We say that the elements of the combinatorial orientation of T are the orientation preserving maps $[k] \to T$. A combinatorial orientation of T determines an orientation of the manifold $|T|$ by requiring that the induced map $|\alpha| : \Delta_k \to |T|$ be orientation preserving for any orientation preserving map $\alpha : [k] \to T$. The chosen orientation of Δ_k is the one such that $\mathrm{d}x_1 \wedge \cdots \wedge \mathrm{d}x_k$ is positive, or equivalently the one it obtains as the boundary of the simplex:

$$\{(x_0, \cdots, x_k) \ : \ x_i \geq 0 \ \text{and} \ \sum x_i \leq 1\},$$

equipped with the orientation of $\mathbb{R}^{k+1}$.

Let T be a $(k+1)$-simplex and T' a k-face, equipped with combinatorial orientations. Choose an orientation preserving $\alpha : [k] \to T'$. Then $\varepsilon(T, T') = 1$ iff the bijection $[k+1] \to T$ defined by:

$$0 \mapsto T \setminus T' \text{ and } i+1 \mapsto \alpha_i,$$

is orientation preserving.

Example 3.4. Let M be a topological space and $(U_i)_{i \in I}$ be an open cover. The *nerve* of this cover is the simplicial complex consisting of the following subsets of I:

$$\{\alpha_0, \cdots, \alpha_k\} \subseteq I \ : \ U_{\alpha_0} \cap \cdots \cap U_{\alpha_k} \neq \emptyset.$$

In what follows we shall frequently identify T with $|T|$.

Let M denote a compact metric space.

To us, a k-dimensional cell in M is a closed subset T of M which is homeomorphic by a bi-Lipschitz map to the closed unit ball of $\mathbb{R}^k$. If a cell T is both k and l dimensional then $k = l$. We denote by ∂T its boundary, the image of the unit sphere in $\mathbb{R}^k$ by the chosen homeomorphism. Different homeomorphisms to the ball give the same boundary. The interior of T is by definition $T \setminus \partial T$ (it is open in T but not necessarily in M). A cellular complex on M is a finite set $\mathcal{T}$ of cells in M such that the following conditions hold:

- The union of all cells in $\mathcal{T}$ is M.
- Distinct cells in $\mathcal{T}$ have disjoint interiors.
- The boundary of any cell in $\mathcal{T}$ is a union of cells in $\mathcal{T}$.
- The topology of M is such that a subset F of M is closed iff $F \cap T$ is closed in T for any cell T in $\mathcal{T}$.

A simplicial complex is a cellular complex in which the intersection of any two cells is a cell (not just a union of cells), and such that the boundary of any cell is split into subcells in the same way as the boundary of a reference simplex is usually split into subsimplices. The reference simplex of dimension k is:

$$\{(x_0, \cdots, x_k) \in \mathbb{R}^{k+1} \ : \ x_i \geq 0 \text{ and } \sum_i x_i = 1\},$$

and its subsimplices are defined by the equations $x_i = 0$ for all $i \in J$, where $J \subseteq \{0, \cdots, k\}$.

We put for each $k \in \mathbb{N}$:

$$\mathcal{T}^k = \{T \in \mathcal{T} : \dim T = k\}.$$

The boundary of any cell T of $\mathcal{T}$ can be naturally equipped with a cellular complex, namely:

$$\{T' \in \mathcal{T} : T' \subseteq T, \, T' \neq T\}.$$

A *refinement* of a cellular complex $\mathcal{T}$ on M is a cellular complex $\mathcal{T}'$ on M such that each element of $\mathcal{T}$ is the union of elements of $\mathcal{T}'$. We will be particularly interested in simplicial refinements of a cellular complex.

A *cellular subcomplex* of a cellular complex $\mathcal{T}$ on M, is a cellular complex $\mathcal{T}'$ on some closed part M' of M such that $\mathcal{T}' \subseteq \mathcal{T}$. For instance, we have seen that the boundary of any cell T in $\mathcal{T}$ can be equipped with a cellular complex which is a subcomplex of $\mathcal{T}$.

Fix a cellular complex $(M, \mathcal{T})$. In the following we suppose that for each $T \in \mathcal{T}$ of dimension ≥ 1, the manifold T has been oriented. *Relative orientations*, also called *incidence numbers*, are defined as follows:

For any edge T, its vertices are ordered by the orientation of T, call them α_0 and α_1 in increasing order, and define $\varepsilon(T, \{\alpha_1\}) = 1$ and $\varepsilon(T, \{\alpha_0\}) = -1$. This definition is extended to higher dimensional cells as follows. Fix $k \geq 1$. Given $T \in \mathcal{T}^{k+1}$ and $T' \in \mathcal{T}^k$, if $T' \subseteq T$ we define $\varepsilon(T, T') = 1$ if T' is outward oriented compared with T, and $\varepsilon(T, T') = -1$ if it is inward oriented. For all $T, T' \in \mathcal{T}$ not covered by this definition we put $\varepsilon(T, T') = 0$.

For each k let $C^k(\mathcal{T})$ denote the set of maps $c : \mathcal{T}^k \to \mathbb{R}$. Such maps are called *k-cochains*. The *coboundary* operator $\delta : C^k(\mathcal{T}) \to C^{k+1}(\mathcal{T})$ is defined by:

$$(\delta c)_T = \sum_{T'} \varepsilon(T, T') c_{T'}.$$

In other words, it is the operator whose canonical matrix is ε. We remark that the coefficients in the sum can be non-zero only when $T' \in \partial T \cap \mathcal{T}^k$.

Lemma 3.6. *We have that $\delta\delta = 0$ as a map $C^k(\mathcal{T}) \to C^{k+2}(\mathcal{T})$.*

Proof. Pick $u \in C^k(\mathcal{T})$. Pick $T \in \mathcal{T}^{k+2}$. Let T'' be a k-face of T; it is shared by two $(k+1)$-faces of T denoted T'_0 and T'_1.

Suppose T'_0 and T'_1 have the same orientation compared with T. Then T'' is oppositely oriented relative to T'_0 and T'_1. If T'_0 and T'_1 have opposite orientations compared with T, then T'' has the same orientation relative to T'_0 and T'_1.

In both cases the contribution of $u_{T''}$ to $(\delta\delta u)_T$ is 0.

In other words, the family $C^\bullet(\mathcal{T})$ is a complex, called the cochain complex and represented by:

$$0 \to C^0(\mathcal{T}) \to C^1(\mathcal{T}) \to C^2(\mathcal{T}) \to \cdots$$

Suppose $\mathcal{T}$ is a cellular complex equipped with an orientation (of the cells) and $\mathcal{T}'$ is a cellular refinement also equipped with an orientation (for instance the same complex but with different orientations). For each cell $T \in \mathcal{T}^k$ and each $T' \in \mathcal{T}'^k$, define $\iota(T,T') = \pm 1$ if $T' \subseteq T$ and they have the same/different orientation, and $\iota(T,T') = 0$ in all other cases.

Proposition 3.5. *The map $\iota : C^\bullet(\mathcal{T}') \to C^\bullet(\mathcal{T})$ defined by:*

$$(\iota u)_T = \sum_{T'} \iota(T,T') u_{T'}$$

is a morphism of complexes.

Proof. We view ι^k as a matrix indexed by $\mathcal{T}^k \times \mathcal{T}'^k$. We also consider the compositions $\delta\iota$ and $\iota\delta'$ as matrices. Pick now $T \in \mathcal{T}^{k+1}$ and $T' \in \mathcal{T}'^k$.

Suppose T' is not a subset of T. Any $(k+1)$-cell of $\mathcal{T}'$ containing T' is not included in T, so $(\iota\delta')_{TT'} = 0$. Any k-cell of $\mathcal{T}$ containing T' is not included in T so $(\delta\iota)_{TT'} = 0$.

Suppose T' is a subset of T, but not of $|\partial T|$. Then T' is not included in any k-cell of $\mathcal{T}$ so $(\delta\iota)_{TT'} = 0$. On the other hand, T' is included in exactly two $(k+1)$-cells of $\mathcal{T}'$ inside T. If they have equal orientations compared with T, they induce opposite orientations on T', and vice-versa. Hence $(\iota\delta')_{TT'}$ is the sum of two opposite terms.

Suppose T' is a subset of $|\partial T|$ and let T_1 be the k-cell of $\mathcal{T}$ containing T' and T_1' the $(k+1)$-cell of $\mathcal{T}'$ included in T and containing T'. We have:

$$(\iota\delta')_{TT'} = \iota(T,T_1')\varepsilon'(T_1',T'),$$

and:

$$(\delta\iota)_{TT'} = \varepsilon(T,T_1)\iota(T_1,T').$$

These terms are both equal to the relative orientation of T' with respect to T. This completes the proof.

3.2.3 Relating continuous and discrete geometry

Let M be a manifold and $\mathcal{T}$ a cellular complex on M. The complex $C^\bullet(\mathcal{T})$ can be related to $\Omega^\bullet(M)$ as follows. For each $T \in \mathcal{T}^k$, let μ_T be the linear form on Ω^k defined by:

$$\mu_T : u \mapsto \int_T u|_T.$$

For each k we denote by $\mu^k : \Omega^k(M) \to C^k(\mathcal{T})$ the map defined by:

$$\mu^k : u \mapsto (\mu_T u)_{(T \in \mathcal{T}^k)}.$$

In the setting of finite element computations, the linear forms μ_T are called *degrees of freedom*.

Proposition 3.6. *For each k the following diagram commutes:*

$$\begin{array}{ccc} \Omega^k(M) & \xrightarrow{\mathrm{d}} & \Omega^{k+1}(M) \\ \downarrow \mu^k & & \downarrow \mu^{k+1} \\ C^k(\mathcal{T}) & \xrightarrow{\delta} & C^{k+1}(\mathcal{T}) \end{array}$$

Proof. This is an application of Stokes' theorem.

It is easy to check that the maps $\mu^k : \Omega^k(M) \to C^k(\mathcal{T})$ are onto, so that the above diagram uniquely determines $\delta : C^k(\mathcal{T}) \to C^{k+1}(\mathcal{T})$.

Suppose $\mathcal{T}$ is a simplicial complex, with oriented simplices. We suppose that $(\lambda_\alpha)_{\alpha \in \mathcal{T}^0}$ is a partition of unity such that the support of λ_α is in the closed star of α (the union of the simplices containing the vertex α). Thus if $\alpha \notin T$, we have $\lambda_\alpha|_T = 0$.

If $T \in \mathcal{T}^k$, and $\alpha : [k] \to T$ is orientation preserving, we define a k-form λ_T attached to T by:

$$\lambda_T = k! \sum_{i=0}^{k} (-1)^i \lambda_{\alpha_i} \mathrm{d}\lambda_{\alpha_0} \wedge \cdots (\mathrm{d}\lambda_{\alpha_i}) \wedge \cdots \wedge \mathrm{d}\lambda_{\alpha_k}.$$

The following notation will be used. For any functions $u_0, \cdots, u_k$ we put:

$$\begin{aligned} [u_0, \cdots, u_k] = & \sum_{i=0}^{k} (-1)^i u_i \mathrm{d}u_0 \wedge \cdots (\mathrm{d}u_i) \wedge \cdots \wedge \mathrm{d}u_k, \\ = & \frac{1}{k!} \sum_{\sigma} \varepsilon(\sigma) u_{\sigma_0} \mathrm{d}u_{\sigma_1} \wedge \cdots \wedge \mathrm{d}u_{\sigma_k}, \end{aligned}$$

where the sum extends over permutations of $[k]$. In particular the definition of λ_T does not depend on the choice of α as long as it is orientation preserving. Note also that if two of the u_i are equal one obtains 0. Suppose that $\sigma : [k] \to [k]$ is a map. If σ is not bijective we have:

$$[u_{\sigma_0}, \cdots, u_{\sigma_k}] = 0.$$

If σ is bijective then:

$$[u_{\sigma_0}, \cdots, u_{\sigma_k}] = \varepsilon(\sigma)[u_0, \cdots, u_k].$$

Note furthermore that if $\alpha : [k] \to \mathcal{T}^0$ is a map whose range is not a simplex, then:

$$[\lambda_{\alpha_0}, \cdots, \lambda_{\alpha_k}] = 0.$$

Proposition 3.7. *For any $T, T' \in \mathcal{T}^k$ we have $\mu_{T'}\lambda_T = \delta_{TT'}$.*

Proof. If $T' \neq T$, there is $\alpha \in T \setminus T'$. Since $\lambda_\alpha|_{T'} = 0$, we have $\lambda_T|_{T'} = 0$, hence $\mu_{T'}\lambda_T = 0$.

To compute $\mu_T \lambda_T$ we choose an orientation preserving $\alpha : [k] \to T$ and notice that on T:

$$\begin{aligned}\lambda_T &= k!\,\mathrm{d}\lambda_{\alpha_1} \wedge \cdots \wedge \lambda_{\alpha_k}\\ &= \varepsilon(T, T')\mathrm{d}\lambda_{T'},\end{aligned}$$

where $T' = \{\alpha_1, \cdots, \alpha_k\}$ is the face of T opposite to α_0. The identity then follows by an induction argument using Stokes' theorem.

We define:

$$\Lambda^k(\mathcal{T}) = \mathrm{span}\ \{\lambda_T \ : \ T \in \mathcal{T}^k\}.$$

By the preceding proposition, $(\lambda_T)_{T \in \mathcal{T}^k}$ is a basis of $\Lambda^k(\mathcal{T})$.

Proposition 3.8. *The exterior derivative* d *maps $\Lambda^k(\mathcal{T})$ into $\Lambda^{k+1}(\mathcal{T})$, and the matrix of $\mathrm{d} : \Lambda^k(\mathcal{T}) \to \Lambda^{k+1}(\mathcal{T})$ in the bases $(\lambda_{T'})_{T' \in \mathcal{T}^k} \to (\lambda_T)_{T \in \mathcal{T}^{k+1}}$ has entry $\varepsilon(T, T')$ at the indices $(T, T') \in \mathcal{T}^{k+1} \times \mathcal{T}^k$.*

Proof. Choose $T' \in \mathcal{T}^k$. Let $\alpha : [k] \to T'$ be orientation preserving. First we remark that:

$$\begin{aligned}\mathrm{d}\lambda_{T'} &= (k+1)!\mathrm{d}\lambda_{\alpha_0} \wedge \cdots \wedge \mathrm{d}\lambda_{\alpha_k}\\ &= (k+1)!(\sum_{\alpha \notin T'} \lambda_\alpha \mathrm{d}\lambda_{\alpha_0} \wedge \cdots \wedge \mathrm{d}\lambda_{\alpha_k} + \sum_{i \in [k]} \lambda_{\alpha_i} \mathrm{d}\lambda_{\alpha_0} \wedge \cdots \wedge \mathrm{d}\lambda_{\alpha_k}).\end{aligned}$$

The last sum can be replaced by:

$$-\sum_{i \in [k]} \lambda_{\alpha_i} \mathrm{d}\lambda_{\alpha_0} \wedge \cdots (\mathrm{d} \sum_{\alpha \notin T'} \lambda_\alpha)_{\text{at } i} \cdots \wedge \mathrm{d}\lambda_{\alpha_k}.$$

Moving $\mathrm{d}\lambda_\alpha$ to the front we get:

$$\begin{aligned}\mathrm{d}\lambda_{T'} &= (k+1)! \sum_{\alpha \notin T'} (\lambda_\alpha \mathrm{d}\lambda_{\alpha_0} \wedge \cdots \wedge \mathrm{d}\lambda_{\alpha_k} - \mathrm{d}\lambda_\alpha \wedge [\lambda_{\alpha_0}, \cdots, \lambda_{\alpha_k}])\\ &= (k+1)! \sum_{\alpha \notin T'} [\lambda_\alpha, \lambda_{\alpha_0}, \cdots, \lambda_{\alpha_k}].\end{aligned}$$

In this sum an α-term is nonzero only if $\{\alpha, \alpha_0, \cdots, \alpha_k\} \in \mathcal{T}^{k+1}$.

By definition of the incidence numbers we have:

$$\mathrm{d}\lambda_{T'} = \sum_{T \in \mathcal{T}^{k+1}} \varepsilon(T,T')\lambda_T. \tag{36}$$

This completes the proof.

Thus $\Lambda^\bullet(\mathcal{T})$ is a subcomplex of $\Omega^\bullet(M)$ which is isomorphic (not just in cohomology) to the cochain complex of $\mathcal{T}$.

We can define projections:

$$I^k : \Omega^k(M) \to \Lambda^k(\mathcal{T}),$$

by requiring that $\mu^k I^k u = \mu^k u$. These projections commute with the exterior derivative.

Corollary 3.1. *Let a compact manifold M be equipped with a simplicial complex $\mathcal{T}$. Then the De Rham map induces surjections in cohomology.*

Proof. We apply Remark 3.2 to the present situation.

Corollary 3.2. *Let M be a starshaped domain, triangulated by a simplicial complex $\mathcal{T}$. Then the simplicial cohomology is trivial in dimension $k > 0$, and has dimension 1 for $k = 0$.*

Proof. We know that $\mathrm{H}^k\Omega^\bullet(M)$ is trivial for $k > 0$ by Poincaré's lemma. Combined with Corollary 3.1 this gives the result.

A significant strengthening of Corollary 3.1 is the following theorem of De Rham:

Theorem 3.4. *Let a compact manifold M be equipped with a simplicial complex. The the De Rham map induces isomorphisms in cohomology.*

Proof. See [13, 54] or [61].

3.3 Finite elements on cellular complexes

The notion of finite element system we shall consider here was introduced in [23]. We develop it in particular in the direction of tensor products.

3.3.1 Finite element systems

Let $\mathcal{T}$ be a cellular complex. For each k and each $T \in \mathcal{T}$, we suppose that we are given a space $A^k(T)$ of differential k-forms on T such that the exterior derivative induces maps $\mathrm{d} : A^k(T) \to A^{k+1}(T)$, and if $T' \subseteq T$ the inclusion map $i : T' \to T$ induces a map $i^\star : A^k(T) \to A^k(T')$. We call such a family of spaces a finite element system. For any subcomplex $\mathcal{T}'$ of $\mathcal{T}$, we define $A^k(\mathcal{T}')$ to consist of families $(u_T)_{T\in\mathcal{T}'}$ such that whenever $T' \subseteq T$ and $i : T' \to T$ is the inclusion map, we have $i^\star u_T = u_{T'}$ (equivalently $u_T|_{T'} = u_{T'}$).

We say that the finite element system A is compatible when, in addition, the following conditions hold:

- *Cohomology.* We require that the following sequence be exact:

$$0 \to \mathbb{R} \to A^0(T) \to A^1(T) \to \cdots \to A^{\dim(T)}(T) \to 0.$$

- *Restrictions.* If we denote by $\rho : |\partial T| \to T$ the inclusion map, we require that the pullback $\rho^\star : A^k(T) \to A^k(\partial T)$ (the restriction to the boundary) be surjective. In other words, compatible forms on the boundary of a cell can be extended to the interior.

The following notation will be handy. For any cellular subcomplex $\mathcal{T}'$ of $\mathcal{T}$ we set:

$$A^\bullet(\mathcal{T}') = \bigoplus_k A^k(\mathcal{T}'),$$

and remark that it is indeed a complex when equipped with the exterior derivative.

We denote by $A^k_0(T)$ the kernel of $\rho^\star : A^k(T) \to A^k(\partial T)$. The following result is about the dimension of $A^k(\mathcal{T})$ when the spaces $A^k(T)$ are finite dimensional.

Proposition 3.9. *Suppose a finite element system has the extension property. Then:*

$$\dim A^k(\mathcal{T}) = \sum_{T\in\mathcal{T}} \dim A^k_0(T).$$

Proof. For any l, recall that $\mathcal{T}^l$ is the subset of $\mathcal{T}$ of l-dimensional cells. We let $\mathcal{T}^{(l)}$ denote the subset of cells of dimension at most l – the so-called l-skeleton of $\mathcal{T}$.

For $l \geq 1$, let ρ^l denote the restriction $A^k(\mathcal{T}^{(l)}) \to A^k(\mathcal{T}^{(l-1)})$. Then ρ^l is a surjection and its kernel can be identified with:

$$\ker \rho^l = \bigoplus_{T\in\mathcal{T}^l} A^k_0(T).$$

Thus we get:

$$\dim A^k(\mathcal{T}^{(l)}) = \dim A^k(\mathcal{T}^{(l-1)}) + \sum_{T\in\mathcal{T}^l} \dim A^k_0(T).$$

Repeating this identity for l ranging from the maximal dimension of cells of $\mathcal{T}$ down to 1, we get the claimed identity.

Remark 3.8. We remark that if the maps $A^k(T) \to A^k(\partial T)$ had not been surjective (but nevertheless well defined) we would have gotten:

$$\dim A^k(\mathcal{T}) \leq \sum_{T\in\mathcal{T}} \dim A^k_0(T).$$

Notice that we have not constructed any explicit isomorphism between the spaces $A^k(\mathcal{T})$ and $\bigoplus_{T\in\mathcal{T}} A^k_0(T)$, even though we proved that such an isomorphism exists.

Example 3.5. Given a cellular complex $\mathcal{T}$ one can fix a simplicial refinement $\mathcal{T}'$ and let $A^k(T)$ denote the Whitney k-forms on T with respect to $\mathcal{T}'$. We proved that the spaces $A^k(T)$ have the right cohomology in Corollary 3.2. The restriction property is trivial. Thus we have a compatible finite element system. It is clear that $\Lambda^k(\mathcal{T}') \subseteq A^k(\mathcal{T})$. Comparing dimensions one finds that the spaces are equal.

Theorem 3.5. *When the finite element system is compatible, the De Rham map $\mu^\bullet : A^\bullet(\mathcal{T}) \to C^\bullet(\mathcal{T})$ induces isomorphisms in cohomology.*

Proof. We suppose that the theorem has been proved for cellular complexes of dimension up to n. We let $\mathcal{T}'$ be a cellular complex of dimension at most $n+1$ for which the theorem is true, and adjoint an $(n+1)$ dimensional cell T with boundary in $\mathcal{T}'$. Let $\mathcal{T}$ denote the cellular complex $\mathcal{T}' \cup \{T\}$.

We have short exact sequences:

$$0 \to A^k(\mathcal{T}) \to A^k(\mathcal{T}') \times A^k(T) \to A^k(\partial T) \to 0,$$

where the second arrow is:

$$u \mapsto (u|_{\mathcal{T}'}, u|_T),$$

and the third arrow is:

$$(u,v) \mapsto u|_{\partial T} - v|_{\partial T}.$$

This provides a long exact sequence. We have similar short exact sequences for $C^\bullet$ and a corresponding long exact sequence. They are related by the De Rham map providing a commuting diagram:

$$\begin{array}{ccc}
\mathrm{H}^k A^\bullet(\mathcal{T}') \times \mathrm{H}^k A^\bullet(T) & \longrightarrow & \mathrm{H}^k C^\bullet(\mathcal{T}') \times \mathrm{H}^k C^\bullet(T) \\
\downarrow & & \downarrow \\
\mathrm{H}^k A^\bullet(\partial T) & \longrightarrow & \mathrm{H}^k C^\bullet(\partial T) \\
\downarrow & & \downarrow \\
\mathrm{H}^{k+1} A^\bullet(\mathcal{T}) & \longrightarrow & \mathrm{H}^{k+1} C^\bullet(\mathcal{T}) \\
\downarrow & & \downarrow \\
\mathrm{H}^{k+1} A^\bullet(\mathcal{T}') \times \mathrm{H}^{k+1} A^\bullet(T) & \longrightarrow & \mathrm{H}^{k+1} C^\bullet(\mathcal{T}') \times \mathrm{H}^{k+1} C^\bullet(T) \\
\downarrow & & \downarrow \\
\mathrm{H}^{k+1} A^\bullet(\partial T) & \longrightarrow & \mathrm{H}^{k+1} C^\bullet(\partial T)
\end{array}$$

The first, second, fourth, and fifth horizontal arrows are isomorphisms by the induction hypothesis (remark that ∂T is n-dimensional), and hence also the third.

Corollary 3.3. *Let $\mathcal{T}$ be a cellular complex and $\mathcal{T}'$ a simplicial refinement. Then the natural map $C^\bullet(\mathcal{T}') \to C^\bullet(\mathcal{T})$ induces isomorphisms in cohomology.*

Proof. We use the finite element system of Example 3.5. By Theorem 3.5 the natural morphism $\Lambda^\bullet(\mathcal{T}') \to C^\bullet(\mathcal{T})$ induces isomorphisms in cohomology. But $\Lambda^\bullet(\mathcal{T}')$ is naturally isomorphic to $C^\bullet(\mathcal{T}')$.

Corollary 3.4. *Let $\mathcal{T}$ be a cellular complex and $\mathcal{T}'$ a cellular refinement. Then the natural map $C^\bullet(\mathcal{T}') \to C^\bullet(\mathcal{T})$ induces isomorphisms in cohomology.*

Proof. Let $\mathcal{T}''$ be a simplicial refinement of $\mathcal{T}'$. Proposition 3.5 provides a commuting diagram of complexes:

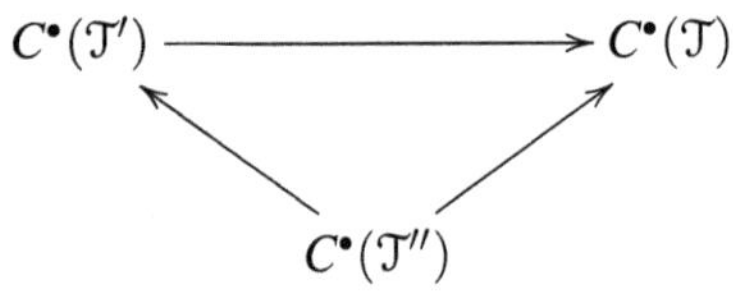

The two diagonal arrows induce isomorphisms in cohomology as noted in the previous corollary. Hence the horizontal arrow also induces isomorphisms.

Corollary 3.5. *Let M be a manifold and $\mathcal{T}$ a cellular complex. The De Rham map $\Omega^\bullet(M) \to C^\bullet(\mathcal{T})$ induces isomorphisms in cohomology.*

Proof. Let $\mathcal{T}'$ be a simplicial refinement of $\mathcal{T}$. We have a commuting diagram of complexes:

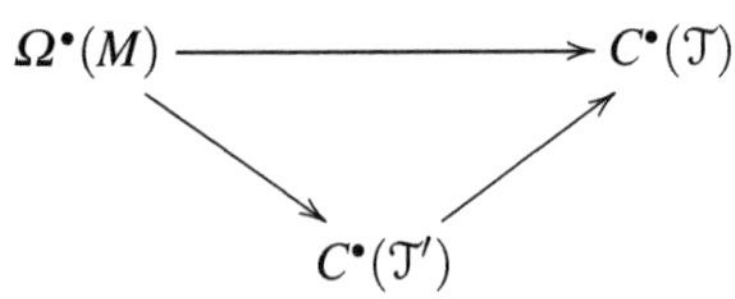

The two diagonal arrows induce isomorphisms in cohomology by Theorem 3.4 and Corollary 3.3. Hence the horizontal arrow also induces isomorphisms.

3.3.2 Harmonic extensions

We now introduce a notion of harmonic extension. It has at least two applications. It yields an isomorphism $\bigoplus_{T\in\mathcal{T}} A_0^k(T) \to A^k(\mathcal{T})$ and also a subcomplex of $A^\bullet(\mathcal{T})$ such that the restriction of the De Rham map to this subcomplex determines isomorphisms to $C^\bullet(\mathcal{T})$ (not just in cohomology). This subcomplex plays the role of lowest order Whitney forms on general cellular complexes.

Let T be a cell of dimension p. Then $C_0^k(T) = 0$ for $k < p$ and $C_0^p(T) \simeq \mathbb{R}$.

Proposition 3.10. *The following sequence is exact:*

$$0 \to A_0^0(T) \to A_0^1(T) \to \cdots \to A_0^{\dim(T)}(T) \to \mathbb{R} \to 0.$$

Proof. We have a diagram of short exact sequences:

$$\begin{array}{ccccccccc} 0 & \longrightarrow & A_0^k(T) & \longrightarrow & A^k(T) & \longrightarrow & A^k(\partial T) & \longrightarrow & 0 \\ & & \downarrow & & \downarrow & & \downarrow & & \\ 0 & \longrightarrow & C_0^k(T) & \longrightarrow & C^k(T) & \longrightarrow & C^k(\partial T) & \longrightarrow & 0 \end{array}$$

The two last vertical arrows induce isomorphisms in cohomology by Theorem 3.5, hence, by applying the five lemma to the long exact sequences, so does the first. But the cohomology of $C_0^\bullet(T)$ is trivial to compute.

From now on we suppose that a scalar product a is given on each $A^k(T)$.

Lemma 3.7. *For each k, $T \in \mathcal{T}^k$ and $\alpha \in \mathbb{R}$ there is a unique element u of $A^{\dim T}(T)$ such that:*

$$\int_T u = \alpha \text{ and } \forall v \in A_0^{\dim T-1}(T) \quad a(u, \mathrm{d}v) = 0.$$

Proof. The orthogonal of $\mathrm{im}\,\mathrm{d} : A_0^{\dim T-1}(T) \to A_0^{\dim T}(T)$ with respect to a is one-dimensional and contains an element with nonzero integral.

With the above notations, the element with integral 1 will be denoted ω_T.

Lemma 3.8. *Pick a k. For each $T \in \mathcal{T}$ such that* $\dim T > k$ *if $u \in A^k(\partial T)$, there is a unique extension $u \in A^k(T)$ such that:*

$$\forall v \in A_0^k(T) \quad a(\mathrm{d}u, \mathrm{d}v) = 0 \text{ and } \forall v \in A_0^{k-1}(T) \quad a(u, \mathrm{d}v) = 0.$$

Proof. Put $K = \mathrm{im}\,\mathrm{d} : A_0^{k-1}(T) \to A_0^k(T)$. Then on K, a is a scalar product but more importantly on its orthogonal $K^\perp$ in $A_0^k(T)$ with respect to a, $a(\mathrm{d}\cdot, \mathrm{d}\cdot)$ is also a scalar product, since $\dim T > k$ and therefore $K = \ker \mathrm{d} : A_0^k(T) \to A_0^{k+1}(T)$. Pick now $u_0 \in A^k(T)$, an arbitrary extension of u. If u_1 and u_2 are in $K^\perp$ and K respectively then $u = u_0 + u_1 + u_2$ solves our problem iff:

$$\forall v \in K^\perp \quad a(\mathrm{d}u_1, \mathrm{d}v) = -a(\mathrm{d}u_0, \mathrm{d}v) \text{ and } \forall v \in K \quad a(u_2, v) = -a(u_0, v).$$

This gives existence and uniqueness.

A differential form $u \in A^k(T)$ such that:

$$\forall v \in A_0^k(T) \quad a(\mathrm{d}u, \mathrm{d}v) = 0 \text{ and } \forall v \in A_0^{k-1}(T) \quad a(u, \mathrm{d}v) = 0.$$

will be called *harmonic*. We say that a differential form $u \in A^k(\mathcal{T})$ is locally harmonic if for each $T \in \mathcal{T}$, $u|_T$ is harmonic.

The construction we propose is the following: Fix a k and a $T \in \mathcal{T}^k$. We will construct a $\lambda_T \in \Omega^k$ attached to T.

- First (using Lemma 3.7) put $\lambda_T|_T = \omega_T$ and for each $T' \in \mathcal{T}^k$ such that $T' \neq T$ we put $\lambda_T|_{T'} = 0$. Then λ_T is set to zero also on cells of dimension $i < k$.

- Second (using Lemma 3.8), supposing λ_T has been defined on all cells of dimension up to some $i \geq k$, we define $\lambda_T|_{T'}$ on a cell $T' \in \mathcal{T}^{i+1}$ to be the unique element $u \in A^k(T')$ with $u|_{\partial T'}$ given by $\lambda_T|_{\partial T'}$ and such that u is harmonic.

We now put:

$$\Lambda^k(\mathcal{T}) = \mathrm{span}\{\lambda_T \ : \ T \in \mathcal{T}^k\}.$$

We have:

Proposition 3.11. *$\Lambda^k(\mathcal{T})$ is the set of locally harmonic elements of $A^k(\mathcal{T})$. The family (λ_T) indexed by $T \in \mathcal{T}^k$ is a basis of $\Lambda^k(\mathcal{T})$, and the induced map $\mu^k : \Lambda^k(\mathcal{T}) \to C^k(\mathcal{T})$ is an isomorphism. The exterior derivative induces maps $\mathrm{d} : \Lambda^k(\mathcal{T}) \to \Lambda^{k+1}(\mathcal{T})$.*

Proof. Since for any T of dimension k, ω_T is harmonic (on T), any element of $\Lambda^k(\mathcal{T})$ is locally harmonic. Since for any T of dimension k, any element of $A^k(T)$, which is harmonic, is proportional to ω_T, locally harmonic forms are in $\Lambda^k(\mathcal{T})$.

We notice that for $T, T' \in \mathcal{T}^k$ we have $\mu_{T'}\lambda_T = \delta_{TT'}$, where the last symbol is the Kronecker delta, hence (λ_T) is linearly independent. It is therefore a basis of $\Lambda^k(\mathcal{T})$. The De Rham map sends this basis to the canonical basis of $C^k(\mathcal{T})$.

We now prove that $\mathrm{d}\Lambda^k(\mathcal{T}) \subseteq \Lambda^{k+1}$. It is trivial to check that if u is locally harmonic, so is $\mathrm{d}u$, and the result follows.

Remark 3.9. The maps μ^k can be used to define interpolation operators $I^k : \Omega^k \to \Lambda^k(\mathcal{T})$ by requiring for any $u \in \Omega^k$ that $\mu^k I^k u = \mu^k u$. They commute with the exterior derivative.

Proposition 3.12. *We have:*

$$\sum_{i \in \mathcal{T}^0} \lambda_i = 1.$$

Proof. Define $u \in \Lambda^0$ by:

$$u = \sum_{i \in \mathcal{T}^0} \lambda_i.$$

We prove that $u = 1$ by induction on the dimension of the cells. For a 0-dimensional cell T we have that $u|_T = 1$. Suppose now it has been proved that $u|_T = 1$ for all cells T of dimension $\leq k$, and consider a $(k+1)$-dimensional cell T. The constant function on T equal to 1 is an element of $A^0(T)$ whose boundary values are 1 and whose exterior derivative is 0. By the uniqueness proved in Lemma 3.8, we therefore have $u|_T = 1$.

It is therefore tempting to believe that the family $(\lambda_i)_{i \in \mathcal{T}^0}$ is a partition of unity. However, in general, the functions λ_i might take negative values. For instance, one can use finite element functions on a refinement of $\mathcal{T}$ to construct the spaces $A^0(T)$ and use L^2-products for a. Then the constructed functions are discretely harmonic on each cell, but it is known that the (refined) mesh needs to satisfy additional requirements for discrete maximum principles to hold.

When M is a domain in a vector space V, it makes sense to speak about constant forms on M. One might then ask that $\Lambda^k(\mathcal{T})$ contains all compatible differential forms, which are constant on each top-dimensional cell. However, in general this is too much to ask. Indeed the pullbacks to a k-dimensional cell T of elements of $\Lambda^k(\mathcal{T})$ form a one-dimensional space (generated by ω_T), whereas the pullbacks to T of constant k-forms on V might form a space of higher dimension. Notice that this is not a problem of the choice of auxiliary spaces and scalar products but comes from our requirement that for k-forms we have one degree of freedom per k-dimensional cell, whereas more would be required to represent the pullbacks of constant k-forms.

We say that a k-dimensional cell is flat if it is included in a k-dimensional affine space. The next proposition shows that when the cells are all flat the previous problem does not occur.

Proposition 3.13. *Suppose we have a scalar product on V and that we use for scalar product a on a cell the L^2 product on forms associated with the induced Riemannian metric. Suppose that each cell is flat and that for each k and T, $A^k(T)$ contains the pullbacks of constant forms on V. Then $\Lambda^k(\mathcal{T})$ contains all compatible differential forms on M, which are constant on each top-dimensional cell.*

Proof. Indeed for any constant differential form on V, the exterior derivative of its pullback to any cell is 0 (by commutation of the exterior derivative with pullbacks). Moreover, if u is the pullback to a cell T of a constant differential form, then $\mathrm{d}^\star u = 0$ by the flatness of T.

Therefore compatible differential forms on M, which are constant on each top-dimensional cell, are piecewise harmonic.

When the hypothesis of Proposition 3.13 is satisfied, the canonical interpolation operator defined by Remark 3.9 can be used to obtain basic error estimates. The only problem is that the canonical interpolation operator is not defined on rough forms, say with mere L^2-regularity, but this can be corrected by using the techniques of [29] or [56].

One application of this construction is the following. Suppose a simplicial mesh $\mathcal{T}$ is given for an n-dimensional oriented manifold. One equips it with the Whitney forms $\Lambda^\bullet(\mathcal{T})$ and wishes to construct finite element spaces of k-forms Γ^k forming a complex such that the bilinear forms:

$$\begin{cases} \Lambda^k(\mathcal{T}) \times \Gamma^{n-k} \to \mathbb{R}, \\ \qquad\qquad (u,v) \mapsto \int u \wedge v. \end{cases} \tag{37}$$

are non-degenerate. One can perform the barycentric refinement of $\mathcal{T}$. It yields a simplicial refinement $\mathcal{T}'$ of $\mathcal{T}$ whose vertices are all the barycenters of all cells of $\mathcal{T}$. Then one can reassemble the elements of $\mathcal{T}'$ into a new cellular complex $\mathcal{U}$ called the dual complex of $\mathcal{T}$, whose k-cells are transverse to the $(n-k)$-cells of $\mathcal{T}$. The locally harmonic forms on $\mathcal{U}$ constructed from Whitney forms on $\mathcal{T}'$ provide a good candidate for $\Gamma^\bullet$. We proved that the bilinear forms (37) are non-degenerate in dimension $n = 2$ in [18], but the case of general dimension remains open. Such dual Whitney forms are useful for preconditioning integral operators discretized on (primal) Whitney forms [2].

Note that on simplicial complexes we now have two definitions of λ_T, as (lowest order) Whitney forms and as locally harmonic ones constructed from some finite element system. We proceed to show that Whitney forms are locally harmonic with respect to any piecewise constant metric, so that in fact there is hardly any ambiguity. Denote by λ_i the barycentric coordinate map associated with any vertex $i \in \mathcal{T}^0$. Recall that for $k \geq 1$, if $T \in \mathcal{T}^k$ and we label (a bit abusively) its vertices $0, 1, \cdots, k$ in a manner consistent with the orientation of T, the Whitney-form associated with T is:

$$\lambda_T = \sum_\sigma \varepsilon(\sigma)\lambda_{\sigma(0)} \mathrm{d}\lambda_{\sigma(1)} \wedge \cdots \wedge \mathrm{d}\lambda_{\sigma(k)},$$

where the sum extends over all permutations of $\{0, 1, \cdots, k\}$, and ε is the signature morphism (see [61]). All we have to check is that, on any cell T', $\mathrm{d}^\star \mathrm{d}\lambda_T = 0$ and $\mathrm{d}^\star \lambda_T = 0$. We have:

$$\mathrm{d}\lambda_T = (k+1)!\mathrm{d}\lambda_0 \wedge \mathrm{d}\lambda_1 \wedge \cdots \wedge \mathrm{d}\lambda_k,$$

which is constant on any simplex, so the first condition is true. Concerning the second condition, we remark that for $k = 0$ it is trivial. For $k \geq 1$ we use the Hodge star operator. When g is a metric on some n-dimensional oriented space E and ω is the corresponding volume form, the Hodge star maps k-forms to $(n-k)$-forms and is characterized by the property that for all k-forms u and v:

$$u \wedge \star v = g(u, v)\omega.$$

One then checks that $\mathrm{d}^\star = \pm \star \mathrm{d} \star$. We have, since g is constant on T':

$$\mathrm{d}\star\lambda_T = \sum_\sigma \varepsilon(\sigma) \mathrm{d}\lambda_{\sigma(0)} \wedge \star(\mathrm{d}\lambda_{\sigma(1)} \wedge \cdots \wedge \mathrm{d}\lambda_{\sigma(k)}).$$

This is a certain constant $(n-k+1)$-form on T' and we wish to show that it is zero. This is the object of the following lemma.

Lemma 3.9. *Let V be an oriented Euclidean space. For $k \geq 1$, an integer, and consider $k+1$ linear forms on V denoted $v_0, \cdots, v_k$. We have:*

$$\sum_\sigma \varepsilon(\sigma) v_{\sigma(0)} \wedge \star(v_{\sigma(1)} \wedge \cdots \wedge v_{\sigma(k)}) = 0.$$

Proof. Let $(\cdot|\cdot)$ denote the scalar product and ω be the volume form.

If $k = 1$ we have:

$$v_0 \wedge \star v_1 - v_1 \wedge \star v_0 = (v_0|v_1)\omega - (v_1|v_0)\omega = 0,$$

which is the identity we wanted to prove.

Suppose now $k \geq 2$. If $u \in \mathrm{L}_a^{k-1}(V)$ we have:

$$u \wedge \sum_\sigma \varepsilon(\sigma) v_{\sigma(0)} \wedge \star(v_{\sigma(1)} \wedge \cdots \wedge v_{\sigma(k)}) = \sum_\sigma \varepsilon(\sigma)(u \wedge v_{\sigma(0)} | v_{\sigma(1)} \wedge \cdots \wedge v_{\sigma(k)})\omega.$$

Denote by S the scalar coefficient in front of ω. We have:

$$S = \sum_{\sigma} \varepsilon(\sigma)(u \wedge v_{\sigma(0)} | v_{\sigma(1)} \otimes \cdots \otimes v_{\sigma(k)}).$$

Now if u is decomposable, i.e., of the form $u = u_1 \wedge \cdots \wedge u_{k-1}$, with $u_i \in V^\star$, we have:

$$u \wedge v_{\sigma(0)} = (1/k!) \sum_{i=1}^{k} \sum_{\tau} (-1)^i \varepsilon(\tau) u_{\tau(1)} \otimes \cdots \otimes u_{\tau(i-1)} \otimes \\ v_{\sigma(0)} \otimes u_{\tau(i)} \otimes \cdots \otimes u_{\tau(k-1)},$$

where the second sum is over all permutations τ of the set $\{1, \cdots, k-1\}$. Inserting this into the expression for S we get:

$$S = (1/k!) \sum_{i,\tau,\sigma} (-1)^i \varepsilon(\tau)\varepsilon(\sigma)(u_{\tau(1)} \otimes \cdots \otimes u_{\tau(i-1)} \otimes \\ v_{\sigma(0)} \otimes u_{\tau(i)} \otimes \cdots \otimes u_{\tau(k-1)} | v_{\sigma(1)} \otimes \cdots \otimes v_{\sigma(k)}).$$

We now fix i and τ and remark that an uneven permutation of $\{0, \cdots, k\}$ can be uniquely written $\sigma\pi$, where π is the permutation exchanging 0 and i. Thus in the above sum we can sum over all *even* permutations τ, two terms - one like above, and one, where $v_{\sigma(0)}$ and $v_{\sigma(i)}$ are switched places:

$$(u_{\tau(1)} \otimes \cdots \otimes u_{\tau(i-1)} \otimes v_{\sigma(0)} \otimes u_{\tau(i)} \otimes \cdots \otimes u_{\tau(k-1)} | v_{\sigma(1)} \otimes \cdots \otimes v_{\sigma(k)}) \\ -(u_{\tau(1)} \otimes \cdots \otimes u_{\tau(i-1)} \otimes v_{\sigma(i)} \otimes u_{\tau(i)} \otimes \cdots \otimes u_{\tau(k-1)} | \\ v_{\sigma(1)} \otimes \cdots \otimes v_{\sigma(i-1)} \otimes v_{\sigma(0)} \otimes v_{\sigma(i+1)} \otimes \cdots \otimes v_{\sigma(k)}).$$

Here the sign reflects that $\sigma\pi$ is an uneven permutation. The above sum is 0 by symmetry of the scalar product. Since this holds for all decomposable u, it is true for all $u \in \mathrm{L}_a^{k-1}(V)$. This proves the lemma.

Finally notice that an element of $A_0^k(T)$ can be extended by 0 to an element of $A^k(\mathcal{T}^{(l)})$, where $l = \dim T$ and $\mathcal{T}^{(l)}$ is the l-skeleton of $\mathcal{T}$. It can thereafter be extended recursively to cells of higher and higher dimension in a unique way by requiring harmonicity at each stage. This yields linear maps $E : A_0^k(T) \to A^k(\mathcal{T})$ which can be thought of as a global map:

$$E : \bigoplus_T A_0^\bullet(T) \to A^\bullet(\mathcal{T}),$$

respecting the degree of differential forms. Notice that for $k < \dim T$, if $u \in A_0^k(T)$ then $\mathrm{d}Eu = E\mathrm{d}u$. When $k = \dim T$, if $u \in A_0^k(T)$ then its harmonic extension will have non-zero exterior derivative if u has non-zero integral.

Proposition 3.14. *The map* $E : \oplus_T A_0^\bullet(T) \to A^\bullet(\mathcal{T})$ *is an isomorphism.*

Proof. The extension operator $E : A_0^k(T) \to A^k(\mathcal{T})$ has the property that for any $u \in A_0^k(T)$ and any T' such that $T \not\subseteq T'$, $(Eu)|_{T'} = 0$.

If now $u \in \bigoplus_T A_0^\bullet(T)$ is such that $Eu = 0$, one deduces that $u|_T = 0$, starting from cells T of dimension k and inductively increasing dimension.

Injectivity yields bijectivity by equality of dimension.

Another extension operator defined for smooth differential forms was used in [22] to prove that compatible piecewise smooth differential forms can be extended from the boundary of a simplex to its interior, showing that the spaces of "all" compatible piecewise smooth differential forms constitute a finite element system. An axiomatic treatment of extension operators is given in [6]. It covers two notable other cases: extension using the degrees of freedom of finite elements and extension using Bernstein polynomials.

Such extension operators reduce the problem of constructing a basis for the global space $A^k(\mathcal{T})$ to that of constructing bases for the local spaces $A_0^k(T)$. See for instance [57].

Notice furthermore that the construction of locally harmonic forms can be applied recursively. Given a (fine) cellular complex $\mathcal{T}_0$ for which one has a compatible finite element system (e.g., a simplicial complex), one can assemble the cells into a coarser complex $\mathcal{T}_1$, whose cells can again be assembled into a coarser complex $\mathcal{T}_2$ etc. At each stage one can construct the space of locally harmonic forms on $\mathcal{T}_{k+1}$ from the finite element system provided by $\mathcal{T}_k$. This provides nested spaces:

$$\Lambda^\bullet(\mathcal{T}_{k+1}) \subseteq \Lambda^\bullet(\mathcal{T}_k) \subseteq \cdots \subseteq \Lambda^\bullet(\mathcal{T}_0),$$

which can be used for instance for preconditioning. See [51].

3.3.3 Constructions with finite element systems

We now show that the notion of a finite element system we introduced behaves naturally with respect to tensor products, that degrees of freedom can be readily defined and that it can also be used to describe *hp*- finite elements.

Suppose we have two manifolds M and N, equipped with cellular complexes $\mathcal{T}$ and $\mathcal{U}$, and auxiliary spaces $A^k(T)$ for $T \in \mathcal{T}$ and $B^k(U)$ for $U \in \mathcal{U}$ subject to the above conditions on cohomology and restrictions.

Let $\mathcal{T} \times \mathcal{U}$ denote the product cellular complex, whose cells are of the form $T \times U$ for $T \in \mathcal{T}$ and $U \in \mathcal{U}$. We equip $\mathcal{T} \times \mathcal{U}$ with auxiliary spaces:

$$Z^\bullet(T \times U) = A^\bullet(T) \otimes B^\bullet(U).$$

Explicitly we put:

$$Z^k(T \times U) = \bigoplus_l A^l(T) \otimes B^{k-l}(U),$$

where the tensor product is that of differential forms.

Proposition 3.15. *$Z^\bullet$ has the right cohomological properties on any product cell.*

Proof. This follows from Theorem 3.2.

We now prove in several steps that the restrictions to boundaries in $Z^\bullet$ is surjective, so that $Z^\bullet$ fulfils the requirements to be a compatible finite element system.

Lemma 3.10. *We have:*

$$Z_0^\bullet(T \times U) = A_0^\bullet(T) \otimes B_0^\bullet(U).$$

Proof. Let $(u_i)_{0 \le i < m}$ denote a basis of $A_0^\bullet(T)$ and extend it to a basis $(u_i)_{0 \le i < m'}$ of $A^\bullet(T)$. Similarly let $(v_j)_{0 \le j < n}$ denote a basis of $B_0^\bullet(U)$ and extend it to a basis $(v_j)_{0 \le j < n'}$ of $B^\bullet(U)$.

Now pick $z \in Z^\bullet(T \times U)$ and write it in the form:

$$z = \sum_{\substack{0 \le i < m' \\ 0 \le j < n'}} \lambda_{ij} u_i \otimes v_j,$$

and suppose that z restricted to the boundary of $T \times U$ is 0.

Restricting z to $\partial T \times \partial U$ we get:

$$\sum_{\substack{m \le i < m' \\ n \le j < n'}} \lambda_{ij} u_i|_{\partial T} \otimes v_j|_{\partial U} = 0.$$

We note that $(u_i|_{\partial T})_{m \le i < m'}$ is linearly independent, as well as $(v_j|_{\partial U})_{n \le j < n'}$. It follows that $\lambda_{ij} = 0$ for $m \le i < m'$ and $n \le j < n'$.

Next one restricts z to $\partial T \times U$ and $T \times \partial U$ (both included in the boundary of $T \times U$) and gets by the same remark that $\lambda_{ij} = 0$ for $m \le i < m'$ and $0 \le j < n$ as well as for $0 \le i < m$ and $n \le j < n'$.

Thus we get:

$$z = \sum_{\substack{0 \le i < m \\ 0 \le j < n}} \lambda_{ij} u_i \otimes v_j,$$

and we are done.

Proposition 3.16. *We have:*

$$Z^\bullet(\mathcal{T} \times \mathcal{U}) = A^\bullet(\mathcal{T}) \otimes B^\bullet(\mathcal{U}).$$

Proof. It is clear that:

$$A^\bullet(\mathcal{T}) \otimes B^\bullet(\mathcal{U}) \subseteq Z^\bullet(\mathcal{T} \times \mathcal{U}).$$

We prove that the spaces have the same dimension.

By Remark 3.8 we have:

$$\begin{aligned}\dim Z^\bullet(\mathcal{T}\times\mathcal{U}) &\leq \Sigma_{TU}\dim Z_0^\bullet(T\times U)\\ &\leq \Sigma_{TU}\dim(A_0^\bullet(T)\otimes B_0^\bullet(U))\\ &\leq \Sigma_{TU}\dim A_0^\bullet(T)\dim B_0^\bullet(U)\\ &\leq \Sigma_T\dim A_0^\bullet(T)\,\Sigma_U\dim B_0^\bullet(U)\\ &\leq \dim A^\bullet(\mathcal{T})\dim B^\bullet(\mathcal{U}).\end{aligned}$$

Thus we get:

$$\dim Z^\bullet(\mathcal{T}\times\mathcal{U}) \leq \dim(A^\bullet(\mathcal{T})\otimes B^\bullet(\mathcal{U})).$$

This completes the proof.

Proposition 3.17. *For any cell $T\in\mathcal{T}$, $U\in\mathcal{U}$, the restriction $Z^\bullet(T\times U)\to Z^\bullet(\partial(T\times U))$ is onto.*

Proof. Pick $z\in Z^\bullet(\partial(T\times U))$.

We have:

$$z|_{\partial T\times\partial U}\in Z^\bullet(\partial T\times\partial U) = A^\bullet(\partial T)\otimes B^\bullet(\partial U).$$

Such a form may therefore be extended from $\partial T\times\partial U$ to $T\times U$ (by extending each component in the sum of tensor products it can be represented by on $\partial T\times\partial U$).

We are therefore left with the problem of extending a $z\in Z^\bullet(\partial(T\times U))$ such that $z|_{\partial T\times\partial U}=0$.

We remark that we have a short exact sequence:

$$0\to A_0^\bullet(T)\to A^\bullet(T)\to A^\bullet(\partial T)\to 0.$$

Taking the tensor products with $B^\bullet(\partial U)$ and using Proposition 3.16, we get a short exact sequence:

$$0\to A_0^\bullet(T)\otimes B^\bullet(\partial U)\to Z^\bullet(T\times\partial U)\to Z^\bullet(\partial T\times\partial U)\to 0.$$

Similarly we have a short exact sequence:

$$0\to A^\bullet(\partial T)\otimes B_0^\bullet(U)\to Z^\bullet(\partial T\times U)\to Z^\bullet(\partial T\times\partial U)\to 0.$$

Pick now $z\in Z^\bullet(\partial(T\times U))$ such that $z|_{\partial T\times\partial U}=0$.

We have:

$$z|_{T\times\partial U}\in A_0^\bullet(T)\otimes B^\bullet(\partial U).$$

Thus we can find $z'\in A_0^\bullet(T)\otimes B^\bullet(U)$ such that $z'|_{T\times\partial U}=z|_{T\times\partial U}$.

Similarly we have:

$$z|_{\partial T\times U}\in A^\bullet(\partial T)\otimes B_0^\bullet(U).$$

Hence we can find $z''\in A^\bullet(T)\otimes B_0^\bullet(U)$ such that $z''|_{\partial T\times U}=z|_{\partial T\times U}$.

We then have:

$$(z'+z'')|_{T\times\partial U}=z'|_{T\times\partial U}=z|_{T\times\partial U},$$

and similarly :

$$(z' + z'')|_{\partial T \times U} = z''|_{\partial T \times U} = z|_{\partial T \times U}.$$

Since:

$$(T \times \partial U) \cup (\partial T \times U) = \partial (T \times U),$$

the form $z' + z''$ is a suitable extension of z.

Thus when A and B are compatible finite element systems, Z is also. There are now two ways of constructing a complex of spaces isomorphic to $C^{\bullet}(\mathcal{T} \times \mathcal{U})$: either construct first the spaces of locally harmonic forms on M and N from A and B, and then take the tensor product, or construct the locally harmonic forms on $M \times N$ from Z. If the scalar product chosen for each product cell $T \times U$ is the product of the scalar product of T and the scalar product of U, then these two constructions yield the same spaces. It suffices to check that the tensor product of two locally harmonic forms is locally harmonic.

So far the only example of a finite element system we have considered on a simplicial complex is the one provided by Whitney forms. One can also construct high order Whitney forms by taking products with polynomials [5, 38, 49]. Specifically, for a simplex T and integer $p \geq 1$, let $A^{k,p}(T)$ denote the space spanned by k-forms of the form $u\lambda_{T'}$, where u is a polynomial of degree at most $p-1$, T' is a k-face of T and $\lambda_{T'}$ the associated Whitney form. As we noted in [22], the wedge product provides a map $A^{k,p}(T) \times A^{k',p'}(T) \to A^{k+k',p+p'}(T)$ (one says that the wedge product respects the filtration), which is the reason for letting the lowest order Whitney forms be indexed by $p = 1$ rather than $p = 0$, which was usual before. We also denote by $B^{k,p}(T)$ the space of polynomial differential k-forms of degree at most p.

Suppose a simplicial complex is given as well as a polynomial degree p. Each simplex is supposed to be oriented. That the system $A^{\bullet,p}(\cdot)$ for given p constitutes a compatible finite element system follows immediately from results proved for instance in [5]. Exactness is a theorem whereas the extension property follows from the existence of degrees of freedom. We will also use that the bilinear forms:

$$\begin{cases} A_0^{k,p}(T) \times B^{n-k,p+k-n-1}(T) \to \mathbb{R}, \\ \qquad\qquad\qquad\qquad (u,v) \mapsto \int u \wedge v. \end{cases}$$

are non-degenerate.

Consider p to be fixed and denote $A^k(T) = A^{k,p}(T)$ and $B^l(T) = B^{l,p-l-1}(T)$. Consider the map Φ^k, which to a compatible differential form u associates with each $T \in \mathcal{T}$ the linear form $v \mapsto \int_T u|_T \wedge v$ on $B^{\dim T - k}(T)$.

Proposition 3.18. *Φ^k induces an isomorphism $A^k(\mathcal{T}) \to \bigoplus_{T \in \mathcal{T}} B^{\dim T - k}(T)^{\star}$.*

Proof. If $\Phi^k u = 0$ then u is zero on k-dimensional cells, therefore also on $(k+1)$-dimensional cells, etc., using the above non-degeneracy at each level. This gives injectivity. Moreover we have equality of dimension.

The map Φ^k is said to associate with u its degrees of freedom. One also says that Φ^k is unisolvent on $A^k(\mathcal{T})$.

To any compatible k-form u we can associate the unique $v \in A^k(\mathcal{T})$ such that $\Phi^k v = \Phi^k u$. We define an interpolation operator I^k by $I^k u = v$.

Proposition 3.19. *Interpolation thus defined commutes with the exterior derivative.*

Proof. Let u be a compatible k-form. Let T be an n-dimensional simplex and $v \in B^{n-k-1}(T)$:

$$\begin{aligned}\int_T \mathrm{d}I^k u \wedge v &= \int_{\partial T} I^k u \wedge v \pm \int_T I^k u \wedge \mathrm{d}v \\ &= \int_{\partial T} u \wedge v \pm \int_T u \wedge \mathrm{d}v \\ &= \int_T \mathrm{d}u \wedge v \\ &= \int_T I^{k+1} \mathrm{d}u \wedge v.\end{aligned}$$

The important properties are that $\mathrm{d}B^{n-k-1}(T) \subseteq B^{n-k}(T)$ (notice that the polynomial degree decreases), and that B^{n-k-1} is stable upon taking traces.

We see that the construction of an interpolator carries over whenever we have a cellular complex equipped with a compatible finite element system $A^k(T)$ together with spaces $B^l(T)$ forming a finite element system such that:

– The form $\int \cdot \wedge \cdot$ is non degenerate on $A_0^k(T) \times B^{\dim T - k}(T)$.

We call such a system of spaces a mirror system. We also say that $B^{\dim T-k}(T)$ is a mirror for $A_0^k(T)$. Thus, in a mirror system, instead on focusing on cohomology and extension, we focus on duality.

Supposing we now have two manifolds M and M' and corresponding cellular complexes $\mathcal{T}$ and $\mathcal{T}'$, suppose that for both we have a finite element system A and a mirror system B. Let A denote also the tensor product finite element system and B the tensor product of the mirror systems. That is, for each product cell $T \times T'$:

$$B^\bullet(T \times T') = B^\bullet(T) \otimes B^\bullet(T').$$

Proposition 3.20. *The tensor product of two mirror systems is a mirror system.*

Proof. Let $\dim T = n$ and $\dim T' = n'$. We have:

$$A_0^k(T \times T') = \bigoplus_l A_0^l(T) \otimes A_0^{k-l}(T').$$

On the other hand:

$$B^{n+n'-k}(T \times T') = \bigoplus_l B^{n-l}(T) \otimes B^{n'-k+l}(T').$$

Duality then follows from the formula:

$$(u \otimes u') \wedge (v \otimes v') = \pm (u \wedge v) \otimes (u' \wedge v'),$$

where u and v are forms on T, and u' and v' are forms on T', pulled back to $T \times T'$.

This provides degrees of freedom for high order Whitney forms on products such as prisms and cubes, of possibly different degree in each direction. Remark that the construction can be used for cellular complexes whose cells are each a product of simplexes; the product need not be global.

Finally we want to consider finite element systems corresponding to the hp-setting, where the polynomial degree is allowed to vary throughout the domain. We will require the following result:

Proposition 3.21. *Suppose we have a cellular complex equipped with a finite element system A having the extension property and the following property on cohomology, namely that for each T the following sequence is exact:*

$$0 \to A_0^0(T) \to A_0^1(A) \to \cdots \to A_0^{\dim T}(T) \to \mathbb{R} \to 0,$$

the second to last arrow being integration. Then the finite element system is compatible.

Proof. Suppose the required cohomology property (that is, without homogeneous boundary conditions) holds for all cells of dimension up to l. Then, following the proof of Theorem 3.5, one checks that the cohomology of $A^\bullet(\mathcal{T}')$ for subcomplexes $\mathcal{T}'$ of $\mathcal{T}$ of dimension at most l is given by the corresponding cochain complex. Letting T be an $(l+1)$ dimensional cell, we note that ∂T is such a subcomplex and we just do a variant of the proof of Proposition 3.10.

Suppose $\mathcal{T}$ is a cellular complex and that for each integer p in some range, a compatible finite element system $A^{k,p}(T)$ has been chosen. We suppose that $A^{k,p}(T) \subseteq A^{k,p+1}(T)$. Choose now some function $p : \mathcal{T} \to \mathbb{N}$ with the property that if $T' \subseteq T$, then $p(T') \leq p(T)$. Define:

$$A^k(T) = \{u \in A^{k,p(T)}(T) \ : \ \forall T' \subseteq T \quad u|_{T'} \in A^{k,p(T')}(T')\}.$$

Proposition 3.22. *These spaces constitute a compatible finite element system.*

Proof. The surjectivity of restriction follows from the monotonicity of p, whereas the cohomological requirement follows from the preceding proposition using the fact that $A_0^k(T) = A_0^{k,p(T)}(T)$.

Now even if we have mirror-systems for each $A^{\bullet,p}(\cdot)$, this does not give a mirror-system for the constructed $A^\bullet(\cdot)$. Choosing the mirror $B^{\dim T - k}(T)$ for each $A_0^k(T)$ given by $p(T)$ does not guarantee that the restrictions map $B^k(T)$ to $B^k(T')$ when $T' \subseteq T$, so the corresponding interpolator does not commute with the exterior derivative. Instead we introduce a generalization to any compatible finite element system of the projection based interpolation described, for instance, in [34].

We suppose a scalar product a has been chosen for each $\Omega^k(T)$. This is similar to the choice of scalar products on $A^k(T)$ we required for the notion of local harmonicity. Let Ψ^k be the map, which to any compatible differential k-form u, associates with each $T \in \mathcal{T}$ the two linear forms $a(u,\cdot)$ on $\mathrm{d}A_0^{k-1}(T)$ and $a(\mathrm{d}u,\cdot)$ on $\mathrm{d}A_0^k(T)$ when $k < \dim T$, and when $k = \dim T$, the linear form $a(u,\cdot)$ on $\mathrm{d}A_0^{k-1}(T)$ and the number $\int u$ (which can be considered as a linear form on $\mathbb{R}$). We let $B^k(T)$ be the space $\mathrm{d}A_0^{k-1}(T) \oplus \mathrm{d}A_0^k(T)$ in the first case and $\mathrm{d}A_0^{k-1}(T) \oplus \mathbb{R}$ in the second. For $k > \dim T$ put $B^k(T) = 0$. Thus, with the help of the scalar products, in each case an element of $\Omega^k(T)$ gives a linear form on $B^k(T)$.

Proposition 3.23. *Ψ^k induces an isomorphism $A^k(\mathcal{T}) \to \bigoplus_{T\in\mathcal{T}} B^k(T)^\star$.*

Proof. Injectivity follows from the uniqueness of Lemmas 3.7 and 3.8 applied in a recursive manner from cells of dimension k and upwards. Then note that we have equality of dimension by Propositions 3.9 and 3.10.

As before we can deduce an interpolation operator I^k, which to a k-form u associates the unique $v \in A^k(\mathcal{T})$ such that $\Psi^k v = \Psi^k u$.

Proposition 3.24. *Interpolation thus defined commutes with the exterior derivative.*

Proof. Note that for any compatible k-form u, if $\Psi^k u = 0$ then $\Psi^{k+1}\mathrm{d}u = 0$. This is nothing but the property already used that locally harmonic forms have locally harmonic exterior derivative, in addition to Stokes theorem. From this remark the proposition follows.

We note also that there is an obvious isomorphism $\oplus_{T\in\mathcal{T}} A_0^k(T) \to \oplus_{T\in\mathcal{T}} B^k(T)^\star$, determined on each cell by the chosen scalar products. Composing with the inverse of Ψ^k gives an isomorphism $\oplus_{T\in\mathcal{T}} A_0^k(T) \to A^k(\mathcal{T})$. It is the map E of Proposition 3.14.

As scalar products one can use, for instance, the L^2-product associated with some Riemannian structure. In [35] fractional order Sobolev spaces are also used. Working with arbitrary scalar products a allows also for scalar products obtained by transport from reference cells.

4 Conclusion

We indicate some directions in which to pursue the present work.

We reviewed homological algebra and tensor products for real vector spaces rather than modules over an arbitrary ring, since this leads to significant simplifications. However, module theory might play a role in future investigations on finite elements, letting the rings be those of multivariate polynomials. This could, for instance, shed light on linear dependence relations in the natural spanning sets for high order Whitney forms. This is of consequence for the choice of bases, for which there are several suggestions [1, 6, 37, 39, 52, 60].

Another interesting line of investigation is finite elements for elasticity. Symmetric tensor fields have algebraic properties not covered by the present study. The linear elasticity complex [4] is related to but different from the De Rham complex.

Finally, some non-linear wave equations such as the Yang-Mills and Einstein equations also have a rich algebraic structure that should be reflected in discretizations [23, 25, 26, 27, 28].

Acknowledgements This work, conducted as part of the award "Numerical analysis and simulations of geometric wave equations" made under the European Heads of Research Councils and European Science Foundation EURYI (European Young Investigator) Awards scheme, was supported by funds from the Participating Organizations of EURYI and the EC Sixth Framework Program.

References

1. Ainsworth, M., Coyle, J.: Hierarchic finite element bases on unstructured tetrahedral meshes. Int. J. Numer. Meth. Eng. **58**, 2103–2130 (2003)
2. Andriulli, F.P., Cools, K., Bagci, H., Olyslager, F., Buffa, A., Christiansen, S.H., Michielssen, E.: A multiplicative Calderon preconditioner for the electric field integral equation. IEEE Trans. Antennas and Propagation **56**(8), 2398–2412 (2008)
3. Arnold, D.N., Falk, R.S., Winther, R.: Differential complexes and stability of finite element methods I. The De Rham complex. In: Arnold, D.N. et al. (eds.) Compatible Spatial Discretizations, IMA Volumes in Mathematics and its Applications 142, pp. 24–46. Springer, New York (2006)
4. Arnold, D.N., Falk, R.S., Winther, R.: Differential complexes and stability of finite element methods II. The elasticity complex. In: Arnold, D.N. et al. (eds.) Compatible Spatial Discretizations, IMA Volumes in Mathematics and its Applications 142, pp. 47–67. Springer, New York (2006)
5. Arnold, D.N., Falk, R.S., Winther, R.: Finite element exterior calculus, homological techniques, and applications. Acta Numerica **15**, 1–155 (2006)
6. Arnold, D.N., Falk, R.S., Winther, R.: Geometric decompositions and local bases for spaces of finite element differential forms. Computer Methods in Appl. Mech. and Eng. To appear
7. Babuska, I.: Error bounds for the finite element method. Numer. Math. **16**, 322–333 (1971)
8. Babuska, I., Osborn, J.: Eigenvalue problems. In: Handbook of Numerical Analysis, Vol. II, pp. 641–787. North-Holland, Amsterdam (1991)
9. Boffi, D., Brezzi, F., Gastaldi, L.: On the problem of spurious eigenvalues in the approximation of linear elliptic problems in mixed form. Math. Comp. **69**(229), 121–140 (2000)
10. Boffi, D., Fernandes, P., Gastaldi, L., Perugia, I.: Computational models of electromagnetic resonators: analysis of edge element approximation. SIAM J. Numer. Anal. **36**(4), 1264–1290 (1999)
11. Bossavit, A.: Mixed finite elements and the complex of Whitney forms. In: The Mathematics of Finite Elements and Applications VI (Uxbridge, 1987), pp. 137–144. Academic Press, London (1988)
12. Bossavit, A.: Discretization of electromagnetic problems: the "generalized finite differences" approach. In: Handbook of Numerical Analysis, Vol. XIII, pp. 105–197. North-Holland, Amsterdam (2005)
13. Bott, R., Tu, L.-W.: Differential Forms in Algebraic Topology. Graduate Texts in Mathematics No. 82. Springer, Berlin (1982)
14. Brezzi, F.: On the existence, uniqueness and approximation of saddle-point problems arising from Lagrangian multipliers. RAIRO Anal. Numér. **8**(R-2), 129–151 (1974)

15. Brezzi, F., Douglas, J., Durán, R., Fortin, M.: Mixed finite elements for second order elliptic problems in three variables. Numer. Math. **51**(2), 237–250 (1987)
16. Brezzi, F., Fortin, M.: Mixed and Hybrid Finite Element Methods. Springer, Berlin (1991)
17. Buffa, A., Christiansen, S.H.: The Electric Field Integral Equation on Lipschitz screens: definitions and numerical approximation. Numer. Math. **94**(2), 229–267 (2003)
18. Buffa, A., Christiansen, S.H.: A dual finite element complex on the barycentric refinement. Math. Comp. **76**, 1743–1769 (2007)
19. Caorsi, S., Fernandes, P., Raffetto, M.: Spurious-free approximations of electromagnetic eigenproblems by means of Nédélec-type elements. Math. Model. Numer. Anal. **35**(2), 331–354 (2001)
20. Christiansen, S.H.: Discrete Fredholm properties and convergence estimates for the Electric Field Integral Equation. Math. Comp. **73**(245), 143–167 (2004)
21. Christiansen, S.H.: A div-curl lemma for edge elements. SIAM J. Numer. Anal. **43**(1), 116–126 (2005)
22. Christiansen, S.H.: Stability of Hodge decompositions in finite element spaces of differential forms in arbitrary dimension. Numer. Math. **107**(1), 87–106 (2007)
23. Christiansen, S.H.: A construction of spaces of compatible differential forms on cellular complexes. Math. Models Methods Appl. Sci. **18**(5), 739–757 (2008)
24. Christiansen, S.H.: On the linearization of Regge calculus. University of Oslo, Department of Mathematics, Preprint Pure Mathematics, ISSN 0806-2439, No. 13, May 2008
25. Christiansen, S.H.: Constraint preserving schemes for some gauge invariant wave equations. SIAM J. Sci. Comp. To appear
26. Christiansen, S.H., Halvorsen, T.G.: Solving the Maxwell-Klein-Gordon equation in the Lattice Gauge Theory formalism. University of Oslo, Department of Mathematics, Preprint Pure Mathematics, ISSN 0806-2439, No. 17, June 2008
27. Christiansen, S.H., Halvorsen, T.G.: Convergence of lattice gauge theory for Maxwell's equations. University of Oslo, Department of Mathematics, Preprint Pure Mathematics, ISSN 0806-2439, No. 21, September 2008
28. Christiansen, S.H., Winther, R.: On constraint preservation in numerical simulations of Yang-Mills equations. SIAM J. Sci. Comp. **28**(1), 75–101 (2006)
29. Christiansen, S.H., Winther, R.: Smoothed projections in finite element exterior calculus. Math. Comp. **77**(262), 813–829 (2008)
30. Ciarlet, P.G.: Basic error estimates for elliptic problems. In: Handbook of Numerical Analysis, Vol. II, pp. 17–351. North-Holland, Amsterdam (1991)
31. Ciarlet, P.G., Zou, J.: Fully discrete finite element approaches for time-dependent Maxwell's equations. Numer. Math. **82**(2), 193–219 (1999)
32. Cohen, G., Joly, P., Roberts, J.E., Tordjman, N.: Higher order triangular finite elements with mass lumping for the wave equation. SIAM J. Numer. Anal. **38**(6), 2047–2078 (2001)
33. Costabel, M., McIntosh, A.: On Bogovskii and regularized Poincaré integral operators for De Rham complexes on Lipschitz domains. Université de Rennes, Prépublication 08–40 (2008)
34. Demkowicz, L., Buffa, A.: H^1, $H(\mathrm{curl})$ and $H(\mathrm{div})$-conforming projection-based interpolation in three dimensions. Quasi-optimal p-interpolation estimates. Comput. Methods Appl. Mech. Eng. **194**(2-5), 267–296 (2005)
35. Demkowicz, L., Kurtz, J.: Projection-based interpolation and automatic hp-adaptivity for finite element discretizations of elliptic and Maxwell problems. In: ESAIM Proceedings Vol. 21, pp. 1–15 EDP Science, Les Ulis (2007)
36. Gelfand, S.I., Manin, Y.I.: Methods of Homological Algebra, 2nd edition. Springer Monographs in Mathematics. Springer, Berlin (2003)
37. Gopalakrishnan, J., Garcia-Castillo, L.E., Demkowicz, L.F.: Nédélec spaces in affine coordinates. Comput. Math. Appl. **49**(7–8), 1285–1294 (2005)
38. Hiptmair, R.: Canonical construction of finite elements. Math. Comp. **68**(228), 1325–1346 (1999)
39. Hiptmair, R.: High order Whitney forms. Prog. Electr. Res. (PIER) **32**, 271–299 (2001)
40. Hiptmair, R.: Finite elements in computational electromagnetism. Acta Numerica **11**, 237–339 (2002)

41. Hiptmair, R.: Discrete compactness for p-version of tetrahedral edge elements. Seminar for Applied Mathematics, ETH Zürich, Research Report 2008–31 (2008)
42. Joly, P.: Variational methods for time-dependent wave propagation problems. In: Ainsworth, M., et al. (eds.) Topics in Computational Wave Propagation, Lecture Notes in Comput. Sci. Eng. **31**, pp. 201–264. Springer, Berlin (2003)
43. Kikuchi, F.: On a discrete compactness property for the Nédélec finite elements. J. Fac. Sci. Univ. Tokyo, Sect. 1A Math. **36**, 479–490 (1989)
44. Lang, S.: Differential and Riemannian Manifolds, 3rd edition. Springer, New York (1995)
45. Lang, S.: Algebra, revised 3rd edition. Graduate Texts in Mathematics, Vol. 211. Springer, New York (2002)
46. Monk, P.: Analysis of a finite element method for Maxwell's equations. SIAM J. Numer. Anal. **29**(3), 714–729 (1992)
47. Monk, P.: An analysis of Nédélec's method for the spatial discretization of Maxwell's equations. J. Comput. Appl. Math. **47**(1), 101–121 (1993)
48. Monk, P.: Finite Element Methods for Maxwell's Equations. Oxford University Press (2003)
49. Nédélec, J.-C.: Mixed finite elements in $\mathbb{R}^3$. Numer. Math. **350**(3), 315–341 (1980)
50. Nédélec, J.-C.: A new family of mixed finite elements in $\mathbb{R}^3$. Numer. Math. **50**(1), 57–81 (1986)
51. Pasciak, J., Vassilevski, P.S.: Exact de Rham sequences of spaces defined on macro-elements in two and three spatial dimensions. SIAM J. Sci. Comput. **30**(5), 2427–2446 (2008)
52. Rapetti, F.: High order edge elements on simplicial meshes. Math. Model. Numer. Anal. **41**(6), 1001–1020 (2007)
53. Raviart, P.A., Thomas, J.-M.: A mixed finite element method for 2nd order elliptic problems. In Galligani, I., Magenes, E. (eds.) Mathematical Aspects of the Finite Element Method. Lecture Notes in Mathematics Vol. 606, pp. 292–315. Springer, Berlin (1977)
54. DeRham, G.: Variétés différentiables. Formes, courants, formes harmoniques (3rd edition). Publications de l'Institut de Mathématique de l'Université de Nancago III. Actualités Scientifiques et Industrielles, No. 1222b. Hermann, Paris (1973)
55. Roberts, J.E., Thomas, J.-M.: Mixed and hybrid methods. In: Handbook of Numerical Analysis, Vol. II, Finite Element Methods (Part 1), pp. 523–640. North-Holland, Amsterdam (1991)
56. Schöberl, J.: A posteriori error estimates for Maxwell equations. Math. Comp. **77**(262), 633–649 (2008)
57. Schöberl, J., Zaglmayr, S.: High order Nédélec elements with local complete sequence properties. COMPEL, Vol. **24**(2), 374–384 (2005)
58. Strang, G., Fix, G.J.: An Analysis of the Finite Element Method. Prentice-Hall Series in Automatic Computation. Prentice-Hall, Englewood Cliffs (1973)
59. Tonti, E.: Finite formulation of the electromagnetic field. Prog. Electr. Res. (PIER) **32**, 1–44 (2001)
60. Webb, J.P.: Hierarchal vector basis functions of arbitrary order for triangular and tetrahedral finite elements. IEEE Trans. Ant. Prop. **47**(8), 1244–1253 (1999)
61. Whitney, H.: Geometric Integration Theory. Princeton University Press (1957)
62. Yee, K.S.: Numerical solution of initial boundary value problems involving Maxwell's equations in isotropic media. IEEE Trans. Ant. Prop. **14**, 302–307 (1966)
63. Zhao, J.: Analysis of finite element approximation for time-dependent Maxwell problems. Math. Comp. **73**(247), 1089–1105 (2004)

An Introduction to the Theory of Scalar Conservation Laws with Spatially Discontinuous Flux Functions

Nils Henrik Risebro

Abstract The aim of these notes is to give a brief introduction to scalar conservation laws with a space-dependent flux function, where the spatial dependence of the flux can have discontinuities.

1 Introduction

We shall restrict ourselves to one spatial dimension, both for reasons of simplicity, and because the theory is more complete in one dimension.

In this case, a conservation law with a space-dependent flux can be written

$$u_t + f(x,u)_x = 0, \quad x \in \mathbb{R}, \quad t > 0, \tag{1}$$

where the flux function f is given, and we aim to determine the unknown $u = u(x,t)$. The conservation law (1) is usually augmented by an initial condition, stating that $u(x,0) = u_0(x)$ in some suitable sense. Formally, integrating (1) between x_1 and x_2 gives

$$\frac{d}{dt}\int_{x_1}^{x_2} u(x,t)\,dx = f\left(x_1, u(x_1,t)\right) - f\left(x_2, u(x_2,t)\right).$$

Hence the interpretation of f is that it gives the rate of flow, i.e., the flux, of u at the point x.

Next we give some examples where such conservation laws arise.

Example 1.1. Traffic flow is a simple model where conservation laws with space-dependent coefficients arise naturally. Let ρ denote the density of cars on a long "one way" road. We normalize units, so that $\rho = 1$ if the cars are packed bumper to bumper. Assume that the speed of the cars is a decreasing function of the density

Centre of Mathematics for Applications, University of Oslo, PO Box 1053 Blindern, NO-0316 Oslo, Norway, e-mail: nilshr@math.uio.no

E. Quak, T. Soomere (eds.), *Applied Wave Mathematics*,
DOI 10.1007/978-3-642-00585-5_18, © Springer-Verlag Berlin Heidelberg 2009

$v = v(\rho)$. The speed of the cars on an empty road ($\rho = 0$) is governed by the road conditions and the speed limits, so that $v(0) = \gamma$, where γ is a function of the position on the road. Furthermore, it is reasonable to assume that $v(1) = 0$. For simplicity we can then define v as $v(\rho) = \gamma(1-\rho)$. In general, conservation of a quantity with density ρ and a velocity v is modeled by

$$\rho_t + (\rho v)_x = 0.$$

If the road conditions, and thereby γ, vary with the position x, then we end up with the conservation law

$$\rho_t + (\gamma(x)\rho(1-\rho))_x = 0, \tag{2}$$

which is an example of a conservation law with an x dependent flux function. On the scale of continuum traffic, where the natural lengths are *many* times that of a single car, the road conditions often vary discontinuously.

Example 1.2. When extracting oil from an oil reservoir, one often injects water in order to maintain the pressure, and thereby to force out more oil. Assuming that we have two phases, oil and water, present, the mass conservation of oil and water reads,

$$\phi s_t + u_x = 0, \quad \text{and} \quad \phi(1-s)_t - v_x = 0,$$

where (the unknown) s denotes the saturation, i.e., the fraction of the available pore space occupied by oil, and u and v are the Darcy velocities of oil and water, respectively. The factor ϕ denotes the fraction of the void space in the material, commonly called the *porosity*. One often assumes that Darcy's law holds,

$$u = -k\lambda_{\text{oil}}P'_{\text{oil}} - g\rho_{\text{oil}}, \quad \text{and} \quad v = -k\lambda_{\text{water}}P'_{\text{water}} - g\rho_{\text{water}},$$

where k denotes the absolute permeability of the medium, g the gravitational constant, and λ_{phase} the mobility, P_{phase} the pressure and ρ_{phase} the density of the phase "phase". If we assume that the two pressures are the same, and that the *total velocity* $q = u + v$ is constant (incompressibilty), we can add the two conservation equations to obtain

$$\phi s_t + \left(\frac{\lambda_{\text{oil}}(s)}{\lambda_{\text{oil}}(s) + \lambda_{\text{water}}(s)} \left(q - k(x)g\lambda_{\text{water}}(s)\Delta\rho\right) \right)_x = 0, \tag{3}$$

where $\Delta\rho = \rho_{\text{water}} - \rho_{\text{oil}}$. The absolute permeability of the rock is often modeled as a piecewise constant function of x, and therefore this is another example of a conservation law where the flux function varies discontinuously.

Example 1.3. Since oil is much more viscous than water, water injection can lead to the formation of thin "fingers" of water. In order to prevent this, one sometimes injects a mixture of polymer and water instead of water only. This polymer is passively transported with the water. In a "one-dimensional" homogeneous oil reservoir, conservation of water and polymer is expressed through the system of conservation laws

$$\begin{aligned} s_t + f(s,c)_x &= 0, \\ (sc)_t + (cf(s,c))_x &= 0, \end{aligned} \tag{4}$$

where c denotes the concentration of the polymer in the water, and the flux function $f(s,c)$ is a known function of the type in (3), where λ_{water} is now a function of both s and c. Introducing new coordinates (y,τ) by

$$\frac{\partial y}{\partial x} = s, \quad \frac{\partial y}{\partial t} = -f(s,c), \quad \frac{\partial \tau}{\partial x} = 0, \quad \text{and} \quad \frac{\partial \tau}{\partial t} = 1,$$

the system (4) reads

$$\begin{aligned} \left(\frac{1}{s}\right)_\tau - \left(\frac{f(s,c)}{s}\right)_y &= 0, \\ c_\tau &= 0. \end{aligned} \tag{5}$$

This change of independent variables is only valid for differentiable (classical) solutions. We shall see that solutions of conservation laws are not even continuous. Therefore we must interpret solutions in the *weak* sense. Nevertheless, by [18, Thm. 2] there is a one-to-one correspondence between weak solutions of (4) and weak solutions of (5). Hence if the initial polymer concentration is discontinuous, expression (5) is another example of a conservation law with a flux function depending discontinuously on the spatial location.

We can always view an x-dependent flux as a flux function depending on a parameter γ which in turn depends on x. In this way we write (1) as a system

$$u_t + f(\gamma, u)_x = 0, \quad \gamma_t = 0. \tag{6}$$

This is a hyperbolic system with a Jacobian matrix

$$\begin{pmatrix} \frac{\partial f}{\partial u} & \frac{\partial f}{\partial \gamma} \\ 0 & 0 \end{pmatrix},$$

which has the eigenvalues

$$\lambda_1 = \frac{\partial f}{\partial u}, \quad \lambda_2 = 0.$$

So if $\partial f/\partial u = 0$ for some values of γ and u, the system is not strictly hyperbolic. This is the cause of many difficulties when working with conservation laws with x dependent fluxes. The classical theory for scalar conservation laws and strictly hyperbolic systems, see [5, 12], uses approximations, which have uniformly bounded spatial variation, to show existence of a weak solution. In [15], Temple exhibited a simple example of a sequence of approximate solutions without any bound on the variation. This means that when studying conservation laws of the type (6), one must use more powerful (and complicated) tools. The "z-transform" used in these notes is perhaps the simplest (and least powerful) example of such tools. Recently, compensated compactness and variants of the "div-curl" lemma have taken the place of the "z-transform" when proving convergence of approximations, see [8] for a recent example.

I emphasize that these notes are meant to be an introduction to (the difficulties of) this field, and they do not contain the most general results.

I have chosen to work with front tracking, but I could well have chosen to use difference methods instead. This choice is a result of personal taste and experience. One can argue that difference methods are simpler to use, but I find that this comes at a price when trying to prove convergence. As mentioned above, the typical tools needed are more complicated. Furthermore, front tracking gives a valuable qualitative insight into the typical behaviour of solutions.

For those readers that are unfamiliar with front tracking, this is a name given to a class of numerical methods where one tries to construct approximate solutions by following (*tracking*) discontinuities in the solution. See the book [6] for an extensive introduction.

Over the last 20 years, conservation laws with spatially discontinuous flux functions have been studied in several papers, a very incomplete list includes [1, 2, 3, 4, 7, 10, 14, 16, 17] and other references therein.

The remainder of the notes is organized as follows. In section 2 we study, and propose a solution to the Riemann problem for conservation laws with discontinuous fluxes. In the following section, section 3, we use a front tracking approximation to establish existence of a (entropy) solution to the Cauchy problem for a model equation. Finally, in section 4, we discuss the question of wellposedness.

2 The Riemann problem

In this section we shall study the Riemann problem, that is the initial value problem where the initial values consist of two constants separated by a jump discontinuity. Concretely this is the problem

$$\begin{cases} u_t + f(\gamma_l, u)_x = 0, & u(x,0) = u_l, \quad \text{for } x < 0, \\ u_t + f(\gamma_r, u)_x = 0, & u(x,0) = u_r, \quad \text{for } x > 0, \end{cases} \tag{7}$$

where γ_l, γ_r, u_l and u_r are constants. Riemann problems for conservation laws lead to the simplest solutions that are not constant. Furthermore, by studying the solution of Riemann problems, we gain insight into the local behavior of typical solutions. It turns out that solutions of Riemann problems can be used as building blocks in many numerical methods, in particular front tracking.

By a solution of (7) we mean a weak solution in the usual sense, i.e., $u \in L^1_{\mathrm{loc}}(\mathbb{R} \times \mathbb{R}^+)$ is called a weak solution if for every test function $\varphi \in C_0^\infty(\mathbb{R} \times \mathbb{R}_0^+)$

$$\int_0^\infty \left(\int_{-\infty}^0 u\varphi_t + f(\gamma_l, u)\,\varphi_x\,dx + \int_0^\infty u\varphi_t + f(\gamma_r, u)\,\varphi_x\,dx \right) dt \\ + \int_{\mathbb{R}} u(x,0)\varphi(x,0)\,dx = 0. \tag{8}$$

Now we shall first show that under reasonable assumptions on f, weak solutions exist, and that if we require that weak solutions satisfy an additional entropy condition, then there exists only one weak solution.

2.1 Existence of a solution

To show existence of a solution, we start by observing that for $x < 0$, $u(x,t)$ must be the solution of a scalar conservation law

$$v_t + f(\gamma_l, v)_x = 0, \tag{9}$$

with $v(x,0)$ given by

$$v(x,0) = \begin{cases} u_l & \text{for } x < 0, \\ u_l' & \text{for } x = 0, \end{cases}$$

where u_l' is a value to be determined. Similarly, for $x > 0$, u must be the solution of the scalar initial value problem

$$w_t + f(\gamma_r, w)_x = 0, \qquad w(x,0) = \begin{cases} u_r' & \text{for } x = 0, \\ u_r & \text{for } x > 0, \end{cases} \tag{10}$$

where u_r' is to be determined. Setting

$$u(x,t) = \begin{cases} v(x,t) & \text{for } x < 0, \\ w(x,t) & \text{for } x > 0, \end{cases} \tag{11}$$

provided that $v(0-,t)$ and $w(0+,t)$ satisfy some extra condition, we find that this will give a weak solution, since both v and w are weak solutions. Therefore, to find a weak solution, we must find solutions of scalar Riemann problems v and w such that this construction is possible. Therefore, we turn to the solution of the Riemann problem for scalar conservation laws.

2.1.1 Solution of the Riemann problem for scalar conservation laws

Now we briefly describe the solution of scalar Riemann problems, for a more thorough description see [6]. Let for the moment v denote the solution of the scalar

Riemann problem

$$v_t + g(v)_x = 0, \qquad v(x,0) = \begin{cases} v_l & \text{for } x < 0, \\ v_r & \text{for } x > 0, \end{cases} \tag{12}$$

where g is a differentiable function, and v_l and v_r are constants. A weak solution to this problem is a function $v(x,t)$ such that for all test functions $\varphi \in C_0^\infty(\mathbb{R} \times \mathbb{R}_0^+)$

$$\int_{\mathbb{R}^+} \int_{\mathbb{R}} v\varphi_t + g(v)\varphi_x \, dxdt + \int_{\mathbb{R}} v(x,0)\varphi(x,0)\, dx = 0. \tag{13}$$

It is easy to see that if v is continuous and is a classical solution almost everywhere, then v is a weak solution. Furthermore, if a weak solution v has an isolated discontinuity along a smooth curve $x = \sigma(t)$, one can choose a test function with support localized near $(\sigma(t_0), t_0)$, and the Gauss-Green theorem to show that

$$\sigma'(t_0)\left(v(\sigma(t_0)+, t_0) - v(\sigma(t_0)-, t_0)\right) = g\left(v(\sigma(t_0)+, t_0)\right) - g\left(v(\sigma(t_0)-, t_0)\right).$$

This relation expresses conservation of v, and is called the Rankine–Hugoniot condition. Conversely, if a piecewise constant (in x) function v satisfies the Rankine–Hugoniot condition across its discontinuities, then it is a weak solution.

To find a weak solution of (12) is not difficult, if we set

$$s = \frac{g(v_l) - g(v_r)}{v_l - v_r},$$

and

$$v(x,t) = \begin{cases} v_l & \text{for } x < st, \\ v_r & \text{for } x > st, \end{cases}$$

then v is a weak solution. This is so since the Rankine–Hugoniot condition is satisfied across the discontinuity in v. However, the solution defined in this way may not be the only solution. In order to pick a unique solution, we require an additional entropy condition to hold. Let $g_\smile$ denote the lower convex envelope of g between v_l and v_r, i.e.,

$$g_\smile(v; v_l, v_r) = \sup\left\{ h(v) \le g(v) \;\middle|\; h \text{ is convex and continuous} \right\}. \tag{14}$$

Geometrically $g_\smile$ can be pictured as cutting a cardboard such that the lower edge has the shape of g and then stretching a rubber band along the lower boundary from v_l to v_r. This (idealized) rubber band will then have the shape of the graph of $g_\smile(\cdot; v_l, v_r)$. See Fig. 1. Since g is differentiable, we find that $g_\smile(\cdot; v_l, v_r)$ will also be differentiable, also $g'_\smile(\cdot; v_l, v_r)$ will be constant on intervals where $g_\smile(v; v_l, v_r) < g(v)$, and strictly increasing on intervals where $g''(v) > 0$ and $g_\smile(v; v_l, v_r) = g(v)$. Since $g_\smile$ is convex, $v \mapsto g'_\smile(v; v_l, v_r)$ will be nondecreasing. Hence we can form its (generalized, right continuous) inverse denoted by

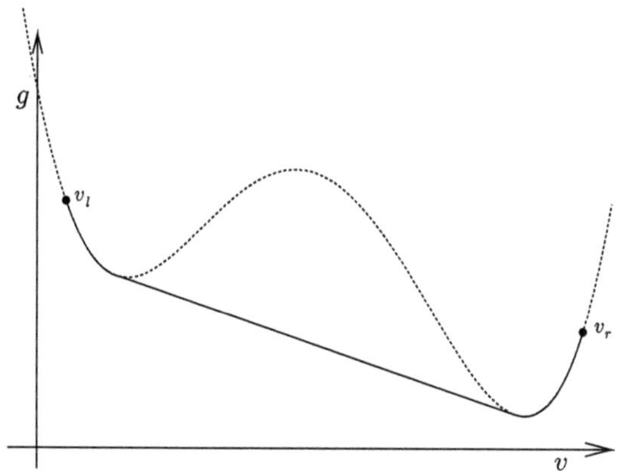

Fig. 1 An example of a flux function g and the lower convex envelope.

$$g'_{\smile}{}^{-1}(\cdot;v_l,v_r).$$

If $g'_{\smile}{}^{-1}(\cdot;v_l,v_r)$ has a jump for some value σ, which means that $g'_{\smile}(\cdot;v_l,v_r)=\sigma$ for some interval (v_1,v_2), we define $g'_{\smile}{}^{-1}(\sigma;v_l,v_r)=v_2$. Since we are looking for weak solutions, isolated point values of $g'_{\smile}{}^{-1}(\cdot;v_l,v_r)$ do not matter for our purposes. If $v_l<v_r$ and σ is not between $g'_{\smile}(v_l;v_l,v_r)$ and $g'_{\smile}(v_r;v_l,v_r)$, we define

$$g'_{\smile}{}^{-1}(\sigma;v_l,v_r)=\begin{cases} v_l & \text{if } \sigma\le g'_{\smile}(v_l;v_l,v_r),\\ v_r & \text{if } \sigma\ge g'_{\smile}(v_r;v_l,v_r).\end{cases}$$

For an illustration of the derivative of the lower envelope and its inverse, see Fig. 2. By scale invariance, the solution of the Riemann problem (12) is a function of x/t,

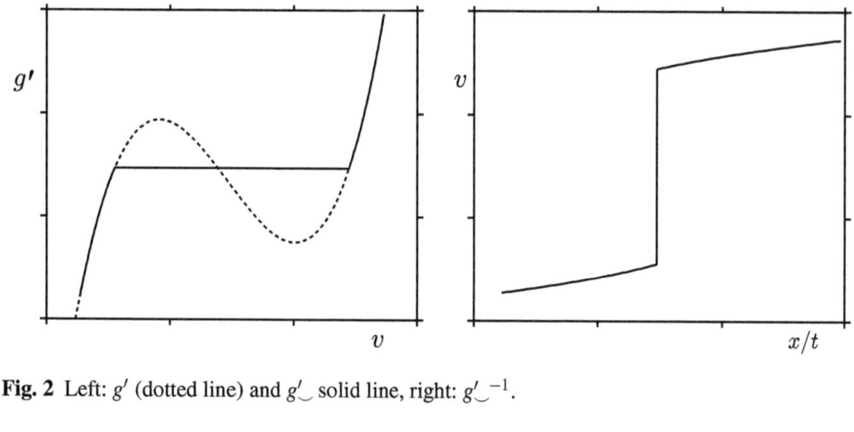

Fig. 2 Left: g' (dotted line) and $g'_{\smile}$ solid line, right: $g'_{\smile}{}^{-1}$.

and if $v_l<v_r$ we claim that the correct solution is found by the formula

$$v(x,t) = g_{\smile}'^{-1}\left(\frac{x}{t}; v_l, v_r\right), \quad t > 0. \tag{15}$$

If $v_l > v_r$, we may transform the problem (12) by $x \mapsto -x$, $g \mapsto -g$, to

$$w_t + \tilde{g}(w)_x = 0, \qquad w(x,0) = \begin{cases} v_r & \text{for } x < 0, \\ v_l & \text{for } x > 0, \end{cases}$$

where $\tilde{g} = -g$. Hence its solution is determined by $\tilde{g}_{\smile}$. This is the *upper concave envelope* of g, defined by

$$g_{\frown}(\cdot; v_l, v_r) = \inf\left\{ h(v) \geq g(v) \;\middle|\; h \text{ is concave and continuous} \right\}. \tag{16}$$

It follows that if $v_l > v_r$ the solution of the Riemann problem (12) is

$$v(x,t) = g_{\frown}'^{-1}\left(\frac{x}{t}; v_l, v_r\right). \tag{17}$$

We call the nonconstant parts of v *waves*. If v is not constant at x/t, then the speed of the wave at this point is given by either $g_{\smile}'(v(x/t); v_l, v_r)$ or $g_{\frown}'(v(x/t); v_l, v_r)$. In general, v will consists of a series of waves adjacent to each other. The continuous non-constant parts of v are called *rarefaction waves*, and the discontinuities are called *shock waves*, or just shocks.

The solution constructed in this way is called the entropy solution of the Riemann problem. It satisfies two important entropy conditions, the first of which is the *Kružkov entropy condition*:

$$\begin{aligned} \int_0^T \int_{\mathbb{R}} |v - c|\, \varphi_t + \operatorname{sign}(u - c)\,[g(v) - g(c)]\, \varphi_x \, dx dt \\ + \int_{\mathbb{R}} |v(x,0) - c|\, \varphi(x,0)\, dx \geq 0, \end{aligned} \tag{18}$$

for all nonnegative test functions $\varphi \in C_0^\infty(\mathbb{R} \times \mathbb{R}_0^+)$, and all constants c. The entropy solution v also satisfies the *vanishing viscosity entropy condition*: If v^ε is the (smooth) solution of the parabolic equation

$$v_t^\varepsilon + g\left(v^\varepsilon\right)_x = \varepsilon v_{xx}^\varepsilon, \qquad v^\varepsilon(x,0) = \begin{cases} v_l & \text{for } x < 0, \\ v_r & \text{for } x \geq 0, \end{cases} \tag{19}$$

then

$$\lim_{\varepsilon \to 0} v^\varepsilon = v,$$

where v is given by either (15) or (17).

In general a (mathematical) *entropy* is a convex smooth function $\eta(v)$. If v is the solution of a scalar conservation law, we are interested in the equation satisfied by

$\eta(v)$. Multiplying the viscous regularization (19) by $\eta'(v)$ we find that

$$\eta(v^\varepsilon)_t + \eta'(v^\varepsilon)g'(v^\varepsilon)v^\varepsilon_x = \varepsilon\eta'(v^\varepsilon)v^\varepsilon_{xx} = \varepsilon\eta(v^\varepsilon)_{xx} - \varepsilon\eta''(v^\varepsilon)(v^\varepsilon_x)^2 \le \varepsilon\eta(v^\varepsilon)_{xx}.$$

If we define the *entropy flux* q by $q' = \eta' g'$, we can thus write

$$\eta(v^\varepsilon)_t + q(v^\varepsilon)_x \le \varepsilon\eta(v^\varepsilon)_{xx}.$$

This means that if v is the limit of v^ε as $\varepsilon \to 0$, then

$$\int_{\mathbb{R}^+}\int_R \eta(v)\varphi_t + q(v)\varphi_x\,dxdt + \int_{\mathbb{R}} \eta(v(x,0))(x,0)\,dx \ge 0, \tag{20}$$

for any convex smooth function η and corresponding entropy flux q. Although $\eta(v) = |v-c|$ is not smooth, we can use an approximation argument to show that (20) implies (18). Conversely, we can approximate any convex smooth function $\eta(v)$ by a sum $\sum_k b_k |v-c_k|$ and use this to show the opposite implication.

To obtain a rather compact formula for the solution of (12), we define

$$\bar{g}(v; v_l, v_r) = \begin{cases} g_\frown(v; v_l, v_r) & \text{if } v_r < v_l, \\ g_\smile(v; v_l, v_r) & \text{if } v_l < v_r. \end{cases} \tag{21}$$

Then the entropy solution v is given by

$$v(x,t) = \bar{g}'^{-1}\left(\frac{x}{t}; v_l, v_r\right), \qquad t > 0. \tag{22}$$

2.1.2 Solution of the Riemann problem

Now that we know how to compute the solution to scalar Riemann problems, we turn again to the Riemann problem (7). The left and right parts of u are v, as given by (9), and w, as given by (10). If we are to form u by gluing together v and w, v must equal u'_l for $x > 0$, and w must equal u'_r for $x < 0$. In other words, v must contain only waves of nonpositive speed, and w only waves with nonnegative speed. To utilize these observations, we introduce the notation

$$f_l(u) = f(\gamma_l, u) \quad \text{and} \quad f_r(u) = f(\gamma_r, u),$$

and define $\bar{f}_l(u; u_l, \tilde{u})$ and $\bar{f}_r(u; \tilde{u}, u_r)$ analogously to (21).

Since v only contains waves of nonpositive speed, we must choose u'_l from the set

$$H_l(u_l) = \left\{\tilde{u} \;\middle|\; \bar{f}'_l(u; u_l, \tilde{u}) \le 0 \quad \text{for all } u \text{ between } u_l \text{ and } \tilde{u}\right\}. \tag{23}$$

Similarly, since w must contain waves of nonnegative speed, we must choose u'_r from the set

$$H_r(u_r) = \left\{\tilde{u} \;\middle|\; \bar{f}'_r(u; \tilde{u}, u_r) \ge 0 \quad \text{for all } u \text{ between } u_r \text{ and } \tilde{u}\right\}. \tag{24}$$

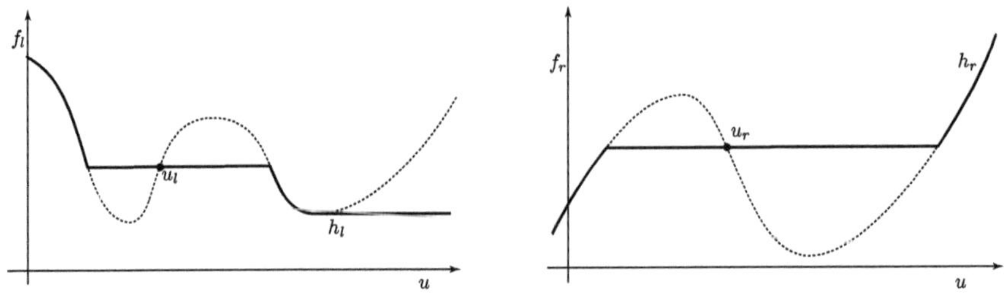

Fig. 3 Left: h_l (solid line) and f_l (dotted line), right: h_r (solid line) and f_r (dotted line).

There is another characterization of the admissible sets H_l and H_r that will be useful. Let h_l be defined by

$$h_l(u;u_l) = \begin{cases} \inf\left\{ h(u) \;\middle|\; \begin{matrix} h(u) \geq f_l(u), & h'(u) \leq 0, \\ \text{and} & h(u_l) = f_l(u_l) \end{matrix} \right\} & \text{if } u \leq u_l, \\ \sup\left\{ h(u) \;\middle|\; \begin{matrix} h(u) \leq f_l(u), & h'(u) \leq 0, \\ \text{and} & h(u_l) = f_l(u_l) \end{matrix} \right\} & \text{if } u \geq u_l, \end{cases} \tag{25}$$

and define h_r by

$$h_r(u;u_r) = \begin{cases} \sup\left\{ h(u) \;\middle|\; \begin{matrix} h(u) \leq f_r(u), & h'(u) \geq 0, \\ \text{and} & h(u_r) = f_r(u_r) \end{matrix} \right\} & \text{if } u \leq u_r, \\ \inf\left\{ h(u) \;\middle|\; \begin{matrix} h(u) \geq f_r(u), & h'(u) \leq 0, \\ \text{and} & h(u_l) = f_l(u_l) \end{matrix} \right\} & \text{if } u \geq u_l. \end{cases} \tag{26}$$

In these definitions, the function h appearing in the infima and suprema is assumed to be continuous. In Fig. 3 we show an example of h_l and h_r. Using h_l and h_r, we can use the following alternative definition of the admissible sets H_l and H_r, namely

$$H_l(u_l) = \left\{ u \;\middle|\; h_l(u;u_l) = f_l(u) \right\}, \tag{27}$$

$$H_r(u_r) = \left\{ u \;\middle|\; h_r(u;u_r) = f_r(u) \right\}. \tag{28}$$

Since the jump in u at $x = 0$ is a discontinuity with zero speed, the Rankine–Hugoniot condition says that for any weak solution we must have

$$f(\gamma_l, u_l') = f(\gamma_r, u_r') =: f^\times. \tag{29}$$

Now $u_l' \in H_l(u_l)$ and $u_r' \in H_r(u_r)$, using (27) and (28), this can be restated as

$$h_l(u_l', u_l) = h_r(u_r', u_r). \tag{30}$$

Now the mapping $u \mapsto h_l(u;u_l)$ is nonincreasing and $u \mapsto h_r(u;u_r)$ is nondecreasing, so the above equality (30) will hold for at most one h value. Therefore, if the graphs of h_l and h_r intersect, the flux value at $x=0$ is determined by the flux value at this intersection point. We label this flux value $f^{\times}$.

From these observations it also follows that if the graph of h_l does *not* intersect the graph of h_r, we cannot hope to find a weak solution to the Riemann problem (7). For instance if

$$f_l(u) = e^{-u^2}, \quad \text{and} \quad f_r(u) = 2 + e^{-u^2},$$

we cannot find any weak solution. Another important example where we cannot find any solution to the Riemann problem is in the case where

$$f_l'(u) \geq 0 \quad \text{and} \quad f_r'(u) \leq 0.$$

In this case $h_l(u;u_l) = f_l(u_l)$, and $h_r(u;u_r) = f_r(u_r)$, so unless these happen to be equal, we cannot find any solution.

Furthermore, even if the flux value at the intersection is uniquely determined, the actual values u_l' and u_r' need not be. This is so since for $u \notin H_l(u_l)$, we have $h_l'(u;u_l) = 0$, and likewise if $u \notin H_r(u_r)$, we have $h_r'(u;u_r) = 0$. In other words, $h_{l,r}$ may both be constant on the interval where their graphs cross. In order to resolve this nonuniqueness problem, we propose that u_l' and u_r' are chosen so that the variation of the solution u is minimal, subject to the above restrictions.

To be more concrete, we choose u_l' to be the unique value such that

$$\left|u_l - u_l'\right| \text{ is minimized provided } u_l' \in H_l(u_l) \text{ and } h_l(u_l';u_l) = f^{\times}. \tag{31}$$

Similarly, we choose u_r' to be the unique value such that

$$\left|u_r - u_r'\right| \text{ is minimized provided } u_r' \in H_r(u_r) \text{ and } h_l(u_r';u_r) = f^{\times}. \tag{32}$$

These criteria for choosing u_l' and u_r' are called *the minimal jump entropy condition*.

It is perhaps instructive to examine this condition in a little more detail. If the graphs of h_l and h_r intersect in a single point $u^{\times}$, then $u^{\times} \in H_l(u_l)$ or $u^{\times} \in H_r(u_r)$. If $u^{\times} \in H_l(u_l)$, then $u_l' = u^{\times}$, and if $u^{\times} \in H_r(u_r)$ then $u_r' = u^{\times}$. Assuming for definiteness that $u_l < u^{\times}$ and $u^{\times} \notin H_l(u_l)$, then there will be a smallest point $\tilde{u}$ in the interval $[u_l, u^{\times}]$ such that the interval $(\tilde{u}, u^{\times}] \not\subset H_l(u_l)$, and $\tilde{u} \in H_l(u_l)$. It is clear that according to (31), we must choose $u_l' = \tilde{u}$.

In Fig. 4 we show how the Riemann problem from Fig. 3 is solved in this way. Here $u^{\times} \in H_l(u_l)$. so $u_l' = u^{\times}$. Also the point minimizing $|u_r' - u_r|$ is clearly u_r, so that $u_r' = u_r$. Finally the Riemann problem is solved by a shock of negative speed from u_l to u_l', and then by a discontinuity at $x=0$ from u_l' to u_r. There is some more important information to be extracted from the minimal jump entropy condition. Since the Riemann problem with $u_l = u_l'$ and $u_r = u_r'$ is solved by a single stationary discontinuity, in the interval spanned by u_l' and u_r', we must have

$$h_l\left(u;u_l'\right) = f^{\times}, \text{ or } h_r\left(u;u_r'\right) = f^{\times}. \tag{33}$$

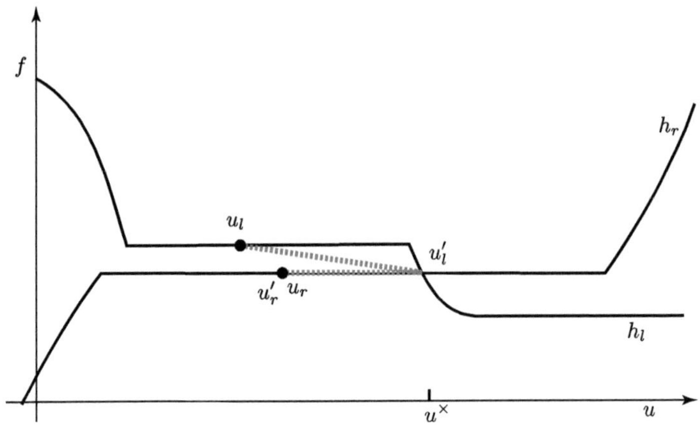

Fig. 4 An example on how to solve a Riemann problem of the type (7).

If $u_l' < u_r'$, since $h_l(\cdot;u_l')$ is the largest nonincreasing continuous function less than or equal to f_l such that $h_l(u_l';u_l') = f_l(u_l')$,

$$h_l\left(u;u_l'\right) = f^\times \quad \Rightarrow \quad f_l(u) > f^\times \text{ for } u \in (u_l', u_r')$$

and

$$h_r\left(u;u_r'\right) = f^\times \quad \Rightarrow \quad f_r(u) > f^\times \text{ for } u \in (u_l', u_r'),$$

since $h_r(\cdot;u_r')$ is the largest continuous nondecreasing function smaller than or equal to f_r. Similarly, if $u_r' < u_l'$, then

$$h_l\left(u;u_l'\right) = f^\times \quad \Rightarrow \quad f_l(u) < f^\times \text{ for } u \in (u_r', u_l'),$$

and

$$h_r\left(u;u_r'\right) = f^\times \quad \Rightarrow \quad f_r(u) < f^\times \text{ for } u \in (u_r', u_l').$$

Summing up, we have

$$u_l' \leq u_r' \quad \Rightarrow \quad \begin{cases} f_l(u) \geq f_l(u_l') & \text{for all } u \in [u_l', u_r'], \text{ or} \\ f_r(u) \geq f_r(u_r') & \text{for all } u \in [u_l', u_r'], \end{cases} \tag{34}$$

$$u_r' \leq u_l' \quad \Rightarrow \quad \begin{cases} f_l(u) \leq f_l(u_l') & \text{for all } u \in [u_r', u_l'], \text{ or} \\ f_r(u) \leq f_r(u_r') & \text{for all } u \in [u_r', u_l']. \end{cases} \tag{35}$$

Furthermore, implications (34) and (35) actually imply that u_l' and u_r' are chosen according to the minimal jump entropy condition.

Lemma 2.1. *If the values u_l' and u_r' are chosen according to the minimal jump entropy condition* (31), (32) *then, for any constant c,*

$$F_r\left(u_r',c\right) - F_l\left(u_l',c\right) \le \left|f_r(c) - f_l(c)\right|, \tag{36}$$

where F_l and F_r are the Kružkov entropy fluxes,

$$\begin{aligned} F_l(u,c) &= \operatorname{sign}(u-c)\left(f_l(u) - f_l(c)\right), \\ \text{and} \quad F_r(u,c) &= \operatorname{sign}(u-c)\left(f_r(u) - f_r(c)\right). \end{aligned}$$

Proof. If $\operatorname{sign}\left(u_l'-c\right) = \operatorname{sign}\left(u_r'-c\right)$ then the left-hand side of (36) equals

$$\begin{aligned} &\operatorname{sign}\left(u_l'-c\right)\left(f_r\left(u_r'\right) - f_r(c) - f_l\left(u_l'\right) + f_l(c)\right) \\ &\qquad\qquad = \operatorname{sign}\left(u_l'-c\right)\left(f_l(c) - f_r(c)\right), \end{aligned}$$

and the inequality clearly holds.

If $u_l' \le c \le u_r'$ then (36) reads

$$2f^\times - f_l(c) - f_r(c) \le \left|f_r(c) - f_l(c)\right|$$

or

$$\begin{aligned} &2f^\times - \max\left\{f_l(c), f_r(c)\right\} - \min\left\{f_l(c), f_r(c)\right\} \\ &\qquad\qquad \le \max\left\{f_l(c), f_r(c)\right\} - \min\left\{f_l(c), f_r(c)\right\}. \end{aligned}$$

In other words, (36) is the same as

$$f^\times \le \max\left\{f_l(c), f_r(c)\right\},$$

and it is immediate that (34) implies this.

If $u_r' \le c \le u_l'$ then (36) reads

$$f^\times \ge \min\left\{f_l(c), f_r(c)\right\},$$

which is implied by (35).

From the proof of Lemma 2.1 it is also transparent that condition (36) does *not* imply the minimal jump entropy conditions (34) and (35). However, define the pair of "constants" c_l and c_r (these numbers depend on u_l' and u_r') by requiring

$$c_l\left(u_l',u_r'\right) = \begin{cases} \operatorname{minarg}_{[u_l',u_r']} f_l(u) & \text{if } u_l' \le u_r', \\ \operatorname{maxarg}_{[u_r',u_l']} f_l(u) & \text{if } u_l' \ge u_r', \end{cases} \tag{37}$$

$$c_r\left(u_l',u_r'\right)=\begin{cases}\operatorname{minarg}_{[u_l',u_r']} f_r(u) & \text{if } u_l'\le u_r',\\ \operatorname{maxarg}_{[u_r',u_l']} f_r(u) & \text{if } u_l'\ge u_r'.\end{cases} \tag{38}$$

Using the arguments of the proof of Lemma 2.1, it readily follows that the minimal jump entropy condition is equivalent to

$$F_r\left(u_r',c_r\right)-F_l\left(u_l',c_l\right)\le\left|f_r\left(c_r\right)-f_l\left(c_l\right)\right|. \tag{39}$$

Furthermore, for any c between u_l' and u_r', the inequality

$$F_r\left(u_r',c\right)-F_l\left(u_l',c\right)\le F_r\left(u_r',c_r\right)-F_l\left(u_l',c_l\right)$$

holds.

Remark 2.1. In a special case (36) actually implies that the values u_l' and u_r' are chosen according to the minimal jump entropy condition. Assume that there is a value $\hat{u}$ such that both $f_l(u)$ and $f_r(u)$ have a global maximum (minimum) at $\hat{u}$, and that $f_{l,r}$ is increasing (decreasing) for $u<\hat{u}$ and decreasing (increasing) for $u>\hat{u}$. To see this we recall that (36) holds trivially if c is not between u_l' and u_r', if c is between these values (36) reads

$$\begin{cases} f^\times\le\max\left\{f_l(c),f_r(c)\right\}, & \text{if } u_l'<u_r',\\ f^\times\ge\max\left\{f_l(x),f_r(c)\right\}, & \text{if } u_l'>u_r'.\end{cases} \tag{40}$$

By assuming that $f_l(u_l')=f_r(u_r')$, that the above holds, and that the flux functions $f_{l,r}$ have a single common maximum, the reader can check that (40) implies (34) and (35). Actually, this implication holds for more general flux functions as well, c.f. the notorious "crossing condition" in [10].

Although it seemingly has nothing to do with the solution of the Riemann problem, at this point it is convenient to state and prove the following lemma, which will play an important role when proving wellposedness in section 4.

Lemma 2.2. *Assume that the pairs (u_l',u_r') and (v_l',v_r') are both chosen according to the minimal jump entropy condition. Then*

$$Q=F_r\left(u_r',v_r'\right)-F_l\left(u_l',v_l'\right)\le 0. \tag{41}$$

Proof. Since $f_l(v_l')=f_r(v_r')$ and $f_l(u_l')=f_r(u_r')$, if $\operatorname{sign}\left(u_l'-v_l'\right)=\operatorname{sign}\left(u_r'-v_r'\right)$, then $Q=0$. Assume therefore that

$$\operatorname{sign}\left(u_l'-v_l'\right)=-1 \quad\text{and}\quad \operatorname{sign}\left(u_r'-v_r'\right)=1.$$

In this case

$$Q=\left[f_r\left(u_r'\right)-f_r\left(v_r'\right)\right]+\left[f_l\left(u_l'\right)-f_l\left(v_l'\right)\right] \tag{42}$$
$$=2\left(f_r\left(u_r'\right)-f_r\left(v_r'\right)\right), \tag{43}$$

$$= 2\left(f_l\left(u_l'\right) - f_l\left(v_l'\right)\right), \tag{44}$$

since $f_l(v_l') = f_r(v_r')$ and $f_l(u_l') = f_r(u_r')$. Moreover

$$u_l' \leq v_l' \quad \text{and} \quad v_r' \leq u_r'.$$

Then either u_l' and u_r' are both in the interval $[v_r', v_l']$ (case **a**), or v_l' and v_r' are in the interval $[u_l', u_r']$ (case **b**), or $v_r' \leq u_l' \leq v_l' \leq u_r'$ (case **c**) or $u_l' \leq v_r' \leq u_r' \leq v_l'$ (case **d**).

If case **a** holds then (35) for v_l' and v_r' gives that either

$$f_l\left(u_l'\right) \leq f_l\left(v_l'\right) \quad \text{or} \quad f_r\left(u_r'\right) \leq f_r\left(v_r'\right).$$

It is easy to see that this coupled with either (43) or (44) will give $Q \leq 0$.

If case **b** holds, then (35) for u gives that either

$$f_l\left(v_l'\right) \geq f_l\left(u_l'\right) \quad \text{or} \quad f_r\left(v_r'\right) \geq f_r\left(u_r'\right).$$

So again $Q \leq 0$.

Recall that case **c** is defined to hold if

$$v_r' \leq u_l' \leq v_l' \quad \text{and} \quad u_l' \leq v_l' \leq u_r'.$$

Using the first inequality and (35) for v we find that

$$f_l\left(u_l'\right) \leq f_l\left(v_l'\right) \quad \text{or} \quad f_r\left(u_r'\right) \leq f_r\left(v_r'\right),$$

which both give the desired conclusion.

Finally in case **d**, we have

$$u_l' \leq v_r' \leq u_r' \quad \text{and} \quad v_r' \leq u_r' \leq v_l'.$$

Using the first inequality with (34) gives

$$f_l\left(v_l'\right) \geq f_l\left(u_l'\right) \quad \text{or} \quad f_r\left(v_r'\right) \geq f_r\left(u_r'\right),$$

so that the proof is concluded.

Example 2.1. Now we pause to consider two examples. First consider the Riemann problem for the equation

$$u_t + \left(\frac{1}{2}u^2 + \gamma\right)_x = 0, \tag{45}$$

where

$$u_0(0) = \begin{cases} u_l & \text{for } x < 0, \\ u_r & \text{for } x > 0, \end{cases} \quad \text{and} \quad \gamma(x) = \begin{cases} \gamma_l & \text{for } x < 0, \\ \gamma_r & \text{for } x > 0. \end{cases}$$

If $u_l \leq 0$ then

$$H_l\left(u_l\right) = (-\infty, 0],$$

and if $u_l \geq 0$ then

$$H_l(u_l) = (-\infty, -u_l] \cup \{u_l\}.$$

Similarly if $u_r \leq 0$ then

$$H_r(u_r) = \{-u_r\} \cup [-u_r, \infty),$$

and if $u_r \geq 0$,

$$H_r(u_r) = [0, \infty).$$

Now it is easy to construct the solution for any initial data and any $\gamma(x)$. Assume that $\gamma_l = -1$, $\gamma_r = 1$, $u_l = 1$ and $u_r = 1$. Then

$$h_l(u; -1) = \begin{cases} \frac{1}{2}u^2 - 1 & \text{if } u \leq -1, \\ -\frac{1}{2} & \text{if } u \geq -1, \end{cases} \quad \text{and} \quad h_r(u; 1) = \begin{cases} 1 & \text{if } u \leq 0, \\ \frac{1}{2}u + 1 & \text{if } u \geq 0. \end{cases}$$

The graphs of h_l and h_r intersect in a single point where the flux equals 1, and $u < 0$, thus we find u_l' as the solution of

$$h_l(u_l'; -1) = 1, \quad u_l' < 0.$$

and thus $u_l' = -2$. Following the general construction, we see that $u_r' = 0$. The complete solution thus consists of the solution of a scalar Riemann problem for the equation

$$v_t + \left(\frac{1}{2}v^2\right)_x = 0, \quad v(x,0) = \begin{cases} 1 & \text{for } x \leq 0, \\ -2 & \text{for } x \geq 0, \end{cases}$$

glued together with the solution of the scalar Riemann problem

$$w_t + \left(\frac{1}{2}w^2\right)_x = 0, \quad w(x,0) = \begin{cases} 0 & \text{for } x \leq 0, \\ 1 & \text{for } x \geq 0. \end{cases}$$

From the general solution procedure for scalar Riemann problems, i.e., taking envelopes, we see that

$$v(x,t) = \begin{cases} 1 & \text{for } x \leq -t/2, \\ -2 & \text{for } x > -t/2, \end{cases} \quad \text{and} \quad w(x,t) = \begin{cases} 0 & \text{for } x \leq 0, \\ x/t & \text{for } 0 < x \leq t, \\ 1 & \text{for } t < x. \end{cases}$$

Finally, we set

$$u(x,t) = \begin{cases} v(x,t) & \text{for } x < 0, \\ w(x,t) & \text{for } x > 0. \end{cases}$$

This solution is depicted in Fig. 5, to the left we see the solution path in the (u, f) plane, and to the right $u(x,t)$. Perhaps the most important lesson to be learned from this example is that the variation of the solution u is *not* bounded by the variation

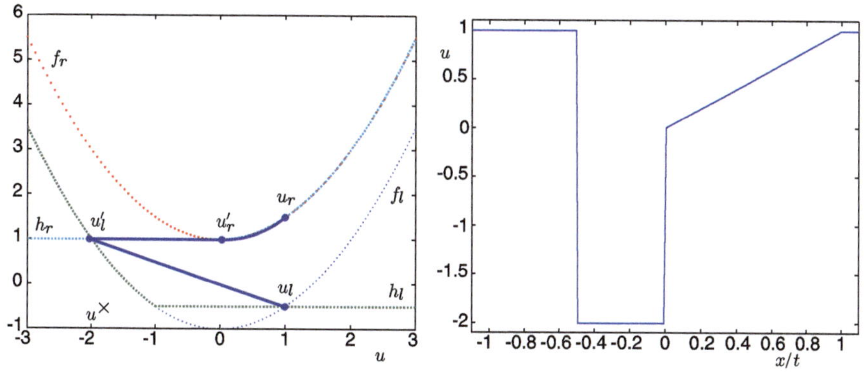

Fig. 5 An example of the solution of a Riemann problem, left: the solution path in (u, f) space, right: $u(x,t)$.

of the initial data $u(x,0)$. Even though this is so, it is natural to ask whether the variation of u is bounded by the variation of u_0 plus the variation of γ. We are not able to show that this is the case, but from the construction of the solution to the Riemann problem, the total variation of $f(u,\gamma)$ is less than the total variation of $f(u_0,\gamma)$. We return to these observations in a later section.

Example 2.2. As a second example we study the (traffic flow) model first considered by Mochon in [13]

$$u_t + (\gamma(x)4u(1-u))_x = 0, \tag{46}$$

where

$$u(x,0) = \begin{cases} u_l & \text{for } x < 0, \\ u_r & \text{for } x \geq 0, \end{cases} \qquad \gamma(x) = \begin{cases} \gamma_l & \text{for } x < 0, \\ \gamma_r & \text{for } x \geq 0. \end{cases}$$

For simplicity, we assume that γ_l and γ_r are positive. Now

$$H_l(u_l) = \begin{cases} \{u_l\} \cup [1-u_l, \infty) & \text{if } u_l \leq 1/2, \\ [1/2, \infty) & \text{if } u_l \geq 1/2, \end{cases}$$

and

$$H_r(u_r) = \begin{cases} (-\infty, 1/2] & \text{if } u_r \leq 1/2, \\ (-\infty, 1-u_r] \cup \{u_r\} & \text{if } u_r \geq 1/2. \end{cases}$$

We shall now detail the complete solution of the Riemann problem in this case, as this is instructive, since (46) exhibits many of the features of Riemann solutions for general flux functions.

We describe the solution by listing what happens in various cases, depending on γ_l, γ_r, u_l and u_r. Note first that $f(\gamma,u)$ has a maximum at $u = 1/2$ for all γ and that $f(\gamma, 1/2) = \gamma$. We start by assuming that

$$u_l \leq \frac{1}{2}. \tag{47}$$

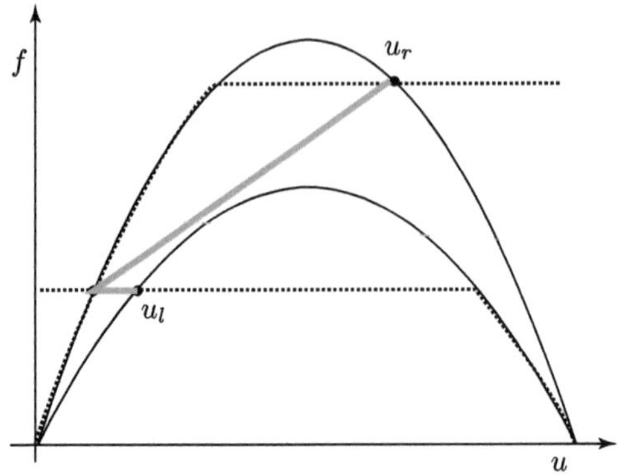

Fig. 6 The solution of the Riemann problem if $u_l < 1/2$, $\gamma_l < \gamma_r$, and $f(\gamma_l, u_l) < f(\gamma_r, u_r)$ or $u_r \leq 1/2$.

In this case the structure of the solution will depend on whether $\gamma_l < \gamma_r$ or not. We start by examining the case where $\gamma_l < \gamma_r$ and $f(\gamma_l, u_l) < f(\gamma_r, u_r)$ or $u_r \leq 1/2$. The situation is depicted in Fig. 6. Here we show the h_l and h_r functions as dotted lines, and the solution path as a gray line. In this case $u_l' = u_l$, and u_r' is the solution of

$$f(\gamma_r, u_r') = f(\gamma_l, u_l), \quad u_r' < \frac{1}{2},$$

in our case, this means that

$$u_r' = \frac{1}{2}\left(1 - \sqrt{1 - \frac{\gamma_l}{\gamma_r} 4u_l(1-u_l)}\right).$$

The solution consists of a stationary discontinuity separating (u_l', γ_l) and (u_r', γ_r), which we shall call a γ-wave, followed by a shock in u moving to the right. This we call a u-wave. For clarity we also show the solution if $u_r \leq 1/2$ in Fig. 7.

Next, we turn to the case where $\gamma_l < \gamma_r$ and $f(\gamma_l, u_l) \geq f(\gamma_r, u_r)$, this case is depicted in Fig. 8. Then the solution consists of a u-wave with negative speed followed by a γ-wave separating u_l' and u_r. In other words, we have $u_r' = u_r$, and u_l' is the solution of

$$f(\gamma_l, u_l') = f(\gamma_r, u_r), \quad u_l' \geq \frac{1}{2}.$$

In the next case we assume that $u_l \geq 1/2$. In this case, if $u_r \leq 1/2$, or $f(\gamma_r, u_r) > f(\gamma_l, 1/2)$ then $u_l' = 1/2$, and u_r' solves

$$f(\gamma_r, u_r') = f(\gamma_l, u_l') = \gamma_l, \quad u_r' < \frac{1}{2}.$$

This is illustrated in Fig. 9. Now the solution consists of a u-wave moving to the left, this u-wave is a rarefaction wave, followed by a γ-wave. The last wave is a u-wave moving to the right, this is a shock wave.

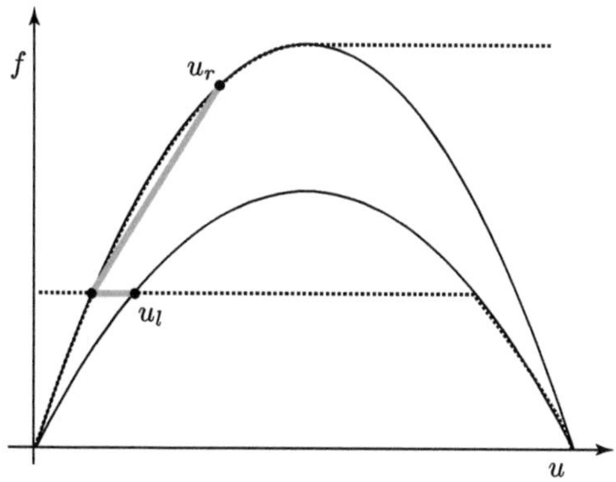

Fig. 7 The solution of the Riemann problem if $u_l < 1/2$, $\gamma_l < \gamma_r$, and $f(\gamma_l, u_l) < f(\gamma_r, u_r)$ or $u_r \leq 1/2$.

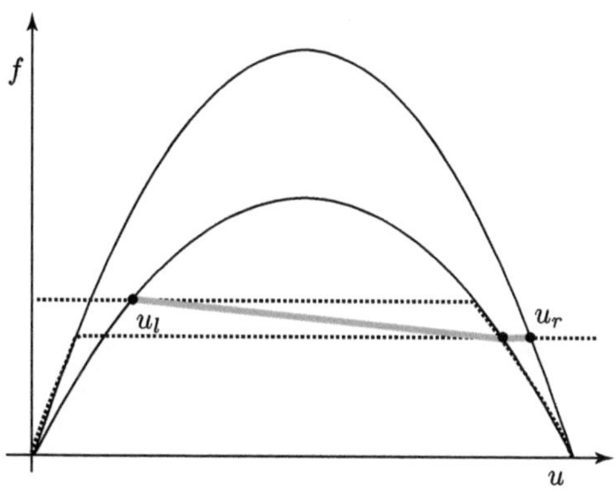

Fig. 8 The solution of the Riemann problem if $u_l < 1/2$, $\gamma_l < \gamma_r$, $f(\gamma_l, u_l) \geq f(\gamma_r, u_r)$ and $u_r \geq 1/2$.

Next, if $u_l \geq 1/2$, $u_r \geq 1/2$ and $f(\gamma_r, u_r) \leq f(\gamma_l, 1/2)$, the solution is shown in Fig. 10. In this case u consists of a leftward moving u-wave followed by a γ-wave. This exhausts the case where $\gamma_l < \gamma_r$.

The case where $\gamma_l > \gamma_r$ is analogous, and we show the four different possibilities in Fig. 11.

In order to determine a particular solution, follow the gray path from u_l to u_r. If the path follows the graph of the f_l or f_r, the wave is a rarefaction wave, and if not it is a shock wave. The horizontal segments joining f_l and f_r are γ-waves. In these figures, the dotted lines indicate the functions h_l and h_r.

From the above diagrams, we observe that if u_l and u_r are in the interval $[0, 1]$, then also the solution $u(x,t)$ will take values in $[0, 1]$. In many applications involving similar conservation laws, u is interpreted as a density, hence it is natural to require that u is between 0 and 1.

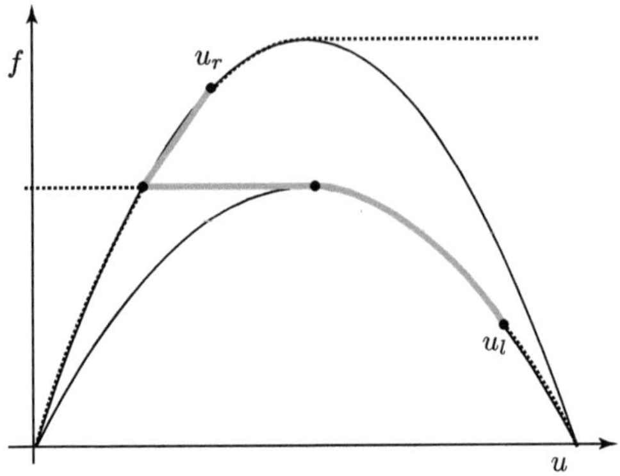

Fig. 9 The solution of the Riemann problem if $u_l \geq 1/2$, $\gamma_l < \gamma_r$, and $f(\gamma_l, 1/2) < f(\gamma_r, u_r)$ or $u_r \leq 1/2$.

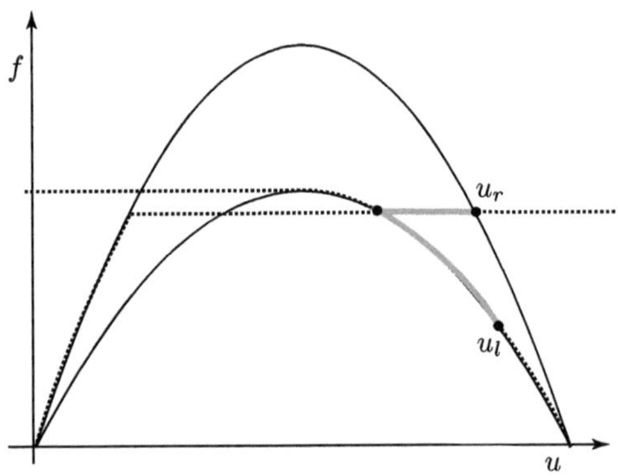

Fig. 10 The solution of the Riemann problem if $u_l \geq 1/2$, $\gamma_l < \gamma_r$, $f(1/2, u_l) \geq f(\gamma_r, u_r)$ and $u_r > 1/2$.

There is another and much more compact way to depict the solution of the general Riemann problem for this conservation law. Let $z = z(\gamma, u)$ be defined as

$$\begin{aligned} z(\gamma,u) &= \text{sign}\left(\frac{1}{2} - u\right)\left[f(\gamma,u) - f\left(\gamma,\frac{1}{2}\right)\right] \\ &= \gamma \text{sign}\left(u - \frac{1}{2}\right)(2u-1)^2 \\ &= \int_{1/2}^{u} \left|\frac{\partial f}{\partial u}(\gamma,\xi)\right| d\xi. \end{aligned} \tag{48}$$

This mapping takes the rectangle $[\gamma_1, \gamma_2] \times [0,1]$ into the region

$$\left\{(z,\gamma) \;\middle|\; \gamma_1 \leq \gamma \leq \gamma_2 \quad \text{and} \quad -\gamma \leq z \leq \gamma\right\}.$$

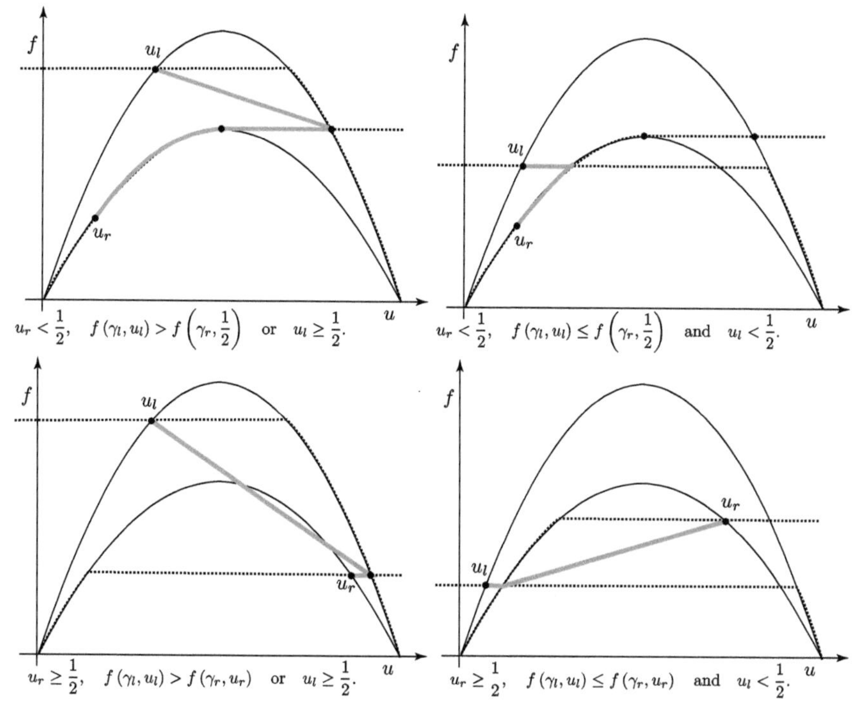

Fig. 11 The different possibilities for a solution of the Riemann problem if $u_r \geq 1/2$. The solution path is the gray line.

Furthermore, $u \mapsto z(\gamma, u)$ is injective, and strictly increasing. In (z, γ) coordinates, γ-waves are straight lines of slope -1 if $u \leq 1/2$ and slope 1 if $u \geq 1/2$, and u-waves are horizontal lines. In Fig. 12 we show what the solutions in the various cases look like in the (z, γ) plane. To read the diagram, start at the point $L = (z(u_l, \gamma_l), \gamma_l)$ and follow the arrows to the right location. The dotted lines mark the boundaries where the solution type is constant. Since we are working with (z, γ) coordinates, we call the u-waves z-waves, and the solution types are $z\gamma$, $z\gamma z$ and γz. If a solution type is e.g., $z\gamma$, this means that the solution consists of a z-wave (u-wave) followed by a γ-wave. This finishes the second example.

Actually, our two examples are more similar than it might seem at first glance. The inverse of the mapping (48) is

$$u = \frac{1}{2}\left(1 + \operatorname{sign}(z)\sqrt{\frac{|z|}{\gamma}}\right),$$

and

$$f(\gamma, u) = |z| + \gamma.$$

Plugging this into the equation (46), we find that

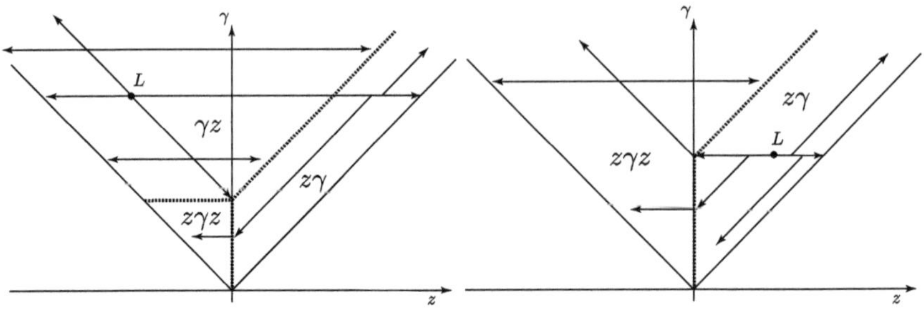

Fig. 12 The solution of the Riemann problem. Left: $z_l \leq 0$, right: $z_l \geq 0$.

$$\left(\frac{1}{2}\left(1+\operatorname{sign}(z)\sqrt{\frac{|z|}{\gamma}}\right)\right)_t + (|z|+\gamma)_x = 0.$$

Since γ is independent of t, we can rearrange this as

$$\left(\operatorname{sign}(z)\sqrt{|z|}\right)_t + 2\sqrt{\gamma}(|z|+\gamma)_x = 0.$$

If we now introduce $w = \operatorname{sign}(z)\sqrt{|z|}$, and a new time coordinate τ such that $\partial/\partial\tau = \sqrt{2\gamma}\partial/\partial t$, then

$$w_\tau + \left(\frac{1}{2}w^2 + \gamma\right)_x = 0.$$

Now we return to the discussion of the Riemann problem for the general conservation law (cf. (7)). We have seen that we cannot always find a weak solution to this problem, but if the graphs of the functions $H_l(\cdot;u_l)$ and $H_r(\cdot;u_r)$ intersect, we can choose a unique pair (u_l', u_r'), which in turn gives us a unique solution of the Riemann problem. We call this solution, satisfying the minimal jump entropy condition, an *entropy solution* of the Riemann problem.

It seems complicated to give a general criterion for f_l and f_r guaranteeing the intersection of h_l and h_r, but for two important classes of flux functions we always have an intersection.

Proposition 2.1. *Consider the Riemann problem*

$$u_t + f(\gamma,u)_x = 0, \quad t > 0,$$
$$u(x,0) = \begin{cases} u_l & \text{for } x < 0, \\ u_r & \text{for } x > 0, \end{cases} \qquad \gamma(x) = \begin{cases} \gamma_l & \text{for } x < 0, \\ \gamma_r & \text{for } x > 0. \end{cases} \tag{49}$$

1. Let $f = f(\gamma,u)$ be a continuously differentiable function on the set

$$(\gamma,u) \in [\gamma_1,\gamma_2] \times [u_1,u_2] = \Omega.$$

Assume that

$$\frac{\partial f}{\partial \gamma}(\gamma, u_1) = \frac{\partial f}{\partial \gamma}(\gamma, u_2) = 0,$$

so that $f(\gamma, u_1) = C_1$ *and* $f(\gamma, u_2) = C_2$ *for some constants* C_1 *and* C_2. *Then the Riemann problem* (49) *has a unique entropy solution for all* (γ_l, u_l) *and* (γ_r, u_r) *in* Ω. *Furthermore* $u(x,t) \in \Omega$ *for all* x *and* t.

2. *Let* $f = f(\gamma, u)$ *be a locally Lipschitz continuous function for* $\gamma \in [\gamma_1, \gamma_2]$ *and* $u \in \mathbb{R}$. *Assume that*

$$\lim_{u \to \pm\infty} f(\gamma, u) = \infty, \quad or \quad \lim_{u \to \pm\infty} f(\gamma, u) = -\infty,$$

for all $\gamma \in [\gamma_1, \gamma_2]$. *Then the Riemann problem* (49) *has a unique entropy solution for all* (γ_l, u_l) *and* (γ_r, u_r) *in* $[\gamma_1, \gamma_2] \times \mathbb{R}$.

Our first example is of the second type mentioned in the proposition, and the second example is of the first type. This proposition is proved simply by constructing the functions h_l and h_r in the two cases.

2.2 Vanishing viscosity and smoothing

We would like to motivate the minimal jump entropy condition. In our construction of the solution of the Riemann problem, it emerged naturally as a candidate for finding a unique solution. In this section we shall give two partial motivations for this entropy condition. Both of these motivations are based on the study of equations, which formally have (7) as a limit, but whose solutions, or the equations themselves, possess more regularity than the conservation law with a discontinuous coefficient. When doing this, we hope that the minimal jump condition will be a consequence of requiring that the solutions to the perturbed equations tend to the solution of the Riemann problem as the size of the perturbations tends to zero.

It is common to motivate entropy conditions for conservation laws by requiring that the solution of Riemann problems are limits of travelling wave solutions to the singularly perturbed equation

$$v_t + f(v)_x = \varepsilon v_{xx},$$

as $\varepsilon \downarrow 0$. For a scalar equation where the flux function does not depend on x, the "lower convex envelope" criterion is indeed a consequence of such an approach. We also say that the weak solution found by taking envelopes satisfies the *vanishing viscosity* entropy condition.

Let now u^ε be a traveling wave solution of the initial value problem

$$u_t^\varepsilon + f(\gamma, u^\varepsilon)_x = \varepsilon u_{xx}^\varepsilon, \quad \gamma(x) = \begin{cases} \gamma_l & \text{for } x < 0, \\ \gamma_r & \text{for } x > 0, \end{cases} \tag{50}$$

and we wish that

$$\lim_{x\to-\infty} u^{\varepsilon}(x,t) = u_l', \text{ and } \lim_{x\to\infty} u^{\varepsilon}(x,t) = u_r'. \tag{51}$$

Since this traveling wave is not really traveling, $u(x,t) = u(x)$, and we can set $\xi = x/\varepsilon$ to obtain,

$$\dot{f}(\gamma,u) = \ddot{u},$$

where $\dot{f} = df/d\xi$. The equation can be integrated once, and if we assume that the limits (51) are reached in a suitable manner, we get

$$\dot{u} = f(\gamma,u) - f\left(\gamma_l,u_l'\right) \tag{52}$$

$$= f(\gamma,u) - f\left(\gamma_r,u_r'\right), \tag{53}$$

which also gives us the Rankine–Hugoniot condition

$$f\left(\gamma_l,u_l'\right) = f\left(\gamma_r,u_r'\right) =: f^{\times}. \tag{54}$$

Summing up, we say that the discontinuity separating (γ_l,u_l') and (γ_r,u_r') admits a viscous profile, or that this discontinuity satisfies the viscous profile entropy conditions, if the ordinary differential equation

$$\frac{du}{d\xi} = \begin{cases} f(\gamma_l,u) - f^{\times} & \text{if } \xi < 0, \\ f(\gamma_r,u) - f^{\times} & \text{if } \xi > 0, \end{cases} \tag{55}$$

has a (at least one) solution $u(\xi)$ such that either

$$\lim_{\xi\to-\infty} u(\xi) = u_l' \quad \text{and} \quad u(\bar{\xi}) = u_r',$$

or

$$u(\bar{\xi}) = u_r' \quad \text{and} \quad \lim_{\xi\to\infty} u(\xi) = u_r',$$

where $\bar{\xi}$ can be finite or infinite. This means that one of two alternatives must hold:

Either the ordinary differential equation

$$\dot{v} = f(\gamma_l,v) - f^{\times} \quad \xi < 0, \quad v(0) = u_r',$$

has a solution such that

$$\lim_{\xi\to-\infty} v(\xi) = u_l'.$$

In this case we say that v is a left viscous profile.

Or the equation

$$\dot{w} = f(\gamma_r,u) - f^{\times}, \quad \xi > 0, \quad w(0) = u_l',$$

has a solution such that

$$\lim_{\xi\to\infty} w(\xi) = u_r',$$

in which case we call w a right viscous profile.

Hence the discontinuity satisfies the viscous profile entropy condition if there exists a left or right viscous profile.

If $u_l' < u_r'$, we will have a left viscous profile if and only if

$$f(\gamma_l, u) > f(\gamma_l, u_l'), \quad \text{for all } u \in (u_l', u_r').$$

Similarly, we will have a right viscous profile if and only if

$$f(\gamma_r, u) > f(\gamma_r, u_r'), \quad \text{for all } u \in (u_l', u_r').$$

Also, if $u_l' > u_r'$, we will have a left viscous profile if

$$f(\gamma_l, u) < f(\gamma_l, u_l'), \quad \text{for all } u \in (u_l', u_r').$$

Similarly, we will have a right viscous profile if and only if

$$f(\gamma_r, u) < f(\gamma_r, u_r'), \quad \text{for all } u \in (u_l', u_r').$$

Summing up, the viscous profile entropy condition is equivalent to

$$u_l' \leq u_r' \quad \Rightarrow \quad \begin{cases} f(\gamma_l, u) > f^\times & \text{for all } u \in (u_l', u_r'), \text{ or} \\ f(\gamma_r, u) > f^\times & \text{for all } u \in (u_l', u_r'), \end{cases} \tag{56}$$

$$u_r' \leq u_l' \quad \Rightarrow \quad \begin{cases} f(\gamma_l, u) < f^\times & \text{for all } u \in (u_r', u_l'), \text{ or} \\ f(\gamma_r, u) < f^\times & \text{for all } u \in (u_r', u_l'). \end{cases} \tag{57}$$

This condition implies the minimal jump entropy condition, and thus provides a motivation for it.

If the coefficient γ is a continuous function of x, then the classical theory of scalar conservation laws applies, and the initial value problem has a unique weak solution. If we let γ^ε denote a smooth approximation to

$$\gamma(x) = \begin{cases} \gamma_l & \text{for } x < 0, \\ \gamma_r & \text{for } x > 0, \end{cases}$$

such that $\gamma^\varepsilon \to \gamma$ as $\varepsilon \to 0$, and let u^ε denote the weak solution to

$$u_t^\varepsilon + f(\gamma^\varepsilon, u^\varepsilon)_x = 0, \quad u^\varepsilon(x,0) = \begin{cases} u_l' & \text{for } x < 0, \\ u_r' & \text{for } x > 0, \end{cases} \tag{58}$$

it is natural to ask whether u^ε tends to the minimal jump entropy solution as $\varepsilon \to 0$.

Example 2.3. We shall consider this in an example. Define

$$f_l(u) = 4 - (u+1)^2,$$

$$f_r(u) = 4 - (u-1)^2,$$
$$f(\gamma,u) = (1-\gamma)f_l(u) + \gamma f_r(u),$$

and consider the Riemann problem

$$u_t + f(\gamma,u)_x = 0, \quad u(x,0) = \begin{cases} -1 & \text{for } x < 0, \\ 1 & \text{for } x > 0, \end{cases} \quad \gamma(x) = \begin{cases} 0 & \text{for } x < 0, \\ 1 & \text{for } x > 0. \end{cases}$$

In this case we find that

$$h_l(u;-1) = \begin{cases} 4 & \text{if } u < -1, \\ 4-(u+1)^2 & \text{if } u \geq -1, \end{cases} \quad h_r(u;1) = \begin{cases} 4-(u+1)^2 & \text{if } u \leq 1, \\ 0 & \text{if } u > 1. \end{cases}$$

Furthermore the discontinuity separating the u and γ values $(-1,0)$ and $(1,1)$ satisfies the minimal jump entropy condition, and hence $u(x,0)$ is a weak solution satisfying the minimal jump entropy condition. Now set

$$\gamma^\varepsilon(x) = \begin{cases} 0 & \text{for } x \leq -\varepsilon, \\ \frac{x+\varepsilon}{2\varepsilon} & \text{for } -\varepsilon < x < \varepsilon, \\ 1 & \text{for } \varepsilon \leq x, \end{cases}$$

and let u^ε denote the stationary solution to (58) with $u_l' = -1$ and $u_r' = 1$. We have that u^ε satisfies

$$f(\gamma^\varepsilon, u^\varepsilon)_x = 0,$$

and thus

$$f(\gamma^\varepsilon, u^\varepsilon) = f(0,-1) = 0.$$

Solving this for u^ε we find that

$$u^\varepsilon(x) = 1 - 2\gamma^\varepsilon(x) \pm \sqrt{(1-2\gamma^\varepsilon(x))^2 + 3}.$$

Since $u^\varepsilon = -1$ for $x \leq -\varepsilon$ and $u^\varepsilon = 1$ for $x \geq \varepsilon$, we can choose the negative sign for x close to $-\varepsilon$ and the positive for x close to ε. Furthermore, since for any (fixed) γ, $f(\gamma,u)$ is concave in u, we can jump from the negative to the positive solution, if this will give a shock with zero speed (recall that u^ε is stationary). But since $f(\gamma^\varepsilon, u^\varepsilon)$ is constant, we can jump at any value of x! For instance, we can choose to jump at $x = 0$, giving

$$u^\varepsilon = \begin{cases} -1 & \text{for } x \leq -\varepsilon, \\ 1 - \frac{2x}{\varepsilon} - \sqrt{\left(1-\frac{2x}{\varepsilon}\right)^2 + 3} & \text{for } -\varepsilon < x < 0, \\ 1 - \frac{2x}{\varepsilon} + \sqrt{\left(1-\frac{2x}{\varepsilon}\right)^2 + 3} & \text{for } 0 < x < \varepsilon, \\ 1 & \text{for } \varepsilon \leq x. \end{cases}$$

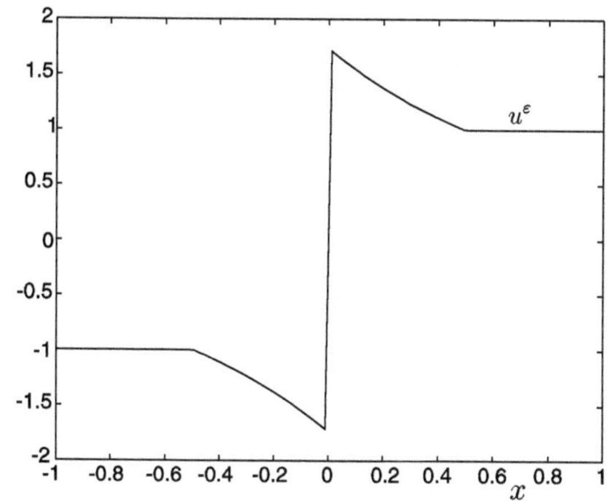

Fig. 13 The stationary solution of (58), $\varepsilon = 1/2$, and the discontinuity at $x = 0$.

We show a plot of this solution in Fig. 13, and we note that although $u^\varepsilon \to u$, the variation of the approximate solution is larger than that of u.

This example readily generalizes to the following case. Assume that the map

$$u \mapsto f(\gamma, u)$$

has a single global maximum for all γ, and

$$\lim_{u\to-\infty} f(\gamma,u) = -\infty \quad \text{and} \quad \lim_{u\to\infty} f(\gamma,u) = -\infty.$$

Let $u^\pm(\gamma,y)$ denote the two solutions of

$$y = f\left(\gamma, u^\pm\right),$$

such that $u^- \leq u^+$. As before, let u^ε denote the stationary solution of (58), where

$$\gamma^\varepsilon(x) = \gamma_l + \frac{x+\varepsilon}{2\varepsilon}\left(\gamma_r - \gamma_l\right), \quad -\varepsilon < x < \varepsilon.$$

Then it is possible to find a weak solution u^ε if and only if

$$u^-\left(\gamma_l, f\left(\gamma_l, u_l'\right)\right) = u_l', \quad \text{or} \quad u^+\left(\gamma_r, f\left(\gamma_r, u_r'\right)\right) = u_r'. \tag{59}$$

Recall that we are always assuming that u_l' and u_r' satisfy the Rankine–Hugoniot condition, i.e., $f(\gamma_l, u_l') = f(\gamma_r, u_r') = f^\times$. If both conditions in (59) hold, then this solution is given by

$$u^{\varepsilon}(x)=\begin{cases}u_l' & \text{for } x<-\varepsilon,\\ u^-\left(\gamma^{\varepsilon}(x),f^{\times}\right) & \text{for } -\varepsilon\le x\le x_J,\\ u^+\left(\gamma^{\varepsilon}(x),f^{\times}\right) & \text{for } x_J<x\le\varepsilon,\\ u_r' & \text{for } \varepsilon<x,\end{cases}\tag{60}$$

for any $x_J\in[-\varepsilon,\varepsilon]$. Since we are jumping from u^- to u^+, this jump is allowed since $u^-\le u^+$ and $f(\gamma,u)>f^{\times}$ in the interval (u^-,u^+). In the case where only one of the conditions in (59) holds, then we stay on u^+ or u^- throughout the interval $[-\varepsilon,\varepsilon]$. If

$$u_l'=u^+\left(\gamma_l,f\left(\gamma_l,u_l'\right)\right),\quad\text{and}\quad u_r'=u^-\left(\gamma_r,f\left(\gamma_r,u_r'\right)\right),$$

we must at some point jump *from u^+ to u^-*, and this will give an entropy violating weak solution. Looking at the shapes of the graphs of $f(\gamma_l,u)$ and $f(\gamma_r,u)$, we see that (59) is equivalent to the minimal jump entropy condition in this case. Hence, if (u_l',u_r') satisfies the minimal jump entropy condition, there exist entropy solutions u^{ε} of (58) such that u^{ε} tends to the minimal jump entropy condition when $\varepsilon\to0$ (if the flux f has the properties assumed above).

Remark 2.2. The minimal jump entropy condition is not always reasonable. Entropy conditions are based on extra information, such as physics or common sense. To illustrate this, consider the equation

$$\begin{gathered}u_t+\left(\frac{1}{2}(u+\gamma)^2\right)_x=0,\\ \gamma(x)=\begin{cases}-1 & \text{for } x<0,\\ 1 & \text{for } x>0,\end{cases}\qquad u(x,0)=0.\end{gathered}\tag{61}$$

In this case

$$h_l(u;0)=\begin{cases}\frac{1}{2}(u-1)^2 & \text{if } u\le1,\\ 0 & \text{if } u>1,\end{cases}\qquad h_r(u;0)=\begin{cases}0 & \text{if } u\le-1,\\ \frac{1}{2}(u+1)^2 & \text{if } u>-1.\end{cases}$$

We see that there is a unique crossing value $f^{\times}=1/2$, and the minimal jump entropy condition gives the solution $u(x,t)=0$.

One can also try to find a solution of (61) by making the substitution $w=u+\gamma$, which turns (61) into

$$w_t+\left(\frac{1}{2}w^2\right)_x=0,\quad w(x,0)=\begin{cases}-1 & \text{for } x<0,\\ 1 & \text{for } x>0.\end{cases}$$

The entropy solution to this, found by taking the lower convex envelope, reads

$$w(x,t)=\begin{cases}-1 & \text{for } x<-t,\\ x/t & \text{for } -t\le x\le t,\\ 1 & \text{for } x>t.\end{cases}$$

Since $u = w - \gamma$, we find the alternative solution

$$\tilde{u}(x,t) = \begin{cases} 0 & \text{for } |x| > t, \\ \frac{x}{t} - \operatorname{sign}(x) & \text{otherwise.} \end{cases} \tag{62}$$

So which of these solutions shall we choose? We have already seen that the minimal jump solution, $u = 0$, is the limit of the viscous approximations u^ε satisfying

$$u_t^\varepsilon + \left(\frac{1}{2}(u^\varepsilon + \gamma)^2\right)_x = \varepsilon u_{xx}^\varepsilon. \tag{63}$$

We know that w is the limit of the viscous approximation w^ε satisfying

$$w_t^\varepsilon + \left(\frac{1}{2}w^\varepsilon\right)_x^2 = \varepsilon w_{xx}^\varepsilon.$$

This means that $\tilde{u}$ is the limit of $\tilde{u}^\varepsilon$, where $\tilde{u}^\varepsilon$ and γ^ε satisfy the viscous approximation for the system (6), i.e.,

$$\begin{aligned} \tilde{u}_t^\varepsilon + \left(\frac{1}{2}(\tilde{u}^\varepsilon + \gamma^\varepsilon)^2\right)_x &= \varepsilon \tilde{u}_{xx}^\varepsilon, \\ \gamma_t^\varepsilon &= \varepsilon \gamma_{xx}^\varepsilon. \end{aligned} \tag{64}$$

Therefore, it is reasonable to choose $u = 0$ if (61) is an approximation of (63) and $\tilde{u}$ if (61) is an approximation of (64).

3 The Cauchy problem

In this section we shall demonstrate the existence of an entropy solution to the conservation law where the flux function depends on a discontinuous coefficient. To be concrete this is the initial value problem

$$\begin{cases} u_t + f(\gamma, u)_x = 0, & x \in \mathbb{R}, \quad t > 0, \\ u(x,0) = u_0(x), \end{cases} \tag{65}$$

where $\gamma = \gamma(x)$ is a function of bounded variation. Fix an arbitrary $T > 0$, and set $\Pi_T = \mathbb{R} \times [0,T)$. By a solution of (65) we mean a weak solution, that is a function u in $L^1_{\text{loc}}(\Pi_T) \cap C([0,T); L^1_{\text{loc}}(\mathbb{R}))$ such that

$$\iint_{\mathbb{R}\times\mathbb{R}^+} u\varphi_t + f(\gamma,u)\varphi_x \, dt\, dx + \int_{\mathbb{R}} u_0(x)\varphi(x,0)\, dx = 0, \tag{66}$$

for all test functions $\varphi \in C_0^1(\Pi_T)$. In order to demonstrate existence we shall assume that f and γ have additional properties, for instance we must be assured that the Riemann problem has a solution for all relevant initial data.

To show that a solution exists, we shall construct it as a limit of a sequence of approximations. This can be done using difference approximations, front tracking approximations or the limits of parabolic regularizations, but we shall use front tracking.

3.1 A model equation

In this section we will restrict ourselves to the model equation, where we have $f(\gamma,u) = 4\gamma u(1-u)$, i.e.,

$$u_t + (4\gamma u(1-u))_x = 0, \quad u(x,0) = u_0(x). \tag{67}$$

We assume that $\gamma : \mathbb{R} \to \mathbb{R}$ is a function of bounded variation, which is continuously differentiable on a finite set of intervals. In particular, we assume that there exists a finite number of intervals

$$I_m = (\xi_m, \xi_{m+1}) \quad \text{for } m = 0, \dots, M,$$

where $\xi_0 = -\infty$, $\xi_{M+1} = \infty$, such that

$$\gamma' \big|_{I_m} \quad \text{is continuous and bounded for } m = 0, \dots, M. \tag{68}$$

For the moment, we also assume that the initial function u_0 is of bounded variation, and such that $u_0(x) \in [0,1]$ for all x.

We shall define a front tracking scheme for (67), and in order to explain this we first start by explaining this scheme in the case where γ is constant, the readers who are already familiar with front tracking for scalar conservation laws may skip ahead to section 3.1.2.

3.1.1 Front tracking for constant γ

Now we aim to construct an entropy solution to the initial value problem

$$v_t + (4v(1-v))_x = 0, \quad v(x,0) = v_0(x), \tag{69}$$

where $v_0(x)$ is a function of bounded variation taking values in the set $[0,1]$. We also demand that $v_0 \in L_1(\mathbb{R})$. Set $g(v) = 4v(1-v)$, then an entropy solution $v(x,t)$ is defined to be the unique function such that

$$\int_0^\infty \int_{\mathbb{R}} |v-c|\,\varphi_t + \operatorname{sign}(v-c)(g(v)-g(c))\varphi_x\,dx\,dt - \int_{\mathbb{R}} |v_0(x)-c|\,\varphi(x,0)\,dx \geq 0, \tag{70}$$

for all nonnegative test functions $\varphi \in C_0^\infty(\mathbb{R} \times \mathbb{R}_0^+)$, and for all constants c.

Let $\mathcal{V}^\delta = \{v_i\}_{i=0}^N$ be a collection of points in $[0,1]$ such that

$$0 = v_0 < v_1 < \cdots < v_{N-1} < v_N = 1.$$

We assume that this collection, and the number of points in it, depend on a positive parameter δ such that

$$\max_{v\in[0,1]} \left(\min_{0\leq i\leq N(\delta)} |v_i - v| \right) \to 0 \quad \text{as } \delta \to 0.$$

Without loss of generality, we can assume that

$$\delta = \max_{0\leq i<N} |v_{i+1} - v_i|, \quad \text{and} \quad \eta = \min_{0\leq i<N} |v_{i+1} - v_i|.$$

We assume that the set $\mathcal{V}^\delta$ is uniform, in the sense that $C\eta \geq \delta$ for some fixed constant $C > 0$. Let g^δ be a piecewise linear approximation to g, obtained by interpolating between the points in $\mathcal{V}^\delta$, i.e.,

$$g^\delta(v) = v_i + (v - v_i)\frac{g_{i+1} - g_i}{v_{i+1} - v_i}, \quad \text{for } v \in [v_i, v_{i+1}],$$

where we have set $g_i = g(v_i)$. Next we let v_0^δ be an approximation to v_0, such that $v_0^\delta(x) \in \mathcal{V}^\delta$ for all x, and

$$\lim_{\delta\to 0} \left\| v_0^\delta - v_0 \right\|_{L^1(\mathbb{R})} = 0.$$

The front tracking approximation v^δ is the entropy solution to the initial value problem

$$v_t^\delta + g^\delta\left(v^\delta\right)_x = 0, \quad v^\delta(x,0) = v_0^\delta(x). \tag{71}$$

We shall now demonstrate that the problem (71) is "easier" than (69), and that the global solution to (71) can be constructed by a finite number of operations. The idea of this is as follows: since the initial function v_0^δ is piecewise constant, the discontinuities define a series of Riemann problems. We shall show that the solution of a Riemann problem for (71) always gives a piecewise constant function of x/t. Hence, for sufficiently small t, $v^\delta(x,t)$ will be a piecewise constant function of x for fixed t. The discontinuities, which we refer to as *fronts*, move, and at some $t > 0$, two (or more) of these will collide. In order to propagate the solution past this collision time, we solve the Riemann problem defined by the constant value to the left of the leftmost colliding front and the constant value to the right of the rightmost one. The solution of this Riemann problem again gives a number of fronts separating constant values. In this way the solution v^δ can be defined until the next collision time, and thus we can define the solution past any finite number of collisions. However, in order to establish that we can define $v^\delta(\cdot,t)$ for any t, we must investigate this construction closer. We start by studying the Riemann problem for (71).

This is the initial value problem for (71), where

$$v_0^\delta(x) = \begin{cases} v_k & \text{for } x < 0, \\ v_l & \text{for } x > 0, \end{cases}$$

where v_k and v_l are in $\mathcal{V}^\delta$. In section 2.1.1, we saw that this solution depends on the envelopes of g^δ between v_k and v_l. Since g^δ is concave, this is not too complicated. By looking up the material containing (12) – (17), we find that the solution reads

$$v^\delta(x,t) = \begin{cases} v_k & \text{for } x < ts_{k,l}, \\ v_l & \text{for } ts_{k,l} \le x, \end{cases} \tag{72}$$

if $v_k < v_l$,

$$v^\delta(x,t) = \begin{cases} v_k & \text{for } x < ts_{k,k-1}, \\ v_i & \text{for } ts_{i,i-1} \le x < ts_{i-1,i-2},\ i = k-1,\dots,l+1, \\ v_l & \text{for } ts_{l+1,l} \le x, \end{cases} \tag{73}$$

if $v_k > v_l$, where we have set

$$s_{j,i} = \frac{g_j - g_i}{v_j - v_i}.$$

Note that, in particular, v^δ is a piecewise constant function taking values in the set $\mathcal{V}^\delta$. Since $g''(u) = -2 < 0$, the map

$$(j,i) \mapsto s_{j,i}$$

is strictly decreasing in both arguments. We need to investigate what happens when two (or more) fronts collide at some time $t_c > 0$. We label the positions of the colliding fronts $x_1(t),\dots,x_n(t)$, such that for $t < t_c$, $x_i(t) < x_{i+1}(t)$ for $i = 1,\dots,n-1$. Set the v value to the left of x_i to be w_i, and of course then w_{i+1} is the state to the right of x_i. Since the fronts are colliding, we also have $x_i'(t) > x_{i+1}'(t)$, and this means that $w_0 < w_1 < \cdots < w_{n+1}$, or that we have at most two colliding fronts and $w_0 \in (w_1, w_2)$ or $w_2 \in (w_0, w_1)$. In order to advance the solution past t_c, we must solve the Riemann problem with a left state w_0 and a right state w_{n+1}. By (72), this solution consists of a single front only, thus at any collision of n fronts, there will only be one resulting front. Hence the number of fronts is strictly decreasing at each collision. Also, this argument shows that the total variation of $v^\delta(\cdot,t)$ is bounded by the total variation of v_0^δ.

Since v_0 has bounded total variation, the approximation v_0^δ would be very unreasonable indeed if its variation was not bounded independently of δ. To keep things simple, we define our approximations so that

$$\left|v_0^\delta\right|_{BV} \le |v_0|_{BV}. \tag{74}$$

Let $C(t)$ denote the number of collisions of fronts for all times less than t, and let $N(t)$ denote the number of fronts in $v^\delta(\cdot,t)$, such that $N(0+)$ is the number of fronts resulting from the solutions of the initial Riemann problems. If a front separates v values v_l and v_r, we call $|v_l - v_r|$ the *strength* of the front. Since the strength of each front is bounded below by η,

$$N(0+) \le \frac{|v_0^\delta|_{BV}}{\eta} \le \frac{|v_0|_{BV}}{\eta}.$$

Also, since the number of fronts is strictly decreasing for each collision, we obtain the bound

$$N(t) \le N(0+) - C(t) \le \frac{|v_0|_{BV}}{\eta} - C(t), \tag{75}$$

and because $N(t) \ge 0$,

$$C(t) \le \frac{|v_0|_{BV}}{\eta},$$

for all t. Concretely, this means that we have only a finite number of collisions for *all* positive time. One consequence of this is that after some *finite* time t_e, the front tracking construction v^δ will consist of some finite number of noninteracting fronts, or be a constant. In particular, we are able to define and calculate $v^\delta(x,t)$ for all $t \in \mathbb{R}^+$ by a finite number of operations. If a numerical approximation has this property, we call it *hyperfast*.

Example 3.1. In Fig. 14 and Fig. 15 we show how this front tracking works[1] in the case where $v_i = i\delta$, $\delta = 0.01$, and v_0^δ approximates

$$v_0(x) = \begin{cases} 0 & \text{for } |x| \ge 1, \\ \frac{1}{2}\left(1 - \sin\left((\pi x)^2\right)\right) & \text{otherwise.} \end{cases} \tag{76}$$

In general, since v^δ is constructed by using entropy solutions of Riemann problems, we have that v^δ is an entropy solution of (71), i.e.,

[1] The code used to produce this example can be found at
www.math.ntnu.no/~holden/FrontBook/Matlab.html

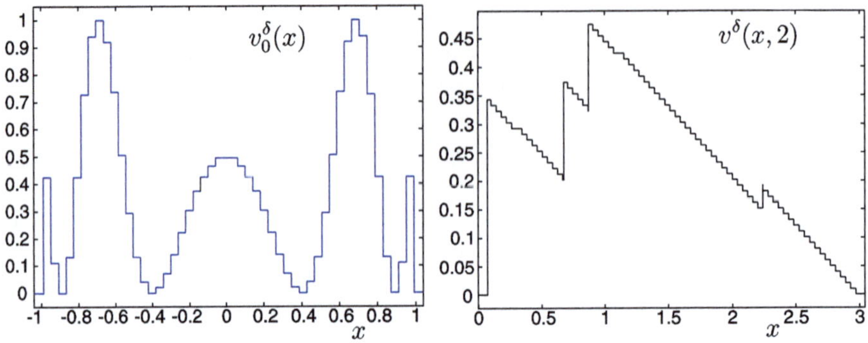

Fig. 14 Left: $v_0^\delta(x)$, right: $v^\delta(x,2)$ for the initial value problem (76).

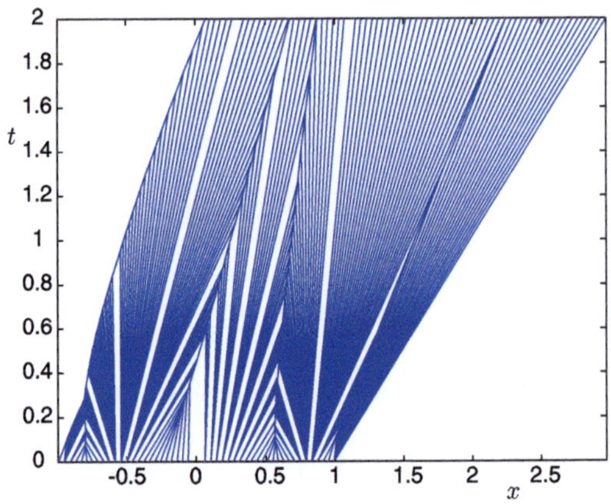

Fig. 15 The fronts in the (x,t) plane for the initial value problem (76).

$$\begin{aligned}\int_0^T \int_{\mathbb{R}} \left|v^\delta - c\right| \varphi_t + \operatorname{sign}\left(v^\delta - c\right)\left(g^\delta\left(v^\delta\right) - g^\delta(c)\right)\varphi_x \, dx dt \\ + \int_{\mathbb{R}} \left|v_0^\delta - c\right| \varphi(x,0)\, dx \geq 0,\end{aligned} \tag{77}$$

for all nonnegative test functions φ and all constants c. Of course, this implies that v^δ is a weak solution.

The inequality (74) shows that the total variation of $v^\delta(\cdot,t)$ is bounded independently of t and δ. Furthermore, the solution procedure for the Riemann problem shows that

$$\min_x v_0^\delta(x) \leq \min_{(x,t)} v^\delta(x,t) \leq \max_{(x,t)} v^\delta(x,t) \leq \max_x v_0^\delta(x).$$

Next, let $t < s$ and let $\alpha_h(t)$ be a smooth approximation to the characteristic function of the interval $[s,t]$, so that

$$\alpha_h \to \chi_{[s,t]}, \quad \text{and} \quad \alpha_h' \to \delta_s - \delta_t,$$

as $h \downarrow 0$, where δ_s denotes the Dirac delta function centered at s. Choose a test function $\varphi(x)$ such that $|\varphi| \leq 1$, and set $\varphi_h(x,t) = \varphi(x)\alpha_h(t)$. Since v^δ is a weak solution we have that

$$\iint_{\Pi_T} v^\delta \partial_t \varphi_h + g^\delta\left(v^\delta\right) \partial_x \varphi_h \, dt\, dx = 0,$$

and sending $h \downarrow 0$ we find that

$$\int_{\mathbb{R}} \varphi(x) \left(v^\delta(x,t) - v^\delta(x,s)\right) dx = \int_s^t \int_{\mathbb{R}} \varphi_x(x) g^\delta\left(v^\delta\right) dt\, dx.$$

Therefore

$$\begin{aligned}
\left\| v^\delta(\cdot,t) - v^\delta(\cdot,s) \right\|_{L^1(\mathbb{R})} &= \sup_{|\varphi|\leq 1} \int \varphi(x) \left(v^\delta(x,t) - v^\delta(x,s)\right) dx \\
&= \sup_{|\varphi|\leq 1} \int_s^t \int_{\mathbb{R}} \varphi_x(x) g^\delta\left(v^\delta\right) dx d\sigma \\
&\leq \|g\|_{\text{Lip}} \int_s^t \left| v^\delta \right|_{BV} d\sigma \\
&\leq (t-s) \left| v_0^\delta \right|_{BV}. \qquad (78)
\end{aligned}$$

Thus $\left\{v^\delta\right\}_{\delta>0}$ is a sequence that satisfies the bounds,

$$\|v^\delta\|_\infty \leq C, \qquad (79)$$

$$\left| v^\delta(\cdot,t) \right|_{BV} \leq C, \qquad (80)$$

$$\left\| v^\delta(\cdot,t) - v^\delta(\cdot,s) \right\|_{L^1} \leq C(t-s), \qquad (81)$$

for some constant C that is independent of δ, t and s. From these bounds it is standard, see e.g., [6], to use Helly's theorem to prove that there is a subsequence of $\left\{v^\delta\right\}$ that converges in $L^1(\mathbb{R} \times [0,\infty))$ to some function v. With a slight abuse of notation, we have that

$$\lim_{\delta \to 0} v^\delta = v, \quad \text{in} \quad L^1(\mathbb{R} \times [0,\infty)).$$

Using that $v \mapsto |v - c|$ and $v \mapsto \operatorname{sign}(v-c)\left(g^\delta(v) - g^\delta(c)\right)$ are continuous, and $g^\delta(v) \to g(v)$ for all v, it is not hard, by taking limits in (77), to show that the limit v is the unique entropy solution to (69). This finishes our description of front tracking for scalar conservation laws.

3.1.2 Front tracking in general

The goal now is to extend the above front tracking scheme to the case where γ is not constant.

In order to prove convergence of the front tracking approximations in the scalar case, we used that the variation of $\{v^\delta\}$ was uniformly bounded. As Example 3.2 will show, such a bound does not exist if γ is not constant.

In order to circumvent this obstacle, we shall work with the variable z defined by (48), and why this is a good idea is outlined in the remark below.

Remark 3.1. Assume that u^ε and v^ε are smooth solutions of the regularized equations

$$u_t^\varepsilon + f(\gamma, u^\varepsilon)_x = \varepsilon u_{xx}^\varepsilon,\ v_t^\varepsilon + f(\gamma, v^\varepsilon)_x = \varepsilon v_{xx}^\varepsilon,$$

with smooth initial data u_0^ε and v_0^ε, respectively. Let η be a smooth convex function. Then we subtract these equations and multiply the result by $\eta'(u^\varepsilon - v^\varepsilon)$ to find

$$\begin{aligned}
\eta(u^\varepsilon - v^\varepsilon)_t &= -\eta'(u^\varepsilon - v^\varepsilon)\left[f(\gamma, u^\varepsilon) - f(\gamma, v^\varepsilon)\right]_x \\
&\quad + \varepsilon\eta(u^\varepsilon - v^\varepsilon)_{xx} - \varepsilon\eta''(u^\varepsilon - v^\varepsilon)\left((u^\varepsilon - v^\varepsilon)_x^2\right) \\
&\le -\left[\eta'(u^\varepsilon - v^\varepsilon)(f(\gamma, u^\varepsilon) - f(\gamma, v^\varepsilon))\right]_x \\
&\quad + \varepsilon\eta(u^\varepsilon - v^\varepsilon)_{xx} + \eta'(u^\varepsilon - v^\varepsilon)_x(f(\gamma, u^\varepsilon) - f(\gamma, v^\varepsilon)).
\end{aligned}$$

Now we let $\eta = \eta_\kappa$ be a continuous approximation to $|\cdot|$, concretely

$$\eta_\kappa(u) = \int_0^u \max\left(-1, \min\left(\frac{v}{\kappa}, 1\right)\right)\, dv.$$

Assuming that $u^\varepsilon - v^\varepsilon$ has compact support in x, we can integrate the above inequality over $x \in \mathbb{R}$, and get

$$\begin{aligned}
\frac{d}{dt}\int_{\mathbb{R}} \eta_\kappa(u^\varepsilon - v^\varepsilon)\, dx &\le \int_{\mathbb{R}} \eta_\kappa''(u^\varepsilon - v^\varepsilon)(f(\gamma, u^\varepsilon) - f(\gamma, v^\varepsilon))(u^\varepsilon - v^\varepsilon)_x\, dx \\
&\le L\int_{|u^\varepsilon - v^\varepsilon| < \kappa} \left|(u^\varepsilon - v^\varepsilon)_x\right|\, dx,
\end{aligned}$$

where $L = \sup|f_u|$, since

$$\eta_\kappa''(u) = \begin{cases} \frac{1}{\kappa} & \text{for } |u| \le \kappa, \\ 0 & \text{otherwise.} \end{cases}$$

By [6, Lemma B.5],

$$\lim_{\kappa\to 0}\int_{|u^\varepsilon - v^\varepsilon|<\kappa} |(u^\varepsilon - v^\varepsilon)_x|\, dx = 0.$$

Thus we can send κ to zero, and obtain for any two solutions of the regularized equation

$$\|u^\varepsilon(\cdot,t) - v^\varepsilon(\cdot,t)\|_{L^1(\mathbb{R})} \le \|u_0^\varepsilon - v_0^\varepsilon\|_{L^1(\mathbb{R})}\,. \tag{82}$$

Now we can set $v^\varepsilon(\cdot,t) = u^\varepsilon(\cdot,t+\tau)$ in (82), then divide by τ and let $\tau \to 0$, to deduce that

$$\|u_t^\varepsilon(\cdot,t)\|_{L^1(\mathbb{R})} \le \|u_t^\varepsilon(\cdot,0+)\|_{L^1(\mathbb{R})} = |f(\gamma,u_0^\varepsilon)|_{BV}\,. \tag{83}$$

Without loss of generality we can construct u_0^ε, so that $|f(\gamma,u_0^\varepsilon)|_{BV} \le |f(\gamma,u_0)|_{BV}$. This means that the total variation of $f(\gamma,u^\varepsilon)$ is bounded independently of ε, i.e.,

$$|f(\gamma,u^\varepsilon(\cdot,t))|_{BV} \le |f(\gamma,u_0)|_{BV}\,. \tag{84}$$

If $f_u(\gamma,u) \ge c > 0$ for all[2] γ and u, then this would imply that also u^ε had uniformly bounded variation. For the flux function in our example $f_u(\gamma,1/2) = 0$, so we cannot deduce that u^ε is of bounded variation. This is precisely where the z-mapping comes to the rescue. We write (48) as

$$z(\gamma,u) = \operatorname{sign}(u-1/2)\,(f(\gamma,u) - f(\gamma,1/2))\,.$$

Now

$$\begin{aligned}|z(\gamma,u^\varepsilon)|_{BV} &\le |f(\gamma,u^\varepsilon)|_{BV} + \|f_\gamma\|_{L^\infty} |\gamma|_{BV}\\ &\le |f(\gamma,u_0)|_{BV} + \|f_\gamma\|_{L^\infty} |\gamma|_{BV}\,.\end{aligned}$$

Thus $z^\varepsilon = z(\gamma,u^\varepsilon)$ has uniformly bounded variation, and the mapping $u \mapsto z(\gamma,u)$ is continuous and invertible. With a little extra work we can then show that $\{z^\varepsilon\}_{\varepsilon>0}$ is compact in $L^1(\mathbb{R}\times\mathbb{R}_0^+)$, and thus (for a subsequence) $z^\varepsilon \to \bar{z}$ as $\varepsilon \to 0$. Then we define

$$u = z^{-1}(\gamma,\bar{z}) = \lim_{\varepsilon\to 0} z^{-1}(\gamma,z^\varepsilon) = \lim_{\varepsilon\to 0} u^\varepsilon.$$

With any luck, we can conclude the argument by showing that the limit u is a weak solution.

This remark is meant to indicate how the z-mapping could be used to show the existence via viscous regularizations, and to motivate the use of the z-mapping also for front tracking approximations.

As in the case without a coefficient, we start with a discussion of an approximate solution to the Riemann problem, or rather with the exact solution of the Riemann problem for an approximate equation. In the simple scalar case, we saw that the exact solution of the Riemann problem was piecewise constant in x/t if the flux function was piecewise linear. We shall now define an approximate flux function f^δ

[2] This excludes resonance.

such that $f^\delta(\gamma,u) \approx 4\gamma u(1-u)$, and the solution of the Riemann problem with flux f^δ is piecewise constant.

From section 2 we saw that the solution of the Riemann problem consisted of a sequence of straight lines in the (z,γ) plane, where

$$z(\gamma,u) = \operatorname{sign}\left(u-\frac{1}{2}\right)\gamma(1-2u)^2. \tag{85}$$

There were z-waves, over which γ is constant, and γ waves, over which γ was not constant. Now fix a (small) positive number δ, and set

$$\gamma_i = i\delta, \quad i>0, \quad i\in\mathbb{N}, \tag{86}$$

and for integers j, such that $-i \le j \le i$, $z_{i,j} = j\delta$, and

$$u_{i,j} = z^{-1}(\gamma_i, z_{i,j}) = \frac{1}{2}\left(1+\operatorname{sign}(z_{i,j})\sqrt{\frac{|z_{i,j}|}{\gamma_i}}\right). \tag{87}$$

Note that the set $\{(z_{i,j},\gamma_i)\}$ defines a grid in the (z,γ) plane, and we define f^δ to be the linear interpolation to f on this grid, i.e.,

$$f^\delta(\gamma_i,u) = f_{i,j} + (u-u_{i,j})\frac{f_{i,j+1}-f_{i,j}}{u_{i,j+1}-u_{i,j}}, \quad \text{for } u\in[u_{i,j},u_{i,j+1}], \tag{88}$$

where $f_{i,j} = f(\gamma_i,u_{i,j}) = 4\gamma_i u_{i,j}(1-u_{i,j})$. For each fixed i, $f^\delta(\gamma_i,u)$ will be a concave function with a maximum for $u=1/2$. Therefore the solution of the Riemann problem

$$\begin{aligned} &u_t + f^\delta(\gamma(x),u)_x = 0,\\ &u(x,0) = \begin{cases} u_{i,j} & \text{for } x<0,\\ u_{m,n} & \text{for } x>0,\end{cases} \quad \gamma(x) = \begin{cases} \gamma_i & \text{for } x<0,\\ \gamma_m & \text{for } x>0,\end{cases}\end{aligned} \tag{89}$$

can be found from the diagrams in Fig. 12. Furthermore, since f^δ is piecewise linear in u, this solution will be piecewise constant in x/t. Also, by our choice of interpolation points when constructing f^δ, *all* the intermediate values of $u(x,t)$ will be grid points, i.e.,

$$z(\gamma(x),u(x,t)) = (z_{i',j'},\gamma_{i'}), \quad \text{where } i'=i \text{ or } i'=m.$$

We label the grid points in the (u,γ) plane, or when there is no danger of misunderstanding, in the (z,γ) plane, as $\mathcal{U}^\delta$. Hence, the solution of the Riemann problem takes pointwise values in $\mathcal{U}^\delta$ if the "initial" states $(u(x,0),\gamma(x))$ take values in $\mathcal{U}^\delta$.

Once we have the solution of the approximate Riemann problem (89), we can use this to design a front tracking scheme. To this end, let $\{u_0^\delta\}_{\delta>0}$ and $\{\gamma^\delta\}_{\delta>0}$ be two sequences of piecewise constant functions such that

$$\left(u_0^\delta(x), \gamma^\delta(x)\right) \in \mathcal{U}^\delta \quad \text{for all but a finite number of } x \text{ values.}$$

Furthermore, we demand that

$$\lim_{\delta\to 0} \left\| u_0^\delta - u_0 \right\|_{L^1} = 0, \tag{90}$$

$$\lim_{\delta\to 0} \left\| \gamma^\delta - \gamma \right\|_{L^1} = 0. \tag{91}$$

We label the discontinuity points of γ^δ as $y_1 < \cdots < y_N$. Of course, these depend on δ, but we suppress this dependency in our notation. At each point of discontinuity of either u_0^δ or γ^δ, we have a Riemann problem whose solution will give a sequence of z-waves and γ-waves. We define the front tracking approximation as in the scalar case, by following discontinuities, called fronts, and solve the Riemann problems (using the approximate flux f^δ) defined by their collisions. We call the resulting piecewise constant function u^δ. As in the scalar case, in order to show that we can define $u^\delta(\cdot,t)$ for any $t > 0$, we must study the interaction of fronts.

The front tracking solution u^δ has two types of fronts, namely z-fronts and γ-fronts, where z-fronts are those fronts whose left and right γ-values are equal. Regarding the collision of two or more z-fronts, we have seen that such collision always results in *one* z-front. Hence, the number of fronts in u^δ decreases when z-fronts collide.

Moreover, γ-fronts have zero speed (recall that these are the discontinuities of γ^δ), and therefore two γ-fronts will never collide. It remains to study collisions between z-fronts and γ-fronts. This turns out to be complicated, and simple examples show that we can have such collisions which result in three outgoing fronts. Furthermore, even if such collisions always result in two outgoing fronts, it is in general not possible to bound the total variation of u^δ independently of δ, as the next example shows.

Example 3.2. Assume for the moment that

$$u_0(x) = \frac{1}{2}, \quad \gamma(x) = \begin{cases} 1 & \text{for } x \le 0, \\ 1+x & \text{for } 0 < x \le 2, \\ 2 & \text{for } 2 < x. \end{cases} \tag{92}$$

In this case $z(\gamma(x), u_0(x)) = 0$, and we can set

$$\gamma^\delta(x) = \begin{cases} 1 & \text{for } x \le 0, \\ 1+i\delta & \text{for } i\delta < x \le (i+1)\delta \; i = 0,\ldots,2/(\delta-1), \\ 2 & \text{for } 2 < x. \end{cases}$$

The z component of the solution of each of the Riemann problems defined by $(u_0^\delta, \gamma^\delta)$ at $x = i\delta$ reads

$$(z,\gamma) = \begin{cases} (0,1+(i-1)\delta) & \text{for } x < i\delta, \\ (-\delta, 1+i\delta) & \text{for } i\delta \le x < ts_i + i\delta, \\ (0,1+i\delta) & \text{for } i\delta + ts_i \le x, \end{cases}$$

where

$$s_i = \sqrt{\delta(1+i\delta)}.$$

This follows from the diagram in Fig. 12. Hence, before any interaction of fronts, the total variation of u^δ reads

$$\begin{aligned} \left|u^\delta\right|_{BV} &= \sum_{i=1}^{1/\delta} \sqrt{\frac{\delta}{1+i\delta}} \\ &\ge \sum_{i=1}^{1/\delta} \sqrt{\frac{\delta}{2}} \\ &= \frac{1}{\delta}\sqrt{\frac{\delta}{2}} \\ &= \frac{1}{\sqrt{2\delta}} \to \infty \quad \text{as } \delta \to 0. \end{aligned}$$

Despite this, since $\gamma(x)$ is Lipschitz continuous, the total variation of the exact solution to this problem is uniformly bounded for $t < T$ for any finite time T, see e.g., Kružkov [11] or Karlsen and Risebro [9]. As an indication of things to come, in passing we observe that

$$\left|z^\delta\right|_{BV} = \sum_{i=1}^{1/\delta} |\delta| = 1,$$

where $z^\delta = z(\gamma^\delta, u^\delta)$. So the total variation of the transformed variable z is uniformly bounded for this example, at least until the first interaction.

For reasons outlined in the above example and in Remark 3.1, we shall work with the z variable instead of u. In the above example, it was trivial to show that the variation of z was bounded independently of δ, but this becomes more cumbersome in general, so to help us we use the Temple[3] functional. For a single front, which we label $\mathfrak{f}$, this is defined as

$$T(\mathfrak{f}) = \begin{cases} |\Delta z| & \text{if } \mathfrak{f} \text{ is a } z\text{-front}, \\ 4|\Delta z| & \text{if } \mathfrak{f} \text{ is a } \gamma\text{-front, and } z_l < z_r, \\ 2|\Delta z| & \text{if } \mathfrak{f} \text{ is a } \gamma\text{-front, and } z_l > z_r, \end{cases} \tag{93}$$

where z_l is the z value to the left of the front, z_r the value to the right, and $\Delta z = z_r - z_l$. Fig. 16 will perhaps be useful later, the figure shows the weights given to $|\Delta z|$ in the various cases. Recall also that if $\mathfrak{f}$ is a γ-front, then

[3] This, or rather a similar, functional was first used in the paper of Temple [15].

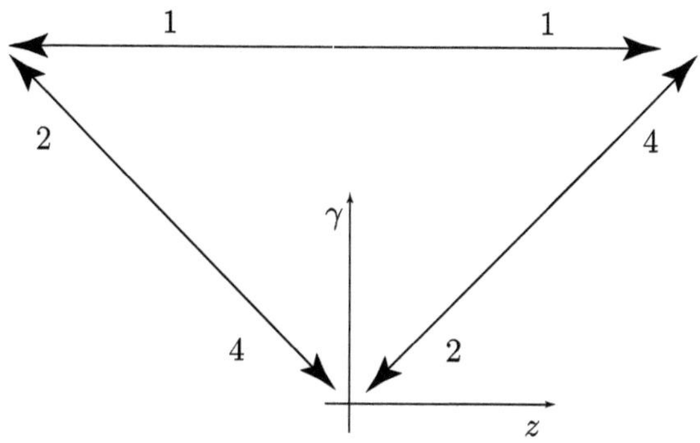

Fig. 16 The weights in the Temple functional (93).

$$|\Delta z| = |\Delta \gamma|,$$

thus an alternative definition of T is

$$T(\mathfrak{f}) = \begin{cases} |\Delta z| & \text{if } \mathfrak{f} \text{ is a } z\text{-front,} \\ 4|\Delta \gamma| & \text{if } \mathfrak{f} \text{ is a } \gamma\text{-front, and } z_l < z_r, \\ 2|\Delta \gamma| & \text{if } \mathfrak{f} \text{ is a } \gamma\text{-front, and } z_l > z_r. \end{cases}$$

For a sequence of fronts, we define T additively, and with a slight abuse of notation we write

$$T\left(u^\delta\right) = \sum_{\mathfrak{f} \in u^\delta} T(\mathfrak{f}).$$

With this definition of T we have the obvious inequalities

$$\left|z^\delta\right|_{BV} \leq T\left(u^\delta\right) \leq 4\left(\left|z^\delta\right|_{BV} + \left|\gamma^\delta\right|_{BV}\right). \tag{94}$$

We also have for any front $\mathfrak{f} \in u^\delta$ that

$$T(\mathfrak{f}) \geq \delta.$$

With a further abuse of notation we shall write $T(t) = T(u^\delta(\cdot,t))$.

Lemma 3.1. *If* $0 < s < t$, *then*

$$T(t) \leq T(s). \tag{95}$$

Hence $\left|z^\delta(\cdot,t)\right|_{BV} \leq T(0+)$.

Proof. The value of T will only change when fronts collide. From the analysis of collisions of z-fronts, we have established that T does not increase at such collisions. To prove the lemma, it therefore remains to study collisions between z-fronts and γ-fronts. We say that a γ-front is nonpositive if it connects points in the half plane $z \leq 0$, similarly we say that it is nonnegative if it connects points in the half plane $z \geq 0$.

We shall study the collision between z-fronts and a γ-fronts, and we thus have three points in the (z,γ) plane; (z_l,γ_l), (z_m,γ_m) and (z_r,γ_r), which lie to the left of, in between, and to the right of the colliding fronts respectively. If we have more than one z-front colliding with the γ front, we can reduce to the two front collision type as follows. If we have several z-fronts colliding with the γ-front from the same side, then we can resolve the collision between the z-fronts first, and then the collision between the (single) resulting z-front and the γ-front.

Therefore we consider the case where we have two z-fronts colliding with one γ front. One z front collides from the left, the other from the right. We label the states to the left of the left z-front $L=(z_l,\gamma_l)$, the one to the left of the γ-front $M_-=(z_-,\gamma_l)$, the state to the left of the right z-front $M_+=(z_+,\gamma_r)$ and finally the state to the right of this z-front $R=(z_r,\gamma_r)$. Of course we may have $z_l=z_-$ or $z_+=z_r$, in which case we have only two colliding fronts. In order to study how T changes by this collision, we study a number of cases. These are distinguished by whether the γ-front lies in the left (it is *nonpositive*) or the right (it is *nonnegative*) half spaces, and by whether $\gamma_l<\gamma_r$ or not.

Case 1: <u>the γ-front is nonpositive, and $\gamma_l>\gamma_r$.</u> Consult Fig. 17 in what follows. Now we regard the z-front, and hence M_- and M_+, as fixed. Since the γ front

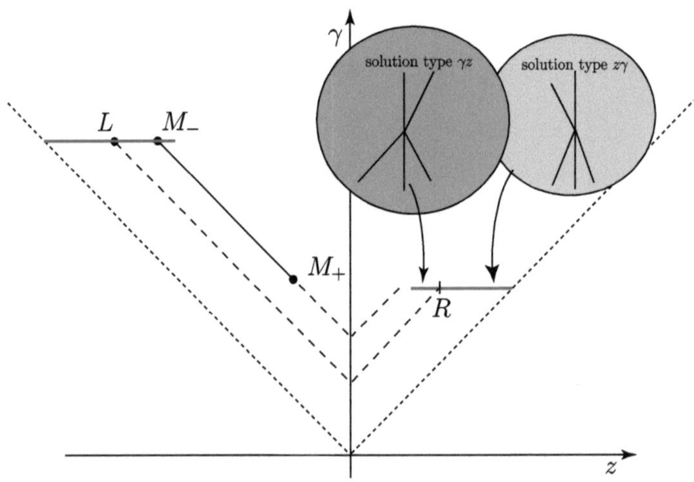

Fig. 17 The possible locations of L and R if the γ-front is nonpositive, and $\gamma_l>\gamma_r$.

is negative, $z_+\le 0$, and since $\gamma_l>\gamma_r$, $z_-\le -\delta$. The z-front between z_l and z_- moves with positive speed, and it is the solution of the Riemann problem defined by these two states with a flux function $f^\delta(\gamma_l,\cdot)$. Hence z_l cannot be larger than "one brakepoint to the right" of z_-. If it was, then the solution would contain more than one front. Furthermore $u_l=z^{-1}(\gamma_l,z_l)\ge 0$, which is the same as $z_l\ge-\gamma_l$. Thus

$$z_l\in[-\gamma_l,z_-+\delta].$$

This interval is indicated by the upper left horizontal gray line in Fig. 17. Reasoning in the same way, the right z-front must have negative speed and thus

$$z_r \in \{z_+\} \cup [-z_+ + \delta, \gamma_r],$$

this interval is indicated by the lower right horizontal gray line in Fig. 17. We have two alternatives. First if $-z_l + \gamma_l \geq z_r + \gamma_r$, then the solution of the Riemann problem defined by (z_l, γ_l) and (z_r, γ_r) is of the type γz, and if $-z_l + \gamma_l < z_r + \gamma_r$, then this Riemann problem has a solution of type $z\gamma$. This is indicated in Fig. 17, where the dashed line passing through the L is the line where $|z| + \gamma = -z_l + \gamma_l$. If $z_l = z_-$, i.e., we have a collision between a γ-front and a z-front from the right, then the solution type is always $z\gamma$. In other words, the wave is transmitted. Consulting Fig. 16, we see that if $z_l \leq z_-$, then T is unchanged by the collision. If $z_l = z_- + \delta$ (which is the maximum value for z_l), and the solution type is $z\gamma$, then T decreases by 2δ. Otherwise T is unchanged. In the special case where $z_r = z_- = 0$ and $z_l = z_- + \delta$, the z-front is reflected. Thus we see that a reflection results in a decrease of T by 2δ. The reader is urged to check these statements.

Case 2: the γ-front is nonpositive, and $\gamma_l < \gamma_r$. Consult Fig. 18 in what follows. Since the fronts are colliding, the speed of the left z-front is positive and that

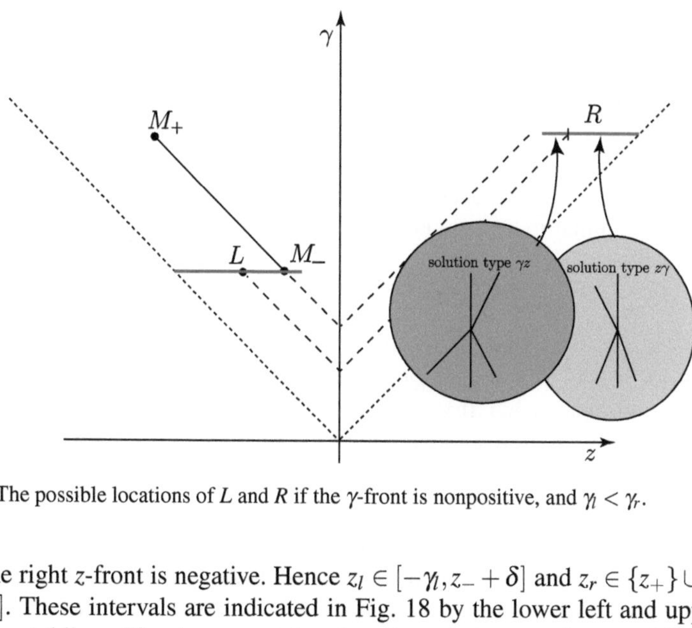

Fig. 18 The possible locations of L and R if the γ-front is nonpositive, and $\gamma_l < \gamma_r$.

of the right z-front is negative. Hence $z_l \in [-\gamma_l, z_- + \delta]$ and $z_r \in \{z_+\} \cup [-z_+ + \delta, \gamma_r]$. These intervals are indicated in Fig. 18 by the lower left and upper right horizontal lines. If $z_r + \gamma_r < -z_l + \gamma_l$ then the solution type is $z\gamma$, and if $z_r + \gamma_r \geq -z_l + \gamma_l$ the solution type is γz. In both of these cases T is unchanged. If $z_r = z_+$ then the solution type is γz, and if $z_l = z_-$ then the solution type is $z\gamma$. Thus there are no reflected fronts in this case.

Case 3: the γ-front is nonnegative, and $\gamma_l > \gamma_r$. Consult Fig. 19 in what follows. This case is similar to Case 2 above. By considering the speeds of the colliding

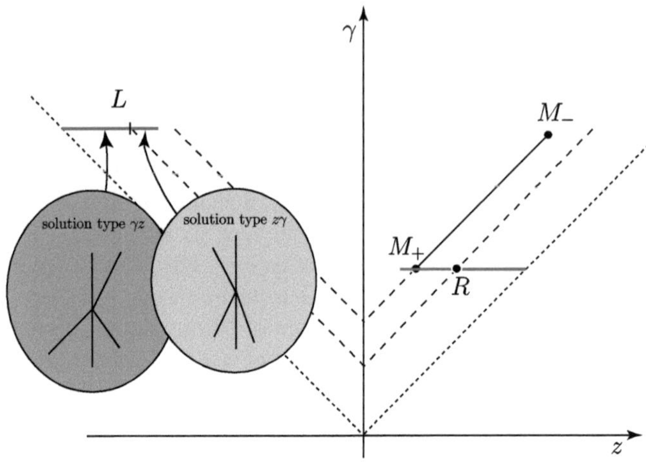

Fig. 19 The possible locations of L and R if the γ-front is nonnegative, and $\gamma_l > \gamma_r$.

fronts, we find that

$$z_l \in [-z_- - \delta, -\gamma_l] \cup \{z_-\}, \quad \text{and} \quad z_r \in [z_+ - \delta, \gamma_r].$$

If $|z_l| + \gamma_l < z_r + \gamma_r$, then the solution is of type γz, and if $|z_l| + \gamma_l \geq z_r + \gamma_r$ the solution is of type $z\gamma$. Note that if $z_r = z_+$, then the solution type is γz, while if $z_l = z_-$, the solution type is $z\gamma$. So also in this case a front cannot be reflected. Furthermore, T is unchanged.

Case 4: the γ-front is nonnegative, and $\gamma_l < \gamma_r$. Consult Fig. 20 in what follows. This case is similar to Case 1 above. We find that

$$z_l \in [-z_- - \delta, -\gamma_l] \cup \{z_-\}, \quad \text{and} \quad z_r \in [z_+ - \delta, \gamma_r].$$

If $|z_l| + \gamma_l > z_r + \gamma_r$, then the solution type is $z\gamma$, while if $|z_l| + \gamma_l \leq z_r + \gamma_r$, the type is γz. If $z_r = z_+ - \delta$ and the solution type is γz, then T decreases by 2δ, otherwise it is unchanged. If $z_+ = z_r$, then the solution type is $z\gamma$, while if $z_l = z_-$ and $z_r = z_+ - \delta$, we have a reflection, and in this case T decreases by 2δ.

This finishes the proof of Lemma 3.1.

Remark 3.2. Recall that we have used the term "reflection" for a collision between a z-front and a γ-front, if the z-front collides from the left, and the solution of the Riemann problem is of type $z\gamma$, or if the z-front collides from the right and the solution type is γz. From the proof of the above lemma, it is clear that whenever we have a reflection, T decreases by 2δ. Hence, if $T(0+)$ is finite, we can only have a finite number of reflections in u^δ.

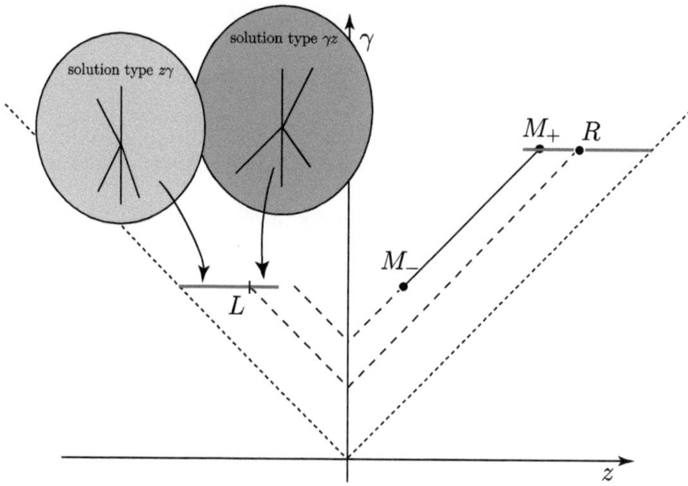

Fig. 20 The possible locations of L and R if the γ-front is nonnegative, and $\gamma_l < \gamma_r$.

One immediate consequence of Lemma 3.1 and (94) is the following result.

Corollary 3.1. *If*

$$\left|\gamma^{\delta}\right|_{BV} \leq |\gamma|_{BV} \quad \textit{and} \quad \left|z(u_0^{\delta},\gamma^{\delta})\right|_{BV} \leq |z(u_0,\gamma)|_{BV}, \tag{96}$$

then for $t \geq 0$,

$$\left|z^{\delta}(\cdot,t)\right|_{BV} \leq |z(u_0,\gamma)|_{BV} + 4\,|\gamma|_{BV},$$

and thus bounded independently of δ and t.

Note that this corollary in itself does not imply that the front tracking construction u^{δ} can be defined up to any time t. In order to show this we have to do some more work. For a z-front $\mathfrak{f}_z$, let $\mathcal{A}(\mathfrak{f}_z)$ be the set of γ-fronts $\mathfrak{f}_\gamma$ that approach $\mathfrak{f}_z$, i.e.,

$$\mathfrak{f}_\gamma \in \mathcal{A}(\mathfrak{f}_z) \quad \text{if} \quad \begin{cases} x(\mathfrak{f}_z) < x(\mathfrak{f}_\gamma) & \text{and } s(\mathfrak{f}_z) \geq 0, \text{ or} \\ x(\mathfrak{f}_z) > x(\mathfrak{f}_\gamma) & \text{and } s(\mathfrak{f}_z) \leq 0, \end{cases}$$

where $x(\mathfrak{f})$ denotes the position of $\mathfrak{f}$ and $s(\mathfrak{f})$ its speed. For any z-front $\mathfrak{f}_z$ define

$$J(\mathfrak{f}_z) = \sum_{\mathfrak{f}_\gamma \in \mathcal{A}(\mathfrak{f}_z)} |\Delta\gamma|, \tag{97}$$

where $\Delta\gamma$ denotes the difference in γ over the front.

Lemma 3.2. *Assume that* (96) *holds. Then for each fixed* δ*, the functional*

$$F(t) = \delta \sum_{\mathfrak{f}_z} J(\mathfrak{f}_z) + T(t)\,|\gamma|_{BV} \tag{98}$$

is nonincreasing, and decreases by at least δ^2 *when a z-front collides with a* γ*-front.*

Proof. Let $N_\mathfrak{f}(t)$ denote the number of fronts in u^δ at a time t. For each front we have $|\Delta z| \geq \delta$, and thus

$$N_\mathfrak{f} \leq \frac{|z^\delta|_{BV}}{\delta}.$$

Recall that T is bounded and $J(\mathfrak{f}_z) \leq |\gamma|_{BV}$. Hence F is bounded by

$$\begin{aligned} F(t) &\leq \delta\,|\gamma|_{BV}\,N_\mathfrak{f} + 2T(0+)\,|\gamma|_{BV} \\ &\leq 4\,|\gamma|_{BV}\left(|z(u_0,\gamma)|_{BV} + 4\,|\gamma|_{BV}\right). \end{aligned} \tag{99}$$

Thus F is bounded independently of δ and t. We must show that F is decreasing by at least δ^2 for collisions between z-fronts and γ-fronts, and nonincreasing when z-fronts collide.

First consider a collision between one (or two) z-front(s) and a γ-front. From the proof of Lemma 3.1, we saw that either (**a**) a z-front "passes through" the γ-front in the collision, or (**b**) we have a reflection, and T decreases by 2δ. If (**a**) holds, then the sum in (98) will "lose" at least one term (two terms if one z-front is lost in the collision) of size $|\Delta\gamma|$, and the second term in (98) does not increase. Thus F decreases by at least $\delta\,|\Delta\gamma| \geq \delta^2$. If (**b**) holds, then T decreases by 2δ, and the sum increases by at most $|\gamma|_{BV}$. Hence F decreases by a least $\delta\,|\gamma|_{BV} \geq \delta^2$.

Next we consider a collision between two (or more) z-fronts. Recall that this collision will result in one z-front. If more than two fronts collide, we can consider this as several collisions between two fronts, occurring at the same point. Therefore, we consider a collision between two z-fronts, $\mathfrak{f}_l$ and $\mathfrak{f}_r$, separating values z_l, z_m and z_r. We label the resulting front $\mathfrak{f}$. If z_m is between z_l and z_r, then T does not change by the collision. However, the speed of $\mathfrak{f}$ is between the speeds of $\mathfrak{f}_l$ and $\mathfrak{f}_r$. If the speed of $\mathfrak{f}$ is different from 0, then $\mathcal{A}(\mathfrak{f}) = \mathcal{A}(\mathfrak{f}_l)$ or $\mathcal{A}(\mathfrak{f}) = \mathcal{A}(\mathfrak{f}_r)$. Hence the sum in (98) looses one term, and F decreases by at least δ^2. If the speed of $\mathfrak{f}$ is 0, then the speed of $\mathfrak{f}_l$ is positive, and the speed of $\mathfrak{f}_r$ negative, thus

$$\mathcal{A}(\mathfrak{f}) = \mathcal{A}(\mathfrak{f}_l) \cup \mathcal{A}(\mathfrak{f}_r),$$

and thus F is constant.

If z_m is not between z_l and z_r, then either $z_r = z_m - \delta$ or $z_l = z_m + \delta$. This is so since f^δ is convex. In this case T decreases by δ, and the first term in equation (98) increases by at most $\delta\,|\gamma^\delta|_{BV}$. This concludes the proof of the lemma.

Note that an immediate consequence of equation (99) and Lemma 3.2 is that for a fixed δ, the number of collisions of z-fronts and γ-fronts is bounded by

$$4\left|\gamma\right|_{BV}\frac{\left|z(u_0,\gamma)\right|_{BV}+4\left|\gamma\right|_{BV}}{\delta^2}.$$

Also, the *smallest* absolute value of the speed of any z-front having a speed different from zero is bounded below by

$$\sqrt{\min\left(\gamma^\delta\right)\delta}.$$

Hence, after some *finite* time T_1, collisions between z-fronts and γ-fronts cannot occur. This means that there must be a time $T_2 \geq T_1$, such that all z-fronts in the interval (y_1, y_N) (recall that γ^δ has discontinuities at $y_1, \ldots, y_N$) have zero speed, that all z-fronts to the left of y_1 have nonpositive speed and that the z-fronts to the right of y_N have nonnegative speed for all $t > T_2$. Outside the interval $[y_1, y_N]$, u^δ is the front tracking approximation to a scalar conservation law with a constant coefficient, and there can only be a finite number of collisions between fronts in u^δ here. Therefore, there exists a finite time $T_3 \geq T_2$, such that there will be no further collisions between fronts in u^δ for $t > T_3$. Thus, the front tracking method is hyperfast.

Example 3.3. Now we pause for a moment in order to exhibit an example of how the front tracking looks in practice. We wish to find the front tracking approximation to the initial value problem

$$\begin{aligned}
&u_t + \left[4\gamma(x)u(1-u)\right]_x = 0, \quad t > 0,\\
&\gamma(x) = \begin{cases} e^{|x|} & \text{for } -1 \leq x \leq 1,\\ \sin(\pi x^2)+2 & \text{for } 1 < |x| < 2,\\ 1 & \text{otherwise,}\end{cases}\\
&u(x,0) = \begin{cases} \frac{1}{2}(1+e^{-|x|}) & \text{for } -1 \leq x \leq 1,\\ 0 & \text{otherwise.}\end{cases}
\end{aligned} \tag{100}$$

In Fig. 21, we show the approximation γ^δ for $\delta = 0.05$, and $u^\delta(\cdot, 3)$. In Fig. 22, we show the fronts in u^δ in the (x,t) plane. Here z-fronts are marked with solid lines, and γ-fronts by dashed lines. We see that the number of fronts decreases rapidly, and there do not seem to be many collisions after $t = 3$.

Returning now to the more general case, we claim that the sequence $\{z^\delta\}_{\delta>0}$ satisfies the following bounds

$$\|z^\delta\|_\infty \leq \|\gamma^\delta\|_\infty \leq C \tag{101}$$

$$\left\|z^\delta(\cdot,t)\right\|_{L^1_{\text{loc}}} \leq C, \quad t < T, \tag{102}$$

$$\|z(\cdot,t) - z(\cdot,s)\|_{L^1} \leq C(t-s), \tag{103}$$

where the constant C does not depend on t or on δ. The first bound (101) follows by the definition of z, (85), and the fact that u^δ takes values in the interval $[0,1]$. Regarding (102), we have that u^δ is a weak solution of

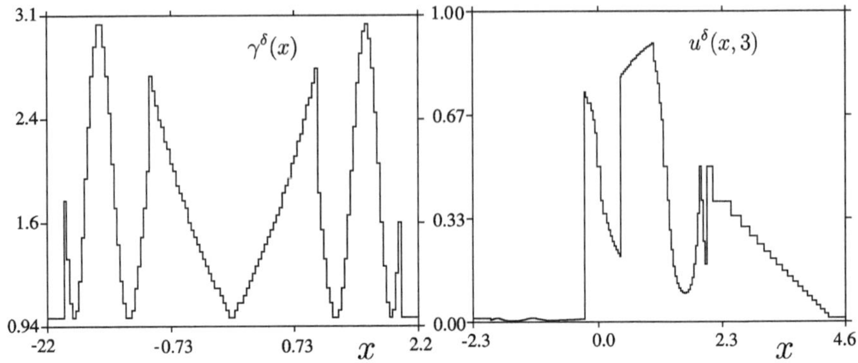

Fig. 21 Left: $\gamma^{\delta}(x)$, right: $u^{\delta}(x,3)$ for $\delta = 0.05$

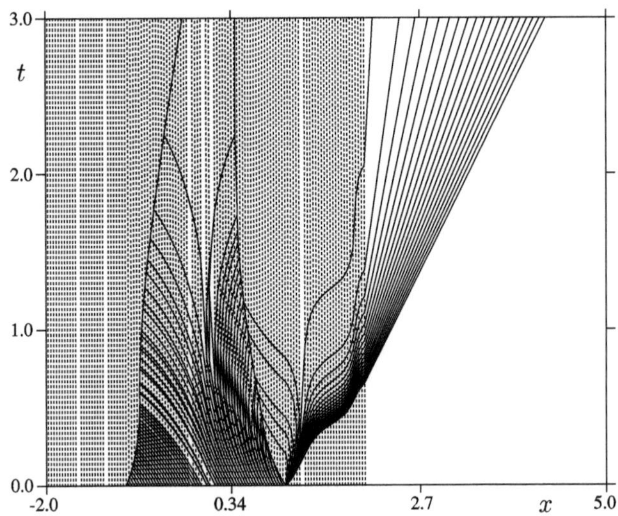

Fig. 22 The fronts in the (x,t) plane for the example (100).

$$u_t^{\delta} + f^{\delta}\left(\gamma^{\delta}, u^{\delta}\right)_x = 0, \quad u^{\delta}(x,0) = u_0^{\delta}(x). \tag{104}$$

Thus we can repeat the argument leading to (78), to obtain

$$\begin{aligned} \left\| u^{\delta}(\cdot,t) - u^{\delta}(\cdot,s) \right\|_{L^1} &\leq \max_{\tau\in[s,t]} \left| f^{\delta}\left(\gamma^{\delta}, u^{\delta}(\cdot,\tau)\right)\right|_{BV} (t-s) \\ &\leq \max_{\tau\in[s,t]} \left| z^{\delta}(\cdot,\tau)\right|_{BV} (t-s) \\ &\leq C(t-s), \end{aligned} \tag{105}$$

for some constant not depending on t, s or δ. Setting $s = 0$, we find

$$\left\| u^{\delta}(\cdot,t) \right\|_{L^1} \leq \left\| u_0^{\delta} \right\|_{L^1} + Ct, \tag{106}$$

and thus $u^{\delta}(\cdot,t)$ is in $L^1(\mathbb{R})$ for all finite t. Now

$$\begin{aligned} \left| z\left(u^{\delta},\gamma^{\delta}\right) \right| &= \left| z\left(0,\gamma^{\delta}\right) + z_u\left(\xi,\gamma^{\delta}\right) u^{\delta} \right| \\ &\leq \left| \gamma^{\delta} \right| + C \left| u^{\delta} \right|, \end{aligned}$$

for some positive constant C, where ξ is in the interval $[0,u^{\delta}]$. Since γ^{δ} is in L^1_{loc}, equation (102) follows. Actually, in our case, since $u^{\delta}(x,t) \in [0,1]$, we have that

$$\left\| u^{\delta}(\cdot,t) \right\|_{L^1} = \int_{\mathbb{R}} u^{\delta}(x,t)\,dx = \int_{\mathbb{R}} u_0^{\delta}(x)\,dx = \left\| u_0^{\delta} \right\|_{L^1},$$

which is stronger than (106).

To prove (103) we use the equality

$$\begin{aligned} z^{\delta}(x,t) - z^{\delta}(x,s) &= z\left(u^{\delta}(x,t),\gamma^{\delta}\right) - z\left(u^{\delta}(x,s),\gamma^{\delta}\right) \\ &= z_u\left(\xi,\gamma^{\delta}\right)\left(u^{\delta}(x,t) - u^{\delta}(x,s)\right). \end{aligned}$$

Since z_u is bounded, by (105) the bound (103) holds.

Hence, by standard techniques as in the case of constant coefficients, it follows that there is a subsequence of $\{\delta\}$ (which we also label $\{\delta\}$), and a function $z \in L^1_{\text{loc}}(\mathbb{R} \times \mathbb{R}_0^+) \cap L^{\infty}(\mathbb{R}^+; BV(\mathbb{R}))$ such that

$$\lim_{\delta \to 0} z^{\delta} = z \quad \text{in } L^1_{\text{loc}}(\mathbb{R} \times [0,T]). \tag{107}$$

Since $z^{\delta} = z(u^{\delta},\gamma^{\delta})$, it also follows that there is a function $u \in L^1_{\text{loc}}(\mathbb{R} \times [0,T])$ such that $u^{\delta} \to u$, and $u = z^{-1}(z,\gamma)$. Furthermore, for this subsequence also $f^{\delta}(\gamma^{\delta},u^{\delta}) \to f(\gamma,u)$. Thus

$$\begin{aligned} \lim_{\delta \to 0} \iint u^{\delta}\varphi_t + f^{\delta}(\gamma^{\delta},u^{\delta})\varphi_x\,dxdt \\ = \iint u\varphi_t + f(\gamma,u)\varphi_x\,dxdt, \end{aligned}$$

and by construction

$$\lim_{\delta \to 0} \int_{\mathbb{R}} u^{\delta}(x,0)\varphi(x,0)\,dx = \int_{\mathbb{R}} u_0(x)\varphi(x,0)\,dx.$$

Since u^{δ} is a weak solution to (104), it follows from this that u is a weak solution to (65).

Furthermore, it is transparent that although we performed the analysis for $f(\gamma,u) = 4\gamma u(1-u)$, our results could be (slightly) extended to include flux functions that are similar to f. To be precise, assume that:

(A.1) There is an interval $[a,b]$ such that $f(\gamma,a) = f(\gamma,b) = C$ for all γ.
(A.2) There is a point $u^\star(\gamma) \in (a,b)$ such that $f_u(\gamma,u) > 0$ for $a < u < u^\star(\gamma)$ and $f_u(\gamma,u) < 0$ for $u^\star(\gamma) < u < b$.
(A.3) The map $\gamma \mapsto f(\gamma,u)$ is strictly monotone for all $u \in (a,b)$.
(A.4) The flux function satisfies $f \in C^2(\mathbb{R} \times [a,b])$.

If f satisfies these assumptions, we can define the mapping z as

$$z(\gamma,u) = \operatorname{sign}(u-u^\star(\gamma))\,(f(\gamma,u^\star(\gamma))-f(\gamma,u)), \tag{108}$$

and use this to show that the front tracking approximation is well defined. This analysis is only a slight modification of the analysis in the case where $f(\gamma,u) = 4\gamma u(1-u)$. Hence, *mutatis mutandis*, we have proved the following theorem.

Theorem 3.1. *Let f be a function satisfying (A.1)–(A.4), and assume that $u_0(x)$ is a function in L^1_{loc} taking values in the interval $[a,b]$, and that γ is a function in $BV(\mathbb{R}) \cup L^1_{\text{loc}}(\mathbb{R})$. Then there exists a weak solution to the initial value problem*

$$u_t + f(\gamma,u)_x = 0, \quad x \in \mathbb{R}, \quad t > 0, \qquad u(x,0) = u_0(x).$$

Furthermore, this solution is the limit of a sequence of front tracking approximations.

3.1.3 An entropy inequality

Now we shall show that the limit of any front tracking approximation to the general conservation law (65) satisfies a Kružkov type entropy condition. Thus we let u^δ be a weak solution to the approximate problem

$$\begin{cases} u^\delta_t + f^\delta\left(\gamma^\delta,u^\delta\right)_x = 0, & x \in \mathbb{R}, \quad t > 0, \\ u^\delta(x,0) = u^\delta_0(x), & x \in \mathbb{R}, \end{cases} \tag{109}$$

where $f^\delta(\gamma,\cdot)$ is a piecewise linear continuous approximation of $f(\gamma,u)$, such that $f^\delta \to f$ as $\delta \to 0$. Here γ^δ is a piecewise constant approximation to γ, such that $\gamma^\delta \to \gamma$ in L^1 as $\delta \to 0$. We *assume* that u^δ can be constructed by front tracking, and that for each fixed $T > 0$

$$u^\delta \to u \text{ in } L^1(\mathbb{R} \times [0,T]) \text{ as } \delta \to 0. \tag{110}$$

Furthermore, we let

$$z(\gamma,u) = \int_0^u |f_u(\gamma,v)|\,dv, \tag{111}$$

and set $z^\delta = z(\gamma^\delta, u^\delta)$. We shall also assume that for each t the family $\{z^\delta(\cdot,t)\}$ is a sequence of uniformly bounded variation in x and satisfies the three basic estimates (101), (102) and (103), so that we have convergence of z^δ along a subsequence.

Using that u^δ is a weak solution to (109), it is not hard to show that u is a weak solution to (65) if $u_0^\delta \to u$ as $\delta \to 0$. We would like to show that the limit u satisfies a generalization of the Kružkov entropy condition. Recall that if γ is continuous, then an entropy solution to (65) in the strip $\Pi_T = \mathbb{R} \times [0,T]$ satisfies the inequality

$$\begin{aligned} \iint_{\Pi_T} |u-c|\,\varphi_t + F(\gamma,u,c)\varphi_x\,dxdt \qquad\qquad (112) \\ -\iint_{\Pi_T} \operatorname{sign}(u-c)\,\partial_x f(\gamma,c)\varphi\,dxdt + \int_{\mathbb{R}} |u_0(x)-c|\,\varphi(x,0)\,dx \geq 0, \end{aligned}$$

for all constants c and all nonnegative test functions φ such that $\varphi(\cdot,T)=0$. Here F is the Kružkov entropy flux defined by

$$F(\gamma,u,c) = \operatorname{sign}(u-c)\left(f(\gamma,u) - f(\gamma,c)\right).$$

We would like to show that the front tracking limit u satisfies (112) if γ is continuous, and if γ has discontinuities, find a suitable generalization that is satisfied by the front tracking limit. The condition (112) does not make sense for discontinuous γ's, since the second integral is undefined.

We shall assume that γ is piecewise continuous on a *finite* number of intervals, i.e., that γ has a finite number of discontinuities. We call this set of discontinuities $\mathcal{D}_\gamma = \{\xi_0,\dots,\xi_N\}$, and we assume that $\gamma(x)$ is continuously differentiable for $x \notin \mathcal{D}_\gamma$. Thus γ and γ' have left and right limits at each discontinuity point $\xi_i \in \mathcal{D}_\gamma$.

Next, we shall require that the approximation $\gamma^\delta(x)$ also has discontinuity points for all $x \in \mathcal{D}_\gamma$ for all relevant δ. In addition to these discontinuities, for a fixed δ, γ^δ has discontinuities at $\{y_{i,j}\}$. These are ordered so that

$$\xi_i = y_{i,0} < y_{i,1} < \cdots < y_{i,N_i} < y_{i,N_i+1} = \xi_{i+1},$$

for $i = 0,\dots,N$. Let $\gamma_{i,j+1/2}$ denote the value of γ^δ in the interval $(y_{i,j}, y_{i,j+1})$, and set

$$\Delta x_{i,j} = \frac{1}{2}\left(y_{i,j+1} - y_{i,j-1}\right), \quad j = 1,\dots,N_i.$$

Of course, these quantities all depend on δ, but for simplicity we omit this in our notation. We also assume that

$$\lim_{\delta\to 0} \frac{f^\delta\left(\gamma_{i,j+1/2},c\right) - f^\delta\left(\gamma_{i,j-1/2},c\right)}{\Delta x_{i,j}} \chi_{I_{i,j}}(x) = \frac{\partial f(\gamma(x),c)}{\partial x}, \qquad (113)$$

where $\chi_{I_{i,j}}$ denotes the characteristic function of the interval

$$I_{i,j} = \left(\frac{y_{i,j-1}+y_{i,j}}{2}, \frac{y_{i,j}+y_{i,j+1}}{2}\right).$$

This is not unreasonable since γ is continuously differentiable in (ξ_i, ξ_{i+1}). In what follows, we let $u_i^{\mp}$ and $u_{i,j}^{\mp}$ denote the left and right limits of u^δ at the points ξ_i and $y_{i,j}$, respectively. Since $u^\delta(\cdot,t)$ is piecewise constant, these limits exist.

In each interval $(y_{i,j}, y_{i,j+1})$ the function u^δ is an entropy solution of the conservation law

$$u_t^\delta + f^\delta(\gamma_{i,j+1/2}, u^\delta)_x = 0,$$

and hence

$$\begin{aligned}
-\int_0^T \int_{y_{i,j}}^{y_{i,j+1}} & \left|u^\delta - c\right| \varphi_t + F^\delta\left(\gamma_{i,j+1/2}, u^\delta, c\right)\varphi_x \, dx\,dt \\
&+ \int_0^T F^\delta\left(\gamma_{i,j+1/2}, u_{i,j+1}^-, c\right)\varphi\left(y_{i,j+1},t\right) - F^\delta\left(\gamma_{i,j+1/2}, u_{i,j}^+, c\right)\varphi\left(y_{i,j},t\right) dt \\
&\qquad - \int_{y_{i,j}}^{y_{i,j+1}} \left|u^\delta(x,0) - c\right| \varphi(x,0)\, dx \le 0,
\end{aligned} \tag{114}$$

where

$$F^\delta(\gamma,u,c) = \operatorname{sign}(u-c)\left(f^\delta(\gamma,u) - f^\delta(\gamma,c)\right).$$

Summing this for $j = 0, \dots, N_i$ we find that

$$\begin{aligned}
-\int_0^T \int_{\xi_i}^{\xi_{i+1}} & \left|u^\delta - c\right| \varphi_t + F^\delta\left(\gamma^\delta, u^\delta, c\right)\varphi_x \, dx\,dt - \int_{\xi_i}^{\xi_{i+1}} \left|u^\delta(x,0) - c\right| \varphi(x,0)\, dx \\
&+ \int_0^T F^\delta\left(\gamma_{i,N_i+1/2}, u_{i+1}^-, c\right)\varphi(\xi_i,t) - F^\delta\left(\gamma_{i,1/2}, u_i^+, c\right)\varphi(\xi_{i+1},t)\, dt \\
&- \int_0^T \sum_{j=1}^{N_i} \left[F^\delta\left(\gamma_{i,j+1/2}, u_{i,j}^+, c\right) - F^\delta\left(\gamma_{i,j-1/2}, u_{i,j}^-, c\right)\right]\varphi(y_{i,j},t)\, dt \\
&\le 0.
\end{aligned} \tag{115}$$

Regarding the terms in the integrand in the last term in (115), we can write

$$\begin{aligned}
&F^\delta\left(\gamma_{i,j+1/2}, u_{i,j}^+, c\right) - F^\delta\left(\gamma_{i,j-1/2}, u_{i,j}^-, c\right) \\
&\quad = \begin{cases}
-\operatorname{sign}\left(u_{i,j}^+ - c\right)\left[f\left(\gamma_{i,j+1/2}, c\right) - f\left(\gamma_{i,j-1/2}, c\right)\right] \\
\qquad + \left\{\operatorname{sign}\left(u_{i,j}^+ - c\right) - \operatorname{sign}\left(u_{i,j}^- - c\right)\right\}\left(f_{i,j}^\times - f\left(\gamma_{i,j-1/2}\right)\right), \\
\text{or} \\
-\operatorname{sign}\left(u_{i,j}^- - c\right)\left[f\left(\gamma_{i,j+1/2}, c\right) - f\left(\gamma_{i,j-1/2}, c\right)\right] \\
\qquad + \left\{\operatorname{sign}\left(u_{i,j}^+ - c\right) - \operatorname{sign}\left(u_{i,j}^- - c\right)\right\}\left(f_{i,j}^\times - f\left(\gamma_{i,j+1/2}\right)\right),
\end{cases}
\end{aligned}$$

where $f^{\times}_{i,j} = f(\gamma_{i,j+1/2}, u^+_{i,j}) = f(\gamma_{i,j-1/2}, u^-_{i,j})$. If $\operatorname{sign}\left(u^+_{i,j} - c\right) = \operatorname{sign}\left(u^-_{i,j} - c\right)$, the last terms in the above expressions are zero, while if $u^-_{i,j} \leq c \leq u^+_{i,j}$, since these values are chosen according to the minimal jump entropy condition (34), we then have that

$$\operatorname{sign}\left(u^+_{i,j} - c\right) - \operatorname{sign}\left(u^-_{i,j} - c\right) = 2 \quad \text{and} \quad \begin{cases} f(\gamma_{i,j-1/2}, c) \geq f^{\times}_{i,j}, & \text{or} \\ \quad f(\gamma_{i,j+1/2}, c) \geq f^{\times}_{i,j}, \end{cases}$$

and thus in this case one of the last terms must be nonpositive. If $u^+_{i,j} < c < u^-_{i,j}$ we use (35), to conclude that

$$\operatorname{sign}\left(u^+_{i,j} - c\right) - \operatorname{sign}\left(u^-_{i,j} - c\right) = -2 \quad \text{and} \quad \begin{cases} f(\gamma_{i,j-1/2}, c) \leq f^{\times}_{i,j}, & \text{or} \\ \quad f(\gamma_{i,j+1/2}, c) \leq f^{\times}_{i,j}, \end{cases}$$

and again we find that one of the last terms is nonpositive. If the first of these last terms is nonpositive for c between $u^-_{i,j}$ and $u^+_{i,j}$, we define $u_{i,j} = u^\delta(y_{i,j}, t) = u^+_{i,j}$, otherwise we define $u_{i,j} = u^\delta(y_{i,j}, t) = u^-_{i,j}$. Using these observations, we find that

$$\begin{aligned} -\int_0^T \int_{\xi_i}^{\xi_{i+1}} & \left|u^\delta - c\right| \varphi_t + F^\delta\left(\gamma^\delta, u^\delta, c\right) \varphi_x \, dxdt - \int_{\xi_i}^{\xi_{i+1}} \left|u^\delta(x,0) - c\right| \varphi(x,0)\, dx \\ & + \int_0^T F^\delta\left(\gamma_{i,N_i+1/2}, u^-_{i+1}, c\right) \varphi\left(\xi_i, t\right) - F^\delta\left(\gamma_{i,1/2}, u^+_i, c\right) \varphi\left(\xi_{i+1}, t\right) dt \\ & + \int_0^T \sum_{j=1}^{N_i} \operatorname{sign}\left(u_{i,j} - c\right) \left[f\left(\gamma_{i,j+1/2}, c\right) - f\left(\gamma_{i,j-1/2}, c\right)\right] \varphi\left(y_{i,j}, t\right) dt \\ \leq 0. & \end{aligned} \tag{116}$$

Now $u_{i,j} = u^\delta(y^-_{i,j}, \cdot)$ or $u_{i,j} = u^\delta(y^+_{i,j}, \cdot)$, hence if we define $\bar{u}^\delta(x,t) = u_{i,j}(t)\chi_{I_{i,j}}(x)$, and set $\bar{z}^\delta = z^\delta_{i,j}(t)\chi_{I_{i,j}}$, we have that

$$\bar{z}^\delta\left(y_{i,j}, t\right) = z^\delta\left(y_{i,j}, t\right).$$

Now we claim that the sequence $\left\{\bar{z}^\delta\right\}$ is compact in $L^1_{\text{loc}}(\mathbb{R} \times [0,T])$. Trivially we have that

$$\|\bar{z}^\delta\|_\infty \leq \|z^\delta\|_\infty < C, \tag{117}$$

and

$$\left|\bar{z}^\delta(\cdot,t)\right|_{BV} \leq \left|z^\delta(\cdot,t)\right|_{BV} \leq C. \tag{118}$$

Furthermore,

$$\begin{aligned}
\left\|\bar{z}^\delta(\cdot,t)-z^\delta(\cdot,t)\right\|_{L^1} &= \int_{\mathbb{R}} \left|\bar{z}^\delta(x,t)-z^\delta(x,t)\right|\,dx \\
&= \sum_{i,j} \int_{y_{i,j-1/2}}^{y_{i,j+1/2}} \left|z^\delta\left(y_{i,j},t\right)-z^\delta(y,t)\right|\,dy \\
&\le \sum_{i,j} \int_{y_{i,j-1/2}}^{y_{i,j+1/2}} \int_{y}^{y_{i,j}} \left|z^\delta_x(x,t)\right|\,dxdy \\
&\le \max_{i,j}\left|\Delta x_{i,j}\right| \left|z^\delta(\cdot,t)\right|_{BV}.
\end{aligned}$$

Setting $\Delta x = \max_{i,j} \Delta x_{i,j}$, we therefore find that

$$\begin{aligned}
\left\|\bar{z}^\delta(\cdot,t)-\bar{z}^\delta(\cdot,s)\right\|_{L^1} &\le \left\|z^\delta(\cdot,t)-z^\delta(\cdot,s)\right\|_{L^1} + 2\Delta x\left|z^\delta(\cdot,t)\right|_{BV} \\
&\le C((t-s)+\Delta x).
\end{aligned} \tag{119}$$

By the bounds (117), (118) and (119), the sequence $\left\{\bar{z}^\delta\right\}$ converges along a subsequence (also labeled δ) and

$$\lim_{\delta\to 0}\bar{z}^\delta = \lim_{\delta\to 0} z^\delta = z.$$

Therefore, also $\lim_{\delta\to 0}\bar{u}^\delta = u$. Now define

$$\Delta_x f^\delta(x,c) = \frac{f\left(\gamma_{i,j+1/2},c\right)-f\left(\gamma_{i,j-1/2},c\right)}{\Delta x_{i,j}}, \quad \text{for } x\in I_{i,j}.$$

Using this notation, the inequality (116) reads

$$\begin{aligned}
&-\int_0^T\int_{\xi_i}^{\xi_{i+1}} \left|u^\delta-c\right|\varphi_t + F^\delta\left(\gamma^\delta,u^\delta,c\right)\varphi_x\,dxdt - \int_{\xi_i}^{\xi_{i+1}} \left|u^\delta(x,0)-c\right|\varphi(x,0)\,dx \\
&\quad -\int_0^T F^\delta\left(\gamma_i^+,u_i^+,c\right)\varphi\left(\xi_i,t\right) - F^\delta\left(\gamma_{i+1}^-,u_{i+1}^-,c\right)\varphi\left(\xi_{i+1},t\right)dt \\
&\quad +\int_0^T\int_{\xi_i}^{\xi_{i+1}} \operatorname{sign}\left(\bar{u}^\delta-c\right)\Delta_x f^\delta(y,c)\sum_{j=1}^{N_i}\varphi\left(y_{i,j},t\right)\chi_{I,j}(y)\,dydt \\
&\le 0.
\end{aligned} \tag{120}$$

Now we can add for $i=0,\ldots,M$ to obtain

$$
\begin{aligned}
&-\iint\limits_{\Pi_T} \left|u^\delta - c\right| \varphi_t + F^\delta\left(\gamma^\delta, u^\delta, c\right) \varphi_x \, dxdt - \int\limits_{\mathbb{R}} \left|u^\delta(x,0) - c\right| \varphi(x,0)\, dx \\
&\qquad - \int\limits_0^T \sum_{i=1}^{M} \left[F^\delta\left(\gamma_i^+, u_i^+, c\right) - F^\delta\left(\gamma_i^-, u_i^-, c\right)\right] \varphi(\xi_i, t)\, dt \\
&\qquad + \int\limits_0^T \sum_{i=0}^{M} \int\limits_{\xi_i}^{\xi_{i+1}} \operatorname{sign}\left(\bar{u}^\delta - c\right) \Delta_x f^\delta(y,c) \sum_{j=1}^{N_i} \varphi\left(y_{i,j}, t\right) \chi_{I_{i,j}}(y)\, dydt \\
&\le 0.
\end{aligned}
\tag{121}
$$

At this point it is convenient to state the following general lemma.

Lemma 3.3. *Let $\Omega \in \mathbb{R}$ be a bounded open set, $g \in L^1(\Omega)$, and suppose that $g_n(x) \to g(x)$ almost everywhere. Then there exists a set Θ, which is at most countable, such that for any $c \in \mathbb{R} \setminus \Theta$,*

$$\operatorname{sign}(g_n(x) - c) \to \operatorname{sign}(g(x) - c) \quad \textit{a.e. in } \Omega.$$

Furthermore, let $c \in \Theta$ and define

$$\mathcal{E}_c = \left\{x \in \Omega \mid g(x) = c\right\},$$

then it is possible to define sequences $\{\underline{c}_m\}_{m=1}^{\infty} \subset \mathbb{R} \setminus \Theta$ and $\{\bar{c}_m\}_{m=1}^{\infty} \subset \mathbb{R} \setminus \Theta$, such that

$$\underline{c}_m \uparrow c \quad \textit{and} \quad \operatorname{sign}(g(x) - \underline{c}_m) \to \operatorname{sign}(g(x) - c) \quad \textit{a.e. in } \Omega \setminus \mathcal{E}_c, \tag{122}$$

$$\bar{c}_m \downarrow c \quad \textit{and} \quad \operatorname{sign}(g(x) - \bar{c}_m) \to \operatorname{sign}(g(x) - c) \quad \textit{a.e. in } \Omega \setminus \mathcal{E}_c, \tag{123}$$

as $m \to \infty$.

Proof. Fix $c \in \mathbb{R}$ and a point $x \in \Omega$ such that $g_n(x) \to g(x)$ and $g(x) \neq c$. For sufficiently large n, $\operatorname{sign}(g_n(x) - c) = \operatorname{sign}(g(x) - c)$, i.e., $\operatorname{sign}(g_n(x) - c)$ is constant in n, and therefore converges to the correct limit. Thus for each $c \in \mathbb{R}$, $\operatorname{sign}(g_n(x) - c) \to \operatorname{sign}(g(x) - c)$ almost everywhere in $\Omega \setminus \mathcal{E}_c$. It remains to show that all but countably many of the sets $\mathcal{E}_c$ have zero measure. To this end, define

$$\mathcal{C}_k = \left\{c \in \mathbb{R} \mid \operatorname{meas}(\mathcal{E}_c) \ge \frac{1}{k}\right\}.$$

Since Ω is bounded, $\mathcal{C}_k$ contains only a finite number of points. Therefore the set

$$\left\{c \in \mathbb{R} \mid \operatorname{meas}(\mathcal{E}_c) > 0\right\} = \bigcup_{k>0} \mathcal{C}_k.$$

is at most countable.

To prove (122), fix $c \in \Theta$. Since Θ is at most countable, we can find a sequence $\underline{c}_n \uparrow c$ such that $c_n \notin \Theta$. For $x \in \Omega \setminus \mathcal{E}_c$, we have that $g(x) \neq c$, and thus

$\operatorname{sign}(g(x)-\underline{c}_n)=\operatorname{sign}(g(x)-c)$ for n sufficiently large. Thus (122) holds. The existence of $\{\bar{c}_n\}$ and (123) is proved in the same way.

Now clearly

$$\Delta_x f^\delta(y,c)\sum_{j=1}^{N_i}\varphi(y_{i,j},t)\,\chi_{I_{i,j}}(y)\to\partial_x f(\gamma(y),c)\varphi(y,t)\text{ as }\delta\to 0$$

in each interval (ξ_i,ξ_{i+1}). Furthermore, by Lemma 3.3

$$\operatorname{sign}\left(\bar{u}^\delta-c\right)\to\operatorname{sign}(u-c),$$

for almost all (x,t) and all but at most a countable set of c's.

Regarding the middle term of (121), by Lemma 2.1 each summand is bounded by

$$\left|f^\delta(\gamma_i^+,c)-f^\delta(\gamma_i^-,c)\right|,$$

since (u_i^-,u_i^+) satisfies the minimal jump entropy condition. Therefore by sending δ to 0 in (121), we find that

$$\begin{aligned}&-\iint_{\Pi_T}|u-c|\,\varphi_t+F(\gamma,u,c)\varphi_x\,dxdt+\underbrace{\iint_{\Pi_T\setminus\mathcal{D}_\gamma}\operatorname{sign}(u-c)\,\partial_x f(\gamma,c)\varphi\,dxdt}_{I(c)}\\&-\int_0^T\sum_{x\in\mathcal{D}_\gamma}\left|f(\gamma(x^+),c)-f(\gamma(x^-),c)\right|\varphi(x,t)\,dt-\int_{\mathbb{R}}|u_0-c|\,\varphi(x,0)\,dx\\&\le 0\end{aligned}\tag{124}$$

for all but a countable set of c's and all nonnegative test functions φ. This can be rewritten as

$$I(c)\le G(c),$$

where G is a *continuous* function of c. Let Θ denote the set where the convergence of $\operatorname{sign}\left(\bar{u}^\delta-c\right)\to\operatorname{sign}(u-c)$ does not hold. Fix some $c\in\Theta$ and define the two sequences $\{\underline{c}_n\}$ and $\{\bar{c}_n\}$ as in Lemma 3.3. Set

$$\mathcal{E}_c=\left\{(x,t)\;\middle|\;u(x,t)=c\right\}.$$

Since (124) holds for $\underline{c}_n$ and $\bar{c}_n$, we can write $I(c)$ as

$$\begin{aligned}&\iint_{\hat{\Pi}_T\setminus\mathcal{E}_c}\operatorname{sign}(u-\underline{c}_n)\,\partial_x f(\gamma,u)\varphi\,dxdt\\&\qquad+\iint_{\mathcal{E}_c\setminus\mathcal{D}_\gamma}\operatorname{sign}(u-\underline{c}_n)\,\partial_x f(\gamma,u)\varphi\,dxdt\le G(c),\end{aligned}\tag{125}$$

where $\hat{\Pi}_T = \Pi_T \setminus \mathcal{D}_\gamma$. Since $\underline{c}_n < c$, the last integral can be rewritten as

$$\iint_{\mathcal{E}_c \setminus \mathcal{D}_\gamma} \partial_x f(\gamma,u)\varphi\, dxdt.$$

Since f is continuous, by sending n to ∞, we find that

$$\iint_{\hat{\Pi}_T \setminus \mathcal{E}_c} \operatorname{sign}(u-c)\,\partial_x f(\gamma,u)\varphi\, dxdt + \iint_{\mathcal{E}_c \setminus \mathcal{D}_\gamma} \partial_x f(\gamma,u)\varphi\, dxdt \le G(c). \tag{126}$$

Similarly, by using the sequence $\{\bar{c}_n\}$, we arrive at

$$\iint_{\hat{\Pi}_T \setminus \mathcal{E}_c} \operatorname{sign}(u-c)\,\partial_x f(\gamma,u)\varphi\, dxdt - \iint_{\mathcal{E}_c \setminus \mathcal{D}_\gamma} \partial_x f(\gamma,u)\varphi\, dxdt \le G(c). \tag{127}$$

Adding (126) and (127) and dividing by 2, we find that

$$\iint_{\hat{\Pi}_T \setminus \mathcal{E}_c} \operatorname{sign}(u-c)\,\partial_x f(\gamma,u)\varphi\, dxdt \le G(c).$$

Since $\operatorname{sign}(0) = 0$, $\operatorname{sign}(u-c) = 0$ on $\mathcal{E}_c$, therefore we can conclude that

$$\iint_{\Pi_T \setminus \mathcal{D}_\gamma} \operatorname{sign}(u-c)\,\partial_x f(\gamma,u)\varphi\, dxdt \le G(c) \tag{128}$$

for *all* constants c. We have proved the following theorem.

Theorem 3.2. *Assume that the flux function satisfies (A.1)–(A.4), and let u^δ be a weak solution of* (109)*, constructed by front tracking, such that u^δ converges to u in $L^1(\Pi_T)$. Then the entropy condition* (124) *holds for all constants c.*

4 Uniqueness of entropy solutions

Now we shall use the Kružkov entropy formulation, (124) to show that there exists at most one entropy solution. For convenience, we restate this condition,

$$\begin{aligned}\iint_{\Pi_T} |u-c|\,\varphi_t + F(\gamma,u,c)\varphi_x\, dtdx - \iint_{\Pi_T \setminus \mathcal{D}_\gamma} \operatorname{sign}(u-c)\,\partial_x f(\gamma,c)\varphi\, dtdx \\ + \int_0^T \sum_i \left| f(\gamma_i^+,c) - f(\gamma_i^-,c)\right| \varphi(\xi_i,t)\, dt + \int_{\mathbb{R}} |u_0 - c|\,\varphi(x,0)\, dx \ge 0,\end{aligned} \tag{129}$$

for all non-negative test functions $\varphi \in C_0^1(\mathbb{R}\times[0,T))$, and where we write $\gamma_i^\pm = \gamma(\xi_i\pm)$.

In addition to satisfying this entropy inequality, we demand that an entropy solution is a weak solution, i.e., it satisfies (66), and is slightly more regular in the sense described below.

If $w\in L^\infty(\Pi_T)$, by the left and right traces of $w(\cdot,t)$ at a point x_0 we understand functions $t\mapsto w(x_0\pm,t)\in L^\infty([0,T])$ which satisfy a.e. $t\in[0,T)$,

$$\begin{aligned} &\operatorname{ess\,lim}_{x\downarrow x_0}\left|w(x,t)-w(x_0+,t)\right| = 0,\\ &\operatorname{ess\,lim}_{x\uparrow x_0}\left|w(x,t)-w(x_0-,t)\right| = 0. \end{aligned} \tag{130}$$

When comparing two entropy solutions, we shall need that they have traces at the points ξ_i, i.e., if u is an entropy solution, then we assume that the following traces exist

$$u_i^\pm(t) = u(x_i\pm,t), \tag{131}$$

in the sense of (130) for almost all t and for $i=1,\dots,N$.

An entropy solution of (65) is a function in $L^1_{\mathrm{loc}}(\Pi_T)\cap C([0,T);L^1_{\mathrm{loc}}(\mathbb{R}))$, such that (66), (129), and the regularity assumption (131) all hold.

We have already shown that an entropy solution exists for our model problem, since the existence of traces follows by noting that $z(\cdot,t)\in BV(\mathbb{R})$, which implies that z has traces. Since $u=z^{-1}(\gamma,z)$ the same applies to u.

Let now $w=w(x)$ be any function on $\mathbb{R}$, and fixing a point y, we use the following notation

$$\begin{aligned} \operatorname{L\,lim}_{x\uparrow y} w(x) &= \lim_{\varepsilon\downarrow 0}\frac{1}{\varepsilon}\int_y^{y+\varepsilon} w(x)\,dx,\\ \operatorname{L\,lim}_{x\downarrow y} w(x) &= \lim_{\varepsilon\downarrow 0}\frac{1}{\varepsilon}\int_{y-\varepsilon}^{y} w(x)\,dx. \end{aligned}$$

Lemma 4.1. *Let $w\in L^\infty(\Pi_T)$, and fix a point $x_0\in\mathbb{R}$. If the left and right traces $t\to w(x_0\pm,t)$ exist in the sense of* (130), *then for a.e. $t\in[0,t)$ we have that*

$$\operatorname{L\,lim}_{x\downarrow x_0} w(x,t) = w(x_0+,t), \quad \operatorname{L\,lim}_{x\uparrow x_0} w(x,t) = w(x_0-,t).$$

Proof. We prove the first limit as follows,

$$\begin{aligned} &\frac{1}{\varepsilon}\int_{x_0}^{x_0+\varepsilon}\left|w(x,t)-w(x_0+,t)\right|dx\\ &\qquad\le \frac{1}{\varepsilon}\int_{x_0}^{x_0+\varepsilon}\operatorname{ess\,sup}_{y\in(x_0,x_0+\varepsilon)}\left|w(y,t)-w(x_0+,t)\right|dx\\ &\qquad= \operatorname{ess\,sup}_{y\in(x_0,x_0+\varepsilon)}\left|w(y,t)-w(x_0+,t)\right| \quad\to 0 \quad \text{as } \varepsilon\downarrow 0. \end{aligned}$$

As a consequence of this lemma, the following limits exist for any entropy solution u

$$\begin{aligned} \operatorname{L\,lim}_{x\downarrow \xi_i} f\left(\gamma(x), u(x,t)\right) &= f\left(\gamma(\xi_i+), u(\xi_i+,t)\right), \\ \operatorname{L\,lim}_{x\uparrow \xi_i} f\left(\gamma(x), u(x,t)\right) &= f\left(\gamma(\xi_i-), u(\xi_i-,t)\right), \end{aligned} \tag{132}$$

and as a consequence, if v is another entropy solution

$$\begin{aligned} \operatorname{L\,lim}_{x\downarrow \xi_i} F\left(\gamma(x), u(x,t), v(x,t)\right) &= F\left(\gamma(\xi_i+), u(\xi_i+,t), v(\xi_i+,t)\right), \\ \operatorname{L\,lim}_{x\uparrow \xi_i} F\left(\gamma(x), u(x,t), v(x,t)\right) &= F\left(\gamma(\xi_i-), u(\xi_i-,t), v(\xi_i-,t)\right), \end{aligned} \tag{133}$$

where F is the Kružkov entropy flux.

Before we continue, let us define the following compactly supported Lipschitz continuous function

$$\theta_\varepsilon(x) = \begin{cases} \frac{1}{\varepsilon}(\varepsilon + x) & \text{if } x \in (-\varepsilon, 0], \\ \frac{1}{\varepsilon}(\varepsilon - x) & \text{if } x \in [0, \varepsilon), \\ 0 & \text{otherwise.} \end{cases} \tag{134}$$

Lemma 4.2. *Let u be an entropy solution, then a.e. $t \in [0,t)$ and for all constants c*

$$\begin{gathered} f\left(\gamma_i^+, u_i^+(t)\right) = f\left(\gamma_i^-, u_i^-(t)\right), \\ F\left(\gamma_i^+, u_i^+, c\right) - F\left(\gamma_i^-, u_i^-\right) \le \left| f\left(\gamma_i^+, c\right) - f\left(\gamma_i^-, c\right)\right|, \end{gathered}$$

where F is the Kružkov entropy flux.

Proof. Since $u \in L^\infty(\Pi_T)$, a density argument shows that

$$\varphi(x,t) = \theta_\varepsilon\left(x - \xi_i\right)\psi(t),$$

where $\psi \in C_0^1((0,T))$ is an admissible test function and can be used in the weak formulation (66). If $\varepsilon < \min_i\{\xi_{i+1} - \xi_i\}$, we get

$$\begin{aligned} &\iint_{\Pi_T} u\theta_\varepsilon\left(x - \xi_i\right)\psi'(t)\,dxdt \\ &\qquad = \int_0^T \left(\frac{1}{\varepsilon}\int_{\xi_i}^{\xi_i+\varepsilon} f(\gamma(x), u)\,dx - \frac{1}{\varepsilon}\int_{\xi_i-\varepsilon}^{\xi_i} f(\gamma(x), u)\,dx\right)\psi(t)\,dt. \end{aligned}$$

By sending $\varepsilon \downarrow 0$ and using Lemma 4.2,

$$\int_0^T \left(f\left(\gamma_i^+, u_i^+\right) - f\left(\gamma_i^-, u_i^-\right)\right)\psi(t)\,dt = 0.$$

Since this holds for every testfunction ψ, the integrand must be zero.

To prove the inequality in the lemma, we choose the same testfunction, but restrict ψ to be non-negative.

By the entropy condition, (129), we get

$$\begin{aligned}\iint_{\Pi_T} &|u-c|\theta_\varepsilon(x-\xi_i)\,\psi'(t)\,dxdt\\ &-\int_0^T\left(\frac{1}{\varepsilon}\int_{\xi_i}^{\xi_i+\varepsilon}F(\gamma(x),u,c)\,dx-\frac{1}{\varepsilon}\int_{\xi_i-\varepsilon}^{\xi_i}F(\gamma(x),u,c)\,dx\right)\psi(t)\,dt\\ &-\iint_{\Pi_T}\operatorname{sign}(u-c)\,\partial_x f(\gamma(x),c)\theta_\varepsilon(x-\xi_i)\,\psi(t)\,dxdt\\ &+\int_0^T\left|f\left(\gamma_i^+,c\right)-f\left(\gamma_i^-,c\right)\right|\psi(t)\,dt\geq 0.\end{aligned}$$

Again, by sending $\varepsilon\downarrow 0$,

$$\int_0^T\left|f\left(\gamma_i^+,c\right)-f\left(\gamma_i^-,c\right)\right|\psi(t)\,dt\geq\int_0^T\left(F\left(\gamma_i^+,u_i^+,c\right)-F\left(\gamma_i^-,u_i^-,c\right)\right)\psi(t)\,dt,$$

which implies the inequality.

This has an immediate corollary.

Corollary 4.1. *Assume that the flux function f satisfies (A.1)–(A.4). If u is an entropy solution, then the pairs (u_i^-,u_i^+) satisfy the minimal jump entropy condition,* (34)–(35) *for $i=1,\ldots,N$.*

For any test function φ, which has support away from $\mathcal{D}_\gamma$, we can double variables "a la Kružkov".

Lemma 4.3. *For any two entropy solutions u and v and any non-negative test function $\varphi\in C_0^1(\Pi_T\setminus\mathcal{D}_\gamma)$ we have that*

$$\begin{aligned}-\iint_{\Pi_T}&|u-v|\,\varphi_t+F(\gamma,u,v)\varphi_x\,dtdx\\ &\leq C\iint_{\Pi_T}|u-v|\,\varphi\,dtdx+\int_{\mathbb{R}}|u_0-v_0|\,\varphi(x,0)\,dx,\end{aligned}\tag{135}$$

where the constant C is zero if γ is piecewise constant.

Proof. The proof is a classical doubling of variables. It uses exactly the same arguments as in Kružkov's original paper [11], but adapted to our situation. I recommend skipping this proof in a first reading.

Let ϕ be a non-negative test function in $C_0^1(\Pi_T^2)$. We use the notation $u=u(y,s)$, $v=v(x,t)$. Then using $c=u(y,s)$ as the constant in the entropy inequality for v and then integrating over (y,s), we get

$$
\begin{aligned}
-\iiiint_{\Pi_T^2} &|u-v|\,\phi_t + F(\gamma(x),u,v)\phi_x \,dtdxdsdy \\
&+ \iiiint_{(\Pi_T\setminus\mathcal{D}_\gamma)^2} \operatorname{sign}(v-u)\, f(\gamma(x),u)_x\, \phi \,dtdxdyds \\
&\le \iint_{\Pi_T}\int_{\mathbb{R}} |v_0-u|\,\phi(x,0,y,s)\,dxdsdy.
\end{aligned} \tag{136}
$$

Similarly, starting with the entropy inequality for u and integrating over (x,t), we arrive at

$$
\begin{aligned}
-\iiiint_{\Pi_T^2} &|u-v|\,\phi_s + F(\gamma(y),u,v)\phi_y \,dsdydtdx \\
&+ \iiiint_{(\Pi_T\setminus\mathcal{D}_\gamma)^2} \operatorname{sign}(u-v)\, f(\gamma(y),v)_y\, \phi \,dsdydtdx \\
&\le \iint_{\Pi_T}\int_{\mathbb{R}} |u_0-v|\,\phi(x,t,y,0)\,dydtdx.
\end{aligned} \tag{137}
$$

Since γ is differentiable outside $\mathcal{D}_\gamma$ for $(x,t)\in \Pi_T\setminus\mathcal{D}_\gamma$, we have

$$
\begin{aligned}
F(\gamma(x),v,u)\,\phi_x - &\operatorname{sign}(v-u)\, f(\gamma(x),u)_x\,\phi \\
&= \operatorname{sign}(v-u)\,(f(\gamma(x),v)-f(\gamma(y),u))\,\phi_x \\
&\quad - \operatorname{sign}(v-u)\,((f(\gamma(x),u)-f(\gamma(y),u))\,\phi)_x .
\end{aligned}
$$

Using this we find that

$$
\begin{aligned}
-\iiiint_{(\Pi_T\setminus\mathcal{D}_\gamma)^2} &F(\gamma(x),v,u)\,\phi_x - \operatorname{sign}(v-u)\, f(\gamma(x),u)_x\,\phi \,dtdxdsdy \\
= -\iiiint_{(\Pi_T\setminus\mathcal{D}_\gamma)^2} &\operatorname{sign}(v-u)\,(f(\gamma(x),v)-f(\gamma(y),u))\,\phi_x \,dtdxdsdy \\
+ \iiiint_{(\Pi_T\setminus\mathcal{D}_\gamma)^2} &\operatorname{sign}(v-u)\,((f(\gamma(x),u)-f(\gamma(y),u))\,\phi)_x \,dtdxdsdy.
\end{aligned}
$$

We also have a similar equality for u,

$$
\begin{aligned}
-\iiiint_{(\Pi_T\setminus\mathcal{D}_\gamma)^2} &F(\gamma(y),v,u)\,\phi_y - \operatorname{sign}(u-v)\, f(\gamma(y),v)_y\,\phi \,dsdydtdx \\
= -\iiiint_{(\Pi_T\setminus\mathcal{D}_\gamma)^2} &\operatorname{sign}(u-v)\,(f(\gamma(y),u)-f(\gamma(x),v))\,\phi_y \,dsdydtdx \\
+ \iiiint_{(\Pi_T\setminus\mathcal{D}_\gamma)^2} &\operatorname{sign}(u-v)\,((f(\gamma(y),v)-f(\gamma(x),v))\,\phi)_y \,dsdydtdx.
\end{aligned}
$$

Now we introduce the notations

$$
\partial_{t+s} = \partial_t + \partial_s, \quad \partial_{x+y} = \partial_x + \partial_y,
$$

use the above and add (137) and (136) to find

$$\begin{aligned}
-\iiiint_{\Pi_T^2} &|v-u|\,\partial_{t+s}\phi + \operatorname{sign}(v-u)\left(f(\gamma(x),v)-f(\gamma(y),u)\right)\partial_{x+y}\phi\,dtdxdsdy \\
&+\iiiint_{\Pi_T^2} \operatorname{sign}(v-u)\Big[\left(\left(f(\gamma(x),u)-f(\gamma(y),u)\right)\phi\right)_x \\
&\qquad\qquad + \left(\left(f(\gamma(y),v)-f(\gamma(x),v)\right)\phi\right)_y\Big]\,dtdxdsdy \\
&\le \iint_{\Pi_T}\int_{\mathbb{R}} |v_0-u|\,\phi(x,0,y,s)\,dxdsdy + \iint_{\Pi_T}\int_{\mathbb{R}} |u_0-v|\,\phi(x,t,y,0)\,dydtdx.
\end{aligned} \tag{138}$$

Now we shall choose a suitable testfunction. First let $\omega \in C_0^\infty(\mathbb{R})$ be a function such that $\omega(-a)=\omega(a)$, $\omega'(a)\le 0$ for $a>0$, $|\omega'(a)|\le 2$, $\omega(a)=0$ for $|a|\ge 1$, and $\int \omega(a)\,da = 1$. For positive ε, set

$$\omega_\varepsilon(a) = \frac{1}{\varepsilon}\omega\left(\frac{a}{\varepsilon}\right).$$

Let $\varphi(x,t)$ be a test function such that

$$\varphi(x,t)=0 \quad \text{for } |x-\xi_i|\le\varepsilon_0,\ i=1,\dots,N,$$

for some positive ε_0. Then we define for $\varepsilon<\varepsilon_0$

$$\phi(x,t,y,s) = \varphi\left(\frac{x+y}{2},\frac{t+s}{2}\right)\omega_\varepsilon\left(\frac{x-y}{2}\right)\omega_\varepsilon\left(\frac{t-s}{2}\right). \tag{139}$$

We can easily check that $\phi \in C_0^1((\Pi_T\setminus\mathcal{D}_\gamma)^2)$. Furthermore, we have the useful identities

$$\begin{aligned}
\partial_{t+s}\phi(x,t,y,s) &= \partial_{t+s}\varphi\left(\frac{x+y}{2},\frac{t+s}{2}\right)\omega_\varepsilon\left(\frac{x-y}{2}\right)\omega_\varepsilon\left(\frac{t-s}{2}\right),\\
\partial_{x+y}\phi(x,t,y,s) &= \partial_{x+y}\varphi\left(\frac{x+y}{2},\frac{t+s}{2}\right)\omega_\varepsilon\left(\frac{x-y}{2}\right)\omega_\varepsilon\left(\frac{t-s}{2}\right).
\end{aligned}$$

If we use these identities in (138) we find that

$$\begin{aligned}
-\iiiint_{\Pi_T^2} &\left(I_{\text{time}}(x,t,y,s)+I_{\text{conv}}(x,t,y,s)\right)\omega_\varepsilon\left(\frac{x-y}{2}\right)\omega_\varepsilon\left(\frac{t-s}{2}\right)dtdxdsdy \\
&\le \iiiint_{\Pi_T^2} I^1_{\text{flux}}(x,t,y,s)+I^2_{\text{flux}}(x,t,y,s)+I^3_{\text{flux}}(x,t,y,s)\,dtdxdsdy \\
&\quad + \underbrace{\iint_{\Pi_T}\int_{\mathbb{R}} |v_0-u|\,\phi(x,0,y,s)\,dxdsdy + \iint_{\Pi_T}\int_{\mathbb{R}} |u_0-v|\,\phi(x,t,y,0)\,dydtdx}_{J_{\text{init}}},
\end{aligned} \tag{140}$$

where

$$
\begin{aligned}
I_{\mathrm{time}}(x,t,y,s) &= |v-u|\,\partial_{t+s}\varphi\left(\frac{x+y}{2},\frac{t+s}{2}\right),\\
I_{\mathrm{conv}}(x,t,y,s) &= \operatorname{sign}(v-u)\left[f(\gamma(x),v)-f(\gamma(y),u)\right]\\
&\qquad \times \partial_{x+y}\varphi\left(\frac{x+y}{2},\frac{t+s}{2}\right),\\
I^1_{\mathrm{flux}}(x,t,y,s) &= -\operatorname{sign}(v-u)\,\omega_\varepsilon\left(\frac{x-y}{2}\right)\omega_\varepsilon\left(\frac{t-s}{2}\right)\varphi\left(\frac{x+y}{2},\frac{t+s}{2}\right)\\
&\qquad \times\left[\gamma'(x)f_\gamma(\gamma(x),u)-\gamma'(y)f_\gamma(\gamma(y),v)\right],\\
I^2_{\mathrm{flux}}(x,t,y,s) &= -\operatorname{sign}(v-u)\,\omega_\varepsilon\left(\frac{x-y}{2}\right)\omega_\varepsilon\left(\frac{t-s}{2}\right)\\
&\qquad \times\left[\partial_x\varphi\left(\frac{x+y}{2},\frac{t+s}{2}\right)\left(f(\gamma(x),u)-f(\gamma(y),u)\right)\right.\\
&\qquad \left.+\partial_y\varphi\left(\frac{x+y}{2},\frac{t+s}{2}\right)\left(f(\gamma(x),v)-f(\gamma(y),v)\right)\right],\\
I^3_{\mathrm{flux}}(x,t,y,s) &= \left[F(\gamma(x),v,u)-F(\gamma(y),v,u)\right]\\
&\qquad \times\varphi\left(\frac{x+y}{2},\frac{t+s}{2}\right)\omega_\varepsilon\left(\frac{t-s}{2}\right)\partial_x\omega_\varepsilon\left(\frac{x-y}{2}\right).
\end{aligned}
$$

Introduce the new variables

$$
\tilde{x}=\frac{x+y}{2},\quad z=\frac{x-y}{2},\quad \tilde{t}=\frac{t+s}{2},\tau=\frac{t-s}{2},
$$

which maps Π_T^2 into

$$
\Omega_T=\left\{(\tilde{x},\tilde{t},z,\tau)\in\mathbb{R}^4\;\middle|\;0\le\tilde{t}\pm\tau\le T\right\},
$$

and $(\Pi_T\setminus\mathcal{D}_\gamma)^2$ into

$$
\Omega_{T,\gamma}=\left\{(\tilde{x},\tilde{t},z,\tau)\in\Omega_T\;\middle|\;\tilde{x}\pm z\neq\xi_i,\quad i=1,\dots,N\right\}.
$$

We start by estimating the terms in J_{init},

$$
\begin{aligned}
&\iint_{\Pi_T}\int_{\mathbb{R}}|v_0(x)-u(y,s)|\,\varphi\left(\frac{x+y}{2},\frac{s}{2}\right)\omega_\varepsilon\left(\frac{x-y}{2}\right)\omega_\varepsilon\left(\frac{-s}{2}\right)dx\,ds\,dy\\
&\quad=\int_0^\varepsilon\int_{\mathbb{R}}\int_{-\varepsilon}^{\varepsilon}|v_0(\tilde{x}+z)-u(\tilde{x}-z,\tilde{t}-\tau)|\,\varphi(\tilde{x},\tau)\omega_\varepsilon(z)\omega_\varepsilon(\tau)\,dz\,d\tilde{x}\,d\tau\\
&\quad\to\frac{1}{2}\int_{\mathbb{R}}|v_0(x)-u(x,0)|\,\varphi(x,0)\,dx
\end{aligned}
$$

as $\varepsilon\to 0$. Since $t\mapsto u(x,t)$ is L^1 continuous, we can replace $u(x,0)$ by $u_0(x)$. Similarly we find that

$$\iint_{\Pi_T} \int_{\mathbb{R}} |u_0 - v| \, \phi(x,t,y,0) \, dy dt dx \to \frac{1}{2} \int_{\mathbb{R}} |u_0 - v_0| \, \varphi(x,0) \, dx$$

as $\varepsilon \to 0$, and thus

$$\lim_{\varepsilon \to 0} J_{\text{init}} = \int_{\mathbb{R}} |u_0 - v_0| \, \varphi(x,0) \, dx. \tag{141}$$

In the transformed variables, we have

$$\begin{aligned}
I_{\text{time}}(\tilde{x},\tilde{t},z,\tau) &= |v(\tilde{x}+z,\tilde{t}+\tau) - u(\tilde{x}-z,\tilde{t}-\tau)| \, \partial_{\tilde{t}} \varphi(\tilde{x},\tilde{t}), \\
I_{\text{conv}}(\tilde{x},\tilde{t},z,\tau) &= \operatorname{sign}(v(\tilde{x}+z,\tilde{t}+\tau) - u(\tilde{x}-z,\tilde{t}-\tau)) \, \partial_{\tilde{x}} \varphi(\tilde{x},\tilde{t}) \\
&\quad \times \Big[f(\gamma(\tilde{x}+z), v(\tilde{x}+z,\tilde{t}+\tau)) \\
&\qquad - f(\gamma(\tilde{x}-z), u(\tilde{x}-z,\tilde{t}-\tau)) \Big], \\
I^1_{\text{flux}}(\tilde{x},\tilde{t},z,\tau) &= \operatorname{sign}(v(\tilde{x}+z,\tilde{t}+\tau) - u(\tilde{x}-z,\tilde{t}-\tau)) \, \omega_\varepsilon(z) \, \omega_\varepsilon(\tau) \\
&\quad \times \Big[\gamma'(\tilde{x}+z) f_\gamma(\gamma(\tilde{x}+z), u(\tilde{x}-z,\tilde{t}-\tau)) \\
&\qquad - \gamma'(\tilde{x}-z) f_\gamma(\gamma(\tilde{x}-z), v(\tilde{x}+z,\tilde{t}+\tau)) \Big] \varphi(\tilde{x},\tilde{t}), \\
I^2_{\text{flux}}(\tilde{x},\tilde{t},z,\tau) &= \operatorname{sign}(v(\tilde{x}+z,\tilde{t}+\tau) - u(\tilde{x}-z,\tilde{t}-\tau)) \, \omega_\varepsilon(z) \, \omega_\varepsilon(\tau) \, \partial_{\tilde{x}} \varphi(\tilde{x},\tilde{t}) \\
&\quad \times \Big[(f(\gamma(\tilde{x}+z), u(\tilde{x}-z,\tilde{t}-\tau)) - f(\gamma(\tilde{x}-z), u(\tilde{x}-z,\tilde{t}-\tau))) \\
&\qquad + (f(\gamma(\tilde{x}+z), v(\tilde{x}+z,\tilde{t}+\tau)) - f(\gamma(\tilde{x}-z), v(\tilde{x}+z,\tilde{t}+\tau))) \Big], \\
I^3_{\text{flux}}(\tilde{x},\tilde{t},z,\tau) &= \Big[F(\gamma(\tilde{x}+z), v(\tilde{x}+z,\tilde{t}+\tau), u(\tilde{x}-z,\tilde{t}-\tau)) \\
&\qquad - F(\gamma(\tilde{x}-z), v(\tilde{x}+z,\tilde{t}+\tau), u(\tilde{x}-z,\tilde{t}-\tau)) \Big] \\
&\quad \times \varphi(\tilde{x},\tilde{t}) \, \omega_\varepsilon(t) \, \partial_z \omega_\varepsilon(z).
\end{aligned}$$

It is straightforward to deduce the limits

$$\lim_{\varepsilon \to 0} \iiiint_\Omega I_{\text{time}}(\tilde{x},\tilde{t},z,\tau) \omega_\varepsilon(z) \, \omega_\varepsilon(t) \, d\tau dz d\tilde{t} d\tilde{x} = \iint_{\Pi_T} |u-v| \, \varphi_t \, dt dx, \tag{142}$$

$$\lim_{\varepsilon \to 0} \iiiint_\Omega I_{\text{conv}}(\tilde{x},\tilde{t},z,\tau) \omega_\varepsilon(z) \, \omega_\varepsilon(t) \, d\tau dz d\tilde{t} d\tilde{x} = \iint_{\Pi_T} F(\gamma(x),u,v) \, \varphi_x \, dt dx. \tag{143}$$

Since γ is C^1 outside $\mathcal{D}_\gamma$, we deduce that

$$\begin{aligned}
\lim_{\varepsilon \to 0} \iiiint_{\Omega_\gamma} I^1_{\text{flux}}(\tilde{x},\tilde{t},z,\tau) \, d\tilde{t} d\tilde{x} d\tau dz &= \iint_{\Pi_T \setminus \mathcal{D}_\gamma} \gamma'(x) F_\gamma(\gamma(x),u,v) \, dt dx \\
&\le C \iint_{\Pi_T} |u-v| \, dt dx,
\end{aligned} \tag{144}$$

where

$$C = \left\|\gamma'\right\|_{L^\infty(\mathbb{R}\setminus\mathcal{D}_\gamma)} \left\|f_{u\gamma}\right\|_{L^\infty}.$$

In particular, we observe that C can be chosen as zero if γ is a piecewise constant function.

Next we consider I^2_{flux}. Since φ vanishes near $\mathcal{D}_\gamma$, I^2_{flux} also vanishes near $\mathcal{D}_\gamma$. Hence γ is uniformly C^1 where $I^2_{\text{flux}} \neq 0$. Therefore

$$\begin{aligned}
&\left|I^2_{\text{flux}}(\tilde{x},\tilde{t},z,\tau)\right| \\
&\quad \leq \omega_\varepsilon(z)\omega_\varepsilon(\tau)\left|\partial_{\tilde{x}}\varphi(\tilde{x},\tilde{t})\right| \\
&\qquad \times \Big(\left|f(\gamma(\tilde{x}+z),u(\tilde{x}-z,\tilde{t}-\tau)) - f(\gamma(\tilde{x}-z),u(\tilde{x}-z,\tilde{t}-\tau))\right| \\
&\qquad\quad + \left|f(\gamma(\tilde{x}+z),v(\tilde{x}+z,\tilde{t}+\tau)) - f(\gamma(\tilde{x}-z),v(\tilde{x}+z,\tilde{t}+\tau))\right|\Big) \\
&\quad \leq \omega_\varepsilon(z)\omega_\varepsilon(\tau)\left|\partial_{\tilde{x}}\varphi(\tilde{x},\tilde{t})\right| 2\left\|f_\gamma\right\|_{L^\infty}\left|\gamma(\tilde{x}+z)-\gamma(\tilde{x}-z)\right| \\
&\quad \leq 4\left\|f_\gamma\right\|_{L^\infty}\left\|\gamma'\right\|_{L^\infty(\mathbb{R}\setminus\mathcal{D}_\gamma)}\omega_\varepsilon(z)\omega_\varepsilon(\tau)\left|\partial_{\tilde{x}}\varphi(\tilde{x},\tilde{t})\right|\left|z\right|.
\end{aligned}$$

From this we conclude

$$\begin{aligned}
&\lim_{\varepsilon\to 0}\left|\iiiint_\Omega I^2_{\text{flux}}(\tilde{x},\tilde{t},z,\tau)\,d\tilde{t}\,d\tilde{x}\,d\tau\,dz\right| \\
&\qquad\qquad \leq \lim_{\varepsilon\to 0} C\int_{-\varepsilon}^{\varepsilon}|z|\,\omega_\varepsilon(z)\,dz = 0.
\end{aligned} \tag{145}$$

Finally we turn to I^3_{flux}, where

$$\begin{aligned}
\left|I^3_{\text{flux}}(\tilde{x},\tilde{t},z,\tau)\right| &\leq \varphi(\tilde{x},\tilde{t})\,\omega_\varepsilon(\tau)\left|\partial_z\omega_\varepsilon(z)\right| \\
&\quad \times \Big|F(\gamma(\tilde{x}+z),v(\tilde{x}+z,\tilde{t}+\tau),u(\tilde{x}-z,\tilde{t}-\tau)) \\
&\qquad - F(\gamma(\tilde{x}-z),v(\tilde{x}+z,\tilde{t}+\tau),u(\tilde{x}-z,\tilde{t}-\tau))\Big| \\
&\leq \varphi(\tilde{x},\tilde{t})\,\omega_\varepsilon(\tau)\left|\partial_z\omega_\varepsilon(z)\right| 2\left\|\gamma'\right\|_{L^\infty(\mathbb{R}\setminus\mathcal{D}_\gamma)}|z| \\
&\quad \times \left\|f_{\gamma u}\right\|_{L^\infty}\left|v(\tilde{x}+z,\tilde{t}+\tau)-u(\tilde{x}-z,\tilde{t}-\tau)\right| \\
&\leq \left\|f_{\gamma u}\right\|_{L^\infty}\left\|\gamma'\right\|_{L^\infty(\mathbb{R}\setminus\mathcal{D}_\gamma)}\varphi(\tilde{x},\tilde{t})\,\omega_\varepsilon(\tau)\frac{8}{2\varepsilon}\mathbf{1}_{|z|\leq\varepsilon} \\
&\qquad \times \left|v(\tilde{x}+z,\tilde{t}+\tau)-u(\tilde{x}-z,\tilde{t}-\tau)\right|.
\end{aligned}$$

Now set

$$h_\varepsilon(\tilde{x},\tilde{t}) = \frac{1}{2\varepsilon}\int_{-\varepsilon}^{\varepsilon}\int_{-\varepsilon}^{\varepsilon}\left|v(\tilde{x}+z,\tilde{t}+\tau)-u(\tilde{x}-z,\tilde{t}-\tau)\right|\varphi(\tilde{x},\tilde{t})\,\omega_\varepsilon(\tau)\,d\tau\,dz.$$

By Lebesgue's differentiation theorem,

$$\lim_{\varepsilon\to 0} h_\varepsilon(x,t) = \left|v(x,t)-u(x,t)\right| \quad \text{a.e. } (x,t).$$

Therefore

$$\lim_{\varepsilon\to 0}\left|\iiiint_{\Omega} I^3_{\text{flux}}(\tilde{x},\tilde{t},z,\tau)\,d\tilde{t}\,d\tilde{x}\,d\tau\,dz\right| \le \lim_{\varepsilon\to 0} C\iint_{\Pi_T} |u-v|\,\varphi\,dt\,dx, \tag{146}$$

where the constant C is zero if γ is piecewise constant.

Combining (142), (143), (144), (145) and (146) we get (135).

Equipped with Lemma 4.3 we can continue to prove the uniqueness of entropy solutions. Define

$$\psi_\varepsilon(x) = \begin{cases} \frac{2}{\varepsilon}(\varepsilon + x) & \text{if } x\in[-\varepsilon,-\varepsilon/2],\\ 1 & \text{if } -\varepsilon/2 < x < \varepsilon/2,\\ \frac{2}{\varepsilon}(\varepsilon - x) & \text{if } x\in[\varepsilon/2,\varepsilon],\\ 0 & \text{otherwise,}\end{cases}$$

and set $\Psi_\varepsilon(x) = 1-\sum_i^N \psi_\varepsilon(x-\xi_i)$. Observe that $\Psi_\varepsilon \to 1$ in $L^1_{\text{loc}}(\mathbb{R})$ as $\varepsilon\to 0$, and we only consider ε that are smaller than $\min_i\{\xi_{i+1}-\xi_i\}$. Let φ be a non-negative testfunction in $C^1_0(\Pi_T)$, then $\phi = \varphi\Psi_\varepsilon$ is an admissible testfunction as a density argument will show. Furthermore ϕ has support away from $\mathcal{D}_\gamma$. With this testfunction, (135) takes the form

$$\begin{aligned} -\iint_{\Pi_T} |u-v|\,\Psi_\varepsilon\varphi_t + F(\gamma,u,v)\Psi_\varepsilon\varphi_x\,dt\,dx - \iint_{\Pi_T} F(\gamma,u,v)\Psi'_\varepsilon\varphi\,dt\,dx \\ \le C\iint_{\Pi_T} |u-v|\,\Psi_\varepsilon\varphi\,dt\,dx + \int_{\mathbb{R}} |u_0-v_0|\,\Psi_\varepsilon\varphi(x,0)\,dx. \end{aligned}$$

Set

$$I_\varepsilon = \iint_{\Pi_T} F(\gamma,u,v)\Psi'_\varepsilon\varphi\,dt\,dx,$$

and let $\varepsilon\downarrow 0$. This yields

$$\begin{aligned} -\iint_{\Pi_T} |u-v|\,\varphi_t + F(\gamma,u,v)\varphi_x\,dt\,dx \\ \le C\iint_{\Pi_T} |u-v|\,\varphi\,dt\,dx + \int_{\mathbb{R}} |u_0-v_0|\,\varphi(x,0)\,dx + \lim_{\varepsilon\downarrow 0} I_\varepsilon. \end{aligned}$$

Now we use that (u_i^-,u_i^+) and (v_i^-,v^+) satisfy the minimal jump entropy condition, and thus Lemma 2.2 applies at each discontinuity in γ. Remembering this we calculate

$$\begin{aligned} &\lim_{\varepsilon\downarrow 0} I_\varepsilon \\ &= \sum_i^N \lim_{\varepsilon\downarrow 0}\int_0^T \left(\frac{2}{\varepsilon}\int_{\xi_i+\varepsilon/2}^{\xi_i+\varepsilon} F(\gamma(x),u,v)\varphi\,dx - \frac{2}{\varepsilon}\int_{\xi_i-\varepsilon}^{\xi_i-\varepsilon/2} F(\gamma(x),u,v)\,\varphi\,dx\right)dt \\ &= \lim_{\varepsilon\downarrow 0}\sum_i^N \int_0^T \left(F(\gamma_i^+,u_i^+,v_i^+) - F(\gamma_i^-,u_i^-,v_i^-)\right)\varphi(\xi_i,t)\,dt \le 0. \end{aligned}$$

Hence for any non-negative testfunction

$$-\iint_{\Pi_T} |u-v|\,\varphi_t + F(\gamma,u,v)\varphi_x\,dtdx \\ \leq C\iint_{\Pi_T} |u-v|\,\varphi\,dtdx + \int_{\mathbb{R}} |u_0-v_0|\,\varphi(x,0)\,dx. \quad (147)$$

Now let $\alpha_r(x)$ be a smooth function taking values in $[0,1]$ such that

$$\alpha_r(x) = \begin{cases} 1 & \text{if } |x| \leq r, \\ 0 & \text{if } |x| \geq r+1, \end{cases}$$

and $\max|\alpha_r'(x)| \leq 2$. Then fix s_0 and s so that $0 < s_0 < s < T$. For any positive κ and τ, such that $s_0+\tau < s+\kappa < T$, let $\beta_{\kappa,\tau}(t)$ be a Lipschitz function which is linear on $[s_0, s_0+\kappa]$ and on $[s, s+\tau]$, and satisfies

$$\beta_{\kappa,\tau}(t) = \begin{cases} 0 & \text{if } t < s_0 \text{ or } t > s+\kappa, \\ 1 & \text{if } s \in [s_0+\tau, s]. \end{cases}$$

By density arguments, $\varphi = \alpha_r\beta_{\kappa,\tau}$ is an admissible test function, and using this in (147) gives

$$\frac{1}{\kappa}\int_s^{s+\kappa}\int_{\mathbb{R}} |u-v|\,\alpha_r\,dxdt - \frac{1}{\tau}\int_{s_0}^{s_0+\tau} |u-v|\,\alpha_r\,dxdt \\ \leq C\int_{s_0}^{s+\kappa}\int_{\mathbb{R}} |u-v|\,\alpha_r\,dxdt \\ +2\int_{s_0}^{s+\kappa}\int_{r<|x|<r+1} |F(\gamma,u,v)|\,\beta_{\kappa,\tau}\,dxdt.$$

Next, we let $s_0 \downarrow 0$, and use the triangle inequality to get

$$\frac{1}{\kappa}\int_s^{s+\kappa}\int_{\mathbb{R}} |u-v|\,\alpha_r\,dxdt \leq \int_{\mathbb{R}} |u_0-v_0|\,\alpha_r\,dx \\ +\frac{1}{\tau}\int_0^{\tau}\int_{\mathbb{R}} |v(x,t)-v_0(x)|\,\alpha_r(x)\,dxdt \\ +\frac{1}{\tau}\int_0^{\tau}\int_{\mathbb{R}} |u(x,t)-u_0(x)|\,\alpha_r(x)\,dxdt \\ +C\int_{s_0}^{s+\kappa}\int_{\mathbb{R}} |u-v|\,\alpha_r\,dxdt \\ +2\int_{s_0}^{s+\kappa}\int_{r<|x|<r+1} |F(\gamma,u,v)|\,\beta_{\kappa,\tau}\,dxdt.$$

We shall now send $\tau \downarrow 0$ and prove later that for any entropy solution u

$$\lim_{\tau\downarrow 0} \frac{1}{\tau} \int_0^\tau \int_{\mathbb{R}} |u(x,t) - u_0(x)|\, \alpha_r(x)\, dxdt = 0. \tag{148}$$

Furthermore, by finite speed of propagation, if $u_0(x) = v_0(x)$ for large $|x|$, then also $u(x,t) = v(x,t)$ for large $|x|$. Hence $F(\gamma(x), u(x,t), v(x,t)) = 0$ for large $|x|$. Hence

$$\lim_{r\to\infty} \int_{s_0}^{s+\kappa} \int_{r<|x|<r+1} |F(\gamma, u, v)|\, \beta_{\kappa,\tau}\, dxdt = 0.$$

Set

$$\mathcal{E}(t) = \int_{\mathbb{R}} |u(x,t) - v(x,t)|\, dx.$$

By sending $\tau \downarrow 0$ and then $r \uparrow \infty$, we obtain

$$\frac{1}{\kappa} \int_s^{s+\kappa} \mathcal{E}(t)\, dt \le \mathcal{E}(0) + C \int_0^{s+\kappa} \mathcal{E}(t)\, dt. \tag{149}$$

Let s be a Lebesgue point for the L^1-function $\mathcal{E}$. Sending $\kappa \downarrow 0$ yields

$$\mathcal{E}(s) \le \mathcal{E}(0) + C \int_0^s \mathcal{E}(t)\, dt.$$

Since the set of Lebesgue points has full measure, we can use Gronwall's inequality to conclude that

$$\mathcal{E}(t) \le e^{Ct} \mathcal{E}(0),$$

for almost every $t > 0$.

It remains to prove (148). To this end, define

$$\beta_\tau(t) = d \begin{cases} \frac{1}{\tau}(\tau - t) & \text{if } 0 \le t \le \tau, \\ 0 & \text{otherwise.} \end{cases}$$

We then use the testfunction $\omega_\varepsilon(x-y)\beta_\tau(t)\alpha_r(x)$ and the constant $c = u_0(y)$ in the entropy formulation (129). The result of this is

$$\begin{aligned}
&\iint_{\Pi_\tau} |u(x,t) - u_0(y)|\, \omega_\varepsilon(x-y)\alpha_r(x)\beta_\tau'(t)\, dtdx \\
&\quad + \iint_{\Pi_\tau} F(\gamma(x), u, u_0(y))\, (\omega_\varepsilon(x-y)\alpha_r(x))_x\, \beta_\tau(t)\, dtdx \\
&\quad - \iint_{\Pi_\tau \setminus \mathcal{D}_\gamma} \operatorname{sign}(u - u_0(y))\, \partial_x f(\gamma(x), u_0(y))\, \omega_\varepsilon(x-y)\alpha_r(x)\beta_\tau(t)\, dtdx \\
&\quad + \int_0^\tau \sum_i \left| f(\gamma_i^+, u_0(y)) - f(\gamma_i^+, u_0(y)) \right| \omega_\varepsilon(\xi_i - y)\, \alpha_r(\xi_i)\, \beta_\tau(t)\, dt \\
&\quad + \int_{\mathbb{R}} |u_0(x) - u_0(y)|\, \omega_\varepsilon(x-y)\alpha_r(x)\, dx \ge 0.
\end{aligned}$$

Since $u \in L^1_{\mathrm{loc}}$, when sending $\tau \downarrow 0$, all terms above containing β_τ will vanish. Recalling that $\beta_\tau'(t) = -1/\tau$ for $t \in (0,\tau)$, after an application of the triangle inequality and an integration over $y \in \mathbb{R}$, we find that

$$\lim_{\tau \downarrow 0} \frac{1}{\tau} \int_0^\tau \int_{\mathbb{R}} |u(x,t) - u_0(x)|\, \alpha_r(x)\, dx dt \\ \leq 2 \int_{\mathbb{R}} \int_{\mathbb{R}} |u_0(x) - u_0(y)|\, \omega_\varepsilon(x-y) \alpha_r(x)\, dx dy.$$

Since $u_0 \in L^1_{\mathrm{loc}}(\mathbb{R})$, we can send $\varepsilon \downarrow 0$ to prove (148).

We have now proved that the initial value problem (65) is well posed in L^1.

Theorem 4.1. *Assume that the flux function f satisfies the assumptions (A.1)–(A.4), and that the initial value u_0 is in $L^1(\mathbb{R})$ and $f(\gamma, u_0) \in BV(\mathbb{R})$. Then there exists a weak entropy solution, in the sense of* (66) *and* (129)*, to the initial value problem* (65).

If v is another entropy solution with initial data v_0, then

$$\|v(\cdot,t) - u(\cdot,t)\|_{L^1(\mathbb{R})} \leq e^{Ct} \|v_0 - u_0\|_{L^1(\mathbb{R})},$$

where the constant C depends on $\gamma'(x)$ for $x \notin \mathcal{D}_\gamma$ and is zero if γ is piecewise constant.

References

1. Adimurti, Gowda, G.V., Mishra, S.: Optimal entropy solutions for conservation laws with discontinuous flux. Journal of Hyp. Diff. Eqns. **2**, 1–56 (2005)
2. Bürger, R., Karlsen, K., Risebro, N.H., Towers, J.: Well posedness in bv_t and convergence of a difference scheme for continuous sedimentation in ideal clarifier thickener units. Numer. Math. **97**(1), 25–65 (2004)
3. Diehl, S.: A conservation law with point source and discontinuous flux function modeling continuous sedimentation. SIAM J. Appl. Math. **78**(3), 385–390 (1995)
4. Gimse, T., Risebro, N.H.: Solution of the Cauchy problem for a conservation law with a discontinuous flux function. SIAM J. Math. Anal. **23**, 635–648 (1992)
5. Glimm, J.: Solutions in the large for nonlinear hyperbolic systems of equations. Comm. Pure Appl. Math. **18**, 697–715 (1965)
6. Holden, H., Risebro, N.H.: Front Tracking for Hyperbolic Conservation Laws. Applied Mathematical Sciences 152. Springer, New York (2002)
7. Karlsen, K., Towers, J.: Convergence of the Lax-Friedrichs scheme and stability for conservation laws with a discontinuous space-time dependent flux. Chinese Ann. Math. Ser. B **25**(3), 287–318 (2004)
8. Karlsen, K.H., Rascle, M., Tadmor, E.: On the existence and compactness of a two-dimensional resonant system of conservation laws. Commun. Math. Sci. **5**(2), 253–265 (2007)
9. Karlsen, K.H., Risebro, N.H.: Convergence of finite difference schemes for viscous and inviscid conservation laws with rough coeffiients. Math. Model. and Numer. Anal. **35**, 239–270 (2001)
10. Karlsen, K.H., Risebro, N.H., Towers, J.D.: L^1 stability for entropy solutions of nonlinear degenerate parabolic convection-diffusion equations with discontinuous coefficients. Skr. Norske. Vitensk. Selsk. **3**, 1–49 (2003)

11. Kružkov, S.N.: First order quasi-linear equations in several independent variables. Math. USSR Sbornik **10**, 217–243 (1970)
12. Kuznetsov, N.N.: On stable methods of solving a quasi-linear equation of first order in a class of discontinuous functions. Soviet Math. Dokl. **16**, 1569–1573 (1975)
13. Mochon, S.: An analysis of the traffic on highways with changing surface conditions. Math. Modelling **9**(1), 1–11 (1987)
14. Seguin, N., Vovelle, J.: Analysis and approximation of a scalar conservation law with a flux function with discontinuous coefficients. Math. Models Methods Appl. Sci. **13**(2), 221–257 (2003)
15. Temple, B.: Global solution of the Cauchy problem for a class of 2×2 non-strictly hyperbolic conservation laws. Adv. Appl. Math. **3**, 335–375 (1982)
16. Towers, J.: Convergence of a difference scheme for a conservation law with a discontinuous flux. SIAM J. Numer. Anal. **38**(2), 681–698 (2000)
17. Towers, J.: A difference scheme for conservation laws with a discontinuous flux, the nonconvex case. SIAM J. Numer. Anal. **39**(4), 1197–1218 (2001)
18. Wagner, D.: Equivalence of the Euler and Lagrangian equations of gas dynamics for weak solutions. J. Diff. Eqn. **68**(1), 118–136 (1987)

Index

DOI 10.1007/978-3-642-00585-5,

MIX
Papier aus verantwortungsvollen Quellen
Paper from responsible sources
FSC® C105338

If you have any concerns about our products,
you can contact us on
ProductSafety@springernature.com

In case Publisher is established outside the EU,
the EU authorized representative is:
Springer Nature Customer Service Center GmbH
Europaplatz 3, 69115 Heidelberg, Germany

Printed by Libri Plureos GmbH
in Hamburg, Germany